2011

OCR

Biology AS

Student Workbook

OCR *Biology AS* 2011

Student Workbook

First edition 2008
Second edition 2010

ISBN 978-1-877462-64-1

Copyright © **2010** Richard Allan
Published by **BIOZONE International Ltd**

Printed by REPLIKA PRESS PVT LTD using paper
produced from renewable and waste materials

About the Writing Team

Tracey Greenwood joined the staff of Biozone at the beginning of 1993. She has a Ph.D in biology, specialising in lake ecology, and taught undergraduate and graduate biology at the University of Waikato for four years.

Kent Pryor has a BSc from Massey University majoring in zoology and ecology. He was a secondary school teacher in biology and chemistry for 9 years before joining Biozone as an author in 2009.

Lissa Bainbridge-Smith worked in industry in a research and development capacity for eight years before joining Biozone in 2006. Lissa has an M.Sc from Waikato University.

Richard Allan has had 11 years experience teaching senior biology at Hillcrest High School in Hamilton, New Zealand. He attained a Masters degree in biology at Waikato University, New Zealand.

All rights reserved. No part of this publication may be reproduced, stored in a retrieval system, or transmitted in any form or by any means, electrical, mechanical, **photocopying**, recording or otherwise, without the permission of BIOZONE International Ltd. This workbook may not be **re-sold**. If the workbook has been purchased at the discounted price for student purchase, then the school must not keep the workbook. The provisions of the 'Sale Agreement' signed by the registered purchaser, specifically prohibit the photocopying of exercises, worksheets, and diagrams from this workbook for any reason.

PHOTOCOPYING PROHIBITED

No part of this workbook may be photocopied under any circumstances. This is a precondition of the sale of the workbook. This specifically excludes any photocopying under any photocopying licence scheme.

Purchases of this workbook may be made direct from the publisher:

BIOZONE   www.biozone.co.uk

UNITED KINGDOM:

BIOZONE Learning Media (UK) Ltd.
Bretby Business Park, Ashby Road, Bretby,
Burton upon Trent, DE15 0YZ, **UK**
Telephone: 01283-553-257
FAX: 01283-553-258
E-mail: sales@biozone.co.uk

AUSTRALIA:

BIOZONE Learning Media Australia
P.O. Box 2841, Burleigh BC,
QLD 4220, **Australia**
Telephone: +61 7-5535-4896
FAX: +61 7-5508-2432
E-mail: info@biozone.com.au

NEW ZEALAND:

BIOZONE International Ltd.
P.O. Box 13-034, Hamilton 3251, **New Zealand**
Telephone: +64 7-856-8104
FAX: +64 7-856-9243
E-mail: sales@biozone.co.nz

Preface to the 2011 Edition

This is the second edition of Biozone's workbook for OCR Biology AS. This aim with this edition was to develop and refine the content and organisation of the workbook in accordance with OCR guidelines. Although it has been specifically written to meet the needs of students enrolled in **OCR Biology**, it is also aligned to the Cambridge International Examinations (**CIE**) and students undertaking this course will find it to be a suitable resource.

OCR Biology AS 2011 builds on the successful features of previous editions, while focusing on scientific literacy and learning within relevant contexts. Much of the content has been substantially revised to improve its relevance, accessibility, and useability by both students and educators. These changes include:

▶ Content reorganisation. The organisation of this edition more closely follows the OCR curriculum document, and concept maps divide the book into each of the three OCR AS units. This reorganisation of content provides a more cohesive coverage of material within each unit, while retaining the flexible nature of the activity worksheet format. Extension material, and additonal material for CIE students, is provided on the **Teacher Resource CD-ROM** (for separate purchase).

▶ A contextual approach. We encourage students to become thinkers by applying their knowledge within appropriate contexts. Some chapters include an account examining a 'biological story' related to the theme of the chapter. This approach provides a context for the material to follow and an opportunity to focus on comprehension and the synthesis of ideas.

▶ Concept maps introduce each unit of the workbook, integrating the content across chapters to encourage linking of ideas.

▶ An easy-to-use chapter introduction comprising succinct learning objectives, a list of key terms, and a short summary of key concepts. The learning objectives are based on the learning outcomes of the OCR scheme, but in some cases include additional detail to assist with comprehension.

▶ There is an emphasis on acquiring skills in scientific literacy. Each chapter includes a comprehension and/or literacy activity, and the appendix (a new feature) includes references for works cited throughout the text.

▶ Web links and Related Activities support the material provided on each activity page.

▶ This workbook will be regularly updated to keep abreast of new developments in biology and to reflect changes to the OCR and CIE curricula. Biozone continues to be committed to providing up-to-date, relevant, interesting, and accurate information.

A Note to the Teacher

This workbook is a student-centred resource, and benefits students by facilitating independent learning and critical thinking. This workbook is just that; a place for your answers notes, asides, and corrections. It is not a textbook and annual revisions are our commitment to providing a current, flexible, and engaging resource. The low price is a reflection of this commitment. Please do not photocopy the activities. If you think it is worth using, then we recommend that the students themselves own this resource and keep it for their own use. I thank you for your support. **Richard Allan**

Acknowledgements

We would like to thank those who have contributed towards this edition:
• Will Robinson and Gemma Conn for artwork and layout review • Sue FitzGerald, Mary McDougall, and Gwen Gilbert for their efficient handling of the office • Deborah Ripley (©deborahripley.com)and her mother jenny, for their generosity in allowing us to write about Jenny's struggle with COPD • TechPool Studios, for their clipart collection of human anatomy: Copyright ©1994, TechPool Studios Corp. USA (some of these images were modified by Richard Allan and Tracey Greenwood) • Totem Graphics, for their clipart collection of plants and animals • John Mahn PLU, for the cross section of a dicot leaf • Corel Corporation, for use of their eps clipart of plants and animals from the Corel MEGAGALLERY collection

Photo Credits

The writing team would like to thank the following individuals and institutions who kindly provided photographs:• Deborah Ripley for the photograph of her mother Jenny • Kimberley Mallady for use of his hedgerow photograph • Joseph E. Armstrong, Professor of Botany,

Head Curator at ISU Herbarium, USA for his permission to use the photo showing a child with kwashiorkor • University of Florida for the photograph of strawberry runners • Alison Roberts for the photograph of plasmodesmata • Dartmouth Electron Microscope Facility for various SEMs • D. Eason (Doc) for the photo of a takahe puppet • I. Flux (Doc) for the photo of kokako release • PASCO for their images of probeware • Alan Sheldon, Sheldon's Nature Photography, Wisconsin for the photo of the lizard without its tail • Stephen Moore for his photos of freshwater insects • Dan Butler for his photograph of a wounded finger • Karl Mueller for the photograph of the Malawi gogo • Walter Siegmund (Flickr) for the photo of the marbled white butterfly • Dept of Biological Sciences, University of Delaware, for the LS of a capillary • www.coastalplanning.net for the image of a marine quadrat • PIER digital library for the image of the coronary artery • Peng S *et al.* for the photograph of a damaged leaf • Currano *et. al.* PNAS for the photo of crop damage • Marc King for photographs of comb types in chickens

We also acknowledge the photographers that have made their images available through Wikimedia Commons under Creative Commons Licences 2.5. or 3.0: • Y tambe • NIH • USDA • Ute Frevert (Plos) • Dr David Midgley • USDS • US Fish and Wildlife • Andrew Dunn, www.andrewdunnphoto.com • ML5 • FrankDrebin • Bjorn Schulz • M. Betley • Keith Hulbert and Paul Zarucki • Dirk Beyer • André Karwath aka Aka • Cereal Research Centre, AAFC • sevenstar • Hans Hillewaert • New York State Dept of Environmental Conservation.

Royalty free images, purchased by Biozone International Ltd, are used throughout this workbook and have been obtained from the following sources: Corel Corporation from various titles in their Professional Photos CD-ROM collection; IMSI (International Microcomputer Software Inc.) images from IMSI's MasterClips® and MasterPhotosTM Collection, 1895 Francisco Blvd. East, San Rafael, CA 94901-5506, USA; ©1996 Digital Stock, Medicine and Health Care collection; ©Hemera Technologies Inc, 1997-2001; © 2005 JupiterImages Corporation www.clipart. com; ©Click Art, ©T/Maker Company; ©1994., ©Digital Vision; Gazelle Technologies Inc.; PhotoDisc®, Inc. USA, www.photodisc.com.

Photos kindly provided by individuals or corporations and identified by way of coded credits as follows: **BF**: Brian Finerran (Uni. of Canterbury), **BH**: Brendan Hicks (Uni. of Waikato), **BOB**: Barry O'Brien (Uni. of Waikato), **CDC**: Centers for Disease Control and Prevention, Atlanta, USA, **DS**: Digital Stock, **EII**: Education Interactive Imaging, **EW**: Environment Waikato, **FRI**: Forest Research Institute **IF**: I. Flux (DoC), **NASA**: National Aeronautic and Space Administration, **NIH**: National Institute of Health, **RA**: Richard Allan, **RCN**: Ralph Cocklin, **TG**: Tracey Greenwood, **USDA**: United States Department of Agriculture, **VM**: Villa Maria Wines, **WBS**: Warwick Silvester (Uni. of Waikato), **WMU**: Waikato Microscope Unit.

Special thanks to all the partners of the Biozone team for their support.

Cover Photographs

Main photograph: The peregrine falcon (*Falco peregrinus*), is a large bird of prey which mainly hunts other birds, and occasionally small mammals. The peregrine falcon hunts by soaring to locate its prey and then diving steeply at speeds of over 322 km/h (200 mph) to capture its prey. PHOTO: Geoff Kuchera, iStock Photos www.istockphoto.com
Background photograph: Autumn leaves, Image ©2005 JupiterImages Corporation www.clipart.com

Contents

Note to the Teacher & Acknowledgments iii
Getting the Most From This Resource 1
Using the Activities 2
Resources Information 4
Using Biozone's Website 4
Course Guides:
OCR AS and A2 Biology.................................. 5
CIE AS and A2 Biology.................................... 6

ASKING QUESTIONS, FINDING ANSWERS

Skills in Biology

△ *Objectives and Key Concepts* 8
□ Hypotheses and Predictions 9
□ △ Terms and Notation 11
□ ☆ A Qualitative Practical Task 12
□ ☆ A Quantitative Practical Task 13
□ Recording Results 15
□ Variables and Data 16
□ ☆ Manipulating Raw Data 17
□ ☆ Constructing Tables 18
□ ☆ Constructing Graphs 19
□ Drawing Bar Graphs 20
□ Drawing Histograms 21
□ Drawing Pie Graphs 22
□ Drawing Kite Graphs 23
□ Drawing Line Graphs 24
□ Interpreting Line and Scatter Graphs 27
□ Drawing Scatter Plots 28
□ Biological Drawings 29
□ Descriptive Statistics 31
□ ☆ Interpreting Sample Variability 33
□ ☆ Evaluating Your Results 35
□ ☆ KEY TERMS: Mix and Match 37

CELLS, EXCHANGE, AND TRANSPORT

Cell Structure

△ *Objectives and Key Concepts* 39
□ Cell Sizes 40
□ Optical Microscopes 41
□ Electron Microscopes 43
□ Prokaryotic Cells 45
□ Plant Cells 47
□ Animal Cells 48
□ ☆ The Cell's Cytoskeleton 49
□ Cell Structures and Organelles 50
□ ☆ Packaging Proteins 53
□ △ Interpreting Electron Micrographs 54
□ ☆ Identifying Structures in an Animal Cell 55

□ △ Identifying Plant Cell Structures 56
□ ☆ How Do we Know? Membrane Structure 57

Cell Membranes and Transport

△ *Objectives and Key Concepts* 58
□ △ The Structure of Membranes 59
□ △ The Role of Membranes in Cells 61
□ Cell Signalling and Receptors 63
□ ☆ Passive Transport Processes 64
□ △ Ion Pumps 67
□ △ Exocytosis and Endocytosis 68
□ △ Active and Passive Transport Summary 69
□ ☆ KEY TERMS: Mix and Match 70

Cell Division an Organisation

△ *Objectives and Key Concepts* 71
□ △ Cell Division 72
□ Mitosis and the Cell Cycle 73
□ The Genetic Origins of Cancer 75
□ Cell Growth and Cancer 76
□ △ Differentiation of Human Cells 77
□ △ Human Cell Specialisation 78
□ ☆ Differentiation of Plant Cells 79
□ Plant Cell Specialisation 80
□ Root Cell Development 81
□ △ Levels of Organisation 82
□ △ Animal Tissues 83
□ Plant Tissues 84
□ ☆ KEY TERMS: Word Find 85

Exchange Surfaces and Breathing

☆ *Objectives and Key Concepts* 86
□ The Need for Gas Exchange 87
□ ☆ Why a Gas Exchange System? 88
□ Surface Area and Volume 89
□ △ Breathing in Humans 91
□ ☆ Responses to Exercise 92
□ The Human Gas Exchange System 93
□ ☆ Measuring Lung Function 95
□ △ Review of Lung Function 97
□ ☆ KEY TERMS: Mix and Match 98

Transport in Animals

△ *Objectives and Key Concepts* 99
□ Transport and Exchange Systems 100
□ ☆ Open Circulatory Systems 101
□ ☆ Closed Circulatory Systems 102
□ ☆ The Human Circulatory System 104

CODES: △ Upgraded ☆ New activity † CIE only ⬚ to be done ✓ when completed

CONTENTS (continued)

The Human Heart 105
Δ Control of Heart Activity 107
Δ The Cardiac Cycle 108
Δ Review of the Human Heart 109
Arteries 110
Veins 111
Δ Capillaries 112
☆ Capillary Networks 113
☆ Formation of Tissue Fluid 114
Blood 115
Gas Transport in Humans 117
The Effects of High Altitude † 119
Exercise and Blood Flow 120
☆ KEY TERMS: Memory Card Game 121

Transport in Plants

Δ Objectives and Key Concepts 123
Transport in Plants 124
Stems and Roots 125
Leaf Structure 127
Xylem 128
Phloem 129
Uptake at the Root 130
Transpiration 131
Adaptations of Xerophytes 133
Translocation 135
☆ KEY TERMS: Mix and Match 137

Biological Molecules

Δ Objectives and Key Concepts 139
Δ The Biochemical Nature of the Cell 140
☆ The Role of Water 141
Δ Organic Molecules 142
Amino Acids 143
Proteins 145
☆ Modification of Proteins 147
☆ Monosaccharides and Disaccharides 148
☆ Carbohydrate Chemistry 149
☆ Polysaccharides 150
☆ Cellulose and Starch 151
Δ Lipids 153
Δ Enzymes 155
☆ How Enzymes Work 156
Enzyme Reaction Rates 157
☆ Enzyme Cofactors 158
☆ Enzyme Inhibitors 159
Δ Biochemical Tests 161
☆ KEY TERMS: Word Find 162

Nucleic Acids

☆ Objectives and Key Concepts 163
DNA Molecules 164
Eukaryote Chromosome Structure 165
Nucleic Acids 167
Creating a DNA Model 169
DNA Replication 173
Review of DNA Replication 175
Δ The Genetic Code 176
The Simplest Case: Genes to Proteins 177
Analysing a DNA Sample 178
☆ KEY TERMS: What Am I? 179

Food and Health

Δ Objectives and Key Concepts 181
Global Human Nutrition 182
A Balanced Diet 183
Deficiency Diseases 185
Δ Malnutrition and Obesity 187
Δ Cardiovascular Disease 189
☆ Atherosclerosis 190
☆ Cholesterol, Diet, and Heart Health 191
The Green Revolution 193
Selective Breeding in Crop Plants 195
Selective Breeding in Animals 197
Producing Food with Microorganisms 199
Increasing Food Production 201
Food Preservation 203
☆ KEY TERMS: Word Find 204

Pathogens and Human Disease

☆ Objectives and Key Concepts 205
Health vs Disease 206
Δ Infection and Disease 207
Transmission of Disease 208
Bacterial Pathogens 209
Δ Tuberculosis 210
Cholera † 211
Δ Protozoan Diseases 212
Δ Malaria 213
☆ Viral Pathogens 214
Δ HIV and AIDS 215
Δ Epidemiology of AIDS 217
☆ Replication in HIV 219
☆ The Impact of HIV/AIDS in Africa 220
☆ The Global Threat of Disease 221
The Control of Disease 223
☆ Antibiotics 225
Resistance in Pathogens........................... 226
☆ KEY TERMS: Crossword 227

CODES: Δ Upgraded   ☆ New activity   † CIE only   • to be done   ✓ when completed

CONTENTS (continued)

Defence and the Immune System

△ *Objectives and Key Concepts* 228
☆ The First Line of Defence 229
The Body's Defences 230
The Action of Phagocytes 231
Inflammation 232
Fever 233
The Lymphatic System 234
The Immune System 235
Antibodies 237
△ Acquired Immunity 239
☆ Vaccines and Vaccination 241
△ New Medicines 243
☆ KEY TERMS: Memory Card Game 245

Smoking: A Choice Against Health

☆ *Objectives and Key Concepts* 247
☆ Living With Chronic Lung Disease 248
☆ Respiratory Diseases 249
Smoking and the Lungs 251
☆ Smoking and Cardiovascular System 253

Identifying Biodiversity

☆ *Objectives and Key Concepts* 255
Global Biodiversity 256
Britain's Biodiversity 257
△ Measuring Diversity in Ecosystems 259
Loss of Biodiversity 261
Characteristics of Life 262
Types of Living Things 263
Types of Cells 264
Features of Taxonomic Groups 265
Unicellular Eukaryotes 270
Classification System 271
Features of the Five Kingdoms 273
△ Features of Microbial Groups 274
Features of Animal Taxa 275
Features of Macrofungi and Plants 277
The Classification of Life 278
The New Tree of Life 284
Phylogenetic Systematics 285
Classification Keys 287
Keying out Plant Species 289
☆ KEY TERMS: Crossword 290

Evolution

△ *Objectives and Resources* 291
☆ Variation in Species 292
Variation 293
Adaptations and Fitness............................. 295

The Modern Theory of Evolution 297
△ Darwin's Theory 298
Fossil Formation 299
The Fossil Record 301
Dating a Fossil site 303
☆ Evolution of Horses 305
☆ DNA Homologies.................................. 306
☆ Protein Homologies 307
Homologous Structures 309
Vestigial Organs 310
☆ Oceanic Island Colonisers...................... 311
Natural Selection.................................... 313
Selection for Human Birth Weight 314
△ Adaptation & Evolution: Darwin's Finches.... 315
What is Speciation?................................. 317
△ Insecticide Resistance 318
Antibiotic Resistance............................... 319
☆ Antigenic Variability in Pathogens 320
☆ KEY TERMS: Mix and Match 321

Maintaining Biodiversity

☆ *Objectives and Key Concepts* 322
☆ Diversity, Stability, and Key Species 323
☆ Global Warming 325
△ Biodiversity and Global Warming 327
☆ Agriculture and Diversity 329
☆ Hedgerows: An Ancient Tradition 331
☆ *In-Situ* Conservation 332
☆ *Ex-Situ* Conservation 333
National Conservation 335
☆ KEY TERMS: Word Find 337

☆ APPENDIX 1: Periodicals 338
☆ APPENDIX 2: Index of Latin & Greek Roots 342
☆ APPENDIX 3: Multiples and SI Units 343
☆ APPENDIX 4: Command Words 344
△ INDEX 345

CODES: △ Upgraded   ☆ New activity   † CIE only   • to be done   ✓ when completed

Getting The Most From This Resource

This workbook is designed as a resource to increase your understanding and enjoyment of biology. It is suitable for students in their AS year of **OCR** and Cambridge International Examinations (**CIE**). The course guides on pages 5-6 (or on the Teacher Resource CD-ROM) indicate where the material required by your syllabus is covered. This workbook will reinforce and extend the ideas developed by your teacher. It is **not a textbook**; its aim is to complement and reinforce the textbooks written for your course. Each topic in the workbook includes the following useful features:

Features of the Concept Map

Each major section of the workbook has core theme:
Cells, Exchange, and Transport
Molecules, Biodiversity, Food and Health
Practical Skills in Biology

Chapter panels identify and summarise the material covered within each chapter.

Encouraging Key Competencies

Thinking - bringing ideas together
Relating to others - communicating
Using language, symbols, and text
Managing self - independence
Participating and contributing

Each section of the workbook emphasises skills and knowledge to be gained.

A summary of why this material is important and where it fits into your understanding of your course content.

Features of the Chapter Topic Page

The part of the OCR (or CIE) scheme to which this chapter applies. Objectives can be assigned at the discretion of the teacher.

The important key ideas in this chapter. You should have a thorough understanding of the concepts summarised here.

The page numbers for the activities covering the material in this subsection of objectives.

The objectives provide a point by point summary of what you should have achieved by the end of the chapter. They are based on the content of the OCR scheme but include some extra explanatory detail where this is important.

A list of key terms used in the chapter. These terms appear in the chapter's vocab activity and can be used to create a glossary for revision purposes. The list represents the minimum literacy requirement for the chapter.

Periodicals of interest are identified by title on a tab on the activity page to which they are relevant. The full citation appears in the **Appendix** on the page indicated.

You can use the check boxes to mark objectives to be completed (a **dot** to be done; a **tick** when completed).

The Weblinks on many of the activities can be accessed through the web links page at:
www.biozone.co.uk/weblink/OCR-AS-2641.html
See page 4 for more details.

Extra resources for this chapter are available on the Teacher Resource CD-ROM (for separate purchase).

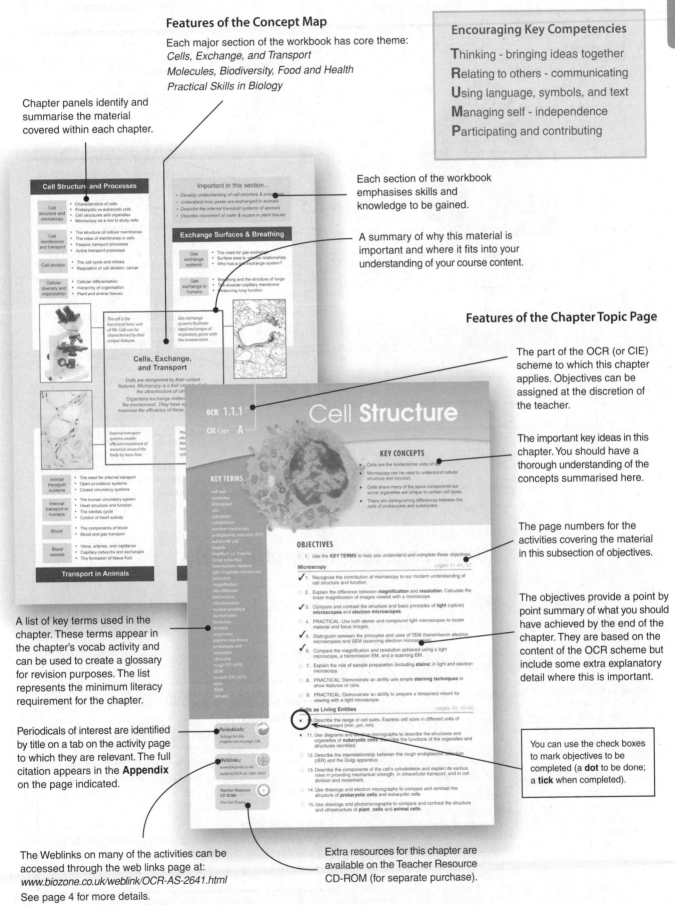

Using the Activities

The activities make up most of the content of this book. Your teacher may use the activity pages to introduce a topic for the first time, or you may use them to revise ideas already covered by other means. They are excellent for use in the classroom, as homework exercises and topic revision, and for self-directed study and personal reference.

Perforations allow easy removal so that pages can be submitted for grading or kept in a separate folder of related work.

Introductory paragraph:
The introductory paragraph provides essential background and provides the focus of the page. Note words that appear in bold, as they are 'key words' worthy of including in a glossary of terms for the topic.

Easy to understand diagrams:
The main ideas of the topic are represented and explained by clear, informative diagrams.

Write-on format:
Your understanding of the main ideas of the topic is tested by asking questions and providing spaces for your answers. Where indicated by the space available, your answers should be concise. Questions requiring more explanation or discussion are spaced accordingly. Answer the questions adequately according to the questioning term used (see Appendix 4).

A tab system at the base of each activity page is a relatively new feature in Biozone's workbooks. With it, we have tagged valuable resources to the activity to which they apply. Use the guide below to help you use the tab system most effectively.

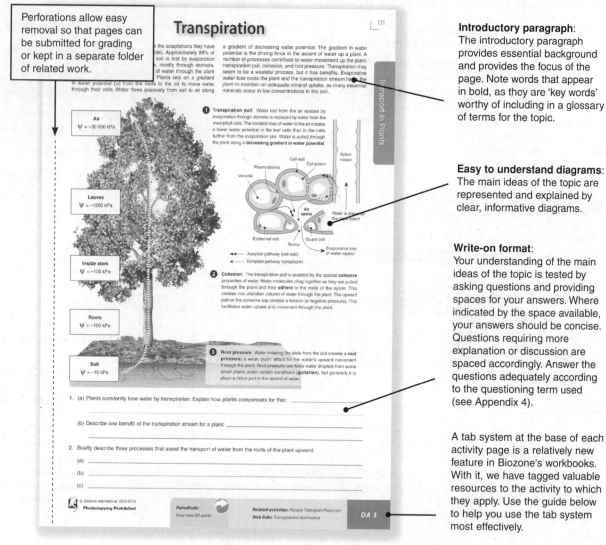

Using page tabs more effectively

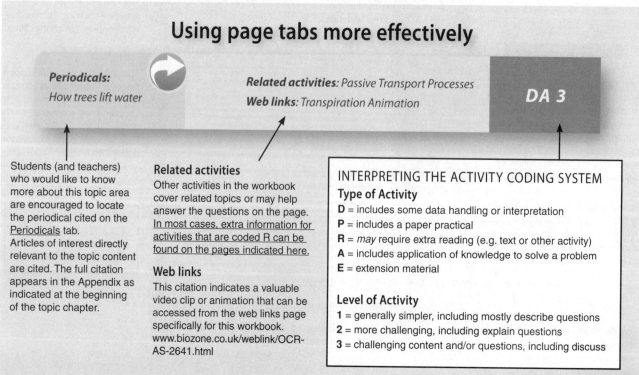

Periodicals:
How trees lift water

Related activities: Passive Transport Processes
Web links: Transpiration Animation

DA 3

Students (and teachers) who would like to know more about this topic area are encouraged to locate the periodical cited on the Periodicals tab.
Articles of interest directly relevant to the topic content are cited. The full citation appears in the Appendix as indicated at the beginning of the topic chapter.

Related activities
Other activities in the workbook cover related topics or may help answer the questions on the page. In most cases, extra information for activities that are coded R can be found on the pages indicated here.

Web links
This citation indicates a valuable video clip or animation that can be accessed from the web links page specifically for this workbook. www.biozone.co.uk/weblink/OCR-AS-2641.html

INTERPRETING THE ACTIVITY CODING SYSTEM

Type of Activity
D = includes some data handling or interpretation
P = includes a paper practical
R = *may* require extra reading (e.g. text or other activity)
A = includes application of knowledge to solve a problem
E = extension material

Level of Activity
1 = generally simpler, including mostly describe questions
2 = more challenging, including explain questions
3 = challenging content and/or questions, including discuss

Resources Information

Your set textbook should always be a starting point for information. There are also many other resources available, including scientific journals, magazine and newspaper articles, supplementary texts covering restricted topic areas, dictionaries, computer software and videos, and the internet.

A synopsis of some of the currently available resources is provided below. Access to the publishers of these resources can be made directly from Biozone's website through our resources hub: **www.biozone.co.uk/resource-hub.html**. Most titles are also available through www.amazon.co.uk. Please note that our listing any product in this workbook does not, in any way, denote Biozone's endorsement of that product.

OCR and CIE Specific Comprehensive Texts

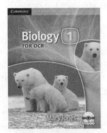

Jones, M., 2007
Biology 1 for OCR
Publisher: Cambridge University Press
Pages: 278
ISBN: 978-0521717632
Comments: *A new resource written for the 2008 OCR specification. Provides HSW themes and summaries. Interactive CD-ROM provides further resources.*

Jones, M., R. Fosbery, D. Taylor, and J. Gregory, 2007
CIE Biology AS and A Level, 2 ed.
Publisher: Cambridge University Press
Pages: 424
ISBN: 0-521-53674-X
Comments: *This text meets the new CIE requirements and covers the complete AS level syllabus, the core A level syllabus, and the new Applications of Biology section.*

Sochaki, F., and Kennedy, P. 2008
OCR Biology
Publisher: Heinemann
Pages: 288
ISBN: 978-0435691806
Comments: *A new book tailored for the new OCR specification. Includes an exam cafe CD-ROM with exam type questions and worked examples.*

Supplementary Texts

Indge, B., 2003
Data and Data Handling for AS and A Level Biology, 128 pp.
Publisher: Hodder Arnold H&S
ISBN: 1340856475
Comments: *Examples and practice exercises to improve skills in data interpretation and analysis.*

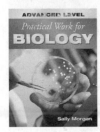

Morgan, S., 2002
Advanced Level Practical Work for Biology, 128 pp.
Publisher: Hodder and Stoughton
ISBN: 0-340-84712-3
Comments: *Caters for the investigative requirements of A level studies: experimental planning, techniques, observation and measurement, and interpretation and analysis.*

Clegg, C.J., 1998.
Mammals: Structure & Function, 96 pp.
ISBN: 0-7195-7551-6
A clearly written supplementary text covering most aspects of basic mammalian anatomy and physiology. Note: This text is now out of print from the publishers, but many schools will still have copies in their collections and it is still available from amazon: www.amazon.co.uk

Clegg, C.J., 2003
Green Plants: The Inside Story, ~. 96 pp.
ISBN: 0-7195-7553-2
The emphasis in this text is on flowering plants. Topics include leaf, stem, and root structure in relation to function, reproduction, economic botany, sensitivity and adaptation.

Periodicals, Magazines and Journals

Biological Sciences Review (Biol. Sci. Rev.)
An excellent quarterly publication for teachers and students. The content is current and the language is accessible. Subscriptions available from Philip Allan Publishers, Market Place, Deddington, Oxfordshire OX 15 OSE.
Tel. 01869 338652 **Fax**: 01869 338803
E-mail: sales@philipallan.co.uk

New Scientist: *Published weekly and found in many libraries. It often summarizes the findings published in other journals. Articles range from news releases to features.*
Subscription enquiries:
Tel. (UK and international): +44 (0)1444 475636. (US & Canada) 1 888 822 3242.
E-mail: ns.subs@qss-uk.com

Scientific American: *A monthly magazine containing mostly specialist feature articles. Articles range in level of reading difficulty and assumed knowledge.*
Subscription enquiries:
Tel. (US & Canada) 800-333-1199.
Tel. (outside North America): 515-247-7631
Web: www.sciam.com

Biology Dictionaries

Clamp, A. **AS/A-Level Biology. Essential Word Dictionary**, 2000, 161 pp. Philip Allan Updates.
ISBN: 0-86003-372-4.
Carefully selected essential words for AS and A2. Concise definitions are supported by further explanation and illustrations where required.

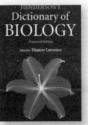

Henderson, E. Lawrence. **Henderson's Dictionary of Biological Terms**, 2008, 776 pp. Benjamin Cummings. **ISBN**: 978-0321505798
This edition has been updated, rewritten for clarity, and reorganised for ease of use. An essential reference and the dictionary of choice for many.

Using Biozone's Website

The current internet address (URL) for the web site is displayed here. You can type a new address directly into this space.

Use Google to search for web sites of interest. The more precise your search words are, the better the list of results. EXAMPLE: If you type in "biotechnology", your search will return an overwhelmingly large number of sites, many of which will not be useful to you. Be more specific, e.g. "biotechnology medicine DNA uses".

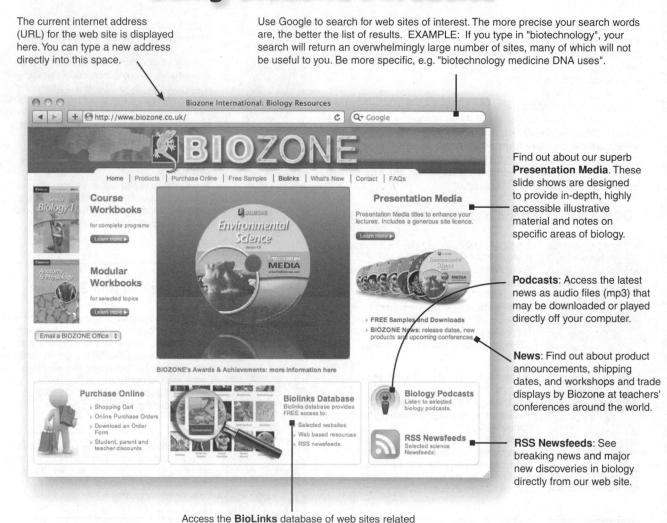

Find out about our superb **Presentation Media**. These slide shows are designed to provide in-depth, highly accessible illustrative material and notes on specific areas of biology.

Podcasts: Access the latest news as audio files (mp3) that may be downloaded or played directly off your computer.

News: Find out about product announcements, shipping dates, and workshops and trade displays by Biozone at teachers' conferences around the world.

RSS Newsfeeds: See breaking news and major new discoveries in biology directly from our web site.

Access the **BioLinks** database of web sites related to each major area of biology. It's a great way to quickly find out more on topics of interest.

Weblinks: www.biozone.co.uk/weblink/OCR-AS-2641.html

BOOKMARK WEBLINKS BY TYPING IN THE ADDRESS: IT IS NOT ACCESSIBLE DIRECTLY FROM BIOZONE'S WEBSITE

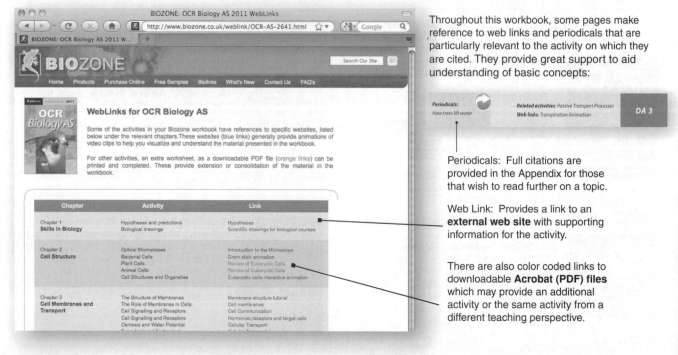

Throughout this workbook, some pages make reference to web links and periodicals that are particularly relevant to the activity on which they are cited. They provide great support to aid understanding of basic concepts:

Periodicals: Full citations are provided in the Appendix for those that wish to read further on a topic.

Web Link: Provides a link to an **external web site** with supporting information for the activity.

There are also color coded links to downloadable **Acrobat (PDF) files** which may provide an additional activity or the same activity from a different teaching perspective.

OCR

Candidates taking the new OCR AS biology course (to be first taught in 2008) are required to complete the units F211, F212 and F213. Candidates taking the OCR A2 course in 2008 will need to complete the requirements for the AS course, as well as 2804, 2805 (one of the five offered options), and 2806 (either components 01 and 02, or components 01 and 03). It should be noted that the new OCR A2 course will be first available for teaching in 2009, at which time the A2 course requirements will be altered.

AS Content — Topics in AS workbook

Unit F211: Cells, Exchange and Transport

Module 1: Cells

1.1.1 Microscopy techniques & stains, organelle structure and function, protein secretion, eukaryote & prokaryote cells characteristics.	Cell Structure
1.1.2 Membrane structure and function. Membrane transport systems. Cell signaling & receptors. Osmosis & water potential.	Cell Membranes & Transport
1.1.3 Mitosis, homologous chromosomes. Stem cells, cell differentiation, and tissue organisation.	Cell Division & Organisation

Module 2: Exchange and Transport

1.2.1 Surface area to volume ratio, diffusion. Mammalian lung & gas exchange system. Breathing in humans, measuring breathing.	Exchange Surfaces and Breathing
1.2.2 Animal transport systems, mammalian heart & the cardiac cycle. Structure and function of arteries, veins and capillaries, tissue fluid and lymph. Haemoglobin and oxygen binding.	Transport in Animals
1.2.3 Plant transport systems. Structure and function of transport tissues. Transpiration, translocation. Water potential, water movement.	Transport in Plants

Unit F212: Molecules, Biodiversity, Food and Health

Module 1: Biological Molecules

2.1.1 Water. Amino acids, peptide bonds. Structure and function of proteins, carbohydrates & lipids. Hydrolysis and condensation. Food testing methods.	Biological Molecules
2.1.2 Nucleic acids, structure of DNA, DNA replication, protein synthesis.	Nucleic Acids
2.1.3 Enzyme structure and function, mode of action, enzyme activity, inhibitors, cofactors.	Biological Molecules

Module 2: Food and Health

2.2.1 Diet & nutrition, nutritional diseases, CHD. Importance of plants. Selective breeding, uses of microorganisms, increasing food production, food preservation.	Food and Health
2.2.2 Health and disease, pathogens. Malaria, HIV, TB. Health effects of smoking. Immune system and the body's defences. Immunity and vaccination. New medicines.	Pathogens & Human Disease, Smoking, Defence & the Immune System

Module 3: Biodiversity and Evolution

2.3.1 Species, habitat and biodiversity. Sampling and measuring biodiversity.	Identifying Diversity
2.3.2 Classification systems. Features of the five kingdoms, identification keys.	Identifying Diversity
2.3.3 Darwin's theory of natural selection, speciation, variation. Mechanisms and evidence for evolution. Pesticide resistance. Antibiotic and other drug resistance.	Evolution
2.3.4 Conservation. Maintaining Biodiversity, global climate change and Biodiversity.	Maintaining Biodiversity

Unit F213: Practical Skills in Biology 1

1. Qualitative Task

Use of appropriate qualitative skills to perform a practical task. Safe techniques, accurate observation and recording.	Skills in Biology, Cell Structure

2. Quantitative Task

Use of appropriate quantitative skills to perform a practical task safely. Accurate measurements and use of the correct degree of precision.	Skills in Biology

3. Evaluation Task

Processing & recording results, evaluating trends & drawing conclusions. Identifying limitations. Reliability and validity of data.	Skills in Biology

A2 Content — Topics in A2 workbook

Unit F214: Communication, Homeostasis, and Energy

Module 1: Communication and Homeostasis

4.1.1 Communication systems in multicellular organisms: nervous and hormonal coordination, feedback, thermoregulation.	Communication and Homeostasis
4.1.2 Sensory receptors, neurones, action potentials, myelination, cholinergic synapses.	Nerves
4.1.3 Endocrine glands and chemical messengers, adrenal glands and pancreas, regulation of blood glucose, diabetes, control of heart rate.	Hormones and Control

Module 2: Excretion

4.2.1 Structure and roles of the liver and role of the kidneys. Control of urine output and composition, urinalysis, renal failure.	Excretion

Module 3: Photosynthesis

4.3.1 Chloroplasts and basic photosynthesis. Photosynthetic pigments LD stage, LI stage, factors affecting photosynthetic rates.	Photosynthesis

Module 4: Respiration

4.4.1 ATP, coenzymes in respiration, glycolysis and aerobic respiration, chemiosmosis, respiratory substrates, alcoholic and lactate fermentation.	Respiration

Unit F215: Control, Genomes, and Environment

Module 1: Cellular Control and Variation

5.1.1 Genes and the genetic code, transcription, translation, mutation, prokaryotic operons, apoptosis and genetic control of development.	Cellular Control
5.1.2 Meiosis and variation, genetic crosses, chi squared, sex linkage, codominance, epistasis. continuous and discontinuous variation, genotype and environment.	Meiosis and Variation
Variation and selection, Hardy-Weinberg principle, calculation of allele frequencies, natural selection, genetic drift, isolation and speciation, species concept, artifical selection.	Variation and Population Change

Module 2: Biotechnology & Gene Technologies

5.2.1 Reproductive and non-reproductive cloning, natural clones in plants, cloning in animals.	Cloning and Biotechnology
5.2.2 Biotechnology, microorganisms in biotechnology, enzyme immobilisation, continuous and batch cultures, metabolites.	Cloning and Biotechnology
5.2.3 Genome sequencing, genetic engineering DNA probes, transgenic organisms, gene therapy, GE xenotransplants, ethical issues.	Genomes and Gene Technologies

Module 3: Ecosystems and Sustainability

5.3.1 Ecosystems: trophic stucture, energy transfers, succession, measuring distribution and abundance, recycling of nitrogen.	Ecosystems
5.3.2 Population growth, species interactions, conservation, preservation, and sustainable management. Human impact (Galapagos).	Populations and Sustainability

Module 4: Responding to the Environment

5.4.1 Tropisms, phytohormones, gibberellins and auxins, commerical uses of plant hormones.	Plant Responses
5.4.2 Animal responses: brain and nervous system, autonomic NS and reflexes, muscle activity and movement, fight or flight responses.	Animal Responses
5.4.3 Innate behaviour, types of learning, sociality in primates, dopamine and behaviour.	Animal Responses

Unit F216: Practical Skills in Biology 1

1 Qualitiative task	OCR AS Skills in Biology
2 Quantitative task	OCR AS Cell Structure
3 Evaluation task	

CIE

The subject content of the **Cambridge International Examinations** (CIE) programme is divided into AS and A2. The A2 includes a core and an Applications of Biology section, which is studied in its entirety, by all A2 candidates. Candidates taking the CIE AS will be assessed on the Learning Outcomes A-K. A level candidates will be assessed on the Learning Outcomes L-U. The *Applications of Biology* section accounts for about 12% of the A level course. The acquisition of practical skills (and their assessment) underpins the course.

AS Content	Topics in OCR Biology AS
Core Syllabus (L-S see OCR Biology A2, T-U see the TRC)	
A Light microscopy, electron microscopy, and cell structure. Eukaryote and prokaryote cells. Functions of organelles.	Cell Structure
B Structure and role of carbohydrates, lipids, and proteins. Water and inorganic ions. Hydrolysis and condensation reactions. Biochemical food testing.	Biological Molecules
C Enzyme action, enzyme activity and enzyme inhibitors.	Biological Molecules
D Fluid mosaic model of membrane structure. Transport across the membrane. Water potential	Cell Membranes and Transport
E Replication and division of nuclei and cells. Role of meiosis in sexual reproduction. Chromosome behaviour during mitosis. Uncontrolled cell division (cancer).	Cell Division & Organisation
F DNA structure and replication. The role of DNA in protein synthesis. Nucleotide base pairing.	Nucleic Acids
G Transport in multicellular plants: structure and distribution of xylem and phloem in dicots. Transpiration. Translocation. Xerophytes. Structure and function of mammalian transport systems. The heart and the cardiac cycle. Haemoglobin and gas transport, gas exchange and altitude.	Transport in Plants / Transport in Animals
H The structure and function of the human gas exchange system. Gas exchange. The effects of smoking on gas exchange. Smoking related diseases.	Exchange Surfaces and Breathing / Smoking: A Choice Against Health
I Causes and transmission of infectious diseases, their control and prevention. HIV/AIDS, TB, cholera and malaria. The use of antibiotics.	Pathogens & Human Disease
J Structure and function of the immune system. Types of immunity. Vaccinations.	Defence and the Immune System
K The ecosystem concept (habitat, niche, populations, communities). Energy transfer, ecological efficiency, the nitrogen cycle.	● TRC: Ecological Principles
Meeting Assessment Objectives	
A Knowledge with understanding, including the use of scientific vocabulary, understanding scientific ideas and concepts, using instruments and scientific measurements, applying of scientific and technological techniques.	Skills in Biology
B Handling information and solving problems: organising and extracting information. Manipulating and presenting data, and drawing conclusions. Applying knowledge to solve problems.	
C Experimental skills and investigations. Following detailed instructions, using techniques and apparatus correctly, making accurate observations, and interpreting the data to form predictions. Designing, planning and carrying out a scientific experiment or investigation.	

A2 Content	Topics in OCR Biology A2 *(unless indicated)*
Core Syllabus (A-K, Q see OCR Biology AS, T-U see the TRC)	
L Energy requirements in living organisms. The structure and function of ATP. Cellular respiration, and energy transfer. Aerobic and anaerobic respiration. Respiratory quotient and the energy value of substrates. Respirometers.	Respiration
M Photosynthesis and energy transfer. The biochemistry of photosynthesis. Limiting factors in photosynthesis. The structure of a dicot leaf.	Photosynthesis
N Principles of homeostasis. Nervous system: sensory receptors, neurones, action potential, and synapses. Endocrine glands (pancreas). Control of blood glucose (diabetes treatment). Role of hormones in flowering plants.	Communication and Homeostasis, Nerves Hormones and Control Excretion Plant Responses
O Meiosis and sources of variation. Genes and alleles. Monohybrid and dihybrid crosses, sex linkage, codominance, multiple alleles. Chi-squared. Environmental effects on phenotype. Mutation.	Meiosis and Variation / Cellular Control
P Natural and artificial selection. The role of natural selection in evolution. The role of environmental factors and isolating mechanisms. The role of artificial selection on livestock improvement. Factors affecting allele frequencies (malaria and sickle cell anaemia).	Variation and Population Change / Also see OCR Biology AS Evolution
Q The five kingdom classification system. The importance of biodiversity. Conservation issues: endangered species and strategies to protect them.	OCR Biology AS Classication Biodiversity & Conservation
R Gene technology, techniques in gene technology. Uses of gene technology (insulin production, DNA sequencing, genetic fingerprinting, and genetic screening). Benefits and hazards of gene technology. Ethical issues.	Genomes and Gene Technologies
S Biotechnology. Industrial use of microorganisms. Large scale production of microorganisms. Enzyme technology. The use of monoclonal antibodies.	Cloning and Biotechnology
T Crop plant reproduction and adaptations. Methods of improving crop production.	● TRC: Crop Plants
U Human reproduction. Gametogenesis (mitosis, growth, meiosis and maturation). The role of hormones in the menstrual cycle. Contraception. In-vitro fertilisation.	● TRC: Aspects of Human Reproduction
Meeting Assessment Objectives	
A Knowledge with understanding. Including the use of scientific vocabulary, understanding scientific ideas and concepts, using instruments and scientific measurements, the application of scientific and technological techniques.	Skills in Biology (OCR AS) Ecosystems (OCR A2)
B Handling information and solving problems. Understand how to organise and extract information. Know how to manipulate and present data, and draw conclusions. Apply knowledge to problem solve.	
C Experimental skills and investigations. Following detailed instructions, using techniques and apparatus correctly, making accurate observations, and interpreting the data to form predictions. Designing, planning and carrying out a scientific experiment or investigation.	

Important in this section...

- *Develop your understanding of how science works*
- *Develop inquiry-based skills to extend and consolidate your biological knowledge*
- *Use vocabulary and knowledge to communicate*

ICT for Data Handling (TRC)

Descriptive statistics
- Entering data in *Excel*® for analysis
- Protocols for entering formulae
- Calculating descriptive statistics

Graphing using Excel®
- Choosing your graph type in *Excel*®
- Labelling axes and adding headings
- Plotting your data

Skills in Biology

Hypotheses
- Applying the scientific questions
- Making a sensible hypothesis

A qualitative practical task
- Procedures for qualitative work
- The limitations of qualitative data
- Presenting qualitative data

A quantitative practical task
- Procedures for quantitative work
- Reliability, accuracy, and precision
- Safety in practical work

Evaluating quantitative work
- Presenting data in graphs and tables
- Descriptive statistics
- Communicating your findings clearly
- Evaluating sources of error

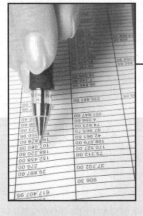

Used properly, computers can help to analyse your data more quickly and understand it better.

Developing skills in practical work and critical evaluation helps you understand how science works

Practical Skills in Biology
Asking Questions, Finding Answers

The scientific method provides framework for investigating the natural world.

Appropriate collection and analysis of data allows us to study biological systems objectively, and to attempt to understand them better.

Field work gives an appreciation of the complexity of community patterns.

Recognising and appreciating biodiversity is an essential part of all biology.

Planning a field study
- Practical considerations for field work
- Collecting and recording data
- Looking for trends and patterns

Population & community ecology
- Considerations for sampling
- Population dynamics
- Quantifying community diversity
- Disturbance and community structure

Biodiversity
- Species' diversity
- Genetic diversity
- Ecosystem diversity

Classifying organisms
- Naming organisms
- Defining species

Ecological sampling is covered in context in OCR Biology A2, but some aspects are introduced in this workbook. The skills developed at AS level can be applied to the study of populations and communities.

The classification of organisms is covered later in this workbook (OCR F212, Module 3). Recognising that evolution is responsible for the diversity of life on Earth is a central theme in all of biology.

Practical Skills in Field Work

Recognising Diversity

Skills in Biology

KEY CONCEPTS

▶ The basis of all science is observation, hypothesis, and investigation.

▶ Data may be quantitative or qualitative.

▶ Data can be analysed and presented in various ways, including in graphs and tables.

▶ Evaluation of quantitative data involves analysis of processed data .and objective assessment of sources of error.

KEY TERMS

accuracy
bibliography
biological drawing
citation
control
controlled variable
data
dependent variable
graph
histogram
hypothesis
independent variable
mean
measurement
median
mode
observation
precision
qualitative data
quantitative data
random sampling
raw data
report
sample
scientific method
standard deviation
statistic
table
transformation (of data)
trend (of data)
variable
X axis
Y axis

OBJECTIVES

☐ 1. Use the **KEY TERMS** to help you understand and complete these objectives.

Understanding the Scientific Method   pages 9-11, 29-30

☐ 2. Describe and explain the basic principles of the scientific method.

☐ 3. Demonstrate an ability or make accurate biological drawings.

A Qualitative Practical Task   page 12

☐ 4. Explain the difference between **qualitative data** and **quantitative data** and give examples of their appropriate use.

☐ 5. A qualitative practical task requires you to carry out a practical investigation for which you collect qualitative (categorical) data (e.g. colour change).

☐ 6. Describe an experiment for which qualitative data are collected and describe its outcome. Describe the limitations of qualitative data and explain why the collection of quantitative data is preferable in most scientific studies.

A Quantitative Practical Task   pages 13-36

☐ 7. You are required to carry out a quantitative investigation in the laboratory. Identify your **dependent** and **independent variables**, their range, and how you will measure them. Identify **controlled variables** and their significance.

☐ 8. Demonstrate an ability to **systematically record** data. Evaluate the **accuracy** and **precision** of any recording or measurements you make.

☐ 9. Demonstrate an ability to process **raw data**. Calculate percentages, rates, and frequencies for raw data (as appropriate) and explain the reason for these manipulations.

☐ 10. Present your of data appropriately in a table, including any calculated values.

☐ 11. Recognise the benefits of graphing data and present your processed data appropriately in a **graph** format.

☐ 12. Considered and objective analysis of your **data** is an essential part of evaluating your quantitative study. Identify and explain the important features of your investigation including:

• trends in your processed data

• conclusions based on analysis of the data and the experimental aims

• discussion of the biological concepts involved

• evaluation, including erroneous results and sources of error

Periodicals:
Listings for this chapter are on page 338

Weblinks:
www.biozone.co.uk/
weblink/OCR-AS-2641.html

Teacher Resource CD-ROM:
Spreadsheets and Statistics

Hypotheses and Predictions

Scientific knowledge grows through a process called the **scientific method**. This process involves observation and measurement, hypothesising and predicting, and planning and executing investigations designed to test formulated **hypotheses**. A scientific hypothesis is a tentative explanation for an observation, which is capable of being tested by experimentation. Hypotheses lead to **predictions** about the system involved and they are accepted or rejected on the basis of findings arising from the investigation. Rejection of the hypothesis may lead to new, alternative explanations (hypotheses) for the observations. Acceptance of the hypothesis as a valid explanation is not necessarily permanent: explanations may be rejected at a later date in light of new findings. This process eventually leads to new knowledge (theory, laws, or models).

Skills In Biology

Making Observations

These may involve the observation of certain behaviours in wild populations, physiological measurements made during previous experiments, or 'accidental' results obtained when seeking answers to completely unrelated questions.

Asking Questions

The observations lead to the formation of questions about the system being studied.

Testing the Predictions

The predictions are tested out in the practical part of an investigation.

Testing predictions may lead to new observations

Forming a Hypothesis

Features of a sound hypothesis:

- It is based on observations and prior knowledge of the system.
- It offers an explanation for an observation.
- It refers to only one independent variable.
- It is written as a definite statement and not as a question.
- It is testable by experimentation.
- It leads to predictions about the system.

Accept or reject the hypothesis

Designing an Investigation

Investigations are planned so that the predictions about the system made in the hypothesis can be tested. Investigations may be laboratory or field based.

Generating a Null Hypothesis

A hypothesis based on observations is used to generate the **null hypothesis (H_0)**; the hypothesis of no difference or no effect. Hypotheses are expressed in the null form for the purposes of statistical testing. H_0 may be rejected in favour of accepting the alternative hypothesis, H_A.

Making Predictions

Based on a hypothesis, **predictions** (expected, repeatable outcomes) can be generated about the behaviour of the system. Predictions may be made on any aspect of the material of interest, e.g. how different variables (factors) relate to each other.

Periodicals:
The truth is out there

Related activities: Experimental Method
Web links: Hypotheses, Terms and Notation

A 2

Useful Types of Hypotheses

A hypothesis offers a tentative explanation to questions generated by observations. Some examples are described below. Hypotheses are often constructed in a form that allows them to be tested statistically. For every hypothesis, there is a corresponding **null hypothesis**; a hypothesis against the prediction. Predictions are tested with laboratory and field experiments and carefully focused observations. For a hypothesis to be accepted it should be possible for anyone to test the predictions with the same methods and get a similar result each time.

Hypothesis involving manipulation
Used when the effect of manipulating a variable on a biological entity is being investigated. **Example**: The composition of applied fertiliser influences the rate of growth of plant A.

Hypothesis of choice
Used when species preference, e.g. for a particular habitat type or microclimate, is being investigated. **Example**: Woodpeckers (species A) show a preference for tree type when nesting.

Hypothesis involving observation
Used when organisms are being studied in their natural environment and conditions cannot be changed. **Example**: Fern abundance is influenced by the degree to which the canopy is established.

1. Generate a prediction for the hypothesis: *"Moisture level of the microhabitat influences woodlouse distribution"*:

_____

2. During the course of any investigation, new information may arise as a result of observations unrelated to the original hypothesis. This can lead to the generation of further hypotheses about the system. For each of the incidental observations described below, formulate a prediction, and an outline of an investigation to test it. *The observation described in each case was not related to the hypothesis the experiment was designed to test:*

(a) **Bacterial cultures**

Prediction: _____

_____

Outline of the investigation: _____

_____

_____

_____

_____

_____

Bacterial Cultures

Observation: During an experiment on bacterial growth, these girls noticed that the cultures grew at different rates when the dishes were left overnight in different parts of the laboratory.

(b) **Plant cloning**

Prediction: _____

_____

_____

Outline of the investigation: _____

_____

_____

_____

_____

_____

_____

Plant Cloning

Observation: During an experiment on plant cloning, a scientist noticed that the root length of plant clones varied depending on the concentration of a hormone added to the agar.

Terms and Notation

The definitions for some commonly encountered terms related to making biological investigations are provided below. Use these as you would use a biology dictionary when planning your investigation and writing up your report. It is important to be consistent with the use of terms i.e. use the same term for the same procedure or unit throughout your study. Be sure, when using a term with a specific statistical meaning, such as sample, that you are using the term correctly.

Questioning Terms

Describe: Provide a detailed account.

Determine: Obtain a value by calculation.

Discuss: Provide a detailed account addressing a range of ideas and arguments.

Explain: Set out reasons using biological background.

Identify: Select and recognise relevant points or characteristics.

Experiments

Data: Facts collected for analysis.

Hypothesis: A tentative explanation of an observation, capable of being tested by experimentation. Hypotheses are written as clear statements, not as questions.

Control treatment (control): A standard (reference) treatment that helps to ensure that responses to other treatments can be reliably interpreted. There may be more than one control in an investigation.

Dependent variable: A variable whose values are determined by another variable (the independent variable). In practice, the dependent variable is the variable representing the biological response.

Independent variable: A variable whose values are set, or systematically altered, by the investigator.

Controlled variables: Variables that may take on different values in different situations, but are controlled (fixed) as part of the design of the investigation.

Experiment: A contrived situation designed to test (one or more) hypotheses and their predictions. It is good practice to use sample sizes that are as large as possible for experiments.

Qualitative: Not quantitative. Described in words or terms rather than by numbers. Includes subjective descriptions in terms of variables such as colour or shape.

Quantitative: Able to be expressed in numbers. Numerical values derived from counts or measurements.

Parameter: A numerical value that describes a characteristic of a population (e.g. the mean height of all 17 year-old males).

Prediction: The prediction of the response (Y) variable on the basis of changes in the independent (X) variable.

Random sample: A method of choosing a sample from a population that avoids any subjective element. It is the equivalent to drawing numbers out of a hat, but using random number tables. For field based studies involving quadrats or transects, random numbers can be used to determine the positioning of the sampling unit.

Repeat / Trial: The entire investigation is carried out again at a different time. This ensures that the results are reproducible. Repeats are not **replicates** in the true sense unless they are run at the same time.

Replicate: A duplication of the entire experimental design run at the same time.

Sample: A sub-set of a whole used to estimate the values that might have been obtained if every individual or response was measured. A sample is made up of **sampling units**, In lab based investigations, the sampling unit might be a test-tube, while in field based studies, the sampling unit might be an individual organism or a quadrat.

Sample size (n): The number of samples taken. In a field study, a typical sample size may involve 20-50 individuals or 20 quadrats. In a lab based investigation, a typical sample size may be two to three sampling units, e.g. two test-tubes held at 10°C.

Sampling unit: Sampling units make up the sample size. Examples of sampling units in different investigations are an individual organism, a test tube undergoing a particular treatment, an area (e.g. quadrat size), or a volume. The size of the sampling unit is an important consideration in studies where the area or volume of a habitat is being sampled.

Statistic: An estimate of a parameter obtained from a sample (e.g. the mean height of all 17 year-old males in your class). A precise (reliable) statistic will be close to the value of the parameter being estimated.

Treatments: Well defined conditions applied to the sample units. The response of sample units to a treatment is intended to shed light on the hypothesis under investigation. What is often of most interest is the comparison of the responses to different treatments.

Variable: A factor in an experiment that is subject to change. Variables may be controlled (fixed), manipulated (systematically altered), or represent a biological response.

Precision and Significance

Accuracy: The correctness of the measurement (the closeness of the measured value to the true value). Accuracy is often a function of the calibration of the instrument used for measuring.

Measurement errors: When measuring or setting the value of a variable, there may be some difference between your answer and the 'right' answer. These errors are often as a result of poor technique or poorly set up equipment.

Objective measurement: Measurement not significantly involving subjective (or personal) judgment. If a second person repeats the measurement they should get the same answer.

Precision (of a measurement): The repeatability of the measurement; the ability to be exact. As there is usually no reason to suspect that a piece of equipment is giving inaccurate measures, making precise measurements is usually the most important consideration.

Reliability: How close the statistic is to the true value of the parameter being estimated. A measure of the confidence in a set of data, as indicated by a measure of variability such as standard error.

Statistical significance: An assigned value that is used to establish

The Expression of Units

The value of a variable must be written with its units where possible. Common ways of recording measurements in biology are: volume in litres, mass in grams, length in metres, time in seconds. The following example shows different ways to express the same term. Note that ml and cm^3 are equivalent.

Oxygen consumption$/cm^3 g^{-1} h^{-1}$

the probability that an observed trend or difference represents a true difference that is not due to chance alone. If a level of significance is less than the chosen value (usually 1-10%), the difference is regarded as statistically significant. In rigorous science, it is the hypothesis of no difference or no effect (the null hypothesis, H_0) that is tested. The alternative hypothesis (your tentative explanation for an observation) can only be accepted through statistical rejection of H_0.

Validity: You are truly measuring the right thing and the outcome is not being distorted by extraneous factors.

A Qualitative Practical Task

Not all the experimental work you carry out in biology will yield quantitative data. It is preferable to collect quantitative data; it is more easily analysed and interpreted and it can be collected without bias more easily. However, some situations do warrant the collection of qualitative data, for example, when recording colour changes in simple biochemical tests for common components of foods. Two common tests for carbohydrates are the iodine/potassium iodide test for starch, and the Benedict's test for reducing sugars such as glucose (specifically, Benedict's reagent detects the presence of an aldehyde functional group, –CHO). These tests indicate the presence of a substance with a colour change. All monosaccharides are reducing sugars as are the disaccharides, lactose and maltose. The monosaccharide fructose is a ketose, but it gives a positive test because it is converted to the aldose glucose in the reagent. When a starchy fruit ripens, the starch is converted to simple reducing sugars.

The Aim
To investigate the effect of ripening on the relative content of starch and simple sugars in bananas.

The Tests

Iodine-potassium iodide test for starch
The sample is covered with the iodine in potassium iodide solution. The sample turns blue-black if starch is present.

Benedict's test for reducing sugars
The sample is heated with the reagent in a boiling water bath. After 2 minutes, the sample is removed and stirred, and the colour recorded immediately after stirring. A change from a blue to a brick red colour indicates a reducing sugar.

Summary of the Method

Two 1 cm thick slices of banana from each of seven stages of ripeness were cut and crushed to a paste. One slice from each stage was tested using the I/KI test for starch, and the other was tested using the Benedict's test.

The colour changes were recorded in a table. Signs (+/–) were used to indicate the intensity of the reaction relative to those in bananas that were either less or more ripe.

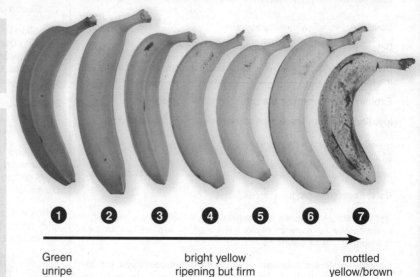

① ② ③ ④ ⑤ ⑥ ⑦

Green
unripe
and hard

bright yellow
ripening but firm
with green tip

mottled
yellow/brown
ripe and soft

Stage of ripeness	Starch-iodine test		Benedict's test	
1	blue-black	+++++	blue clear	–
2	blue-black	++++	blue clear	–
3	blue-black	+++	green	+
4	blue-black	++	yellow cloudy	++
5	slight darkening	+	orange thick	+++
6	no change	–	orangey-red thick	++++
7	no change	–	brick-red thick	+++++

1. Explain why each of the following protocols was important:

    (a) All samples of banana in the Benedict's reagent were heated for 2 minutes: _____

    _____

    (b) The contents of the banana sample and Benedict's reagent were stirred after heating: _____

    _____

2. Explain what is happening to the relative levels of starch and glucose as bananas ripen: _____

    _____

    _____

3. Fructose is a ketose sugar (not an aldose with an aldehyde functional group like glucose).

    (a) Explain why fructose also gives a positive result in a Benedict's test: _____

    _____

    (b) What could this suggest to you about the results of this banana test? _____

    _____

Related activities: Biochemical Tests, Organic Molecules

A Quantitative Practical Task

The middle stage of any investigation (following the planning) is the practical work when the data are collected. Practical work may be laboratory or field based. Typical laboratory based experiments involve investigating how a biological response is affected by manipulating a particular **variable**, e.g. temperature. The data collected for a quantitative practical task should be recorded systematically, with due attention to safe practical techniques, a suitable quantitative method, and accurate measurements to a an appropriate degree of precision. If your quantitative practical task is executed well, and you have taken care throughout, your evaluation of the experimental results will be much more straightforward and less problematic.

Carrying out Your Practical Work

Preparation
Familiarise yourself with the equipment and how to set it up. If necessary, calibrate equipment to give accurate measurements.

Read through the methodology and identify key stages and how long they will take.

Execution
Know how you will take your measurements, how often, and to what degree of precision.

If you are working in a group, assign tasks and make sure everyone knows what they are doing.

Recording
Record your results systematically, in a hand-written table or on a spreadsheet.

Record your results to the apropriate number of significant figures according to the precision of your measurement.

Identifying Variables

A variable is any characteristic or property able to take any one of a range of values. Investigations often look at the effect of changing one variable on another. It is important to identify all variables in an investigation: independent, dependent, and controlled, although there may be nuisance factors of which you are unaware. In all fair tests, only one variable is changed by the investigator.

Dependent variable
- Measured during the investigation.
- Recorded on the y axis of the graph.

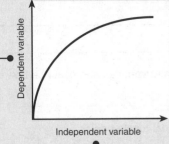

Controlled variables
- Factors that are kept the same or controlled.
- List these in the method, as appropriate to your own investigation.

Independent variable
- Set by the experimenter.
- Recorded on the graph's x axis.

Assumptions

In any experimental work, you will make certain assumptions about the biological system you are working with.

Assumptions are features of the system (and your experiment) that you assume to be true but do not (or cannot) test.

Examples of Investigations

Aim		Variables	
Investigate the effect of varying ...	on the following ...	Independent variable	Dependent variable
Temperature	Leaf width	Temperature	Leaf width
Light intensity	Activity of woodlice	Light intensity	Woodlice activity
Soil pH	Plant height at age 6 months	pH	Plant height

In order to write a sound method for your investigation, you need to determine how the independent, dependent, and controlled variables will be set and measured (or monitored). A good understanding of your methodology is crucial to a successful investigation. You need to be clear about how much data, and what type of data, you will collect. You should also have a good idea about how you will analyse the data. Use the example below to practise your skills in identifying this type of information.

Case Study: Catalase Activity

Catalase is an enzyme that converts hydrogen peroxide (H_2O_2) to oxygen and water. An experiment investigated the effect of temperature on the rate of the catalase reaction. Small (10 cm^3) test tubes were used for the reactions, each containing 0.5 cm^3 of enzyme and 4 cm^3 of hydrogen peroxide. Reaction rates were assessed at four temperatures (10°C, 20°C, 30°C, and 60°C). For each temperature, there were two reaction tubes (e.g. tubes 1 and 2 were both kept at 10°C). The height of oxygen bubbles present after one minute of reaction was used as a measure of the reaction rate; a faster reaction rate produced more bubbles. The entire experiment, involving eight tubes, was repeated on two separate days.

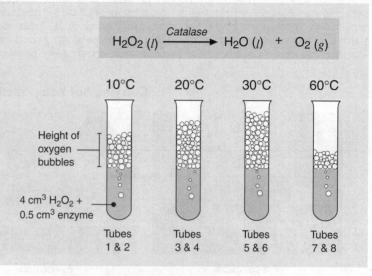

$$H_2O_2 \ (l) \xrightarrow{\text{Catalase}} H_2O \ (l) \ + \ O_2 \ (g)$$

10°C   20°C   30°C   60°C

Height of oxygen bubbles

4 cm^3 H_2O_2 + 0.5 cm^3 enzyme

Tubes 1 & 2   Tubes 3 & 4   Tubes 5 & 6   Tubes 7 & 8

1. Write a suitable aim for this experiment: _____

   _____

2. Write a suitable hypothesis for this experiment: _____

   _____

3. (a) Identify the **independent variable**: _____

   (b) State the range of values for the independent variable: _____

   (c) Name the unit for the independent variable: _____

   (d) List the equipment needed to set the independent variable, and describe how it was used: ____

   _____

4. (a) Identify the **dependent variable**: _____

   (b) Name the unit for the dependent variable: _____

   (c) List the equipment needed to measure the dependent variable, and describe how it was used: ____

   _____

5. (a) Each temperature represents a treatment/sample/trial (circle one):

   (b) State the number of tubes at each temperature: _____

   (c) State the sample size for each treatment: _____

   (d) State how many times the whole investigation was repeated: _____

6. Explain why it would have been desirable to have included an extra tube containing no enzyme: ____

   _____

7. Identify three variables that might have been controlled in this experiment, and how they could have been monitored:

   (a) _____

   (b) _____

   (c) _____

8. Explain why controlled variables should be monitored carefully: _____

   _____

Recording Results

Using a table is the preferred way to record your results systematically, both during the course of your experiment and in presenting your results. A table can also be used to show calculated values, such as rates or means. An example of a table for recording results is shown below. It relates to an investigation of the net growth of plants at three pH levels, but it represents a relatively standardised layout. The labels on the columns and rows are chosen to represent the design features of the investigation. The first column shows the entire range of the independent variable. There are spaces for multiple sampling units, repeats (trials), and calculated mean values. A version of this table would be given in the write-up of the experiment.

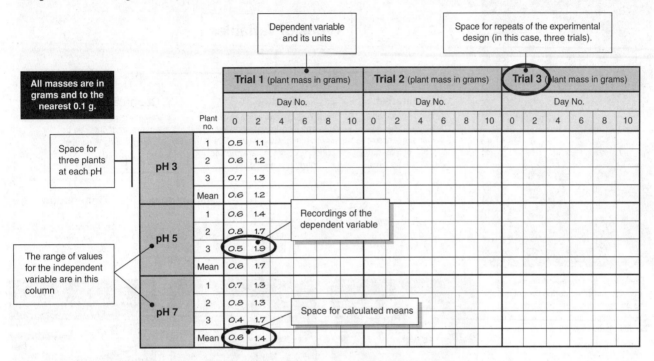

Dependent variable and its units

Space for repeats of the experimental design (in this case, three trials).

All masses are in grams and to the nearest 0.1 g.

Space for three plants at each pH

The range of values for the independent variable are in this column

Recordings of the dependent variable

Space for calculated means

	Plant no.	Trial 1 (plant mass in grams)						Trial 2 (plant mass in grams)						Trial 3 (plant mass in grams)					
		Day No.						Day No.						Day No.					
		0	2	4	6	8	10	0	2	4	6	8	10	0	2	4	6	8	10
pH 3	1	0.5	1.1																
	2	0.6	1.2																
	3	0.7	1.3																
	Mean	0.6	1.2																
pH 5	1	0.6	1.4																
	2	0.8	1.7																
	3	0.5	1.9																
	Mean	0.6	1.7																
pH 7	1	0.7	1.3																
	2	0.8	1.3																
	3	0.4	1.7																
	Mean	0.6	1.4																

Skills In Biology

1. In the space (below) design a table to collect data from the case study below. Include space for individual results and averages from the three set ups (use the table above as a guide).

Carbon dioxide levels in a respiration chamber

A datalogger was used to monitor the concentrations of carbon dioxide (CO_2) in respiration chambers containing five green leaves from one plant species. The entire study was performed in conditions of full light (quantified) and involved three identical set-ups. The CO_2 concentrations were measured every minute, over a period of ten minutes, using a CO_2 sensor. A mean CO_2 concentration (for the three set-ups) was calculated. The study was carried out two more times, two days apart.

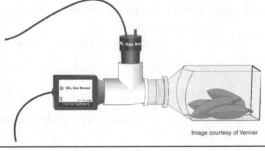

Image courtesy of Vernier

2. Next, the effect of various light intensities (low light, half-light, and full light) on CO_2 concentration was investigated. Describe how the results table for this investigation would differ from the one you have drawn above (for full light only):

_____

_____

Variables and Data

When planning any kind of biological investigation, it is important to consider the type of data that will be collected. It is best, whenever possible, to collect quantitative or numerical data, as these data lend themselves well to analysis and statistical testing. Recording data in a systematic way as you collect it, e.g. using a table or spreadsheet, is important, especially if data manipulation and transformation are required. It is also useful to calculate summary, descriptive statistics (e.g. mean, median) as you proceed. These will help you to recognise important trends and features in your data as they become apparent.

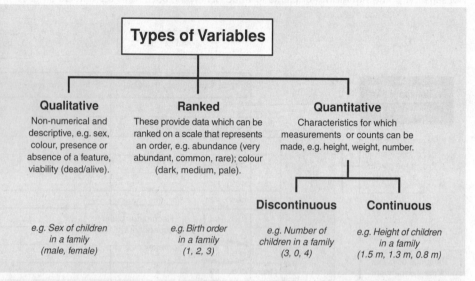

Types of Variables

Qualitative
Non-numerical and descriptive, e.g. sex, colour, presence or absence of a feature, viability (dead/alive).

e.g. Sex of children in a family (male, female)

Ranked
These provide data which can be ranked on a scale that represents an order, e.g. abundance (very abundant, common, rare); colour (dark, medium, pale).

e.g. Birth order in a family (1, 2, 3)

Quantitative
Characteristics for which measurements or counts can be made, e.g. height, weight, number.

Discontinuous
e.g. Number of children in a family (3, 0, 4)

Continuous
e.g. Height of children in a family (1.5 m, 1.3 m, 0.8 m)

The values for monitored or measured variables, collected during the course of the investigation, are called **data**. Like their corresponding variables, data may be quantitative, qualitative, or ranked.

A: Leaf shape

B: Number per litter

C: Fish length

1. For each of the photographic examples (A – C above), classify the variables as quantitative, ranked, or qualitative:

   (a) Leaf shape: _____

   (b) Number per litter: _____

   (c) Fish length: _____

2. Explain clearly why it is desirable to collect quantitative data where possible in biological studies: _____

   _____

   _____

   _____

   _____

3. Suggest how you might measure the colour of light (red, blue, green) quantitatively: _____

   _____

4. (a) Give an example of data that could not be collected in a quantitative manner, explaining your answer:

      _____

      _____

   (b) Sometimes, ranked data are given numerical values, e.g. rare = 1, occasional = 2, frequent = 3, common = 4, abundant = 5. Suggest why these data are sometimes called **semi-quantitative**:

      _____

      _____

Related activities: Descriptive Statistics

Periodicals:
Descriptive statistics

Manipulating Raw Data

The data collected by measuring or counting in the field or laboratory are called **raw data**. They often need to be changed (**transformed**) into a form that makes it easier to identify important features of the data (e.g. trends). Some basic calculations, such as totals (the sum of all data values for a variable), are made as a matter of course to compare replicates or as a prelude to other transformations. The calculation of **rate** (amount per unit time) is another example of a commonly performed calculation,

and is appropriate for many biological situations (e.g. measuring growth or weight loss or gain). For a line graph, with time as the independent variable plotted against the values of the biological response, the slope of the line is a measure of the rate. Biological investigations often compare the rates of events in different situations (e.g. the rate of photosynthesis in the light and in the dark). Other typical transformations include frequencies (number of times a value occurs) and percentages (fraction of 100).

Skills In Biology

Tally Chart

Records the number of times a value occurs in a data set

HEIGHT / cm	TALLY	TOTAL
0 - 0.99	III	3
1 - 1.99	++++ I	6
2 - 2.99	++++ ++++	10
3 - 3.99	++++ ++++ II	12
4 - 4.00	III	3
5 - 5.99	II	2

- A useful first step in analysis; a neatly constructed tally chart doubles as a simple histogram.

- Cross out each value on the list as you tally it to prevent double entries. Check all values are crossed out at the end and that totals agree.

Example: Height of 6d old seedlings

Percentages

Expressed as a fraction of 100

Women	Body mass in kg	Lean body mass	% lean body mass
Athlete	50	38	76.0
Lean	56	41	73.2
Normal weight	65	46	70.8
Overweight	80	48	60.0
Obese	95	52	54.7

- Percentages provide a clear expression of what proportion of data fall into any particular category, e.g. for pie graphs.

- Allows meaningful comparision between different samples.

- Useful to monitor change (e.g. % increase from one year to the next).

Example: Percentage of lean body mass in women

Rates

Expressed as a measure per unit time

Time / minutes	Cumulative sweat loss	Rate of sweat loss / mL min^{-1}
0	0	0
10	50	5
20	130	8
30	220	9
60	560	11.3

- Rates show how a variable changes over a standard time period (e.g. one second, one minute, or one hour).

- Rates allow meaningful comparison of data that may have been recorded over different time periods.

Example: Rate of sweat loss in exercise

---

1. Explain why you might perform basic data transformations: _____

_____

2. (a) Describe a transformation for data relating to the relative abundance of plant species in different habitats:

_____

(b) Explain your answer: _____

_____

3. Complete the transformations on the table (right). The first value is given for you.

Table: *Incidence of cyanogenic clover in different areas*

Working: 120 ÷ 158 = 0.76 = 76%

This is the number of cyanogenic clover out of the total.

Incidence of cyanogenic clover in different areas

Clover plant type	Frost free area		Frost prone area		Totals
	Number	%	Number	%	
Cyanogenic	120	76	22		
Acyanogenic	38		120		
Total	158				

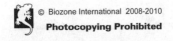
© Biozone International 2008-2010
Photocopying Prohibited

 Periodicals:
Percentages

Related activities: Variables and Data

DA 2

Constructing Tables

Tables provide a convenient way to systematically record and condense a large amount of information for later presentation and analysis. The protocol for creating tables for recording data during the course of an investigation is provided elsewhere, but tables can also provide a useful summary in the results section of a finished report. They provide an accurate record of numerical values and allow you to organise your data in a way that allows you to clarify the relationships and trends that are apparent. Columns can be provided to display the results of any data transformations such as rates. Some basic descriptive statistics (such as mean or standard deviation) may also be included prior to the data being plotted. For complex data sets, graphs tend to be used in preference to tables, although the latter may be provided as an appendix.

Presenting Data in Tables

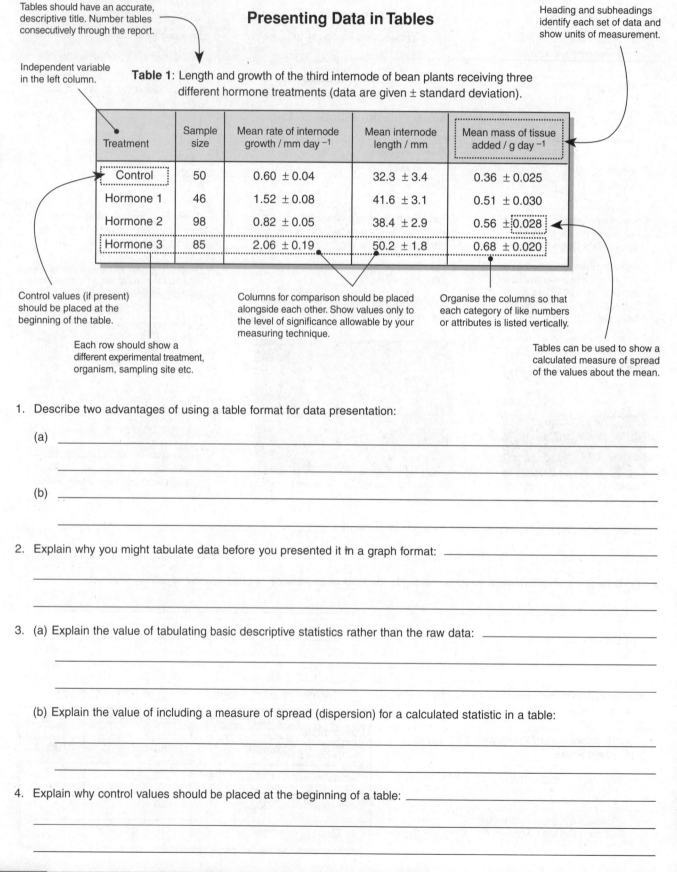

Tables should have an accurate, descriptive title. Number tables consecutively through the report.

Independent variable in the left column.

Heading and subheadings identify each set of data and show units of measurement.

Table 1: Length and growth of the third internode of bean plants receiving three different hormone treatments (data are given ± standard deviation).

Treatment	Sample size	Mean rate of internode growth / mm day^{-1}	Mean internode length / mm	Mean mass of tissue added / g day^{-1}
Control	50	0.60 ± 0.04	32.3 ± 3.4	0.36 ± 0.025
Hormone 1	46	1.52 ± 0.08	41.6 ± 3.1	0.51 ± 0.030
Hormone 2	98	0.82 ± 0.05	38.4 ± 2.9	0.56 ± 0.028
Hormone 3	85	2.06 ± 0.19	50.2 ± 1.8	0.68 ± 0.020

Control values (if present) should be placed at the beginning of the table.

Each row should show a different experimental treatment, organism, sampling site etc.

Columns for comparison should be placed alongside each other. Show values only to the level of significance allowable by your measuring technique.

Organise the columns so that each category of like numbers or attributes is listed vertically.

Tables can be used to show a calculated measure of spread of the values about the mean.

1. Describe two advantages of using a table format for data presentation:

   (a) _____

   _____

   (b) _____

   _____

2. Explain why you might tabulate data before you presented it in a graph format: _____

   _____

   _____

3. (a) Explain the value of tabulating basic descriptive statistics rather than the raw data: _____

   _____

   _____

   (b) Explain the value of including a measure of spread (dispersion) for a calculated statistic in a table:

   _____

   _____

   _____

4. Explain why control values should be placed at the beginning of a table: _____

   _____

   _____

Related activities: *Variables and Data, Manipulating Raw Data, Constructing Graphs*

Periodicals: *Descriptive statistics*

© Biozone International 2008-2010
Photocopying Prohibited

Constructing Graphs

Presenting results in a graph format provides a visual image of trends in data in a minimum of space. The choice between graphing or tabulation depends on the type and complexity of the data and the information that you are wanting to convey. Presenting graphs properly requires attention to a few basic details, including correct orientation and labelling of the axes, and accurate plotting of points. Common graphs include scatter plots and line graphs (for continuous data), and bar charts and histograms (for categorical data). Where there is an implied trend, a line of best fit can be drawn through the data points, as indicated in the figure below. Further guidelines for drawing graphs are provided on the following pages.

Presenting Data in Graph Format

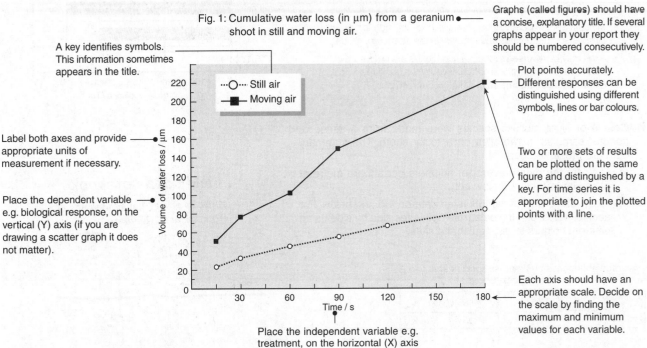

Fig. 1: Cumulative water loss (in μm) from a geranium shoot in still and moving air.

Graphs (called figures) should have a concise, explanatory title. If several graphs appear in your report they should be numbered consecutively.

A key identifies symbols. This information sometimes appears in the title.

Plot points accurately. Different responses can be distinguished using different symbols, lines or bar colours.

Label both axes and provide appropriate units of measurement if necessary.

Two or more sets of results can be plotted on the same figure and distinguished by a key. For time series it is appropriate to join the plotted points with a line.

Place the dependent variable e.g. biological response, on the vertical (Y) axis (if you are drawing a scatter graph it does not matter).

Each axis should have an appropriate scale. Decide on the scale by finding the maximum and minimum values for each variable.

Place the independent variable e.g. treatment, on the horizontal (X) axis

Skills In Biology

1. Describe an advantage of using a graph format for data presentation: _____

_____

2. (a) Explain the importance of using an appropriate scale on a graph: _____

_____

(b) Scales on X and Y axes may sometimes be "floating" (not meeting in the lower left corner), or they may be broken using a double slash and recontinued. Explain the purpose of these techniques:

_____

_____

3. (a) Explain what is wrong with the graph plotted to the right:

_____

_____

_____

_____

(b) Describe the graph's appearance if it were plotted correctly:

_____

_____

_____

_____

Fig. 1: Yeast growth against time

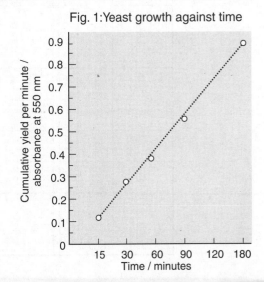

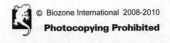
Periodicals:

It's a plot!

Related activities: Descriptive Statistics

DA 2

Drawing Bar Graphs

Guidelines for Bar Graphs

Bar graphs are appropriate for data that are non-numerical and **discrete** for at least one variable, i.e. they are grouped into separate categories. There are no dependent or independent variables. Important features of this type of graph include:

- Data are collected for discontinuous, non-numerical categories (e.g. place, colour, and species), so the bars do not touch.

- Data values may be entered on or above the bars if you wish.

- Multiple sets of data can be displayed side by side for direct comparison (e.g. males and females in the same age group).

- Axes may be reversed so that the categories are on the x axis, i.e. the bars can be vertical or horizontal. When they are vertical, these graphs are sometimes called column graphs.

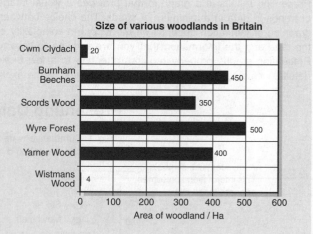

Size of various woodlands in Britain

1. Counts of eight mollusc species were made from a series of quadrat samples at two sites on a rocky shore. The summary data are presented here.

   (a) Tabulate the mean (**average**) numbers per square metre at each site in Table 1 (below left).

   (b) Plot a **bar graph** of the tabulated data on the grid below. For each species, plot the data from both sites side by side using different colours to distinguish the sites.

Average abundance of 8 mollusc species from two sites along a rocky shore.

Species	Average/ no m⁻²	
	Site 1	Site 2

Field data notebook

Total counts at site 1 (11 quadrats) and site 2 (10 quadrats). Quadrats 1 sq. m.

Species	Site 1		Site 2	
	No m⁻²		No m⁻²	
	Total	Mean	Total	Mean
Ornate limpet	232	21	299	30
Radiate limpet	68	6	344	34
Limpet sp. A	420	38	0	0
Cats-eye	68	6	16	2
Top shell	16	2	43	4
Limpet sp. B	628	57	389	39
Limpet sp. C	0	0	22	2
Chiton	12	1	30	3

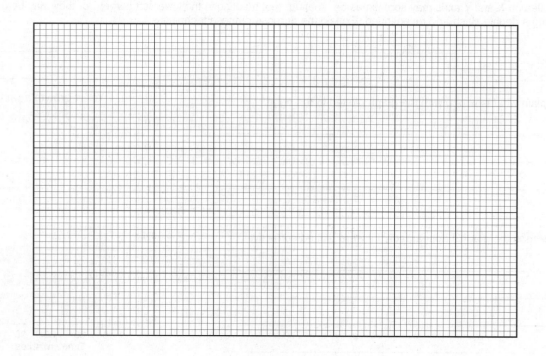

Related activities: Constructing Tables, Descriptive Statistics

Periodicals: Drawing graphs

© Biozone International 2008-2010
Photocopying Prohibited

Drawing Histograms

Guidelines for Histograms

Histograms are plots of **continuous** data and are often used to represent frequency distributions, where the y-axis shows the number of times a particular measurement or value was obtained. For this reason, they are often called frequency histograms. Important features of this type of graph include:

- The data are numerical and continuous (e.g. height or weight), so the bars touch.

- The x-axis usually records the class interval. The y-axis usually records the number of individuals in each class interval (frequency).

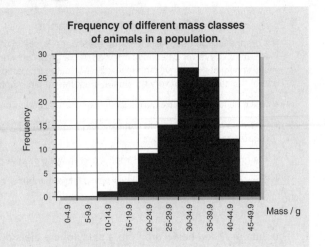

Frequency of different mass classes of animals in a population.

1. The weight data provided below were recorded from 95 individuals (male and female), older than 17 years.

   (a) Create a tally chart (frequency table) in the frame provided, organising the weight data into a form suitable for plotting. An example of the tally for the weight grouping 55-59.9 kg has been completed for you as an example. Note that the raw data values, once they are recorded as counts on the tally chart, are crossed off the data set in the notebook. It is important to do this in order to prevent data entry errors.

   (b) Plot a **frequency histogram** of the tallied data on the grid provided below.

Weight /kg	Tally	Total
45-49.9		
50-54.9		
55-59.9	LHT //	7
60-64.9		
65-69.9		
70-74.9		
75-79.9		
80-84.9		
85-89.9		
90-94.9		
95-99.9		
100-104.9		
105-109.9		

Lab notebook

Weight (in kg) of 95 individuals

63.4	81.2	65
56.5	83.3	75.6
84	95	76.8
81.5	105.5	67.8
73.4	82	68.3
56	73.5	63.5
60.4	75.2	58
83.5	63	58.5
82	70.4	50
61	82.2	92
55.2	87.8	91.5
48	86.5	88.3
53.5	85.5	81
63.8	87	72
69	98	66.5
82.8	71	61.5
68.5	76	66
67.2	72.5	65.5
82.5	61	67.4
83	60.5	73
78.4	67	67
76.5	86	71
83.4	85	70.5
77.5	93.5	65.5
77	62	68
87	62.5	90
89	63	83.5
93.4	60	73
83	71.5	66
80	73.8	57.5
76	77.5	76
56	74	

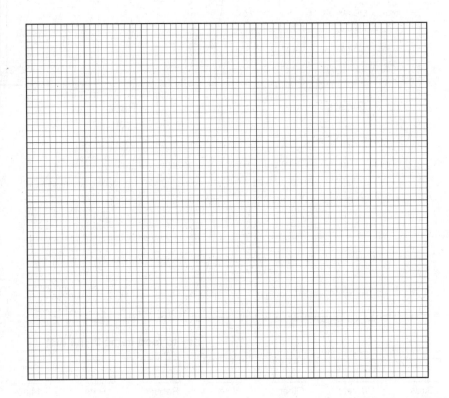

© Biozone International 2008-2010
Photocopying Prohibited

Periodicals:
Drawing graphs

Related activities: Manipulating Raw Data

DA 2

Drawing Pie Graphs

Guidelines for Pie Graphs

Pie graphs can be used instead of bar graphs, generally in cases where there are six or fewer categories involved. A pie graph provides strong visual impact of the relative proportions in each category, particularly where one of the categories is very dominant. Features of pie graphs include:

- The data for one variable are discontinuous (non-numerical or categories).

- The data for the dependent variable are usually in the form of counts, proportions, or percentages.

- Pie graphs are good for visual impact and showing relative proportions.

- They are not suitable for data sets with a large number of categories.

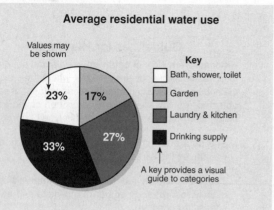

Average residential water use

Values may be shown → 23% | 17% | 33% | 27%

Key
- Bath, shower, toilet
- Garden
- Laundry & kitchen
- Drinking supply

A key provides a visual guide to categories

1. The data provided below are from a study of the diets of three vertebrates.

   (a) Tabulate the data from the notebook in the frame provided. Calculate the angle for each percentage, given that each percentage point is equal to 3.6° (the first example is provided: 23.6 x 3.6 = 85).

   (b) Plot a pie graph for each animal in the circles provided. The circles have been marked at 5° intervals to enable you to do this exercise without a protractor. For the purposes of this exercise, begin your pie graphs at the 0° (= 360°) mark and work in a clockwise direction from the largest to the smallest percentage. Use one key for all three pie graphs.

Field data notebook

% of different food items in the diet

Food item	Stoats	Rats	Cats
Birds	23.6	1.4	6.9
Crickets	15.3	23.6	0
Other insects (not crickets)	15.3	20.8	1.9
Voles	9.2	0	19.4
Rabbits	8.3	0	18.1
Rats	6.1	0	43.1
Mice	13.9	0	10.6
Fruits and seeds	0	40.3	0
Green leaves	0	13.9	0
Unidentified	8.3	0	0

Percentage occurrence of different foods in the diet of stoats, rats, and cats. Graph angle representing the % is shown to assist plotting.

Food item in diet	Stoats		Rats		Cats	
	% in diet	Angle / °	% in diet	Angle / °	% in diet	Angle / °
Birds	23.6	85				

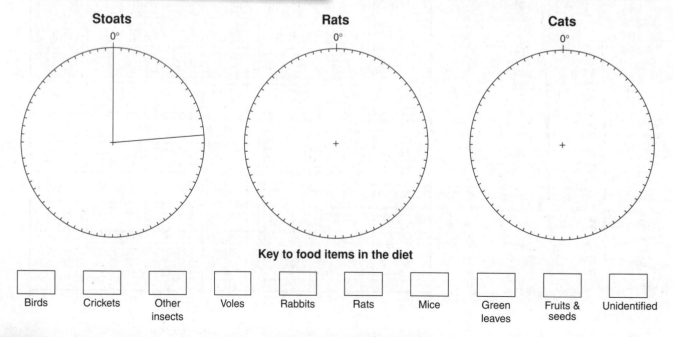

Stoats 0°

Rats 0°

Cats 0°

Key to food items in the diet

Birds | Crickets | Other insects | Voles | Rabbits | Rats | Mice | Green leaves | Fruits & seeds | Unidentified

Related activities: Manipulating Raw Data

Periodicals: Drawing graphs

Drawing Kite Graphs

Guidelines for Kite Graphs

Kite graphs are ideal for representing distributional data, e.g. abundance along an environmental gradient. They are elongated figures drawn along a baseline. Important features of kite graphs include:

- Each kite represents changes in species abundance across a landscape. The abundance can be calculated from the kite width.
- They often involve plots for more than one species; this makes them good for highlighting probable differences in habitat preferences between species.
- A thin line on a kite graph represents species absence.
- The axes can be reversed depending on preference.
- Kite graphs may also be used to show changes in distribution with time, for example, with daily or seasonal cycles of movement.

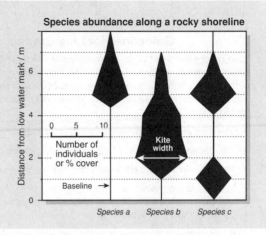

Species abundance along a rocky shoreline

1. The following data were collected from three streams of different lengths and flow rates. Invertebrates were collected at 0.5 km intervals from the headwaters (0 km) to the stream mouth. Their wet weight was measured and recorded (per m²).

   (a) Tabulate the data below for plotting.

   (b) Plot a **kite graph** of the data from all three streams on the grid provided below. Do not forget to include a scale so that the weight at each point on the kite can be calculated.

Wet mass of invertebrates along three different streams

Distance from mouth/ km	Wet weight/g m⁻²		
	Stream A	Stream B	Stream C

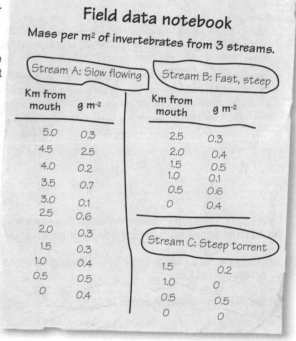

Field data notebook

Mass per m² of invertebrates from 3 streams.

Stream A: Slow flowing

Km from mouth	g m⁻²
5.0	0.3
4.5	2.5
4.0	0.2
3.5	0.7
3.0	0.1
2.5	0.6
2.0	0.3
1.5	0.3
1.0	0.4
0.5	0.5
0	0.4

Stream B: Fast, steep

Km from mouth	g m⁻²
2.5	0.3
2.0	0.4
1.5	0.5
1.0	0.1
0.5	0.6
0	0.4

Stream C: Steep torrent

1.5	0.2
1.0	0
0.5	0.5
0	0

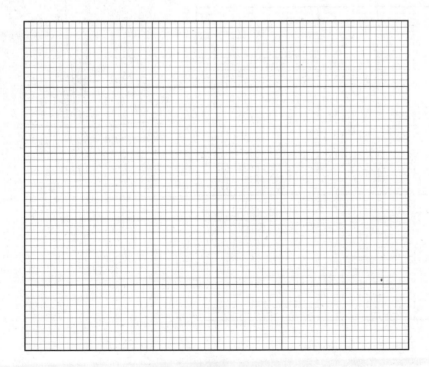

© Biozone International 2008-2010
Photocopying Prohibited

Periodicals:
Drawing graphs

Related activities: Manipulating Raw Data

DA 2

Drawing Line Graphs

Guidelines for Line Graphs

Line graphs are used when one variable (the independent variable) affects another, the dependent variable. Line graphs can be drawn without a measure of spread (top figure, right) or with some calculated measure of data variability (bottom figure, right). Important features of line graphs include:

- The data must be continuous for both variables.

- The dependent variable is usually the biological response.

- The independent variable is often time or the experimental treatment.

- In cases where there is an implied trend (e.g. one variable increases with the other), a line of best fit is usually plotted through the data points to show the relationship.

- If fluctuations in the data are likely to be important (e.g. with climate and other environmental data) the data points are usually connected directly (point to point).

- Line graphs may be drawn with measure of error. The data are presented as points (the calculated means), with bars above and below, indicating a measure of variability or spread in the data (e.g. standard error, standard deviation, or 95% confidence intervals).

- Where no error value has been calculated, the scatter can be shown by plotting the individual data points vertically above and below the mean. By convention, bars are not used to indicate the range of raw values in a data set.

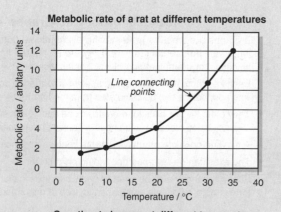

Metabolic rate of a rat at different temperatures

Line connecting points

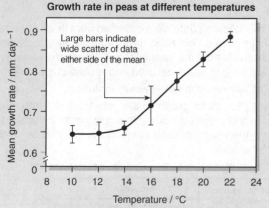

Growth rate in peas at different temperatures

Large bars indicate wide scatter of data either side of the mean

1. The results (shown right) were collected in a study investigating the effect of temperature on the activity of an enzyme.

   (a) Using the results provided in the table (right), plot a line graph on the grid below:

   (b) Estimate the rate of reaction at 15°C: _____

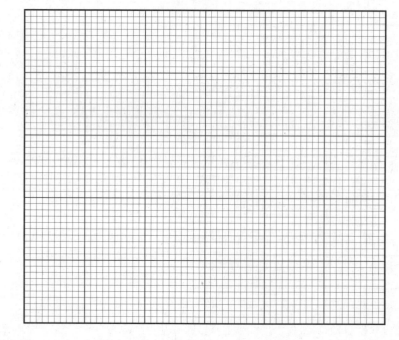

Lab Notebook

An enzyme's activity at different temperatures

Temperature /°C	Rate of reaction /mg of product formed per minute
10	1.0
20	2.1
30	3.2
35	3.7
40	4.1
45	3.7
50	2.7
60	0

Plotting Multiple Data Sets

A single figure can be used to show two or more data sets, i.e. more than one curve can be plotted per set of axes. This type of presentation is useful when you want to visually compare the trends for two or more treatments, or the response of one species against the response of another. Important points regarding this format are:

- If the two data sets use the same measurement units and a similar range of values for the independent variable, one scale on the y axis is used.

- If the two data sets use different units and/or have a very different range of values for the independent variable, two scales for the y axis are used (see example provided). The scales can be adjusted if necessary to avoid overlapping plots

- The two curves must be distinguished with a key.

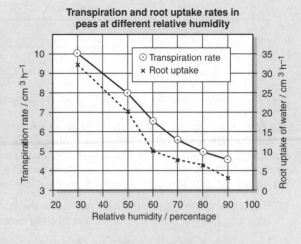

Transpiration and root uptake rates in peas at different relative humidity

2. A census of a deer population on an island indicated a population of 2000 animals in 1960. In 1961, ten wolves (natural predators of deer) were brought to the island in an attempt to control deer numbers. Over the next nine years, the numbers of deer and wolves were monitored. The results of these population surveys are presented in the table, right.

Plot a line graph (joining the data points) for the tabulated results. Use one scale (on the left) for numbers of deer and another scale (on the right) for the number of wolves. Use different symbols or colours to distinguish the lines and include a key.

Field data notebook
Results of a population survey on an island

Time/ year	Wolf numbers	Deer numbers
1961	10	2000
1962	12	2300
1963	16	2500
1964	22	2360
1965	28	2244
1966	24	2094
1967	21	1968
1968	18	1916
1969	19	1952

Skills In Biology

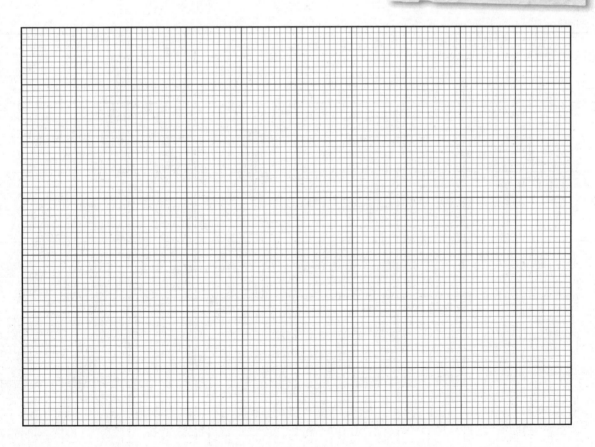

(b) Study the line graph that you plotted for the wolf and deer census on the previous page. Provide a plausible explanation for the pattern in the data, stating the evidence available to support your reasoning:

_____

_____

_____

_____

3. In a sampling programme, the number of perch and trout in a hydro-electric reservoir were monitored over a period of time. A colony of black shag was also present. Shags take large numbers of perch and (to a lesser extent) trout. In 1960-61, 424 shags were removed from the lake during the nesting season and nest counts were made every spring in subsequent years. In 1971, 60 shags were removed from the lake, and all existing nests dismantled. The results of the population survey are tabulated below (for reasons of space, the entire table format has been repeated to the right for 1970-1978).

(a) Plot a line graph (joining the data points) for the survey results. Use one scale (on the left) for numbers of perch and trout and another scale for the number of shag nests. Use different symbols to distinguish the lines and include a key.

(b) Use a vertical arrow to indicate the point at which shags and their nests were removed.

Results of population survey at reservoir

Time / year	Fish number / average per haul		Shag nest numbers	Time / year continued	Fish number / average per haul		Shag nest numbers
	Trout	Perch			Trout	Perch	
1960	–	–	16	1970	1.5	6	35
1961	–	–	4	1971	0.5	0.7	42
1962	1.5	11	5	1972	1	0.8	0
1963	0.8	9	10	1973	0.2	4	0
1964	0	5	22	1974	0.5	6.5	0
1965	1	1	25	1975	0.6	7.6	2
1966	1	2.9	35	1976	1	1.2	10
1967	2	5	40	1977	1.2	1.5	32
1968	1.5	4.6	26	1978	0.7	2	28
1969	1.5	6	32				

Interpreting Line & Scatter Graphs

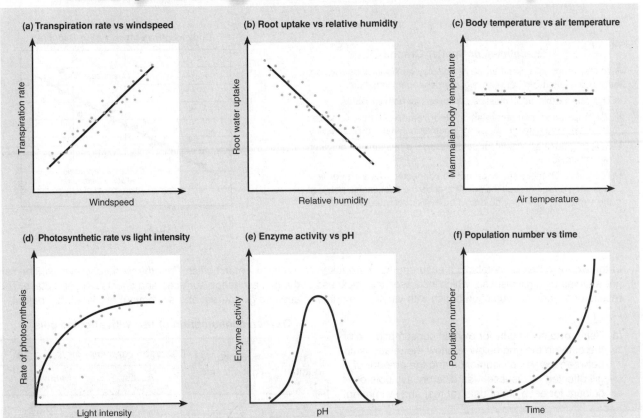

(a) Transpiration rate vs windspeed

Transpiration rate / Windspeed

(b) Root uptake vs relative humidity

Root water uptake / Relative humidity

(c) Body temperature vs air temperature

Mammalian body temperature / Air temperature

(d) Photosynthetic rate vs light intensity

Rate of photosynthesis / Light intensity

(e) Enzyme activity vs pH

Enzyme activity / pH

(f) Population number vs time

Population number / Time

1. For each of the graphs (b-f) above, give a description of the slope and an interpretation of how one variable changes with respect to the other. For the purposes of your description, call the independent variable (horizontal or x-axis) in each example "variable X" and the dependent variable (vertical or y-axis) "variable Y". Be aware that the existence of a relationship between two variables does not necessarily mean that the relationship is causative (although it may be).

(a) Slope: _Positive linear relationship, with constantly rising slope_

Interpretation: _Variable Y (transpiration) increases regularly with increase in variable X (windspeed)_

(b) Slope: _____

Interpretation: _____

(c) Slope: _____

Interpretation: _____

(d) Slope: _____

Interpretation: _____

(e) Slope: _____

Interpretation: _____

(f) Slope: _____

Interpretation: _____

2. Study the line graph of trout, perch and shag numbers that you plotted on the previous page:

(a) Describe the evidence suggesting that the shag population is exercising some control over perch numbers:

_____

(b) Describe evidence that the fluctuations in shag numbers are related to fluctuations in trout numbers:

_____

_____

Periodicals:
Dealing with data

Related activities: Drawing Line Graphs

Drawing Scatter Plots

Guidelines for Scatter Graphs

A scatter graph is a common way to display continuous data where there is a relationship between two interdependent variables.

- The data for this graph must be continuous for both variables.
- There is no independent (manipulated) variable, but the variables are often correlated, i.e. they vary together in some predictable way.
- Scatter graphs are useful for determining the relationship between two variables.
- The points on the graph need not be connected, but a line of best fit is often drawn through the points to show the relationship between the variables (this may be drawn be eye or computer generated).

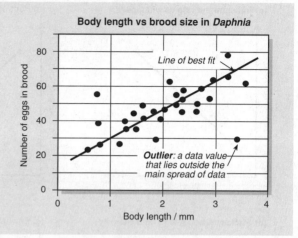

Body length vs brood size in *Daphnia*

1. In the example below, metabolic measurements were taken from seven Antarctic fish *Pagothenia borchgrevinski*. The fish are affected by a gill disease, which increases the thickness of the gas exchange surfaces and affects oxygen uptake. The results of oxygen consumption of fish with varying amounts of affected gill (at rest and swimming) are tabulated below.

   (a) Using **one** scale only for oxygen consumption, plot the data on the grid below to show the relationship between oxygen consumption and the amount of gill affected by disease. Use different symbols or colours for each set of data (at rest and swimming).

   (b) Draw a line of best fit through each set of points.

2. Describe the relationship between the amount of gill affected and oxygen consumption in the fish:

   (a) For the **at rest** data set:

   _____

   _____

   _____

   _____

   _____

   _____

   (b) For the **swimming** data set:

   _____

   _____

   _____

   _____

   _____

   _____

   _____

Oxygen consumption of fish with affected gills

Fish number	Percentage of gill affected	Oxygen consumption/ *cm³ g⁻¹ h⁻¹*	
		At rest	Swimming
1	0	0.05	0.29
2	95	0.04	0.11
3	60	0.04	0.14
4	30	0.05	0.22
5	90	0.05	0.08
6	65	0.04	0.18
7	45	0.04	0.20

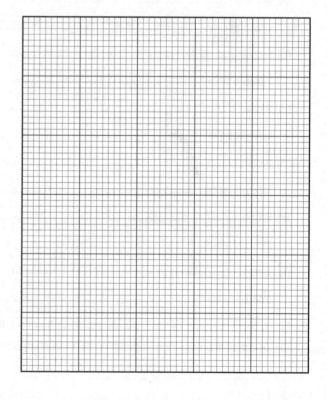

3. Describe how the gill disease affects oxygen uptake in resting fish: _____

_____

Related activities: Interpreting Line Graphs

Biological Drawings

Microscopes are a powerful tool for examining cells and cell structures. In order to make a permanent record of what is seen when examining a specimen, it is useful to make a drawing. It is important to draw **what is actually seen**. This will depend on the **resolution** of the microscope being used. Resolution refers to the ability of a microscope to separate small objects that are very close together. Making drawings from mounted specimens is a skill. Drawing forces you to observe closely and accurately. While photographs are limited to representing appearance at a single moment in time, drawings can be composites of the observer's cumulative experience, with many different specimens of the same material. The total picture of an object thus represented can often communicate information much more effectively than a photograph. Your attention to the outline of suggestions below will help you to make more effective drawings. If you are careful to follow the suggestions at the beginning, the techniques will soon become habitual.

1. **Drawing materials**: All drawings should be done with a clear pencil line on good quality paper. A sharp HB pencil is recommended. A soft rubber of good quality is essential. Diagrams in ballpoint or fountain pen are unacceptable because they cannot be corrected.

2. **Positioning**: Centre your diagram on the page. Do not draw it in a corner. This will leave plenty of room for the addition of labels once the diagram is completed.

3. **Size**: A drawing should be large enough to easily represent all the details you see without crowding. Rarely, if ever, are drawings too large, but they are often too small. Show only as much as is necessary for an understanding of the structure; a small section shown in detail will often suffice. It is time consuming and unnecessary, for example, to reproduce accurately the entire contents of a microscope field.

4. **Accuracy**: Your drawing should be a complete, accurate representation of the material you have observed, and should communicate your understanding of the material to anyone who looks at it. Avoid making "idealised" drawings; your drawing should be a picture of what you actually see, not what you imagine should be there. Proportions should be accurate. If necessary, measure the lengths of various parts with a ruler. If viewing through a microscope, estimate them as a proportion of the field of view, then translate these proportions onto the page. When drawing shapes that indicate an outline, make sure the line is complete. Where two ends of a line do not meet (as in drawing a cell outline) then this would indicate that the structure has a hole in it.

5. **Technique**: Use only simple, narrow lines. Represent depth by stippling (dots close together). Indicate depth only when it is essential to your drawing (usually it is not). Do not use shading. Look at the specimen while you are drawing it.

6. **Labels**: Leave a good margin for labels. All parts of your diagram must be labelled accurately. Labelling lines should be drawn with a ruler and should not cross. Where possible, keep label lines vertical or horizontal. Label the drawing with:
   - A title, which should identify the material (organism, tissues or cells).
   - Magnification under which it was observed, or a scale to indicate the size of the object.
   - Names of structures.
   - In living materials, any movements you have seen.

Remember that drawings are intended as records for you, and as a means of encouraging close observation; artistic ability is not necessary. Before you turn in a drawing, ask yourself if you know what every line represents. If you do not, look more closely at the material. *Take into account the rules for biological drawings and draw what you see, not what you think you see!*

Examples of acceptable biological drawings: The diagrams below show two examples of biological drawings that are acceptable. The example on the left is of a whole organism and its size is indicated by a scale. The example on the right is of plant tissue: a group of cells that are essentially identical in the structure. It is not necessary to show many cells even though your view through the microscope may show them. As few as 2-4 will suffice to show their structure and how they are arranged. Scale is indicated by stating how many times larger it has been drawn. Do not confuse this with what magnification it was viewed at under the microscope. The abbreviation **T.S.** indicates that the specimen was a *cross* or *transverse section*.

Skills In Biology

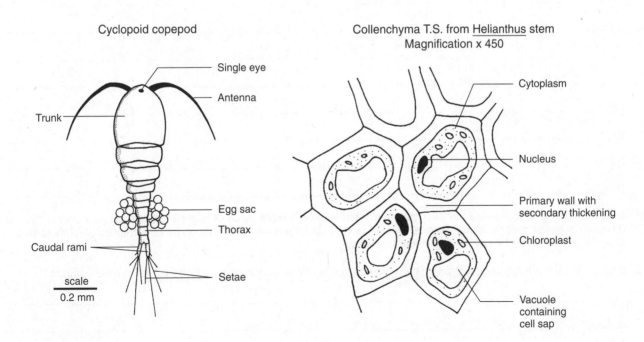

Cyclopoid copepod

Collenchyma T.S. from <u>Helianthus</u> stem
Magnification x 450

© Biozone International 2008-2010
Photocopying Prohibited

Periodicals:
Size does matter

Related activities: Optical Microscopes
Web links: Scientific Drawings for Biological Courses

A 2

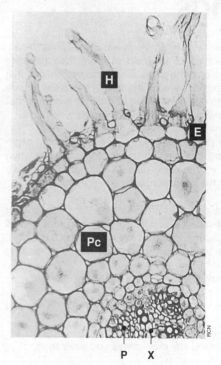

An Unacceptable Biological Drawing

The diagram below is an example of how *not* to produce a biological drawing; it is based on the photograph to the left. There are many aspects of the drawing that are unacceptable. The exercise below asks you to identify the errors in this student's attempt.

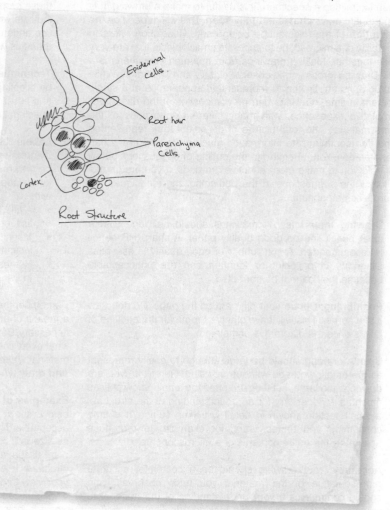

P   X

Specimen used for drawing

The photograph above is a light microscope view of a stained transverse section (cross section) of a root from a *Ranunculus* (buttercup) plant. It shows the arrangement of the different tissues in the root. The vascular bundle is at the centre of the root, with the larger, central xylem vessels (**X**) and smaller phloem vessels (**P**) grouped around them. The root hair cells (**H**) are arranged on the external surface and form part of the epidermal layer (**E**). Parenchyma cells (**Pc**) make up the bulk of the root's mass. The distance from point **X** to point **E** on the photograph (above) is about 0.15 mm (150 μm).

1. Identify and describe eight unacceptable features of the student's biological diagram above:

   (a) _____

   (b) _____

   (c) _____

   (d) _____

   (e) _____

   (f) _____

   (g) _____

   (h) _____

2. In the remaining space next to the 'poor example' (above) or on a blank piece of refill paper, attempt your own version of a biological drawing for the same material, based on the photograph above. Make a point of correcting all of the errors that you have identified in the sample student's attempt.

3. Explain why accurate biological drawings are more valuable to a scientific investigation than an 'artistic' approach:

   _____

   _____

   _____

Descriptive Statistics

For most investigations, measures of the biological response are made from more than one sampling unit. The sample size (the number of sampling units) will vary depending on the resources available. In lab based investigations, the sample size may be as small as two or three (e.g. two test-tubes in each treatment). In field studies, each individual may be a sampling unit, and the sample size can be very large (e.g. 100 individuals). It is useful to summarise the data collected using **descriptive statistics**.

Descriptive statistics, such as mean, median, and mode, can help to highlight trends or patterns in the data. Each of these statistics is appropriate to certain types of data or distributions, e.g. a mean is not appropriate for data with a skewed distribution (see below). Frequency graphs are useful for indicating the distribution of data. Standard deviation and standard error are statistics used to quantify the amount of spread in the data and evaluate the reliability of estimates of the true (population) mean.

Variation in Data

Whether they are obtained from observation or experiments, most biological data show variability. In a set of data values, it is useful to know the value about which most of the data are grouped; the centre value. This value can be the mean, median, or mode depending on the type of variable involved (see schematic below). The main purpose of these statistics is to summarise important trends in your data and to provide the basis for statistical analyses.

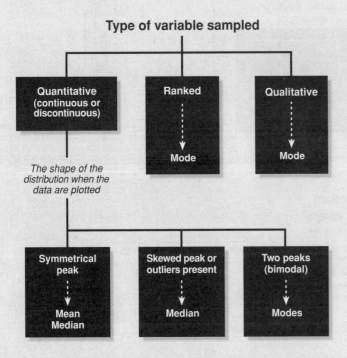

Variability in continuous data is often displayed as a **frequency distribution**. A frequency plot will indicate whether the data have a normal distribution (A), with a symmetrical spread of data about the mean, or whether the distribution is skewed (B), or bimodal (C). The shape of the distribution will determine which statistic (mean, median, or mode) best describes the central tendency of the sample data.

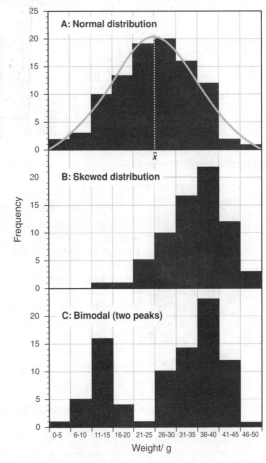

Statistic	Definition and use	Method of calculation
Mean	• The average of all data entries. • Measure of central tendency for normally distributed data.	• Add up all the data entries. • Divide by the total number of data entries.
Median	• The middle value when data entries are placed in rank order. • A good measure of central tendency for skewed distributions.	• Arrange the data in increasing rank order. • Identify the middle value. • For an even number of entries, find the mid point of the two middle values.
Mode	• The most common data value. • Suitable for bimodal distributions and qualitative data.	• Identify the category with the highest number of data entries using a tally chart or a bar graph.
Range	• The difference between the smallest and largest data values. • Provides a crude indication of data spread.	• Identify the smallest and largest values and find the difference between them.

When NOT to calculate a mean:

In certain situations, calculation of a simple arithmetic mean is inappropriate.

Remember:

• *DO NOT* calculate a mean from values that are already means (averages) themselves.

• *DO NOT* calculate a mean of ratios (e.g. percentages) for several groups of different sizes; go back to the raw values and recalculate.

• *DO NOT* calculate a mean when the measurement scale is not linear, e.g. pH units are not measured on a linear scale.

Periodicals: Describing the normal distribution

Related activities: Variables and Data, The Reliability of the Mean
Web links: Taking the Next Step

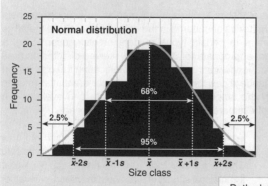

Normal distribution

Measuring Spread

The **standard deviation** is a frequently used measure of the variability (spread) in a set of data. It is usually presented in the form $\bar{x} \pm s$. In a normally distributed set of data, 68% of all data values will lie within one standard deviation (s) of the mean ($\bar{x}$) and 95% of all data values will lie within two standard deviations of the mean (left).

Two different sets of data can have the same mean and range, yet the distribution of data within the range can be quite different. In both the data sets pictured in the histograms below, 68% of the values lie within the range $\bar{x} \pm 1s$ and 95% of the values lie within $\bar{x} \pm 2s$. However, in B, the data values are more tightly clustered around the mean.

Both plots show a normal distribution with a symmetrical spread of values about the mean.

Histogram B has a smaller standard deviation; the values are clustered more tightly around the mean.

Histogram A has a larger standard deviation; the values are spread widely around the mean.

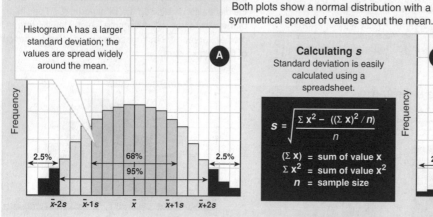

A

Calculating s

Standard deviation is easily calculated using a spreadsheet.

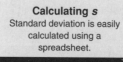
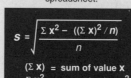

$$ s = \sqrt{\frac{\Sigma x^2 - ((\Sigma x)^2 / n)}{n}} $$

(Σx) = sum of value x
Σx^2 = sum of value x^2
n = sample size

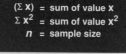

B

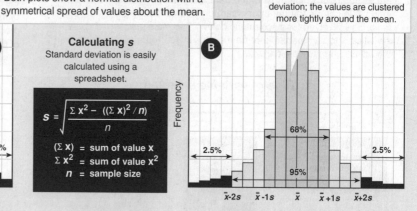

Fern spores

Case Study: Fern Reproduction

Raw data (below) and descriptive statistics (right) from a survey of the number of spores found on the fronds of a fern plant.

Raw data: Number of spores per frond

64	60	64	62	68	66	63
69	70	63	70	70	63	62
71	69	59	70	66	61	70
67	64	63	64			

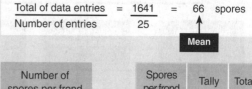

$$ \frac{\text{Total of data entries}}{\text{Number of entries}} = \frac{1641}{25} = 66 \text{ spores} $$

Mean

Number of spores per frond (in rank order)		Spores per frond	Tally	Total
59	66	59	✔	1
60	66	60	✔	1
61	67	61	✔	1
62	68	62	✔✔	2
62	69	63	✔✔✔✔	4
63	69	64	✔✔✔✔	4
63	70	65		0
63	70 **Median**	66	✔✔	2
63	70	67	✔	1
63	70	68	✔	1
64	70	69	✔✔	2
64	70	70	✔✔✔✔✔	5
64	71	71	✔	1
64			**Mode**	

1. Give a reason for the difference between the mean, median, and mode for the fern spore data:

_____

_____

_____

_____

2. Calculate the mean, median, and mode for the data on beetle masses below. Draw up a tally chart and show all calculations:

Beetle masses / g		
2.2	2.1	2.6
2.5	2.4	2.8
2.5	2.7	2.5
2.6	2.6	2.5
2.2	2.8	2.4

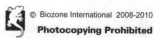

Interpreting Sample Variability

Measures of central tendency, such as mean, attempt to identify the most representative value in a set of data, but the description of a data set also requires that we know something about how far the data values are spread around that central measure. As we have seen in the previous activity, the **standard deviation** (s) gives a simple measure of the spread or **dispersion** in data. The **variance** (s^2) is also a measure of dispersion, but the standard deviation is usually preferred because it is expressed in the original units. Two data sets could have exactly the same mean values, but very different values of dispersion. If we were simply to use the central tendency to compare these data sets, the results would (incorrectly) suggest that they were alike. The assumptions we make about a population will be affected by what the sample data tell us. This is why it is important that sample data are unbiased (e.g. collected by **random sampling**) and that the sample set is as large as practicable. This exercise will help to illustrate how our assumptions about a population are influenced by the information provided by the sample data.

Random Sampling, Sample Size, and Dispersion in Data

Sample size and sampling bias can both affect the information we obtain when we sample a population. In this exercise you will calculate some descriptive statistics for some sample data.

The complete set of sample data we are working with comprises 689 length measurements of year zero (young of the year) perch (column left). Basic descriptive statstics for the data have bee calculated for you below and the frequency histogram has also been plotted.

Look at this data set and then complete the exercise to calculate the same statistics from each of two smaller data sets (tabulated right) drawn from the same population. This excercise shows how random sampling, large sample size, and sampling bias affect our statistical assessment of variation in a population.

Complete sample set n = 689 (random)

Length in mm	Freq
25	1
26	0
27	0
28	0
29	0
30	0
31	0
32	2
33	3
34	3
35	4
36	5
37	10
38	23
39	22
40	33
41	39
42	41
43	41
44	36
45	49
46	32
47	14
48	32
49	27
50	25
51	24
52	17
53	18
54	27
55	21
56	20
57	11
58	18
59	16
60	22
61	13
62	8
63	10
64	5
65	7
66	2
67	3
68	3
69	1
70	0
71	1

Small sample set n = 30 (random)

Length in mm	Freq
25	1
26	0
27	0
28	0
29	0
30	0
31	0
32	0
33	0
34	0
35	2
36	0
37	0
38	3
39	2
40	1
41	3
42	0
43	0
44	0
45	0
46	1
47	0
48	2
49	0
50	0
51	1
52	3
53	0
54	0
55	0
56	0
57	1
58	0
59	3
60	2
61	2
62	0
63	0
64	0
65	0
66	0
67	2
68	1
	30

Small sample set n = 50 (bias)

Length in mm	Freq
46	1
47	0
48	0
49	1
50	0
51	0
52	1
53	1
54	1
55	1
56	0
57	2
58	2
59	4
60	1
61	0
62	8
63	10
64	13
65	2
66	0
67	2
	50

The person gathering this set of data was biased towards selecting larger fish because the mesh size on the net was too large to retain small fish

This population was sampled randomly to obtain this data set

This column records the number of fish of each size

Number of fish in the sample

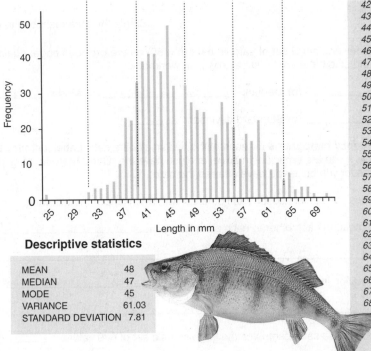

Length of year zero perch

$\bar{x}-2s$ $\bar{x}-1s$ $\bar{x}$ $\bar{x}+1s$ $\bar{x}+2s$

Length in mm

Descriptive statistics

MEAN	48
MEDIAN	47
MODE	45
VARIANCE	61.03
STANDARD DEVIATION	7.81

1. For the complete data set (n = 689) calculate the percentage of data falling within:

   (a) ± one standard deviation of the mean: _____

   (b) ± two standard deviations of the mean: _____

   (c) Explain what this information tells you about the distribution of year zero perch from this site: _____

   _____

2. Give another reason why you might reach the same conclusion about the distribution: _____

   _____

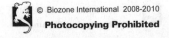

Periodicals:
Estimating the mean
and standard deviation

Related activities: Descriptive Statistics,
Drawing Histograms

DA 3

Skills In Biology

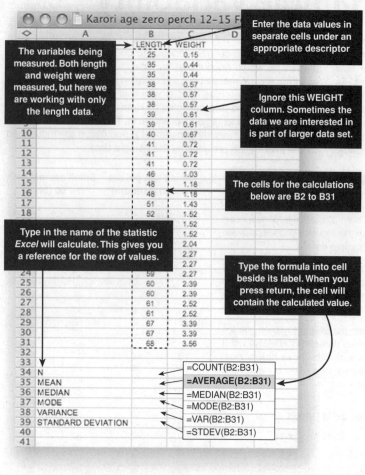

The variables being measured. Both length and weight were measured, but here we are working with only the length data.

Enter the data values in separate cells under an appropriate descriptor

Ignore this WEIGHT column. Sometimes the data we are interested in is part of larger data set.

The cells for the calculations below are B2 to B31

Type in the name of the statistic *Excel* will calculate. This gives you a reference for the row of values.

Type the formula into cell beside its label. When you press return, the cell will contain the calculated value.

	A	B	C	D
◇		LENGTH	WEIGHT	
		25	0.15	
		35	0.44	
		35	0.44	
		38	0.57	
		38	0.57	
		38	0.57	
		39	0.61	
		39	0.61	
10		40	0.67	
11		41	0.72	
12		41	0.72	
13		41	0.72	
14		46	1.03	
15		48	1.18	
16		48	1.18	
17		51	1.43	
18		52	1.52	
			1.52	
			1.52	
			1.52	
			2.04	
			2.27	
			2.27	
24		59	2.27	
25		60	2.39	
26		60	2.39	
27		61	2.52	
28		61	2.52	
29		67	3.39	
30		67	3.39	
31		68	3.56	
32				
33				
34	N	=COUNT(B2:B31)		
35	MEAN	=AVERAGE(B2:B31)		
36	MEDIAN	=MEDIAN(B2:B31)		
37	MODE	=MODE(B2:B31)		
38	VARIANCE	=VAR(B2:B31)		
39	STANDARD DEVIATION	=STDEV(B2:B31)		
40				
41				

Calculating Descriptive Statistics Using *Excel*

You can use *Microsoft Excel* or another similar spreadsheet program to easily calculate descriptive statistics for sample data.

In this first example, the smaller data set ($n = 30$) is shown as it would appear on an *Excel* spreadsheet, ready for the calculations to be made. Use this guide to enter your data into a spreadsheet and calculate the descriptive statistics as described.

When using formulae in *Excel*, = indicates that a formula follows. The cursor will become active and you will be able to select the cells containing the data you are interested in, or you can type the location of the data using the format shown. The data in this case are located in the cells B2 through to B31 (B2:B31).

3. For this set of data, use a spreadsheet to calculate:

    (a)   Mean: _____

    (b)   Median _____

    (c)   Mode: _____

    (d)   Sample variance: _____

    (e)   Standard deviation: _____

Staple the spreadsheet into your workbook.

4. Repeat the calculations for the second small set of sample data (n = 50) on the previous page. Again, calculate the statistics as indicated below and staple the spreadsheet into your workbook:

    (a) Mean: _____ (b) Median: _____ (c) Mode: _____

    (d) Variance: _____ (e) Standard deviation: _____

5. On a separate sheet, plot **frequency histograms** for each of the two small data sets. Label them $n = 30$ and $n = 50$. Staple them into your workbook. If you are proficient in *Excel* and you have the "Data Analysis" plug in loaded, you can use *Excel* to plot the histograms for you once you have entered the data.

6. Compare the descriptive statistics you calculated for each data set with reference to the following:

    (a) How close the median and mean to each other in each sample set: _____

    _____

    (b) The size of the standard deviation in each case: _____

    _____

    (c) How close each small of the sample sets resembles the large sample set of 689 values: _____

    _____

    _____

7. (a) Compare the two frequency histograms you have plotted for the two smaller sample data sets: _____

    _____

    _____

    (b) Explain why you think two histograms look so different: _____

    _____

    _____

Evaluating Your Results

Once you have completed the practical part of an experiment, the next task is to evaluate your results in the light of your own hypothesis and your current biological knowledge. A critical evaluation of any study involves analysing, presenting, and discussing the results, as well as accounting for any deficiencies in your procedures and erroneous results. This activity describes an experiment in which germinating seeds of different ages were tested for their level of catalase activity using hydrogen peroxide solution as the substrate and a simple apparatus to measure oxygen production (see background). Completing this activity, which involves a critical evaluation of the second-hand data provided will help to prepare you for your own evaluative task.

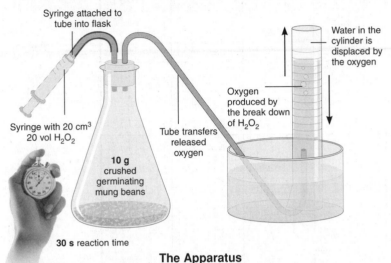

Syringe attached to tube into flask

Syringe with 20 cm³ 20 vol H₂O₂

10 g crushed germinating mung beans

30 s reaction time

Tube transfers released oxygen

Oxygen produced by the break down of H₂O₂

Water in the cylinder is displaced by the oxygen

The Apparatus

In this experiment, 10 g germinating mung bean seeds (0.5, 2, 4, 6, or 10 days old) were ground by hand with a mortar and pestle and placed in a conical flask as above. There were six trials at each of the five seedling ages. With each trial, 20 cm³ of 20 vol H_2O_2 was added to the flask at time 0 and the reaction was allowed to run for 30 seconds. The oxygen released by the decomposition of the H_2O_2 by catalase in the seedlings was collected via a tube into an inverted measuring cylinder. The volume of oxygen produced is measured by the amount of water displaced from the cylinder. The results from all trials are tabulated below:

The Aim

To investigate the effect of germination age on the level of catalase activity in mung beans.

Background

Germinating seeds are metabolically very active and this metabolism inevitably produces reactive oxygen species, including hydrogen peroxide (H_2O_2). H_2O_2 is helps germination by breaking dormancy, but it is also toxic. To counter the toxic effects of H2O2 and prevent cellular damage, germinating seeds also produce **catalase**, an enzyme that catalyses the breakdown of H_2O_2 to water and oxygen.

A class was divided into six groups with each group testing the seedlings of each age. Each group's set of results (for 0.5, 2, 4, 6, and 10 days) therefore represents one trial.

Stage of germination / days	Volume of oxygen collected after 30s / cm³						Mean	Standard deviation	Mean rate / cm³ s⁻¹ g⁻¹
Trial #	1	2	3	4	5	6			
0.5	9.5	10	10.7	9.5	10.2	10.5			
2	36.2	30	31.5	37.5	34	40			
4	59	66	69	60.5	66.5	72			
6	39	31.5	32.5	41	40.3	36			
10	20	18.6	24.3	23.2	23.5	25.5			

1. Write the equation for the catalase reaction with hydrogen peroxide: _____

2. Complete the table above to summarise the data from the six trials:

   (a) Calculate the mean volume of oxygen for each stage of germination and enter the values in the table.

   (b) Calculate the standard deviation for each mean and enter the values in the table (you may use a spreadsheet).

   (c) Calculate the mean rate of oxygen production in cm³ per second per gram. For the purposes of this exercise, assume that the weight of germinating seed in every case was 10.0 g.

3. In another scenario, group (trial) #2 obtained the following measurements for volume of oxygen produced: 0.5 d: 4.8 cm³, 2 d: 29.0 cm³, 4 d: 70 cm³, 6 d: 30.0 cm³, 10 d: 8.8 cm³ (pencil these values in beside the other group 2 data set).

   (a) Describe how group 2's new data accords with the measurements obtained from the other groups: _____

   _____

   (b) Describe how you would approach a reanalysis of the data set incorporating group 2's new data: _____

   _____

   _____

(c) Explain the rationale for your approach _____

_____

4. Use the tabulated data to plot an appropriate graph of the results on the grid provided:

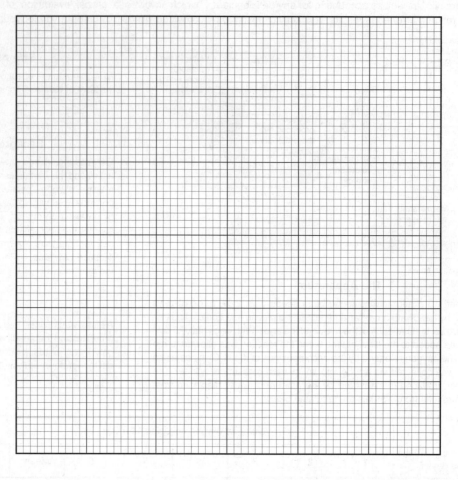

5. (a) Describe the trend in the data: _____

_____

_____

(b) Explain the relationship between stage of germination and catalase activity shown in the data: _____

_____

_____

6. Describe any potential sources of errors in the apparatus or the procedure: _____

_____

_____

7. Describe two things that might affect the validity of findings in this experimental design: _____

_____

_____

8. Describe one improvement you could make to the experiment in order to generate more reliable data: _____

_____

_____

_____

KEY TERMS Mix and Match

ACCURACY

BIBLIOGRAPHY

BIOLOGICAL DRAWING

CITATION

CONTROL

CONTROLLED VARIABLE

DATA

DATALOGGER

DEPENDENT VARIABLE

GRAPH

HISTOGRAM

HYPOTHESIS

INDEPENDENT VARIABLE

MEAN

MEASUREMENT

MEDIAN

MODE

OBSERVATION

PARAMETER

PRECISION

QUALITATIVE DATA

QUANTITATIVE DATA

RAW DATA

RELIABILITY

SAMPLE

SCIENTIFIC METHOD

TREND (OF DATA)

VARIABLE

A Note normally appearing directly after a new fact or data that states the author of the information and the date it was published.

B A variable whose values are set, or systematically altered, by the investigator.

C A diagram drawn to accurately show what has been seen by the observer.

D Facts collected for analysis.

E The value that occurs most often in a data set.

F A pattern observed in processed data showing that data values may be linked.

G A standard (reference) treatment that helps to ensure that the responses to the other treatments can be reliably interpreted.

H The sum of the data divided by the number of data entries (n).

I A variable whose values are determined by another variable.

J Data able to be expressed in numbers. Numerical values derived from counts or measurements.

K Variable that is fixed at a specific amount as part of the design of experiment.

L A type of column graph used to display frequency distributions.

M A tentative explanation of an observation, capable of being tested by experimentation.

N The number that occurs in the middle of a set of sorted numbers. It divides the upper half of the number data set from the lower half.

O How close a statistic is to the value of the parameter being estimated.

P The sampling of an object or substance to record numerical data that describes some aspect of the it, e.g. length or temperature.

Q Data that have not been processed or manipulated in any way.

R A diagram which often displays numerical information in a way that can be used to identify trends in the data.

S The act of seeing and noting an occurrence in the object or substance being studied.

T The use of an ordered, repeatable method to investigate, manipulate, gather, and record data.

U Data described in descriptors or terms rather than by numbers.

V A sub-set of a whole used to estimate the values that might have been obtained if every individual or response was measured.

W Device that is able to record data as it changes, in real time.

X The measure of confidence that can be placed in a set of data, e.g. as indicated by standard deviation or other measures of sample variability.

Y A list displaying the titles and publication information of resources used in the gathering of information.

Z How close a measured value to its true amount. The exactness of a measurement.

AA A factor in an experiment that is subject to change.

BB A quantity that defines a characteristic of a system.

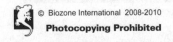

Cell Structure and Processes

Cell structure and microscopy
- Characteristics of cells
- Prokaryotic vs eukaryotic cells
- Cell structures and organelles
- Microscopy as a tool to study cells

Cell membranes and transport
- The structure of cellular membranes
- The roles of membranes in cells
- Passive transport processes
- Active transport processes

Cell division
- The cell cycle and mitosis
- Regulation of cell division; cancer

Cellular diversity and organisation
- Cellular differentiation
- Hierarchy of organisation
- Plant and animal tissues

Important in this section...
- *Develop understanding of cell structure & processes*
- *Understand how gases are exchanged in animals*
- *Describe the internal transport systems of animals*
- *Describe movement of water & sugars in plant tissues*

Exchange Surfaces & Breathing

Gas exchange systems
- The need for gas exchange
- Surface area to volume relationships
- Who has a gas exchange system?

Gas exchange in humans
- Breathing and the structure of lungs
- The alveolar-capillary membrane
- Measuring lung function

The cell is the functional basic unit of life. Cells can be characterised by their unique features.

Gas exchange systems facilitate rapid exchanges of respiratory gases with the environment.

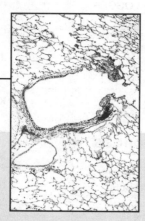

Cells, Exchange, and Transport

Cells are recognised by their unique features. Microscopy is a tool used to study the ultrastructure of cells.

Organisms exchange materials with the environment. They have systems to maximise the efficiency of these exchanges.

Internal transport systems enable efficient movement of materials around the body by mass flow.

Plant transport tissues also provide support. Water and sugars are transported in the xylem and phloem.

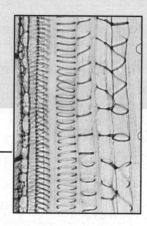

Animal transport systems
- The need for internal transport
- Open circulatory systems
- Closed circulatory systems

Internal transport in humans
- The human circulatory system
- Heart structure and function
- The cardiac cycle
- Control of heart activity

Blood
- The components of blood
- Blood and gas transport

Blood vessels
- Veins, arteries, and capillaries
- Capillary networks and exchanges
- The formation of tissue fluid

Transport in Animals

Plant transport tissues
- Xylem and phloem
- The structure and function of leaves
- The structure and function of stems
- The structure and function of roots

Transport of water and minerals
- Water loss and gas exchange
- The transpiration stream
- Water and mineral uptake in roots
- Adaptations of xerophytes

Transport of sugars
- Translocation in plants
- Sources and sinks
- Hypotheses for phloem transport

Transport in Plants

Cell Structure

KEY CONCEPTS

▶ Cells are the fundamental units of life.

▶ Microscopy can be used to understand cellular structure and function.

▶ Cells share many of the same components but some organelles are unique to certain cell types.

▶ There are distinguishing differences between the cells of prokaryotes and eukaryotes.

KEY TERMS

cell wall
centrioles
chloroplast
cilia
cytoplasm
cytoskeleton
electron microscope
endoplasmic reticulum (ER)
eukaryotic cell
flagella
flagellum (pl. flagella)
Golgi apparatus
intermediate filament
light (=optical) microscope
lysosome
magnification
microfilament
microtubule
mitochondrion
nuclear envelope
nuclear pore
nucleolus
nucleus
organelles
plasma membrane
prokaryotic cell
resolution
ribosome
rough ER (rER)
SEM
smooth ER (sER)
stain
TEM
vacuole

Periodicals:

Listings for this chapter are on page 338

Weblinks:

www.biozone.co.uk/
weblink/OCR-AS-2641.html

Teacher Resource CD-ROM:

The Cell Theory

OBJECTIVES

☐ 1. Use the **KEY TERMS** to help you understand and complete these objectives.

Microscopy
pages 41-44, 57

☐ 2. Recognise the contribution of microscopy to our modern understanding of cell structure and function.

☐ 3. Explain the difference between **magnification** and **resolution**. Calculate the linear magnification of images viewed with a microscope.

☐ 4. Compare and contrast the structure and basic principles of **light** (optical) **microscopes** and **electron microscopes**.

☐ 5. PRACTICAL: Use both stereo and compound light microscopes to locate material and focus images.

☐ 6. Distinguish between the principles and uses of TEM (transmission electron microscopes) and SEM (scanning electron microscopes).

☐ 7. Compare the magnification and resolution achieved using a light microscope, a transmission EM, and a scanning EM.

☐ 8. Explain the role of sample preparation (including **stains**) in light and electron microscopy.

☐ 9. PRACTICAL: Demonstrate an ability use simple **staining techniques** to show features of cells.

☐ 10. PRACTICAL: Demonstrate an ability to prepare a temporary mount for viewing with a light microscope.

Cells as Living Entities
pages 40, 45-56

☐ 11. Describe the range of cell sizes. Express cell sizes in different units of measurement (mm, μm, nm).

☐ 12. Use diagrams and electron micrographs to describe the structures and organelles of **eukaryotic cells**. Describe the functions of the organelles and structures identified.

☐ 13. Describe the interrelationship between the rough endoplasmic reticulum (rER) and the Golgi apparatus.

☐ 14. Describe the components of the cell's cytoskeleton and explain its various roles in providing mechanical strength, in intracellular transport, and in cell division and movement.

☐ 15. Use drawings and electron micrographs to compare and contrast the structure of **prokaryotic cells** and eukaryotic cells.

☐ 16. Use drawings and photomicrographs to compare and contrast the structure and ultrastructure of **plant cells** and **animal cells**.

Cell Sizes

Cells are extremely small and can only be seen properly when viewed through the magnifying lenses of a microscope. The diagrams below show a variety of cell types, together with a virus and a microscopic animal for comparison. For each of these images, note the scale and relate this to the type of microscopy used.

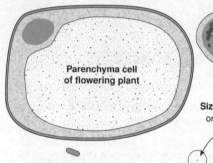

Parenchyma cell of flowering plant

Human white blood cell

Eukaryotic cells
(e.g. plant and animal cells)
Size: 10–100 μm diameter. Cellular organelles may be up to 10 μm.

Prokaryotic cells
Size: Typically 2–10 μm length, 0.2–2 μm diameter. Upper limit, 30 μm long.

Viruses
Size: 0.02–0.25 μm (20–250 nm)

Units of length (International System)

Unit	Metres	Equivalent
1 metre (m)	1 m	= 1000 millimetres
1 millimetre (mm)	10^{-3} m	= 1000 micrometres
1 micrometre (μm)	10^{-6} m	= 1000 nanometres
1 nanometre (nm)	10^{-9} m	= 1000 picometres

Micrometres are sometime referred to as **microns**. Smaller structures are usually measured in nanometres (nm) e.g. molecules (1 nm) and plasma membrane thickness (10 nm).

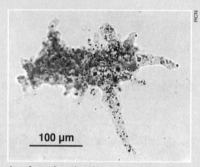

100 μm

An **Amoeba** showing extensions of the cytoplasm called pseudopodia. This protoctist changes its shape, exploring its environment.

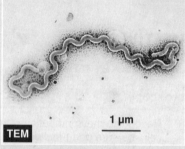

1 μm

TEM

A long thin cell of the spirochete bacterium **Leptospira pomona**, which causes the disease leptospirosis.

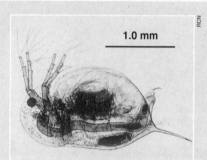

1.0 mm

Daphnia showing its internal organs. These freshwater microcrustaceans are part of the zooplankton found in lakes and ponds.

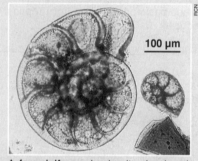

100 μm

A **foraminiferan** showing its chambered, calcified shell. These single-celled protozoans are marine planktonic amoebae.

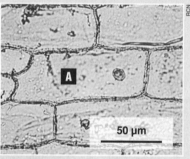

A

50 μm

Epidermal cells (skin) from an onion bulb showing the nucleus, cell walls and cytoplasm. Organelles are not visible at this resolution.

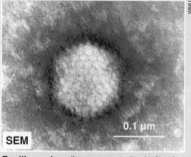

0.1 μm

SEM

Papillomavirus (human wart virus) showing its polyhedral protein coat (20 triangular faces, 12 corners) made of ball-shaped structures.

1. Using the measurement scales provided on each of the photographs above, determine the longest dimension (length or diameter) of the cell/animal/virus in μm and mm (choose the cell marked **A** for epidermal cells):

(a) *Amoeba*: _____ μm _____ mm   (d) Epidermis: _____ μm _____ mm

(b) Foraminiferan: _____ μm _____ mm   (e) *Daphnia*: _____ μm _____ mm

(c) *Leptospira*: _____ μm _____ mm   (f) *Papillomavirus*: _____ μm _____ mm

2. List these six organisms in order of size, from the smallest to the largest: _____

_____

3. Study the scale of your ruler and state which of these six organisms you would be able to see with your unaided eye:

_____

4. Calculate the equivalent length in millimetres (mm) of the following measurements:

(a) 0.25 μm: _____   (b) 450 μm: _____   (c) 200 nm: _____

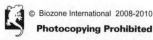

Optical Microscopes

The light microscope is one of the most important instruments used in biology practicals, and its correct use is a basic and essential skill of biology. High power light microscopes use a combination of lenses to magnify objects up to several hundred times. They are called **compound microscopes** because there are two or more separate lenses involved. A typical compound light microscope (bright field) is shown below (top photograph). The specimens viewed with these microscopes must be thin and mostly transparent. Light is focused up through the condenser and specimen; if the specimen is thick or opaque,

little or no detail will be visible. The microscope below has two eyepieces (**binocular**), although monocular microscopes, with a mirror rather than an internal light source, may still be encountered. Dissecting microscopes (lower photograph) are a type of binocular microscope used for observations at low total magnification (x4 to x50), where a large working distance between objective lenses and stage is required. A dissecting microscope has two separate lens systems, one for each eye. Such microscopes produce a 3-D view of the specimen and are sometimes called stereo microscopes for this reason.

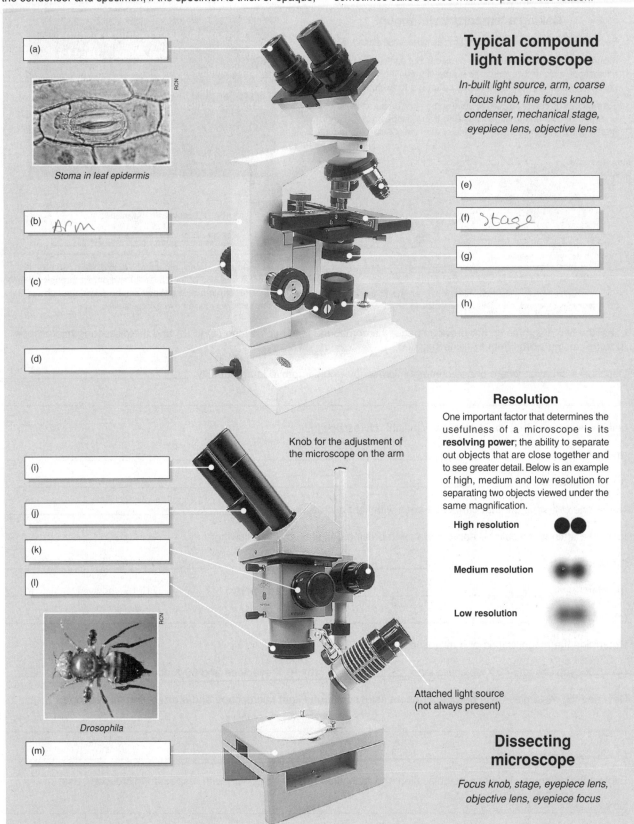

(a)

Stoma in leaf epidermis

(b) Arm

(c)

(d)

(e)

(f) stage

(g)

(h)

Typical compound light microscope

In-built light source, arm, coarse focus knob, fine focus knob, condenser, mechanical stage, eyepiece lens, objective lens

(i)

(j)

(k)

(l)

Knob for the adjustment of the microscope on the arm

Drosophila

(m)

Attached light source (not always present)

Resolution

One important factor that determines the usefulness of a microscope is its **resolving power**; the ability to separate out objects that are close together and to see greater detail. Below is an example of high, medium and low resolution for separating two objects viewed under the same magnification.

High resolution

Medium resolution

Low resolution

Dissecting microscope

Focus knob, stage, eyepiece lens, objective lens, eyepiece focus

Cell Structure

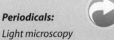

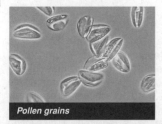

Pollen grains

Phase contrast illumination increases contrast of transparent specimens by producing interference effects.

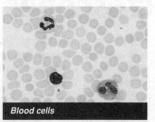

Blood cells

Leishman's stain is used to show red blood cells as red/pink, while staining the nucleus of white blood cells blue.

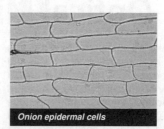

Onion epidermal cells

Standard **bright field** lighting shows cells with little detail; only cell walls, with the cell nuclei barely visible.

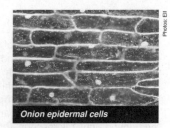

Onion epidermal cells

Dark field illumination is excellent for viewing near transparent specimens. The nucleus of each cell is visible.

Photos: Ell

Making a temporary wet mount

1. **Sectioning**: Very thin sections of fresh material are cut with a razorblade.

2. **Mounting**: The thin section(s) are placed in the centre of a clean glass microscope slide and covered with a drop of mounting liquid (e.g. water, glycerol or stain). A coverslip is placed on top to exclude air (below).

3. **Staining**: Dyes can be applied to stain some structures and leave others unaffected. The stains used in dyeing living tissues are called **vital stains** and they can be applied before or after the specimen is mounted.

Commonly used temporary stains

Stain	Final colour	Used for
Iodine solution	blue-black	Starch
Aniline sulfate	yellow	Lignin
Schultz's solution	blue	Starch
	blue or violet	Cellulose
	yellow	Protein, cutin, lignin, suberin
Methylene blue	blue	Nuclei

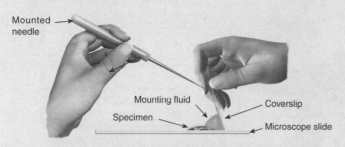

Mounted needle

Mounting fluid

Specimen

Coverslip

Microscope slide

A mounted needle is used to support the coverslip and lower it gently over the specimen. This avoids including air in the mount.

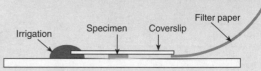

Irrigation   Specimen   Coverslip   Filter paper

If a specimen is already mounted, a drop of stain can be placed at one end of the coverslip and drawn through using filter paper (above). Water can be drawn through in the same way to remove excess stain.

1. Label the two diagrams on the previous page, the compound light microscope (a) to (h) and the dissecting microscope (i) to (m), using words from the lists supplied.

2. Describe a situation where phase contrast microscopy would improve image quality: _____
_____

3. List two structures that could be seen with light microscopy in:

    (a) A plant cell: _cell wall, nuclei_____

    (b) An animal cell: _____

4. Name one cell structure that cannot be seen with light microscopy: _____

5. Identify a stain that would be appropriate for improving definition of the following:

    (a) Blood cells: _____   (d) Lignin: _____

    (b) Starch: _____   (e) Nuclei and DNA: _____

    (c) Protein: _____   (f) Cellulose: _____

6. Determine the magnification of a microscope using:

    (a) 15 X eyepiece and 40 X objective lens: _____   (b) 10 X eyepiece and 60 X objective lens: _____

7. Describe the main difference between a bright field, compound light microscope and a dissecting microscope:
_____
_____

8. Explain the difference between magnification and resolution (resolving power) with respect to microscope use:
_____
_____

Electron Microscopes

Electron microscopes (EMs) use a beam of electrons, instead of light, to produce an image. The higher resolution of EMs is due to the shorter wavelengths of electrons. There are two basic types of electron microscope: **scanning electron microscopes**

(SEMs) and **transmission electron microscopes** (TEMs). In SEMs, the electrons are bounced off the surface of an object to produce detailed images of the external appearance. TEMs produce very clear images of specially prepared thin sections.

Transmission Electron Microscope (TEM)

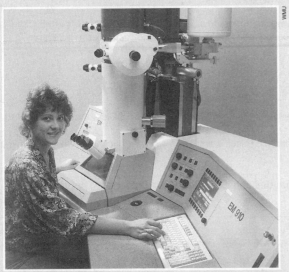

The transmission electron microscope is used to view extremely thin sections of material. Electrons pass through the specimen and are scattered. Magnetic lenses focus the image onto a fluorescent screen or photographic plate. The sections are so thin that they have to be prepared with a special machine, called an **ultramicrotome**, that can cut wafers to just 30 thousandths of a millimetre thick. It can magnify several hundred thousand times.

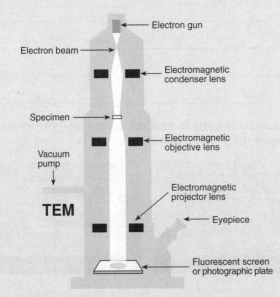

TEM: Electron gun, Electron beam, Electromagnetic condenser lens, Specimen, Electromagnetic objective lens, Vacuum pump, Electromagnetic projector lens, Eyepiece, Fluorescent screen or photographic plate

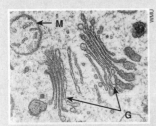

TEM photo showing the Golgi (**G**) and a mitochondrion (**M**).

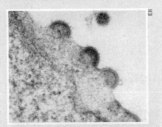

Three HIV viruses budding out of a human lymphocyte (TEM).

Scanning Electron Microscope (SEM)

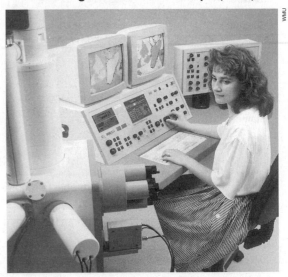

The scanning electron microscope scans a sample with a beam of primary electrons that knock electrons from its surface. These secondary electrons are picked up by a collector, amplified, and transmitted onto a viewing screen or photographic plate, producing a superb 3-D image. A microscope of this power can easily obtain clear pictures of organisms as small as bacteria and viruses. The image produced is of the outside surface only.

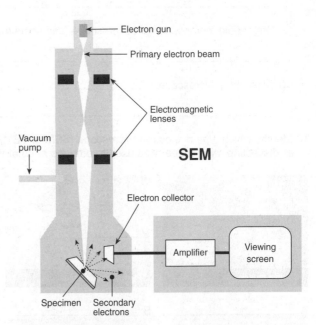

SEM: Electron gun, Primary electron beam, Electromagnetic lenses, Vacuum pump, Electron collector, Amplifier, Viewing screen, Specimen, Secondary electrons

SEM photo of stoma and epidermal cells on the upper surface of a leaf.

Image of hair louse clinging to two hairs on a Hooker's sealion (SEM).

Cell Structure

Periodicals: Transmission electron microscopy

Related activities: Optical Microscopes, Interpreting Electron Micrographs

RA 2

	Light Microscope	Transmission Electron Microscope (TEM)	Scanning Electron Microscope (SEM)
Radiation source:	light	electrons	electrons
Wavelength:	400-700 nm	0.005 nm	0.005 nm
Lenses:	glass	electromagnetic	electromagnetic
Specimen:	living or non-living supported on glass slide	non-living supported on a small copper grid in a vacuum	non-living supported on a metal disc in a vacuum
Maximum resolution:	200 nm	1 nm	10 nm
Maximum magnification:	1500 x	250 000 x	100 000 x
Stains:	coloured dyes	impregnated with heavy metals	coated with carbon or gold
Type of image:	coloured	monochrome (black & white)	monochrome (black & white)

1. Explain why electron microscopes are able to resolve much greater detail than a light microscope:

_____

_____

_____

2. Describe two typical applications for each of the following types of microscope:

   (a) Transmission electron microscope (TEM): _____

   _____

   (b) Scanning electron microscope (SEM): _____

   _____

   (c) Bright field, compound light microscope (thin section): _____

   _____

   (d) Dissecting microscope: _____

   _____

3. Identify which type of electron microscope (SEM or TEM) or optical microscope (bright field, compound light microscope or dissecting microscope) was used to produce each of the images in the photos below (A-H):

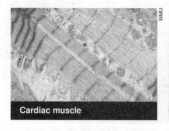
Cardiac muscle

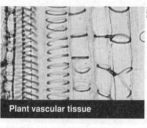
Plant vascular tissue

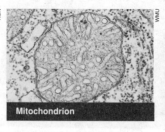
Mitochondrion

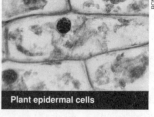
Plant epidermal cells

A _____   B _____   C _____   D _____

Head louse

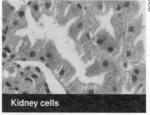
Kidney cells

Alderfly larva

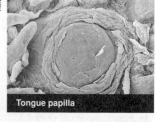
Tongue papilla

E _____   F _____   G _____   H _____

Prokaryotic Cells

Bacterial (prokaryotic) cells are much smaller and simpler than the cells of eukaryotes. They lack many eukaryotic features (e.g. a distinct nucleus and membrane-bound cellular organelles). The bacterial cell wall is an important feature. It is a complex, multi-layered structure and often has a role in virulence. These pages illustrate some features of bacterial structure and diversity.

Structure of a Generalised Bacterial Cell

Plasmids: Small, circular DNA molecules (accessory chromosomes) which can reproduce independently of the main chromosome. They can move between cells, and even between species, by **conjugation**. This property accounts for the transmission of antibiotic resistance between bacteria. Plasmids are also used as vectors in recombinant DNA technology.

Single, circular main chromosome: Makes them haploid for most genes. It is possible for some genes to be found on both the plasmid and chromosome and there may be several copies of a gene on a group of plasmids.

The cell lacks a nuclear membrane, so there is no distinct nucleus and the chromosome is in direct contact with the cytoplasm. It is possible for free ribosomes to attach to mRNA while the mRNA is still in the process of being transcribed from the DNA.

Fimbriae: Hairlike structures that are shorter, straighter, and thinner than flagella. They are used for attachment, not movement. Pili are similar to fimbriae, but are longer and less numerous. They are involved in bacterial conjugation (below) and as phage receptors (opposite).

Cell surface membrane: Similar in composition to eukaryotic membranes, although less rigid.

1 µm

Cytoplasm

Glycocalyx. A viscous, gelatinous layer outside the cell wall. It is composed of polysaccharide and/or polypeptide. If it is firmly attached to the wall, it is called a **capsule**. If loosely attached, it is called a **slime layer**. Capsules may contribute to virulence in pathogenic species, e.g. by protecting the bacteria from the host's immune attack. In some species, the glycocalyx allows attachment to substrates.

Cell wall. A complex, semi-rigid structure that gives the cell shape, prevents rupture, and serves as an anchorage point for flagella. The cell wall is composed of a macromolecule called **peptidoglycan**; repeating disaccharides attached by polypeptides to form a lattice. The wall also contains varying amounts of lipopolysaccharides and lipoproteins. The amount of peptidoglycan present in the wall forms the basis of the diagnostic **gram stain**. In many species, the cell wall contributes to their virulence (disease-causing ability).

Flagellum (pl. flagella). Some bacteria have long, filamentous appendages, called flagella, that are used for locomotion. There may be a single polar flagellum (monotrichous), one or more flagella at each end of the cell, or the flagella may be distributed over the entire cell (peritrichous).

Cell Structure

Bacterial cell shapes

Most bacterial cells range between 0.20-2.0µm in diameter and 2-10µm length. Although they are a very diverse group, much of this diversity is in their metabolism. In terms of gross morphology, there are only a few basic shapes found (illustrated below). The way in which members of each group aggregate after division is often characteristic and is helpful in identifying certain species.

Bacilli
Rod-shape,
e.g. *E. coli*

Bacilli: Rod-shaped bacteria that divide only across their short axis. Most occur as single rods, although pairs and chains are also found. The term bacillus can refer (as here) to shape. It may also denote a genus.

Cocci
Ball-shaped
e.g. *Staphylococcus*

Cocci: usually round, but sometimes oval or elongated. When they divide, the cells stay attached to each other and remain in aggregates e.g. pairs (diplococci) or clusters (staphylococci), that are usually a feature of the genus.

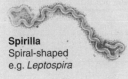

Spirilla
Spiral-shaped
e.g. *Leptospira*

Spirilla and vibrio: Bacteria with one or more twists. Spirilla bacteria have a helical (corkscrew) shape which may be rigid or flexible (as in spirochetes). Bacteria that look like curved rods (comma shaped) are called vibrios.

Bacterial conjugation

The two bacteria below are involved in conjugation: a one-way exchange of genetic information from a donor cell to a recipient cell. The plasmid, which must be of the 'conjugative' type, passes through a tube called a **sex pilus** to the other cell. Which is donor and which is recipient appears to be genetically determined. Conjugation should not be confused with sexual reproduction, as it does not involve the fusion of gametes or formation of a zygote.

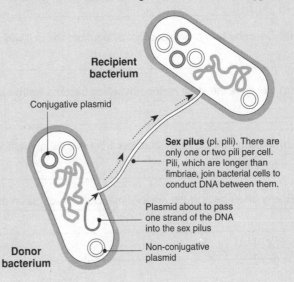

Recipient bacterium

Conjugative plasmid

Sex pilus (pl. pili). There are only one or two pili per cell. Pili, which are longer than fimbriae, join bacterial cells to conduct DNA between them.

Plasmid about to pass one strand of the DNA into the sex pilus

Non-conjugative plasmid

Donor bacterium

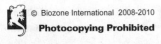
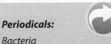

Periodicals:
Bacteria

Related activities: Bacterial Pathogens
Web links: Gram Stain Animation

RA 2

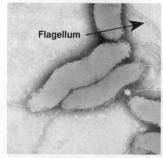
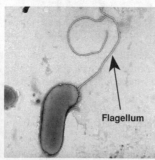

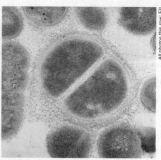

Campylobacter jejuni, a spiral bacterium responsible for foodborne intestinal disease. Note the single flagellum at each end (amphitrichous arrangement).

Helicobacter pylori, a comma-shaped vibrio bacterium that causes stomach ulcers in humans. This bacterium moves by means of multiple polar flagella.

A species of *Spirillum*, a spiral shaped bacterium with a tuft of polar flagella. Most of the species in this genus are harmless aquatic organisms.

Bacteria usually divide by binary fission. During this process, DNA is copied and the cell splits into two cells, as in these gram positive cocci.

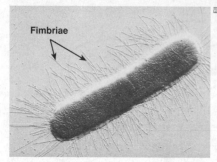
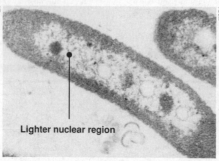

Escherichia coli, a common gut bacterium with **peritrichous** (around the entire cell) **fimbriae**. *E. coli* is a gram negative rod; it does not take up the gram stain but can be counter stained with safranin.

TEM showing *Enterobacter* bacteria, which belong to the family of gut bacteria commonly known as enterics. They are widely distributed in water, sewage, and soil. The family includes motile and non-motile species.

SEM of endospores of ***Bacillus anthracis*** bacteria, which cause the disease anthrax. These heat-resistant spores remain viable for many years and enable the bacteria to survive in a dormant state.

1. Describe three features distinguishing prokaryotic cells from eukaryotic cells:

   (a) ~~Slime~~ layer looped DNA

   (b) no membrane bound organelles

   (c) No nucleus, DNA is free

2. (a) Describe the function of flagella in bacteria: for locomotion, movement of the cell

   (b) Explain how fimbriae differ structurally and functionally from flagella: _____

3. (a) Describe the location and general composition of the bacterial cell wall: It is the outside of the cell, made up of peptidoglycan

   (b) Describe how the glycocalyx differs from the cell wall: _____

4. (a) Describe the main method by which bacteria reproduce: _____

   (b) Explain how conjugation differs from this usual method: _____

5. Briefly describe how the artificial manipulation of plasmids has been used for technological applications: _____

Plant Cells

Plant cells are enclosed in a cellulose cell wall. The cell wall protects the cell, maintains its shape, and prevents excessive water uptake. It does not interfere with the passage of materials into and out of the cell. The diagram below shows the structure and function of a typical plant cell and its organelles.

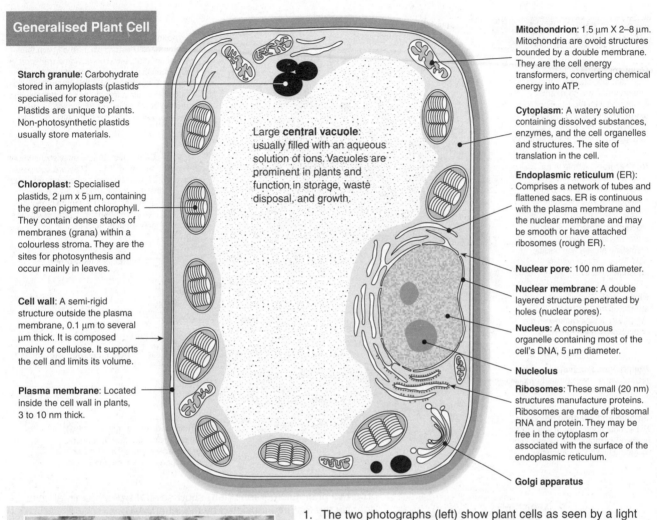

Generalised Plant Cell

Starch granule: Carbohydrate stored in amyloplasts (plastids specialised for storage). Plastids are unique to plants. Non-photosynthetic plastids usually store materials.

Chloroplast: Specialised plastids, 2 µm x 5 µm, containing the green pigment chlorophyll. They contain dense stacks of membranes (grana) within a colourless stroma. They are the sites for photosynthesis and occur mainly in leaves.

Cell wall: A semi-rigid structure outside the plasma membrane, 0.1 µm to several µm thick. It is composed mainly of cellulose. It supports the cell and limits its volume.

Plasma membrane: Located inside the cell wall in plants, 3 to 10 nm thick.

Large **central vacuole**: usually filled with an aqueous solution of ions. Vacuoles are prominent in plants and function in storage, waste disposal, and growth.

Mitochondrion: 1.5 µm X 2–8 µm. Mitochondria are ovoid structures bounded by a double membrane. They are the cell energy transformers, converting chemical energy into ATP.

Cytoplasm: A watery solution containing dissolved substances, enzymes, and the cell organelles and structures. The site of translation in the cell.

Endoplasmic reticulum (ER): Comprises a network of tubes and flattened sacs. ER is continuous with the plasma membrane and the nuclear membrane and may be smooth or have attached ribosomes (rough ER).

Nuclear pore: 100 nm diameter.

Nuclear membrane: A double layered structure penetrated by holes (nuclear pores).

Nucleus: A conspicuous organelle containing most of the cell's DNA, 5 µm diameter.

Nucleolus

Ribosomes: These small (20 nm) structures manufacture proteins. Ribosomes are made of ribosomal RNA and protein. They may be free in the cytoplasm or associated with the surface of the endoplasmic reticulum.

Golgi apparatus

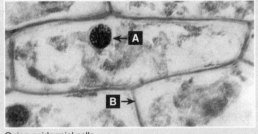

Onion epidermial cells

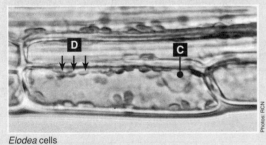

Elodea cells

Photos: RCN

1. The two photographs (left) show plant cells as seen by a light microscope. Identify the basic features labelled **A-D**:

A: _Nucleus_

B: _cell wall_

C: _cytoplasm nucleus_

D: _chloroplasts_

2. Cytoplasmic streaming is a feature of eukaryotic cells, often clearly visible with a light microscope in plant (and algal) cells.

(a) Explain what is meant by cytoplasmic streaming:

_____

_____

(b) For the *Elodea* cell (lower, left), draw arrows to indicate cytoplasmic streaming movements.

3. Describe three structures/organelles present in generalised plant cells but absent from animal cells (also see page 48):

(a) _vacuole_

(b) _cell wall_

(c) _chloroplasts_

 Starch granules

Related activities: Animal Cells, Cell Structure and Organelles
Web links: Review of Eukaryotic Cells

RA 2

Cell Structure

Animal Cells

Animal cells, unlike plant cells, do not have a regular shape. In fact, some animal cells (such as phagocytes) are able to alter their shape for various purposes (e.g. engulfment of foreign material). The diagram below shows the structure and function of a typical animal cell and its organelles. Note the differences between this cell and the generalised plant cell.

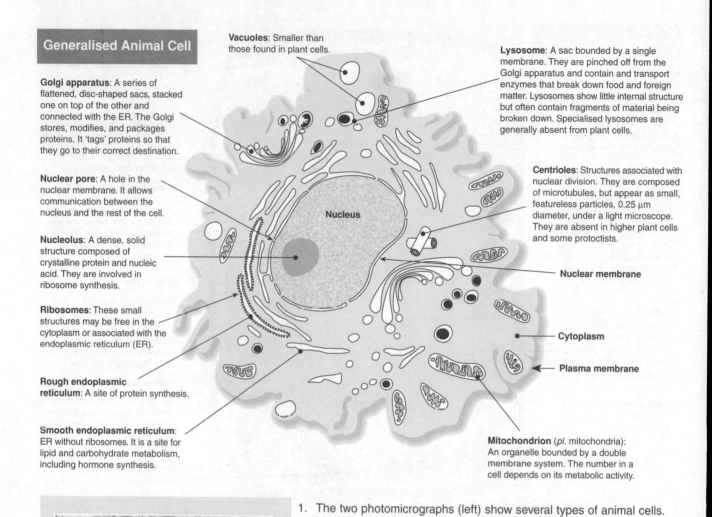

Generalised Animal Cell

Vacuoles: Smaller than those found in plant cells.

Golgi apparatus: A series of flattened, disc-shaped sacs, stacked one on top of the other and connected with the ER. The Golgi stores, modifies, and packages proteins. It 'tags' proteins so that they go to their correct destination.

Nuclear pore: A hole in the nuclear membrane. It allows communication between the nucleus and the rest of the cell.

Nucleolus: A dense, solid structure composed of crystalline protein and nucleic acid. They are involved in ribosome synthesis.

Ribosomes: These small structures may be free in the cytoplasm or associated with the endoplasmic reticulum (ER).

Rough endoplasmic reticulum: A site of protein synthesis.

Smooth endoplasmic reticulum: ER without ribosomes. It is a site for lipid and carbohydrate metabolism, including hormone synthesis.

Lysosome: A sac bounded by a single membrane. They are pinched off from the Golgi apparatus and contain and transport enzymes that break down food and foreign matter. Lysosomes show little internal structure but often contain fragments of material being broken down. Specialised lysosomes are generally absent from plant cells.

Centrioles: Structures associated with nuclear division. They are composed of microtubules, but appear as small, featureless particles, 0.25 μm diameter, under a light microscope. They are absent in higher plant cells and some protoctists.

Nucleus

Nuclear membrane

Cytoplasm

Plasma membrane

Mitochondrion (*pl.* mitochondria): An organelle bounded by a double membrane system. The number in a cell depends on its metabolic activity.

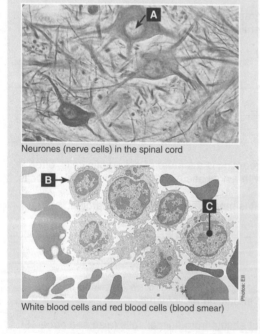

Neurones (nerve cells) in the spinal cord

White blood cells and red blood cells (blood smear)

1. The two photomicrographs (left) show several types of animal cells. Identify the features indicated by the letters **A-C**:

   A: _____

   B: _____

   C: _____

2. White blood cells are mobile, phagocytic cells, whereas red blood cells are smaller than white blood cells and, in humans, lack a nucleus.

   (a) In the photomicrograph (below, left), circle a white blood cell and a red blood cell:

   (b) With respect to the features that you can see, explain how you made your decision.

   _____

   _____

   _____

   _____

3. Name and describe one structure or organelle present in generalised animal cells but absent from plant cells:

   _____

   _____

Related activities: Plant Cells, Cell Structure and Organelles, Human Cell Specialisation **Web links**: Review of Eukaryotic Cells

The Cell's Cytoskeleton

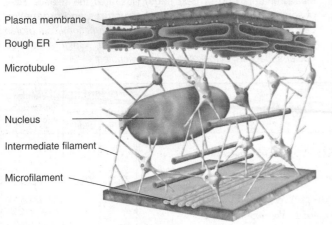

Plasma membrane

Rough ER

Microtubule

Nucleus

Intermediate filament

Microfilament

The cell's cytoplasm is not a fluid filled space; it contains a complex network of fibres called the **cytoskeleton**. The cytoskeleton resists tension and so provides structural support to maintain the cell's shape. The cytoskeleton is made up of three proteinaceous elements: microfilaments, intermediate filaments, and microtubules. Each has a distinct size structure and protein composition, and a specific role in cytoskeletal function. Cilia and flagella are made up of microtubules and for this reason they are considered to be part of the cytoskeleton.

The elements of the cytoskeleton are dynamic; they move and change to alter the cell's shape, move materials within the cell, and move the cell itself. This movement is achieved through the action of motor proteins, which transport material by 'walking' along cytoskeletal 'tracks', hydrolysing ATP at each step.

Microfilaments

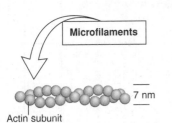

7 nm

Actin subunit

Intermediate filaments

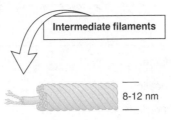

8-12 nm

Microtubules

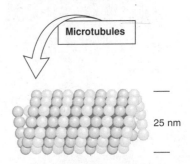

25 nm

	Microfilaments	Intermediate filaments	Microtubules
Protein subunits	Actin	Fibrous proteins, e.g. keratin	α and β tubulin dimers
Structure	Two intertwined strands	Fibres wound into thicker cables	Hollow tubes
Functions	• Maintain cell shape • Motility (pseudopodia) • Contraction (muscle) • Cytokinesis of cell division	• Maintain cell shape • Anchor nucleus and organelles	• Maintain cell shape • Motility (cilia and flagella) • Move chromosomes (spindle) • Move organelles

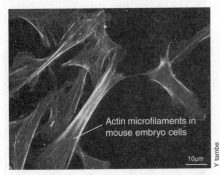

Actin microfilaments in mouse embryo cells

10μm

Y tambe

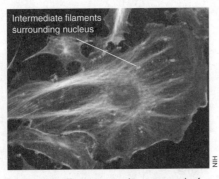

Intermediate filaments surrounding nucleus

NIH

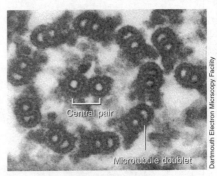

Central pair

Microtubule doublet

Dartmouth Electron Micrscopy Facility

Microfilaments are long polymers of the protein actin. Microfilaments can grow and shrink as actin subunits are added or taken away from either end. Networks of microfilaments form a matrix that helps to define the cell's shape. Actin microfilaments are also involved in cell division (during cytokinesis) and in muscle contraction.

Intermediate filaments can be composed of a number of different fibrous proteins and are defined by their size rather than composition. The protein subunits are wound into cables around 10 nm in diameter. Intermediate filaments form a dense network within and projecting from the nucleus, helping to anchor it in place.

Microtubules are the largest cytoskeletal components and grow or shrink in length as tubulin subunits are added or subtracted from one end. The are involved in movement of material within the cell and in moving the cell itself. This EM shows a cilia from *Chlamydomonas*, with the 9+2 arrangement of microtubular doublets.

1. Describe the role that all components of the cytoskeleton have in common: _____

2. Explain the importance of the cytoskeleton being a dynamic structure: _____

_____

3. Explain how the presence of a cytoskeleton could aid in directing the movement of materials within the cell:

_____

_____

Periodicals:
The beat goes on

Related activities: Cell Structures and Organelles

A 2

Cell Structure

Cell Structures and Organelles

The table below provides a format to summarise information about structures and organelles of typical eukaryotic cells. Complete the table using the list provided and by referring to a textbook and to other pages in this topic. Fill in the final three columns by writing either 'YES' or 'NO'. The first cell component has been completed for you as a guide and the log scale of measurements (top of next page) illustrates the relative sizes of some cellular structures. **List of structures and organelles**: *cell wall, mitochondrion, chloroplast, cell junctions, centrioles, ribosome, flagella, endoplasmic reticulum, Golgi apparatus, nucleus, flagella, cytoskeleton and vacuoles.*

Cell Component	Details	Present in Plant cells	Present in Animal cells	Visible under light microscope
(a) 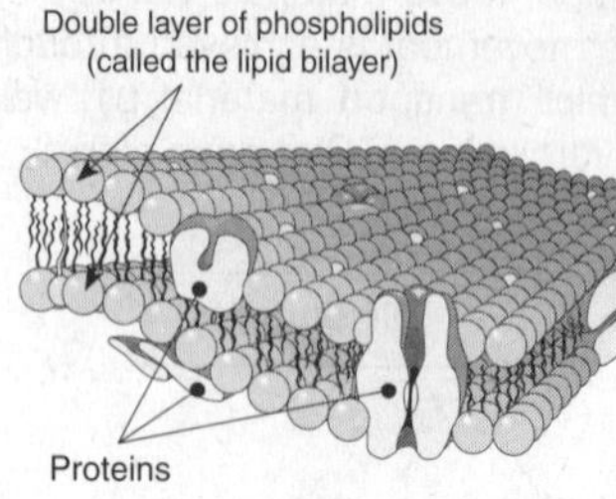 Double layer of phospholipids (called the lipid bilayer); Proteins	Name: Plasma (cell surface) membrane Location: Surrounding the cell Function: Gives the cell shape and protection. It also regulates the movement of substances into and out of the cell.	YES	YES	YES (but not at the level of detail shown in the diagram)
(b) 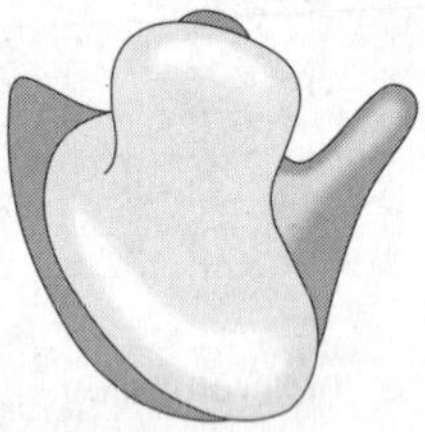	Name: Ribosome Location: free or on RR Function: Synthesis proteins	✓	✓	✗
(c) 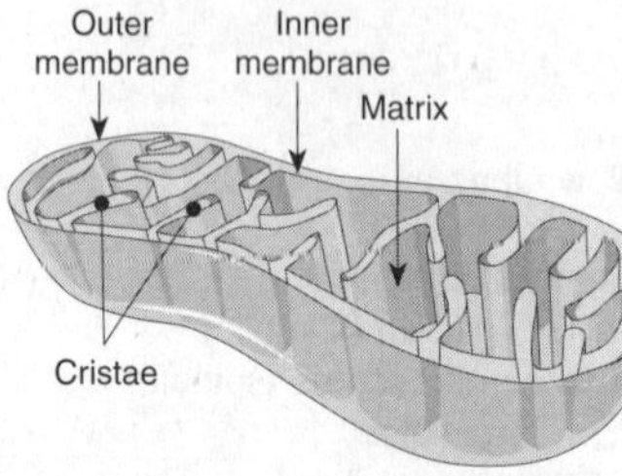 Outer membrane; Inner membrane; Matrix; Cristae	Name: Mitochondria Location: in the cell Function: where respiration takes place, making most of the cells energy in form ATP	✓	✓	✓
(d) 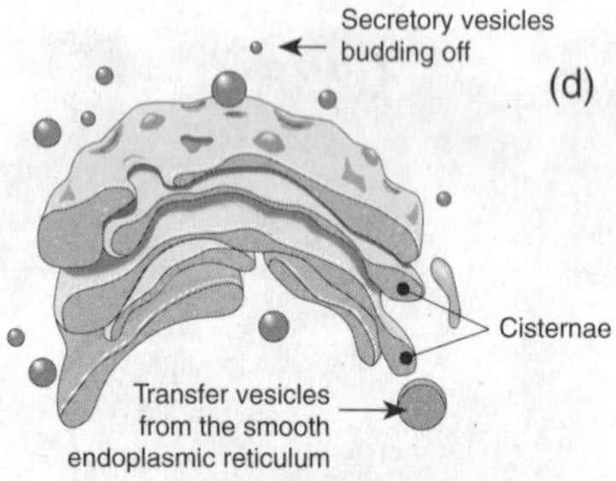 Secretory vesicles budding off; Cisternae; Transfer vesicles from the smooth endoplasmic reticulum	Name: Golgi apparatus Location: In the cell Function: modifies and packages protiens or other substances to transfer them to other places in the cell or out	✓	✓	✗
(e) 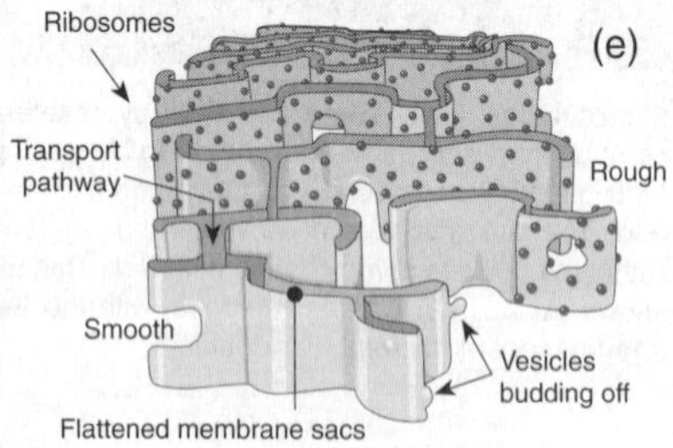 Ribosomes; Transport pathway; Rough; Smooth; Vesicles budding off; Flattened membrane sacs	Name: Rough er Location: In, the cell, next to the nucleus. Function: makes the basic proteins, translation; from the instructions from the nucleus			
(f) 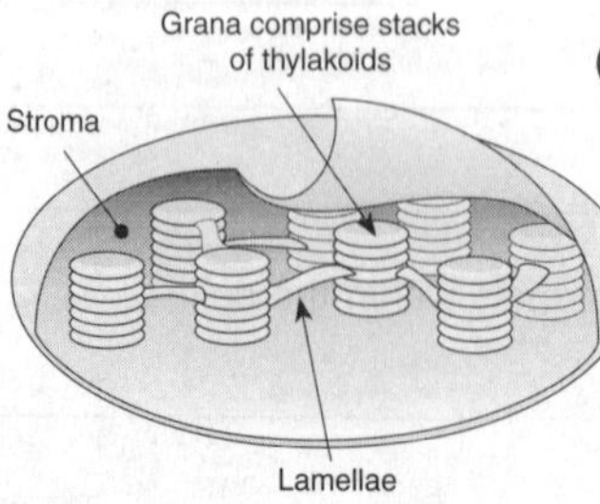 Grana comprise stacks of thylakoids; Stroma; Lamellae	Name: chloroplast Location: in the cell Function: where photosynthesis takes place			

Related activities: Plant Cells, Animal Cells
Web links: Eukaryotic Cells Interactive Animation

Periodicals: Cellular factories, Chloroplasts

© Biozone International 2008-2010
Photocopying Prohibited

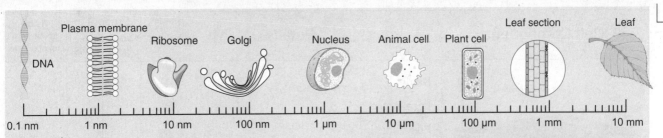

	DNA	Plasma membrane	Ribosome	Golgi	Nucleus	Animal cell	Plant cell	Leaf section	Leaf
	0.1 nm	1 nm	10 nm	100 nm	1 µm	10 µm	100 µm	1 mm	10 mm

Cell Component	Details	Present in		Visible under light microscope
		Plant cells	Animal cells	
(g)	**Name:** Lysosome and food vacuole **Location:** **Function:**			
(h)	**Name:** Nucleus **Location:** in the cell **Function:** houses the DNA and sends instructions to other parts of the cell in the form of gene codes.			
(i)	**Name:** cytoskeleton **Location:** all over the cell **Function:**			
(j)	**Name:** Cilia and flagella (some eukaryotic cells) **Location:** **Function:** used for locomotion.			

Labels on diagram (g): Digestion, Lysosome, Food Vacuole, Phagocytosis of food particle

Labels on diagram (h): Nuclear membrane, Nuclear pores, Nucleolus, Genetic material

Labels on diagram (i): Microtubules

Labels on diagram (j): Two central, single microtubules; 9 doublets of microtubules in an outer ring; Extension of plasma membrane surrounding a core of microtubules in a 9+2 pattern; Basal body anchors the cilium

Cell Structure

Cell Component	Details	Present in		Visible under light microscope
		Plant cells	Animal cells	

(k)

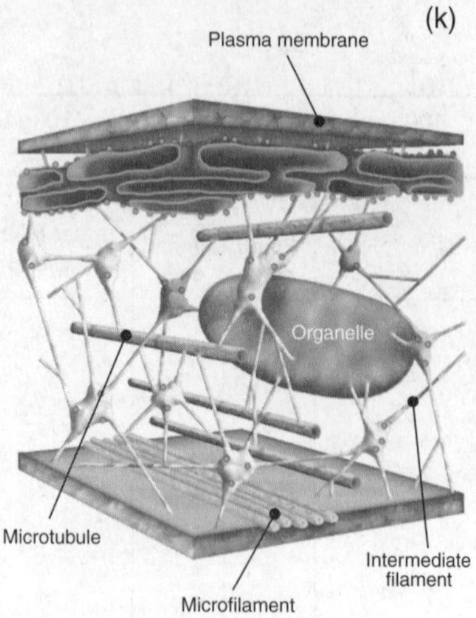

Plasma membrane

Organelle

Microtubule

Intermediate filament

Microfilament

Name:

Location:

Function:

(l)

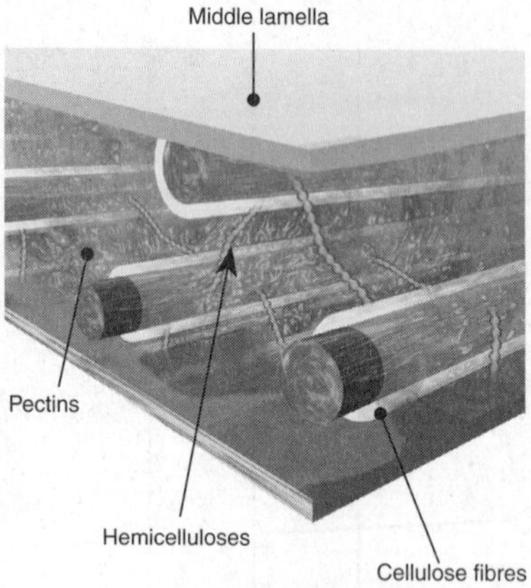

Middle lamella

Pectins

Hemicelluloses

Cellulose fibres

Name: Cellulose cell wall

Location:

Function:

(m)

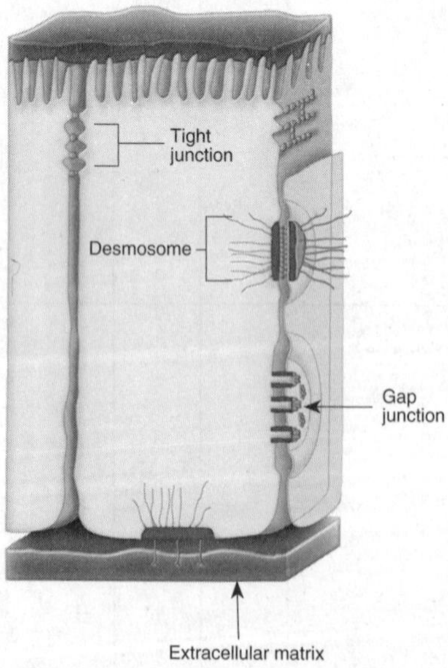

Tight junction

Desmosome

Gap junction

Extracellular matrix

Name: Cell junctions

Location:

Function:

Packaging Proteins

Cells produce a range of organic polymers made up of repeating units of smaller molecules. The synthesis, packaging and movement of these **macromolecules** inside the cell involves a number of membrane bound organelles, as indicated below. These organelles provide compartments where the enzyme systems involved can be isolated.

Golgi apparatus
The Golgi apparatus comprises stacks of flattened membranes in the shape of curved sacs. This organelle receives transport vesicles and the products they contain from smooth ER. They are modified, stored and eventually shipped to the surface of the cell or other destinations.

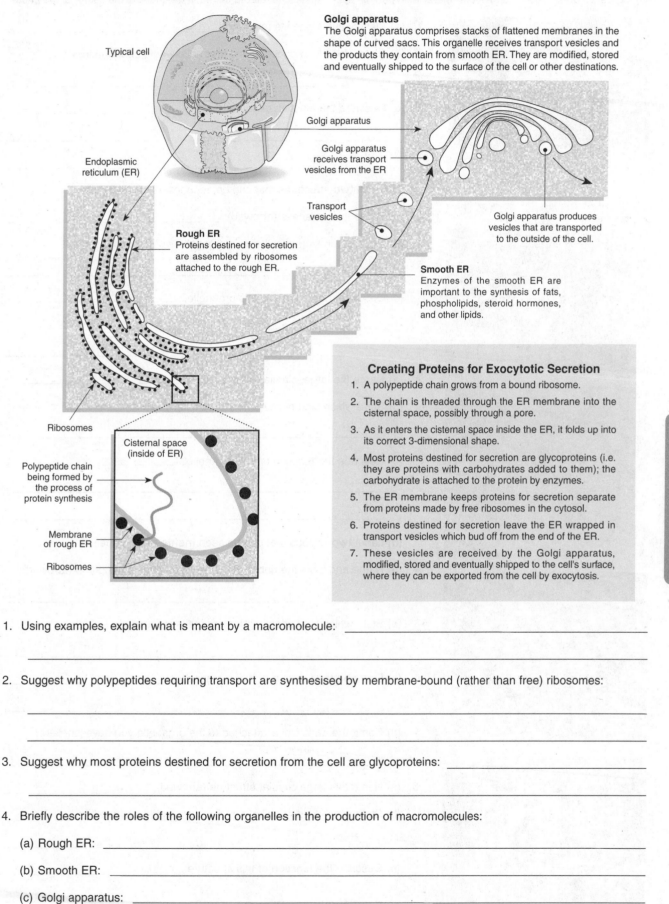

Typical cell

Endoplasmic reticulum (ER)

Golgi apparatus

Golgi apparatus receives transport vesicles from the ER

Transport vesicles

Golgi apparatus produces vesicles that are transported to the outside of the cell.

Rough ER
Proteins destined for secretion are assembled by ribosomes attached to the rough ER.

Smooth ER
Enzymes of the smooth ER are important to the synthesis of fats, phospholipids, steroid hormones, and other lipids.

Ribosomes

Cisternal space (inside of ER)

Polypeptide chain being formed by the process of protein synthesis

Membrane of rough ER

Ribosomes

Creating Proteins for Exocytotic Secretion

1. A polypeptide chain grows from a bound ribosome.

2. The chain is threaded through the ER membrane into the cisternal space, possibly through a pore.

3. As it enters the cisternal space inside the ER, it folds up into its correct 3-dimensional shape.

4. Most proteins destined for secretion are glycoproteins (i.e. they are proteins with carbohydrates added to them); the carbohydrate is attached to the protein by enzymes.

5. The ER membrane keeps proteins for secretion separate from proteins made by free ribosomes in the cytosol.

6. Proteins destined for secretion leave the ER wrapped in transport vesicles which bud off from the end of the ER.

7. These vesicles are received by the Golgi apparatus, modified, stored and eventually shipped to the cell's surface, where they can be exported from the cell by exocytosis.

Cell Structure

1. Using examples, explain what is meant by a macromolecule: _____

_____

2. Suggest why polypeptides requiring transport are synthesised by membrane-bound (rather than free) ribosomes:

_____

_____

3. Suggest why most proteins destined for secretion from the cell are glycoproteins: _____

_____

4. Briefly describe the roles of the following organelles in the production of macromolecules:

(a) Rough ER: _____

(b) Smooth ER: _____

(c) Golgi apparatus: _____

(d) Transport vesicles: _____

Related activities: The Role of Membranes in Cells

RA 2

Interpreting Electron Micrographs

The photographs below were taken using a **transmission electron microscope** (TEM). They show some of the cell organelles in great detail. Remember that these photos are showing only **parts of cells**, **not whole cells**. Some of the photographs show more than one type of organelle. The questions refer to the main organelle in the centre of the photo.

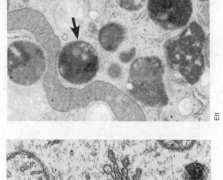

1. (a) Identify this organelle (arrowed): _____

    (b) State which kind of cell(s) this organelle would be found in:

    _____

    (c) Describe the function of this organelle: _____

    _____

    _____

    (d) Label **two** structures that can be seen inside this organelle.

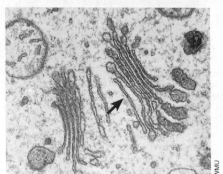
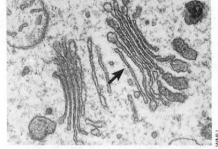

2. (a) Name this organelle (arrowed): _____

    (b) State which kind of cell(s) this organelle would be found in:

    _____

    (c) Describe the function of this organelle: _____

    _____

    _____

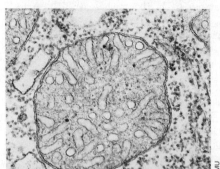
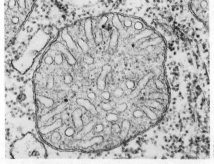

3. (a) Name the large, circular organelle: _____

    (b) State which kind of cell(s) this organelle would be found in:

    _____

    (c) Describe the function of this organelle: _____

    _____

    _____

    (d) Label **two** regions that can be seen **inside** this organelle.

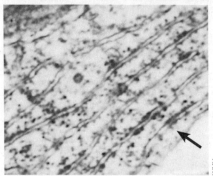
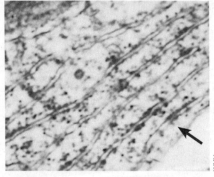

4. (a) Name and label the ribbon-like organelle in this photograph (arrowed):

    _____

    (b) State which kind of cell(s) this organelle is found in:

    _____

    (c) Describe the function of these organelles: _____

    _____

    (d) Name the dark 'blobs' attached to the organelle you have labelled:

    _____

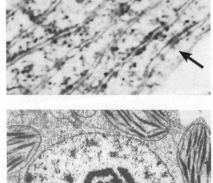

5. (a) Name this large circular structure (arrowed): _____

    (b) State which kind of cell(s) this structure would be found in: _____

    _____

    (c) Describe the function of this structure: _____

    _____

    (d) Label three features relating to this structure in the photograph.

Related activities: *Electron Microscopes, Plant Cells, Animal Cells, Cell Structures and Organelles*

Periodicals:
The power behind an electron microscopist

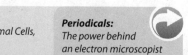

© Biozone International 2008-2010
Photocopying Prohibited

Identifying Structures in an Animal Cell

Our current knowledge of cell ultrastructure has been made possible by the advent of electron microscopy. Transmission electron microscopy is the most frequently used technique for viewing cellular organelles. When viewing TEMs, the cellular organelles may appear to be quite different depending on whether they are in transverse or longitudinal section.

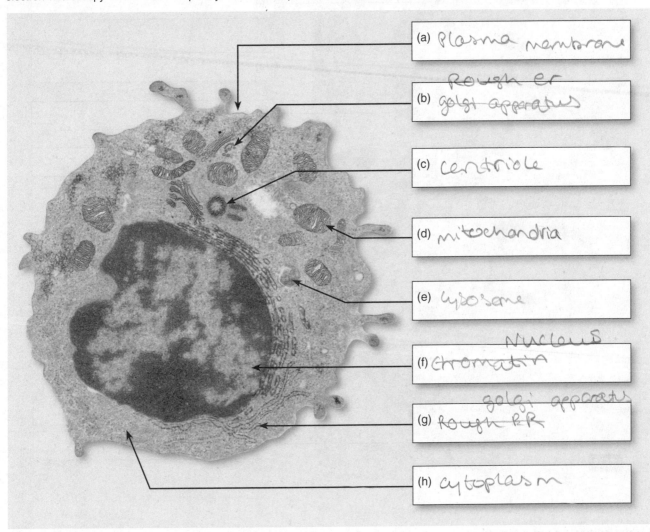

(a) Plasma membrane

(b) Rough er Golgi apparatus

(c) centriole

(d) mitochondria

(e) lysosome

(f) Nucleus ~~Chromatin~~

(g) golgi apparatus Rough ER

(h) cytoplasm

Cell Structure

1. Identify and label the structures in the cell above using the following list of terms: *cytoplasm, plasma membrane, rough endoplasmic reticulum, mitochondrion, nucleus, centriole, Golgi apparatus, lysosome*

2. In the electron micrograph above, identify which of the organelles are shown in both transverse and longitudinal section:

    _____

3. Plants lack any of the mobile phagocytic cells typical of animals. Explain why this is the case: _____

    _____

    _____

4. The animal cell pictured above is a lymphocyte. Describe the features that suggest to you that:

    (a) It has a role in producing and secreting proteins: _____

    _____

    (b) It is metabolically very active: _____

    _____

5. Describe the features of the lymphocyte cell above that identify it as an eukaryotic cell: _____

    _____

    _____

Related activities: Cell Structure and Organelles, Animal Cells, Prokaryotic vs Eukaryotic Cells

RA 2

Identifying Plant Cell Structures

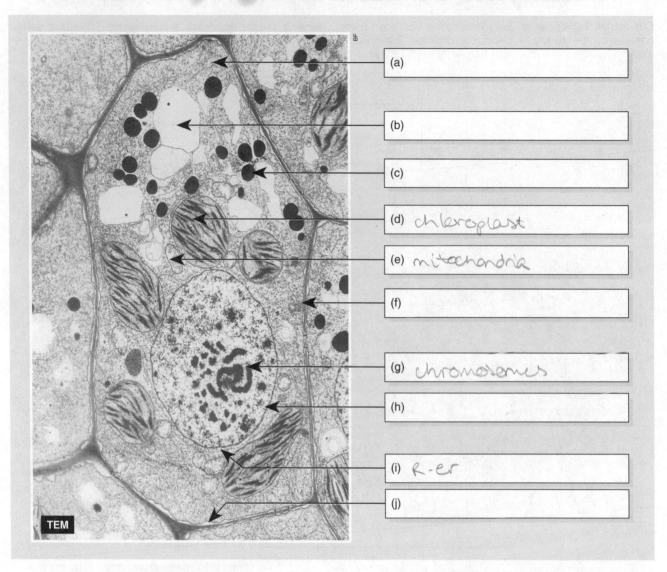

(a)

(b)

(c)

(d) chloroplast

(e) mitochondria

(f)

(g) chromosomes

(h)

(i) R-er

(j)

TEM

1. Identify and label the ten structures in the cell above using the following list of terms: *nuclear membrane, cytoplasm, endoplasmic reticulum, mitochondrion, starch granules, chromosome, vacuole, plasma membrane, cell wall, chloroplast*

2. State how many cells, or parts of cells, are visible in the electron micrograph above: _____

3. Name the features which identify this cell as a plant cell:

   _____

   _____

4. (a) Explain where cytoplasm is found in the cell: _____

   _____

   (b) Describe what cytoplasm is made up of: _____

   _____

5. Describe two structures, pictured in the cell above, that are associated with storage:

   (a) _____

   _____

   (b) _____

   _____

How Do We Know? Membrane Structure

Cellular membranes play many extremely important roles in cells, and understanding their structure is central to understanding cellular function. Moreover, understanding the structure and function of membrane proteins is essential to understanding cellular transport processes, and cell recognition and signalling. Cellular membranes are far too small to be seen clearly using light microscopy, and certainly any detail is impossible to resolve. Since early last century, scientists have known that membranes comprised a lipid bilayer with associated proteins. But how did they elucidate just how these molecules were organised?

The answers were provided with electron microscopy, and one technique in particular – **freeze fracture**. As the name implies, freeze fracture, at its very simplest level, is the freezing of a cell and then cleaving it so that it fractures in a certain way. Scientists can then use electron microscopy to see the indentations and outlines of the structures remaining after cleavage. Membranes are composed of two layers held together by weak intermolecular bonds, so they cleave into two halves when fractured. This provides views of the inner surfaces of the membrane.

The procedure involves several steps:
▶ The tissue is prefixed using a cross linking agent. This alters the strength of the internal and external parts of the membrane.
▶ The cell is fixed to immobilise any mobile macromolecules.
▶ The specimen is passed through a sequential series of glycerol solutions of increasing strength. This protects the cells from bursting when placed into the cryomaterial.
▶ The specimen is frozen using liquid propane cooled by liquid nitrogen. The specimens are mounted on gold supports and cooled briefly before transfer to the freeze-etch machine.
▶ Specimen is cleaved in a helium-vented vacuum at -150°C. A razor blade cooled to -170°C acts as both a cold trap for water and the cleaving instrument.
▶ At this stage the specimen may be evaporated a little to produce some relief in the surface of the fracture (known as etching) so that a 3-dimensional effect occurs.
▶ For viewing under EM, a replica of the specimen is made and coated in gold or platinum to ~3 nm thick. This produces a shadow effect that allows structures to be seen clearly. A layer of carbon around 30 nm thick is used to stabilise the specimen.
▶ The samples are then raised to room temperature and placed into distilled water or digestive enzymes, which allows the replica to separate from the sample. The replica is then rinsed several times in distilled water before it is ready for viewing.

The freeze fracture technique provided the necessary supporting evidence for the current fluid mosaic model of membrane structure. When cleaved, proteins in the membrane left impressions that showed they were embedded into the membrane and not a continuous layer on the outside as earlier models proposed.

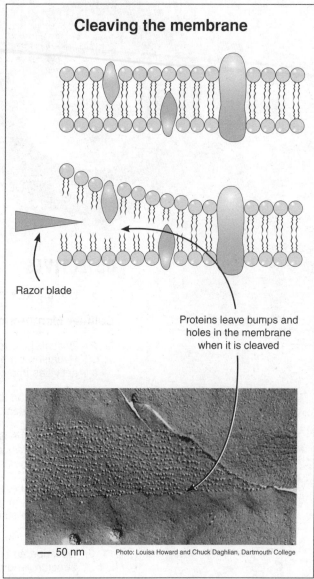

Cleaving the membrane

Razor blade

Proteins leave bumps and holes in the membrane when it is cleaved

— 50 nm   Photo: Louisa Howard and Chuck Daghlian, Dartmouth College

Cell Structure

1. Describe the principle of freeze fracture and explain why it is such a useful technique for studying membrane structure:

_____

_____

_____

2. Explain how this freeze-fracture studies provided evidence for our current model of membrane structure: _____

_____

_____

3. An earlier model of membrane structure was the unit membrane; a phospholipid bilayer with a protein coat. Explain how the freeze-fracture studies showed this model to be flawed:

_____

_____

_____

Related activities: The Structure of Membranes   **RA 2**

Cellular Membranes and Transport

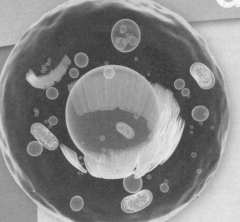

KEY CONCEPTS

▶ Cellular metabolism depends on the transport of substances across cellular membranes.

▶ The plasma membrane forms a partially permeable barrier to entry and exit of substances into the cell.

▶ The fluid mosaic model satisfies the observed characteristics of cellular membranes.

▶ Substances can move by passive or active transport.

▶ Active transport requires energy expenditure.

KEY TERMS

active transport
amphipathic
carrier protein
channel protein
cholesterol
concentration gradient
diffusion
endocytosis
exocytosis
facilitated diffusion
fluid mosaic model
glycolipid
glycoprotein
ion pump
osmosis
partially permeable
passive transport
phagocytosis
phospholipid
pinocytosis
plasma membrane
plasmolysis
pressure potential
receptor mediated phago-
cytosis
solute potential
surface area: volume ratio
turgor
water potential

OBJECTIVES

☐ 1. Use the **KEY TERMS** to help you understand and complete these objectives.

Cellular Membranes
pages 57, 59-63

☐ 2. Describe the roles of membranes (including the plasma membrane) in cells. Recall how membranes serve to compartmentalise regions of the cell's activities into specific organelles.

☐ 3. Describe the role of **plasma membranes** as **partially permeable** barriers to the movement of substances into and out of cells.

☐ 4. Describe the fluid mosaic model of membrane structure. Include reference to the significance of the **amphipathic** character the phospholipids and the role of transmembrane proteins in the integral structure of the membrane.

☐ 5. Describe the functional roles of the phospholipids, cholesterol, glycolipids, proteins, and glycoproteins in the plasma membrane. Include reference to membrane stability and fluidity, cellular communication, and cellular transport.

☐ 6. Describe and explain the effects of temperature changes on membrane structure and permeability.

☐ 7. Explain the term **cell signalling**. Identify some types of signalling molecules and describe their general role in cellular communication. Explain the role of membrane-bound receptors as sites where hormones and drugs can bind.

Cellular Transport Processes
pages 64-69

☐ 8. Summarise the types of movements that occur across membranes. Recall the role of proteins as receptors and carriers in membrane transport.

☐ 9. Describe and explain **passive transport** across membranes by **diffusion** and **facilitated diffusion**.

☐ 10. Explain what is meant by water potential (Ψ) and define its components. With reference to the **concentration gradient** and **water potential**, explain net movements of water by **osmosis**. Describe and explain the effects of solutions of different water potential on plant and animal cells.

☐ 11. Using examples, explain **active transport** processes in cells, including **ion pumps**, **endocytosis**, and **exocytosis**. Recognise categories of endocytosis (**phagocytosis**, **receptor mediated phagocytosis**, and **pinocytosis**) and identify when these operate.

☐ 12. Describe factors affecting the rates of Transport processes in cells.

Periodicals:
Listings for this
chapter are on page 338

Weblinks:
www.biozone.co.uk/
weblink/OCR-AS-2641.html

**Teacher Resource
CD-ROM:** Sucrose
Transport in Phloem

The Structure of Membranes

All cells have a plasma membrane that forms the outer limit of the cell. Membranes are also found inside eukaryotic cells as part of membranous **organelles**. Our knowledge of membrane structure has been built up as a result of many observations and experiments. The now-accepted **fluid-mosaic model** of membrane structure (below) satisfies the observed properties of membranes. The self-orientating properties of the phospholipids allows cellular membranes to reseal themselves when disrupted. The double layer of lipids is also quite fluid, and proteins move quite freely within it. The plasma membrane is more than just a passive envelope; it is a dynamic structure actively involved in cellular activities.

The Fluid Mosaic Model of Membrane Structure

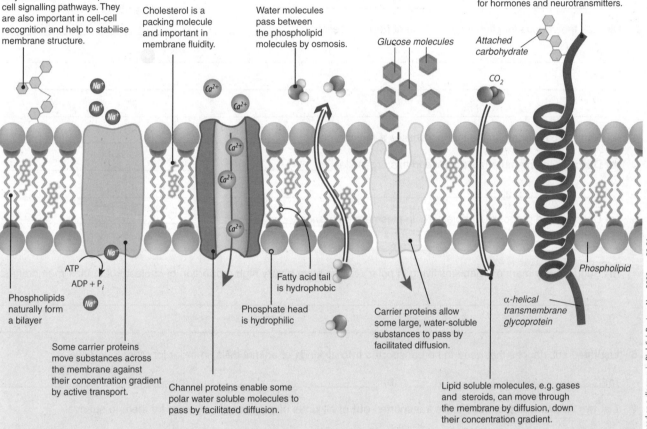

Glycolipids in membranes are phospholipids with attached carbohydrate. Like glycoproteins, they act as surface receptors in cell signalling pathways. They are also important in cell-cell recognition and help to stabilise membrane structure.

Cholesterol is a packing molecule and important in membrane fluidity.

Water molecules pass between the phospholipid molecules by osmosis.

Glucose molecules

Glycoproteins are proteins with attached carbohydrate. They are important in membrane stability, in cell-cell recognition, and in cell signalling, acting as receptors for hormones and neurotransmitters.

Attached carbohydrate

CO_2

ATP

ADP + P_i

Phospholipids naturally form a bilayer

Some carrier proteins move substances across the membrane against their concentration gradient by active transport.

Channel proteins enable some polar water soluble molecules to pass by facilitated diffusion.

Fatty acid tail is hydrophobic

Phosphate head is hydrophilic

Carrier proteins allow some large, water-soluble substances to pass by facilitated diffusion.

Lipid soluble molecules, e.g. gases and steroids, can move through the membrane by diffusion, down their concentration gradient.

α-helical transmembrane glycoprotein

Phospholipid

Based on a diagram in Biol. Sci. Review, Nov. 2009, pp. 20-21

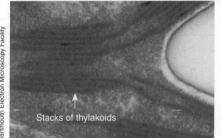

Membranes can be organised into stacks, as in this TEM of the thylakoid membranes in a chloroplast.

Stacks of thylakoids

Dartmouth Electron Microscopy Facility

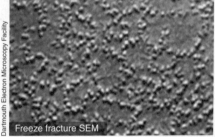

Pits and mounds in the middle of the membrane show that proteins must span the lipid bilayer.

Freeze fracture SEM

Dartmouth Electron Microscopy Facility

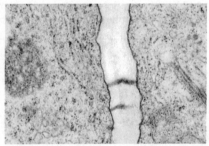

Membranous connections link cells together as in these plasmodesmata between adjacent plant cells.

Alison Roberts

Cellular Membranes and Transport

1. Identify the component(s) of the plasma membrane involved in:

(a) Facilitated diffusion: _channel / carrier_

(c) Cell signalling: _glycolipid/proteins_

(b) Active transport: _channel/carrier ATP proteins_

(d) Regulating membrane fluidity: _cholestrol_

2. (a) Describe the modern fluid mosaic model of membrane structure: _____

_____

_____

Periodicals:
Border control
The fluid mosaic model

Related activities: The Role of Membranes in Cells
Web links: Membrane Structure Tutorial

RA 2

(b) Explain how the fluid mosaic model accounts for the observed properties of cellular membranes:

_____

_____

_____

_____

_____

_____

_____

3. Discuss the various functional roles of membranes in cells: _____

_____

_____

_____

4. (a) Name a cellular organelle that possesses a membrane: _____

(b) Describe the membrane's purpose in this organelle: _____

_____

5. (a) Describe the purpose of cholesterol in plasma membranes: ____ stability ____

_____

(b) Suggest why marine organisms living in polar regions have a very high proportion of cholesterol in their membranes:

_____

_____

6. List three substances that need to be transported **into** all kinds of animal cells, in order for them to survive:

(a) _____ (b) _____ (c) _____

7. List two substances that need to be transported **out** of all kinds of animal cells, in order for them to survive:

(a) _____ (b) _____

8. Use the symbol for a phospholipid molecule (below) to draw a **simple labelled diagram** to show the structure of a plasma membrane (include features such as lipid bilayer and various kinds of proteins):

The Role of Membranes in Cells

Many of the important structures and organelles in cells are composed of, or are enclosed by, membranes. These include: the endoplasmic reticulum, mitochondria, nucleus, Golgi apparatus, chloroplasts, lysosomes, vesicles and the plasma membrane itself. All membranes within eukaryotic cells share the same basic structure as the plasma membrane that encloses the entire cell.

They perform a number of critical functions in the cell: serving to compartmentalise regions of different function within the cell, controlling the entry and exit of substances, and fulfilling a role in recognition and communication between cells. Some of these roles are described below. The role of membranes in the production of macromolecules (e.g. proteins) is shown on the next page:

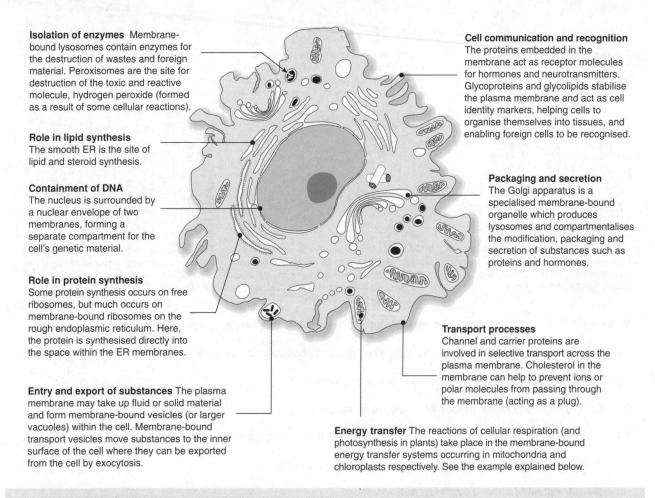

Isolation of enzymes Membrane-bound lysosomes contain enzymes for the destruction of wastes and foreign material. Peroxisomes are the site for destruction of the toxic and reactive molecule, hydrogen peroxide (formed as a result of some cellular reactions).

Role in lipid synthesis
The smooth ER is the site of lipid and steroid synthesis.

Containment of DNA
The nucleus is surrounded by a nuclear envelope of two membranes, forming a separate compartment for the cell's genetic material.

Role in protein synthesis
Some protein synthesis occurs on free ribosomes, but much occurs on membrane-bound ribosomes on the rough endoplasmic reticulum. Here, the protein is synthesised directly into the space within the ER membranes.

Entry and export of substances The plasma membrane may take up fluid or solid material and form membrane-bound vesicles (or larger vacuoles) within the cell. Membrane-bound transport vesicles move substances to the inner surface of the cell where they can be exported from the cell by exocytosis.

Cell communication and recognition
The proteins embedded in the membrane act as receptor molecules for hormones and neurotransmitters. Glycoproteins and glycolipids stabilise the plasma membrane and act as cell identity markers, helping cells to organise themselves into tissues, and enabling foreign cells to be recognised.

Packaging and secretion
The Golgi apparatus is a specialised membrane-bound organelle which produces lysosomes and compartmentalises the modification, packaging and secretion of substances such as proteins and hormones.

Transport processes
Channel and carrier proteins are involved in selective transport across the plasma membrane. Cholesterol in the membrane can help to prevent ions or polar molecules from passing through the membrane (acting as a plug).

Energy transfer The reactions of cellular respiration (and photosynthesis in plants) take place in the membrane-bound energy transfer systems occurring in mitochondria and chloroplasts respectively. See the example explained below.

Compartmentation within Membranes

Membranes play an important role in separating regions within the cell (and within organelles) where particular reactions occur. Specific enzymes are therefore often located in particular organelles. The reaction rate is controlled by controlling the rate at which substrates enter the organelle and therefore the availability of the raw materials required for the reactions.

Example (right): The enzymes involved in cellular respiration are arranged in different parts of the mitochondria. Reactions are localised and separated by membrane systems.

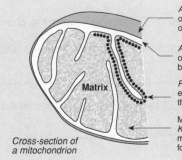

Cross-section of a mitochondrion

Amine oxidases and other enzymes on the outer membrane surface

Adenylate kinase and other phosphorylases between the membranes

Respiratory assembly enzymes embedded in the membrane (ATPase)

Many soluble enzymes of the Krebs cycle floating in the matrix, as well as enzymes for fatty acid degradation.

1. Explain the crucial role of membrane systems and organelles in the following:

(a) Providing compartments within the cell: _____

_____

_____

(b) Increasing the total membrane surface area within the cell: _____

_____

_____

Related activities: Cell Structures and Organelles, Packaging Macromolecules
Web links: Cell Membranes

Functional Roles of Membranes in Cells

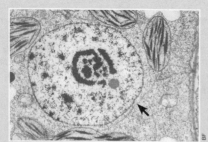

The **nuclear membrane**, which surrounds the nucleus, regulates the passage of genetic information to the cytoplasm and may also protect the DNA from damage.

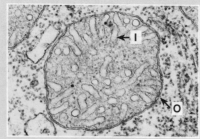

Mitochondria have an outer membrane (**O**) which controls the entry and exit of materials involved in aerobic respiration. Inner membranes (**I**) provide attachment sites for enzyme activity.

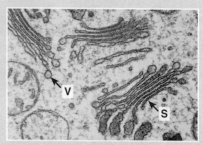

The **Golgi apparatus** comprises stacks of membrane-bound sacs (**S**). It is involved in packaging materials for transport or export from the cell as secretory vesicles (**V**).

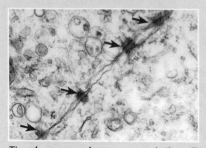

The **plasma membrane** surrounds the cell. In this photo, intercellular junctions called **desmosomes**, which connect neighbouring cells, are indicated with arrows.

Chloroplasts are large organelles found in plant cells. The stacked membrane systems of chloroplasts (grana) trap light energy which is then used to fix carbon into 6-C sugars.

Grana

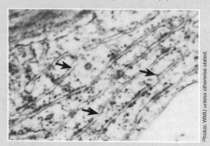

This EM shows stacks of rough endoplasmic reticulum (arrows). The membranes are studded with ribosomes, which synthesize proteins into the intermembrane space.

2. Match each of the following organelles with the correct description of its functional role in the cell:

chloroplast, rough endoplasmic reticulum, lysosome, smooth endoplasmic reticulum, mitochondrion, Golgi apparatus

(a) Active in synthesis, sorting, and secretion of cell products: _____

(b) Digestive organelle where macromolecules are hydrolysed: _____

(c) Organelle where most cellular respiration occurs and most ATP is generated: _____

(d) Active in membrane synthesis and synthesis of secretory proteins: _____

(e) Active in lipid and hormone synthesis and secretion: _____

(f) Photosynthetic organelle converts light energy to chemical energy stored in sugar molecules: _____

3. Explain how the membrane surface area is increased within cells and organelles: _____

_____

4. Discuss the importance of each of the following to cellular function:

(a) High membrane surface area: _____

_____

(b) Channel proteins and carrier proteins in the plasma membrane: _____

_____

5. Non-polar (lipid-soluble) molecules diffuse more rapidly through membranes than polar (lipid-insoluble) molecules:

(a) Explain the reason for this: _____

_____

(b) Discuss the implications of this to the transport of substances into the cell through the plasma membrane:

_____

_____

Cell Signalling and Receptors

Cells use **signals** (chemical messengers) to gather information about, and respond to, changes in their cellular environment and for communication between cells. The reception of the signal by membrane receptors is followed by a **signal transduction pathway**. Signal transduction often involves a number of enzymes and molecules in a **signal cascade**, which results in a large response in the target cell. Cell signalling pathways are categorised primarily on the distance over which the signal molecule travels to reach its target cell, and generally fall into three categories. The **endocrine** pathway involves the transport of hormones varying distances through the circulatory system. During **paracrine** signalling, the signal acts locally upon neighbouring cells. Cells also produce and respond to their own signals in a process called autocrine signalling.

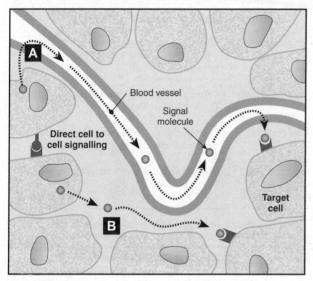

Cells communicate through a variety of mechanisms. Hormone or **endocrine signals** are released by ductless endocrine glands and carried long distances through the body by the circulatory system to the target cells. Examples include sex hormones, growth factors and neurohormones such as dopamine. Signal molecules can also act upon target cells within the immediate vicinity (this is called **paracrine signalling**). The chemical messenger can be transferred through the extracellular fluid (e.g. at synapses) or directly between cells, which is important during embryonic development.

Structure of a Transmembrane Receptor

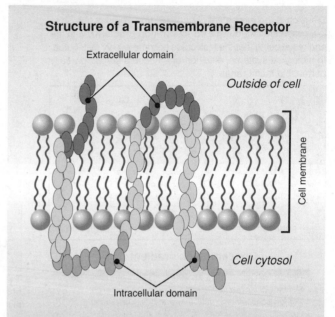

Transmembrane receptors span the cell membrane and bind signal molecules (ligands) which cannot cross the plasma membrane on their own. They have an extra-cellular domain outside the cell, and an intracellular domain within the cell cytosol. Ion channels and protein kinases are examples of transmembrane receptors.

1. Explain what is meant by cell signalling: _____

_____

_____

2. Identify the type of cell communication indicated by the letters A and B in the diagram, above left:

    (a) A: _____

    (b) B: _____

3. Describe the components that all cell signalling mechanisms have in common: _____

_____

_____

4. Explain the role of membrane-bound receptors in:

    (a) Hormonal responses: _____

_____

_____

    (b) Responses to drugs: _____

_____

_____

Related activities: New Medicines
Web links: Cell Communication, Hormones, Receptors and Target Cells

A 2

Passive Transport Processes

The molecules that make up substances are constantly moving about in a random way. This random motion causes molecules to disperse from areas of high to low concentration; a process called **diffusion**. The molecules move down a **concentration gradient**. Diffusion and osmosis (diffusion of water molecules across a partially permeable membrane) are **passive** processes, and use no energy. Diffusion occurs freely across membranes, as long as the membrane is permeable to that molecule (partially permeable membranes allow the passage of some molecules but not others). Each type of molecule diffuses down its own concentration gradient. Diffusion of molecules in one direction does not hinder the movement of other molecules. Diffusion is important in allowing exchanges with the environment and in the regulation of cell water content.

Diffusion is the movement of particles from regions of high to low concentration (down a **concentration gradient**), with the end result being that the molecules become evenly distributed. In biological systems, diffusion often occurs across partially permeable membranes.

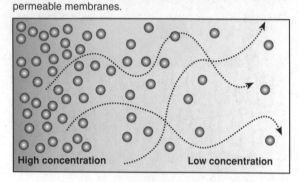

High concentration   **Low concentration**

Concentration gradient

If molecules are free to move, they move from high to low concentration until they are evenly dispersed.

Factors affecting rates of diffusion

Concentration gradient:	Diffusion rates will be higher when there is a greater difference in concentration between two regions.
The distance involved:	Diffusion over shorter distances occurs at a greater rate than diffusion over larger distances.
The area involved:	The larger the area across which diffusion occurs, the greater the rate of diffusion.
Barriers to diffusion:	Thicker barriers slow diffusion rate. Pores in a barrier enhance diffusion.

$$\text{FICK'S LAW} \quad \frac{\text{Surface area of membrane} \times \text{Difference in concentration across the membrane}}{\text{Length of the diffusion path (thickness of the membrane)}}$$

These factors are expressed in **Fick's law**, which governs the rate of diffusion of substances within a system. Temperature also affects diffusion rates; at higher temperatures molecules have more energy and move more rapidly.

Diffusion Through Membranes

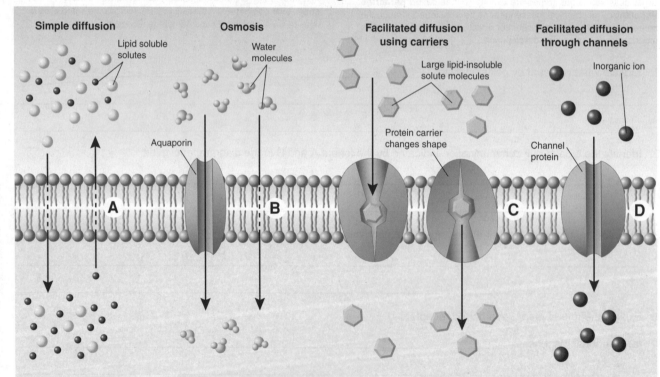

Simple diffusion — Lipid soluble solutes

Osmosis — Water molecules — Aquaporin

Facilitated diffusion using carriers — Large lipid-insoluble solute molecules — Protein carrier changes shape

Facilitated diffusion through channels — Inorganic ion — Channel protein

A: Some molecules (e.g. gases and lipid soluble molecules) diffuse directly across the plasma membrane. Two-way diffusion is common in biological systems, e.g. at the alveolar surface of the lung, CO_2 diffuses out and oxygen diffuses into the blood.

B: Osmosis describes the diffusion of water across a partially permeable membrane (in this case, the plasma membrane). Some water can diffuse directly through the lipid bilayer, but movement is also aided by specific protein channels called **aquaporins**.

C: A lipid-insoluble molecule is aided across the membrane by **carrier mediated facilitated diffusion**. This involves a transmembrane carrier protein specific to the molecule being transported (for example, glucose transport into red blood cells).

D: Small polar molecules and ions molecules diffuse rapidly across the membrane by **channel-mediated facilitated diffusion**. Special channel proteins (sometimes called ionophores) create hydrophilic pores that allow some solutes, usually inorganic ions, to pass through.

Related activities: Active and Passive Transport Summary
Web links: Cellular Transport

Osmosis and Water Potential

Osmosis is the term describing the diffusion of water down its concentration gradient across a partially permeable membrane. It is the principal mechanism by which water enters and leaves cells in living organisms. The direction of this movement can be predicted on the basis of the water potential of the solutions involved. The water potential of a solution (denoted by ψ) is the term given to the tendency for water molecules to enter or leave a solution by osmosis.

Pure water has the highest water potential, set at zero. Dissolving any solute in water lowers ψ (makes it more negative). Water always diffuses from regions of less negative to more negative water potential. Water potential is determined by two components: the **solute potential**, ψs (of the cell sap) and the **pressure potential**, ψp, expressed by:

$$\psi cell = \psi s + \psi p$$

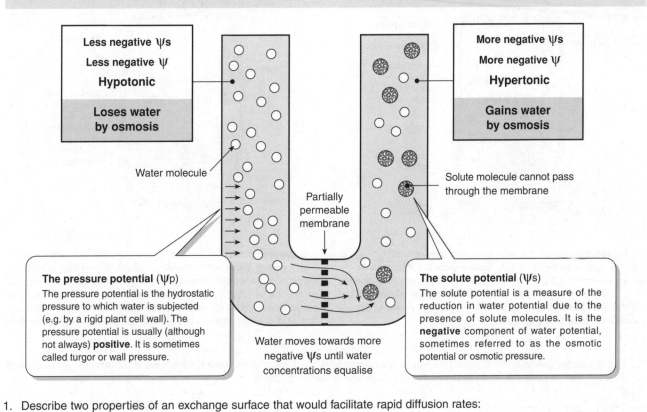

Less negative ψs
Less negative ψ
Hypotonic
Loses water by osmosis

Water molecule

More negative ψs
More negative ψ
Hypertonic
Gains water by osmosis

Solute molecule cannot pass through the membrane

Partially permeable membrane

The pressure potential (ψp)
The pressure potential is the hydrostatic pressure to which water is subjected (e.g. by a rigid plant cell wall). The pressure potential is usually (although not always) **positive**. It is sometimes called turgor or wall pressure.

Water moves towards more negative ψs until water concentrations equalise

The solute potential (ψs)
The solute potential is a measure of the reduction in water potential due to the presence of solute molecules. It is the **negative** component of water potential, sometimes referred to as the osmotic potential or osmotic pressure.

1. Describe two properties of an exchange surface that would facilitate rapid diffusion rates:

   (a) _____ (b) _____

2. Describe two biologically important features of diffusion:

   (a) _____

   (b) _____

3. Describe how facilitated diffusion is achieved for:

   (a) Small polar molecules and ions: _____

   _____

   (b) Glucose: _____

   _____

4. Explain how concentration gradients across membranes are maintained: _____

   _____

5. Explain the role of aquaporins in the rapid movement of water through some cells: _____

   _____

6. (a) State the water potential of pure water at standard temperature and pressure: _____

   (b) Explain what happens if a cell takes up sucrose by active transport: _____

   (c) Describe a situation where this occurs in plants: _____

Water Relations in Plant Cells

The plasma membrane of cells is a partially permeable membrane and osmosis is the principal mechanism by which water enters and leaves the cell. When the external water potential is the same as that of the cell there is no net movement of water. Two systems (cell and environment) with the same water potential are termed **isotonic**. The diagram below illustrates two different situations: when the external water potential is less negative than the cell (**hypotonic**) and when it is more negative than the cell (**hypertonic**).

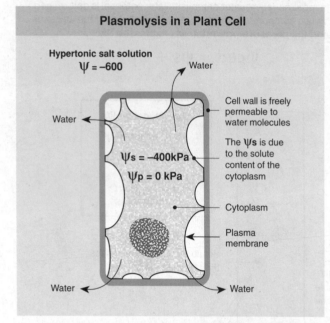

Plasmolysis in a Plant Cell

Hypertonic salt solution
$\Psi = -600$

Water

Cell wall is freely permeable to water molecules

The Ψs is due to the solute content of the cytoplasm

$\Psi s = -400 kPa$
$\Psi p = 0\ kPa$

Cytoplasm

Plasma membrane

Water

Water

In a **hypertonic** solution, the external water potential is more negative than the water potential of the cell ($\Psi cell = \Psi s + \Psi p$). Water leaves the cell and, because the cell wall is rigid, the plasma membrane shrinks away from the cell wall. This process is termed **plasmolysis** and the cell becomes **flaccid** ($\Psi p = 0$). Full plasmolysis is irreversible; the cell cannot recover by taking up water.

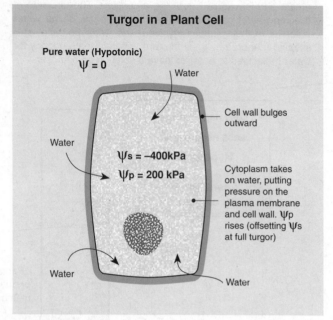

Turgor in a Plant Cell

Pure water (Hypotonic)
$\Psi = 0$

Water

Cell wall bulges outward

Water

$\Psi s = -400 kPa$
$\Psi p = 200\ kPa$

Cytoplasm takes on water, putting pressure on the plasma membrane and cell wall. Ψp rises (offsetting Ψs at full turgor)

Water

Water

In a **hypotonic** solution, the external water potential is less negative than the $\Psi cell$. Water enters the cell causing it to swell tight. A pressure potential is generated when sufficient water has been taken up to cause the cell contents to press against the cell wall. Ψp rises progressively until it offsets Ψs. Water uptake stops when $\Psi cell = 0$. The rigid cell wall prevents cell rupture. Cells in this state are **turgid**.

7. The diagrams below show three hypothetical situations where adjacent cells have different water potentials. Draw arrows on each pair of cells (a)-(c) to indicate the net direction of water movement:

(a)

A	B
$\psi s = -400$ kPa	$\psi s = -500$ kPa
$\psi p = 300$ kPa	$\psi p = 300$ kPa
$\psi = -100$ kPa	$\psi = -200$ kPa

(b)

A	B
$\psi s = -500$ kPa	$\psi s = -600$ kPa
$\psi p = 100$ kPa	$\psi p = 100$ kPa
$\psi = -400$ kPa	$\psi = -400$ kPa

(c)

A	B
$\psi s = -600$ kPa	$\psi s = -500$ kPa
$\psi p = 200$ kPa	$\psi p = 300$ kPa
$\psi = -400$ kPa	$\psi = -200$ kPa

8. Fluid replacements are usually provided for heavily perspiring athletes after endurance events. Suggest what the water potential of these drinks should be, relative to the body fluids, and explain your answer:

_____

_____

9. (a) Explain the role of pressure potential in generating cell turgor in plants: _____

_____

(b) Explain the purpose of cell turgor to plants: _____

_____

10. Explain how animal cells differ from plant cells with respect to the effects of net water movements: _____

_____

11. Describe what would happen to an animal cell (e.g. a red blood cell) if it was placed into:

(a) Pure water: _____

(b) A solution of lower (more negative) water potential: _____

(c) A solution of higher (less negative) water potential: _____

Ion Pumps

Diffusion alone cannot supply the cell's entire requirements for molecules (and ions). Some molecules (e.g. glucose) are required by the cell in higher concentrations than occur outside the cell. Others (e.g. sodium) must be removed from the cell in order to maintain fluid balance. These molecules must be moved across the plasma membrane by active transport mechanisms. **Active transport** requires the expenditure of energy because the molecules (or ions) must be moved **against** their concentration gradient. The work of active transport is performed by specific carrier proteins in the membrane. These transport proteins harness the energy of ATP to pump molecules from a low to a high concentration. When ATP transfers a phosphate group to the carrier protein, the protein changes its shape in such a way as to move the bound molecule across the membrane. Three types of membrane pump are illustrated below. The sodium-potassium pump (below, centre) is almost universal in animal cells and is common in plant cells also. The concentration gradient created by ion pumps such as this and the proton pump (left) is frequently coupled to the transport of molecules such as glucose (e.g. in the intestine) as shown below right.

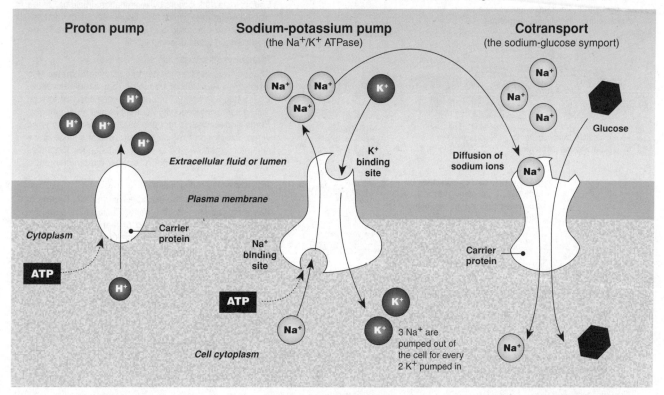

Proton pumps
ATP driven proton pumps use energy to remove hydrogen ions (H+) from inside the cell to the outside. This creates a large difference in the proton concentration either side of the membrane, with the inside of the plasma membrane being negatively charged. This potential difference can be coupled to the transport of other molecules.

Sodium-potassium pump
The sodium-potassium pump is a specific protein in the membrane that uses energy in the form of ATP to exchange sodium ions (Na+) for potassium ions (K+) across the membrane. The unequal balance of Na+ and K+ across the membrane creates large concentration gradients that can be used to drive transport of other substances (e.g. cotransport of glucose).

Cotransport (coupled transport)
In intestinal epithelial cells, a gradient in sodium ions drives the active transport of **glucose**. The specific transport protein couples the return of Na+ down its concentration gradient to the transport of glucose into the intestinal epithelial cell. A low intracellular concentration of Na+ (and therefore the concentration gradient) is maintained by a sodium-potassium pump.

1. Explain why the ATP is required for membrane pump systems to operate: _____

_____

2. (a) Explain what is meant by cotransport: _____

_____

   (b) Explain how cotransport is used to move glucose into the intestinal epithelial cells: _____

_____

_____

   (c) Explain what happens to the glucose that is transported into the intestinal epithelial cells: _____

_____

3. Describe two consequences of the extracellular accumulation of sodium ions: _____

_____

_____

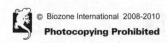
Periodicals: How biological membranes achieve selective transport

Related activities: Active and Passive Transport Summary
Web links: Cellular Transport, Symport

A 2

Cellular Membranes and Transport

Exocytosis and Endocytosis

Most cells carry out **cytosis**: a form of **active transport** involving the in- or outfolding of the plasma membrane. The ability of cells to do this is a function of the flexibility of the plasma membrane. Cytosis results in the bulk transport into or out of the cell and is achieved through the localised activity of microfilaments and microtubules in the cell cytoskeleton. Engulfment of material is termed **endocytosis.** Endocytosis typically occurs in protozoans and certain white blood cells of the mammalian defence system (e.g. neutrophils, macrophages). **Exocytosis** is the reverse of endocytosis and involves the release of material from vesicles or vacuoles that have fused with the plasma membrane. Exocytosis is typical of cells that export material (secretory cells).

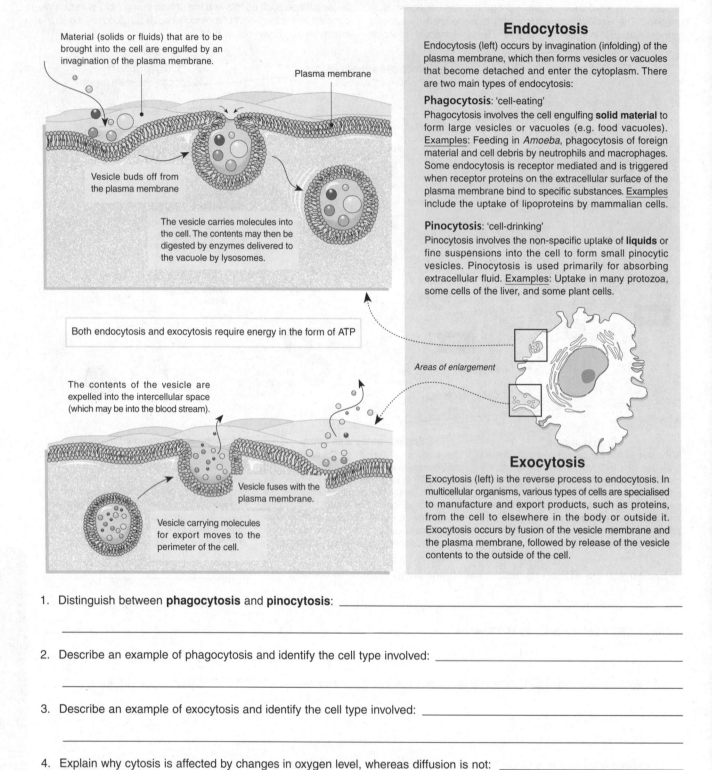

Material (solids or fluids) that are to be brought into the cell are engulfed by an invagination of the plasma membrane.

Plasma membrane

Vesicle buds off from the plasma membrane

The vesicle carries molecules into the cell. The contents may then be digested by enzymes delivered to the vacuole by lysosomes.

Both endocytosis and exocytosis require energy in the form of ATP

The contents of the vesicle are expelled into the intercellular space (which may be into the blood stream).

Vesicle fuses with the plasma membrane.

Vesicle carrying molecules for export moves to the perimeter of the cell.

Areas of enlargement

Endocytosis

Endocytosis (left) occurs by invagination (infolding) of the plasma membrane, which then forms vesicles or vacuoles that become detached and enter the cytoplasm. There are two main types of endocytosis:

Phagocytosis: 'cell-eating'

Phagocytosis involves the cell engulfing **solid material** to form large vesicles or vacuoles (e.g. food vacuoles). Examples: Feeding in *Amoeba*, phagocytosis of foreign material and cell debris by neutrophils and macrophages. Some endocytosis is receptor mediated and is triggered when receptor proteins on the extracellular surface of the plasma membrane bind to specific substances. Examples include the uptake of lipoproteins by mammalian cells.

Pinocytosis: 'cell-drinking'

Pinocytosis involves the non-specific uptake of **liquids** or fine suspensions into the cell to form small pinocytic vesicles. Pinocytosis is used primarily for absorbing extracellular fluid. Examples: Uptake in many protozoa, some cells of the liver, and some plant cells.

Exocytosis

Exocytosis (left) is the reverse process to endocytosis. In multicellular organisms, various types of cells are specialised to manufacture and export products, such as proteins, from the cell to elsewhere in the body or outside it. Exocytosis occurs by fusion of the vesicle membrane and the plasma membrane, followed by release of the vesicle contents to the outside of the cell.

1. Distinguish between **phagocytosis** and **pinocytosis**: _____

   _____

2. Describe an example of phagocytosis and identify the cell type involved: _____

   _____

3. Describe an example of exocytosis and identify the cell type involved: _____

   _____

4. Explain why cytosis is affected by changes in oxygen level, whereas diffusion is not: _____

   _____

5. Identify how each of the following substances enter a living macrophage (for help, see *Passive Transport Processes*):

   (a) Oxygen: _____ (c) Water: _____

   (b) Cellular debris: _____ (d) Glucose: _____

Related activities: *Active and Passive Transport Summary*
Web links: *Cellular Transport*

Periodicals:
What is endocytosis?

© Biozone International 2008-2010
Photocopying Prohibited

Active and Passive Transport Summary

Cells have a need to move materials both into and out of the cell. Raw materials and other molecules necessary for metabolism must be accumulated from outside the cell. Some of these substances are scarce outside of the cell and some effort is required to accumulate them. Waste products and molecules for use in other parts of the body must be 'exported' out of the cell.

Some materials (e.g. gases and water) move into and out of the cell by **passive transport** processes, without the expenditure of energy on the part of the cell. Other molecules (e.g. sucrose) are moved into and out of the cell using **active transport**. Active transport processes involve the expenditure of energy in the form of ATP, and therefore use oxygen.

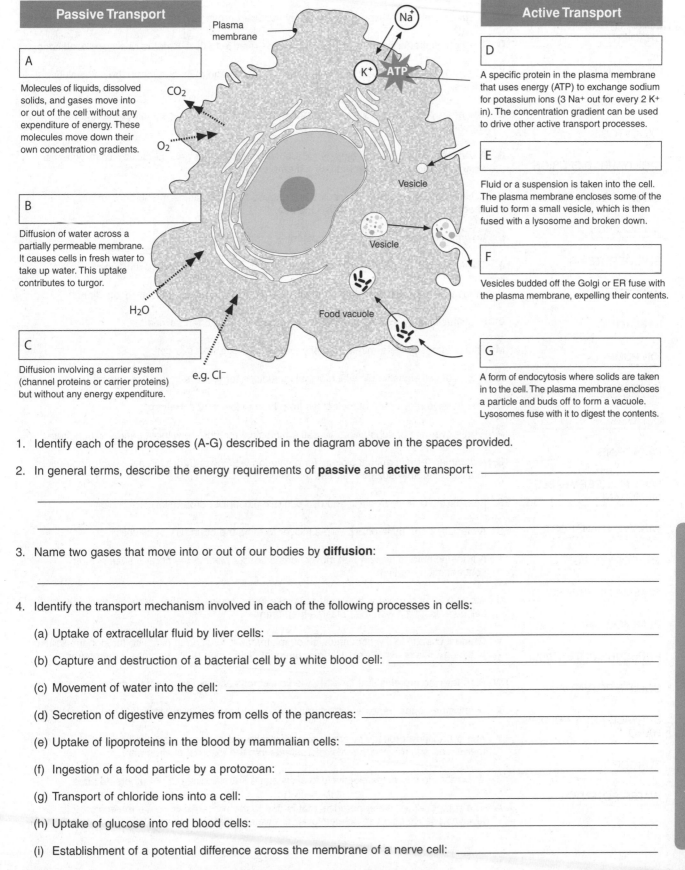

Passive Transport

A

Molecules of liquids, dissolved solids, and gases move into or out of the cell without any expenditure of energy. These molecules move down their own concentration gradients.

B

Diffusion of water across a partially permeable membrane. It causes cells in fresh water to take up water. This uptake contributes to turgor.

C

Diffusion involving a carrier system (channel proteins or carrier proteins) but without any energy expenditure.

Active Transport

D

A specific protein in the plasma membrane that uses energy (ATP) to exchange sodium for potassium ions (3 Na+ out for every 2 K+ in). The concentration gradient can be used to drive other active transport processes.

E

Fluid or a suspension is taken into the cell. The plasma membrane encloses some of the fluid to form a small vesicle, which is then fused with a lysosome and broken down.

F

Vesicles budded off the Golgi or ER fuse with the plasma membrane, expelling their contents.

G

A form of endocytosis where solids are taken in to the cell. The plasma membrane encloses a particle and buds off to form a vacuole. Lysosomes fuse with it to digest the contents.

Labels on diagram: Plasma membrane, Na+, K+, ATP, CO_2, O_2, Vesicle, Vesicle, H_2O, Food vacuole, e.g. Cl−

1. Identify each of the processes (A-G) described in the diagram above in the spaces provided.

2. In general terms, describe the energy requirements of **passive** and **active** transport: _____

_____

_____

3. Name two gases that move into or out of our bodies by **diffusion**: _____

_____

4. Identify the transport mechanism involved in each of the following processes in cells:

   (a) Uptake of extracellular fluid by liver cells: _____

   (b) Capture and destruction of a bacterial cell by a white blood cell: _____

   (c) Movement of water into the cell: _____

   (d) Secretion of digestive enzymes from cells of the pancreas: _____

   (e) Uptake of lipoproteins in the blood by mammalian cells: _____

   (f) Ingestion of a food particle by a protozoan: _____

   (g) Transport of chloride ions into a cell: _____

   (h) Uptake of glucose into red blood cells: _____

   (i) Establishment of a potential difference across the membrane of a nerve cell: _____

Related activities: Passive Transport Processes, Ion Pumps, Exocytosis and Endocytosis

RA 1

Cellular Membranes and Transport

KEY TERMS: Mix and Match

INSTRUCTIONS: *Test your vocab by matching each term to its correct definition, as identified by its preceding letter code.*

ACTIVE TRANSPORT

AMPHIPATHIC

CHANNEL PROTEINS

CONCENTRATION GRADIENT

DIFFUSION

ENDOCYTOSIS

EXOCYTOSIS

FACILITATED DIFFUSION

FLUID MOSAIC MODEL

GLYCOLIPIDS

GLYCOPROTEINS

HYPERTONIC

HYPOTONIC

ION PUMP

ISOTONIC

OSMOSIS

PARACRINE

PARTIALLY PERMEABLE MEMBRANE

PASSIVE TRANSPORT

PHAGOCYTOSIS

PLASMA MEMBRANE

PLASMOLYSIS

PRESSURE POTENTIAL

SOLUTE POTENTIAL

SURFACE AREA: VOLUME RATIO

TURGOR

WATER POTENTIAL

A Passive movement of water molecules across a partially permeable membrane down a concentration gradient.

B The model for membrane structure which proposes a double phospholipid bilayer in which proteins and cholesterol are embedded.

C A type of passive transport, facilitated by transport proteins.

D The potential energy of water per unit volume relative to pure water in reference conditions.

E The process in plant cells where the plasma membrane pulls away from the cell wall due to the loss of water through osmosis.

F The energy-requiring movement of substances across a biological membrane against a concentration gradient.

G A common term in animal biology for a solution with higher (less negative) water potential relative to another solution (across a membrane).

H Active transport in which molecules are engulfed by the plasma membrane, forming a phagosome or food vacuole within the cell.

I An important component of water potential in plant cells.

J The force exerted outward on a plant cell wall by the water contained in the cell.

K Lipids with attached carbohydrates which serve as markers for cellular recognition.

L In animal biology, solutions of equal water potential are often termed this.

M This relationship determines capacity for effective diffusion in a cell.

N A type of cell signalling in which signal molecules target cells in their immediate vicinity.

O The passive movement of molecules from high to low concentration.

P The movement of substances across a biological membrane without energy expenditure.

Q A common term in animal biology for a solution with lower (more negative) water potential relative to another solution (across a membrane).

R Possessing both hydrophilic and hydrophobic (lipophilic) properties.

S A partially-permeable phospholipid bilayer forming the boundary of all cells.

T A transmembrane protein that moves ions across a plasma membrane against their concentration gradient.

U Active transport process by which cells take in molecules (such as proteins) from outside the cell by engulfing it with their plasma membrane.

V Gradual change in the concentration of solutes as a function of distance through the solution. In biology, this is usually results from unequal distribution of ions across a membrane.

W Pore-forming proteins that facilitate the flow of ions across the plasma membrane.

X Membrane-bound proteins with attached carbohydrates, involved in cell to cell interactions.

Y Active transport process by which membrane-bound secretory vesicles fuse with the plasma membrane and release the vesicle contents into the external environment.

Z A membrane that acts selectively to allow some substances, but not others, to pass.

AA The component of water potential that is due to the presence of solute molecules. It is always negative because solutes lower the water potential of the system.

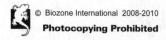

Cellular Division and Organisation

KEY CONCEPTS

▶ Cellular metabolism depends on the transport of substances across cellular membranes.

▶ Cell size is limited by surface area to volume ratio.

▶ New cells arise through cell division.

▶ Cellular diversity arises through specialization from stem cell progenitors.

▶ Regulation of cell division is important in development.

▶ Emergent properties are a feature of increasing complexity in biological systems.

KEY TERMS

anaphase
cancer
cell cycle
cellular differentiation
centrioles
chromosome
cytokinesis
diploid
DNA replication
epithelium
erythrocytes
G1
G2
guard cells
haploid
homologous chromosomes
meiosis
metaphase
mitosis
neutrophil
nuclear envelope
organ
organ system
palisade mesophyll cell
phloem
phloem sieve cell
prophase
reduction division
root hair cell
S phase
stem cell
telophase
tissue
xylem
xylem vessel

OBJECTIVES

☐ 1. Use the **KEY TERMS** to help you understand and complete these objectives.

The Nucleus and Cell Division   pages 72-76

☐ 2. Describe how genetic information is copied and passed to daughter cells during the **cell cycle**. Describe the cell cycle in eukaryotes, including reference to: **mitosis**, **growth** (G1 and G2), and DNA replication (S). Identify the approximate proportion of the cycle spent in each stage.

☐ 3. Use diagrams and photographs to describe the main stages in mitosis: **prophase**, **metaphase**, **anaphase**, and **telophase**. Include reference to the behaviour of the chromosomes, nuclear envelope, plasma membrane, and centrioles (if present).

☐ 4. Contrast **cytokinesis** in plant and animal cells.

☐ 5. Describe and explain the role of mitosis in growth and repair, and asexual reproduction. Use diagrams to describe cell division by budding in yeast.

☐ 6. Explain the terms **homologous chromosomes**, **diploid**, and **haploid**. Compare and contrast **meiosis (reduction division)** and **mitosis** in terms of their cellular and genetic outcomes.

☐ 7. <u>CIE AND OCR EXTENSION</u>: Explain how uncontrolled mitotic division can lead to tumour formation and **cancer**. Describe factors that can increase the chances of tumour growth.

Cellular Differentiation   pages 77-84

☐ 8. Describe the properties of **stem cells** and their role in multicellular organisms.

☐ 9. Recognise mitotic cell division as a prelude to **cellular differentiation**. Describe cell differentiation and specialisation with reference to:

(a) Differentiation of blood cells from stem cells in bone marrow

(b) Differentiation of xylem vessels and phloem sieve tubes from cambium.

☐ 10. Use diagrams and photographs to describe functional specialisation in plant and animal cells. Examples include **erythrocytes**, **neutrophils**, **epithelial cells**, **sperm cells**, **palisade mesophyll cells**, **root hair cells**, **guard cells**.

☐ 11. Describe the hierarchy of organisation in multicellular organisms. Appreciate that each step in the hierarchy of biological order is associated with the emergence of properties not present at simpler levels of organisation.

☐ 12. Describe the organisation of specialised cells into **tissues**. Examples should include **squamous epithelia**, **ciliated epithelia**, **xylem**, and **phloem**. Explain how tissues are grouped together into **organs** and **organ systems** and discuss the role of cooperation between the various components in achieving this.

Periodicals:

Listings for this chapter are on page 338

Weblinks:

www.biozone.co.uk/ weblink/OCR-AS-2641.html

Teacher Resource CD-ROM:

Stem Cell Technology

Cell Division

The life cycle of a diploid sexually reproducing organism, such as a human, with **gametic meiosis** is illustrated below. In this life cycle, **gametogenesis** involves meiotic division to produce male and female gametes for the purpose of sexual reproduction. The life cycle in flowering plants is different in that the gametes are produced through mitosis in haploid gametophytes. The male gametes are produced inside the pollen grain and the female gametes are produced inside the embryo sac of the ovule. The gametophytes develop and grow from haploid spores, which are produced from meiosis.

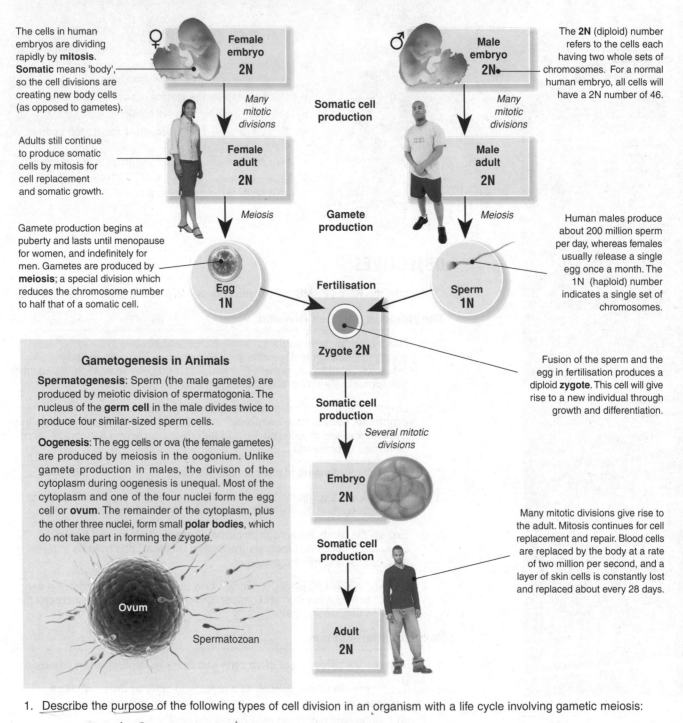

The cells in human embryos are dividing rapidly by **mitosis**. **Somatic** means 'body', so the cell divisions are creating new body cells (as opposed to gametes).

Adults still continue to produce somatic cells by mitosis for cell replacement and somatic growth.

Gamete production begins at puberty and lasts until menopause for women, and indefinitely for men. Gametes are produced by **meiosis**; a special division which reduces the chromosome number to half that of a somatic cell.

The **2N** (diploid) number refers to the cells each having two whole sets of chromosomes. For a normal human embryo, all cells will have a 2N number of 46.

Human males produce about 200 million sperm per day, whereas females usually release a single egg once a month. The 1N (haploid) number indicates a single set of chromosomes.

Fusion of the sperm and the egg in fertilisation produces a diploid **zygote**. This cell will give rise to a new individual through growth and differentiation.

Many mitotic divisions give rise to the adult. Mitosis continues for cell replacement and repair. Blood cells are replaced by the body at a rate of two million per second, and a layer of skin cells is constantly lost and replaced about every 28 days.

Gametogenesis in Animals

Spermatogenesis: Sperm (the male gametes) are produced by meiotic division of spermatogonia. The nucleus of the **germ cell** in the male divides twice to produce four similar-sized sperm cells.

Oogenesis: The egg cells or ova (the female gametes) are produced by meiosis in the oogonium. Unlike gamete production in males, the divison of the cytoplasm during oogenesis is unequal. Most of the cytoplasm and one of the four nuclei form the egg cell or **ovum**. The remainder of the cytoplasm, plus the other three nuclei, form small **polar bodies**, which do not take part in forming the zygote.

1. Describe the purpose of the following types of cell division in an organism with a life cycle involving gametic meiosis:

   (a) Mitosis: _for growth + repair_

   (b) Meiosis: _to produce sex cells_

2. Describe the basic difference between the cell divisions involved in spermatogenesis and oogenesis:

   _____

   _____

3. Explain how gametogenesis differs between humans and flowering plants: _____

   _____

Related activities: Mitosis and the Cell Cycle

Mitosis and the Cell Cycle

Mitosis is part of the 'cell cycle' in which an existing cell (the parent cell) divides into two (the daughter cells). Mitosis does not result in a change of chromosome numbers (unlike meiosis) and the daughter cells are identical to the parent cell. Although mitosis is part of a continuous cell cycle, it is divided into stages (below). The example below illustrates the cell cycle in a plant cell. Note that in animal cells, **cytokinesis** involves the formation of a constriction that divides the cell in two. It is usually well underway by the end of telophase and does not involve the formation of a cell plate.

The Cell Cycle and Stages of Mitosis

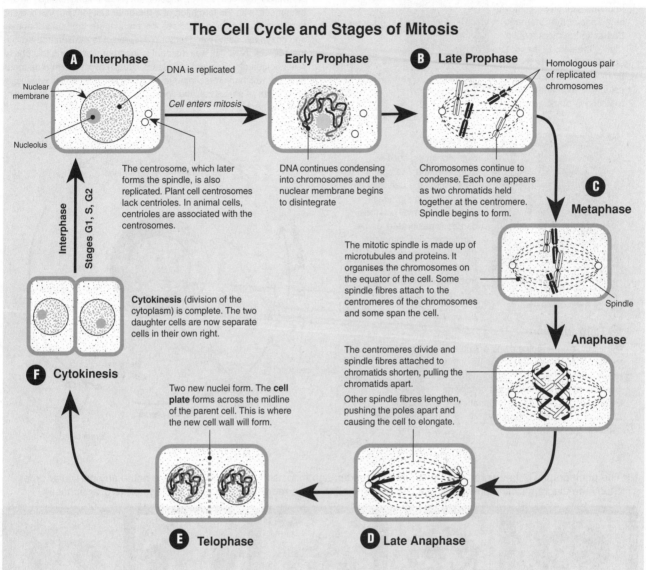

A Interphase

Nuclear membrane
DNA is replicated
Nucleolus

The centrosome, which later forms the spindle, is also replicated. Plant cell centrosomes lack centrioles. In animal cells, centrioles are associated with the centrosomes.

Interphase
Stages G1, S, G2

Early Prophase

DNA continues condensing into chromosomes and the nuclear membrane begins to disintegrate

B Late Prophase

Homologous pair of replicated chromosomes

Chromosomes continue to condense. Each one appears as two chromatids held together at the centromere. Spindle begins to form.

C Metaphase

The mitotic spindle is made up of microtubules and proteins. It organises the chromosomes on the equator of the cell. Some spindle fibres attach to the centromeres of the chromosomes and some span the cell.

Spindle

Anaphase

The centromeres divide and spindle fibres attached to chromatids shorten, pulling the chromatids apart.

Other spindle fibres lengthen, pushing the poles apart and causing the cell to elongate.

F Cytokinesis

Cytokinesis (division of the cytoplasm) is complete. The two daughter cells are now separate cells in their own right.

Two new nuclei form. The **cell plate** forms across the midline of the parent cell. This is where the new cell wall will form.

E Telophase

D Late Anaphase

The Cell Cycle Overview

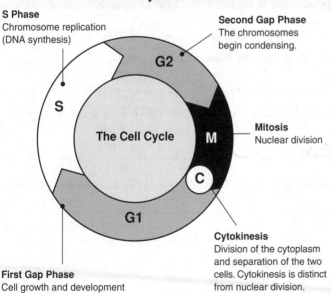

S Phase
Chromosome replication (DNA synthesis)

Second Gap Phase
The chromosomes begin condensing.

G2

S

The Cell Cycle

M

C

Mitosis
Nuclear division

Cytokinesis
Division of the cytoplasm and separation of the two cells. Cytokinesis is distinct from nuclear division.

G1

First Gap Phase
Cell growth and development

Homologous Chromosomes

In sexually reproducing organisms, the chromosomes of most cells are present as **homologous pairs**. One chromosome of a pair is supplied by the female parent and one by the male parent. Each homologue carries an identical assortment of genes, but the version of the gene (allele) from each parent may differ.

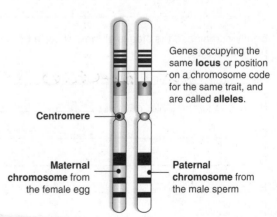

Genes occupying the same **locus** or position on a chromosome code for the same trait, and are called **alleles**.

Centromere

Maternal chromosome from the female egg

Paternal chromosome from the male sperm

Periodicals:
The cell cycle and mitosis

Related activities: The Genetic Origins of Cancer, Root Cell Development
Web links: Mitosis in an Animal Cell

Mitotic cell division has several purposes (below left). In multicellular organisms, mitosis repairs damaged cells and tissues, and produces the growth in an organism that allows it to reach its adult size. In unicellular organisms, and some small multicellular organisms, cell division allows organisms to reproduce asexually (as in the budding yeast cell cycle below).

The Functions of Mitosis

❶ Growth

In plants, cell division occurs in regions of **meristematic tissue**. In the plant root tip (right), the cells in the root apical meristem are dividing by mitosis to produce new cells. This elongates the root, resulting in **plant growth**.

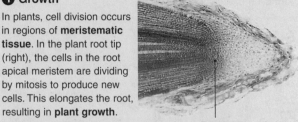

Root apical meristem

Photo: AB Sheldon

❷ Repair

Some animals, such as this skink (left), detach their limbs as a defence mechanism in a process called autotomy. The limbs can be **regenerated** via the mitotic process, although the tissue composition of the new limb differs slightly from that of the original.

❸ Reproduction

Mitotic division enables some animals to reproduce **asexually**. The cells of this Hydra (left) undergo mitosis, forming a 'bud' on the side of the parent organism. Eventually the bud, which is genetically identical to its parent, detaches to continue the life cycle.

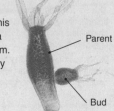

Parent

Bud

The Budding Yeast Cell Cycle

Yeasts can reproduce asexually through **budding**. In *Saccharomyces cerevisiae* (baker's yeast), budding involves mitotic division in the parent cell, with the formation of a daughter cell (or bud). As budding begins, a ring of chitin stabilises the area where the bud will appear and enzymatic activity and turgor pressure act to weaken and extrude the cell wall. New cell wall material is incorporated during this phase. The nucleus of the parent cell also divides in two, to form a daughter nucleus, which migrates into the bud. The daughter cell is genetically identical to its parent cell and continues to grow, eventually separating from the parent cell.

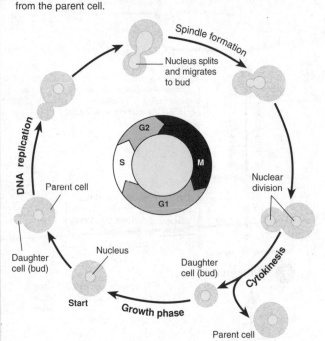

Spindle formation

Nucleus splits and migrates to bud

Nuclear division

Cytokinesis

DNA replication

Parent cell

G2

S

M

G1

Nucleus

Daughter cell (bud)

Daughter cell (bud)

Start

Growth phase

Parent cell

Photos: RCN

1. The photographs below were taken at various stages through mitosis in a plant cell. They are not in any particular order. Study the diagram on the previous page and determine the stage represented in each photograph (e.g. anaphase).

   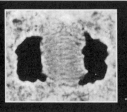 

(a) _Anaphase_   (b) _Prophase_   (c) _Metophase_   (d) _Telophase_   (e) _Interphase_

2. State two important changes that chromosomes must undergo before cell division can take place: _____

_____ Duplicate _____

3. Briefly summarise the stages of the cell cycle by describing what is happening at the points (**A-F**) in the diagram on the previous page:

I A. _DNA double. cell develps, energy store, organelles._

P B. _chromatin condense to chromosom. N.e disintegrates_

M C. _chr. line up at middle c.mares attach it mk_

A D. _chr. move to opp. ends of the cell_

T E. _chr. back to chromatin envelope redevelps_

C F. _p. memb invagination cnar splits._

The Genetic Origins of Cancer

Normal cells do not live forever. Under certain circumstances, cells are programmed to die, particularly during development. Cells that become damaged beyond repair will normally undergo this programmed cell death (called **apoptosis** or **cell suicide**). Cancer cells evade this control and become immortal, continuing to divide regardless of any damage incurred. **Carcinogens** are agents capable of causing cancer. Roughly 90% of carcinogens are also mutagens, i.e. they damage DNA. Chronic exposure to carcinogens accelerates the rate at which dividing cells make errors. Susceptibility to cancer is also influenced by genetic make-up. Any one or a number of cancer-causing factors (including defective genes) may interact to induce cancer.

Cancer: Cells out of Control

Cancerous transformation results from changes in the genes controlling normal cell growth and division. The resulting cells become immortal and no longer carry out their functional role. Two types of gene are normally involved in controlling the cell cycle: proto-oncogenes, which start the cell division process and are essential for normal cell development, and **tumour-suppressor** genes, which switch off cell division. In their normal form, both kinds of genes work as-a team, enabling the body to perform vital tasks such as repairing defective cells and replacing dead ones. But mutations in these genes can disrupt these finely tuned checks and balances. Proto-oncogenes, through mutation, can give rise to **oncogenes**; genes that lead to uncontrollable cell division. Mutations to tumour-suppressor genes initiate most human cancers. The best studied tumour-suppressor gene is **p53**, which encodes a protein that halts the cell cycle so that DNA can be repaired before division.

The panel, right, shows the mutagenic action of some selected carcinogens on four of five codons of the **p53 gene**.

Features of Cancer Cells

The diagram right shows a single **lung cell** that has become cancerous. It no longer carries out the role of a lung cell, and instead takes on a parasitic lifestyle, taking from the body what it needs in the way of nutrients and contributing nothing in return. The rate of cell division is greater than in normal cells in the same tissue because there is no *resting phase* between divisions.

A mutation in one or two of the controlling genes causes a **benign** (nonmalignant) **tumour**. As the number of controlling genes with mutations increases, so too does the loss of control until the cell becomes cancerous.

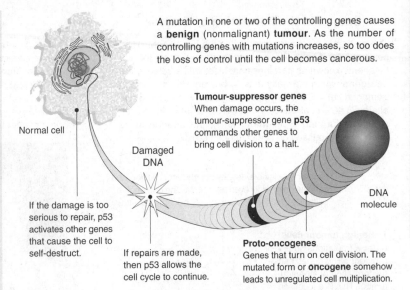

Normal cell

Damaged DNA

If the damage is too serious to repair, p53 activates other genes that cause the cell to self-destruct.

If repairs are made, then p53 allows the cell cycle to continue.

Tumour-suppressor genes
When damage occurs, the tumour-suppressor gene **p53** commands other genes to bring cell division to a halt.

DNA molecule

Proto-oncogenes
Genes that turn on cell division. The mutated form or **oncogene** somehow leads to unregulated cell multiplication.

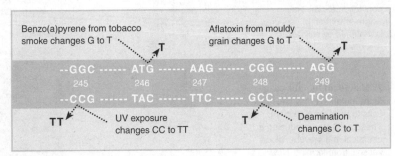

Benzo(a)pyrene from tobacco smoke changes G to T

Aflatoxin from mouldy grain changes G to T

--GGC	ATG	AAG	CGG	AGG
245	246	247	248	249
--CCG	TAC	TTC	GCC	TCC

UV exposure changes CC to TT

Deamination changes C to T

Given a continual supply of nutrients, cancer cells can go on dividing indefinitely and are said to be immortal.

Cancer cells may have unusual numbers of chromosomes.

The bloated, lumpy shape is readily distinguishable from a healthy cell, which has a flat, scaly appearance.

Metabolism is disrupted and the cell ceases to function constructively.

Cancerous cells lose their attachments to neighbouring cells.

1. Explain how cancerous cells differ from normal cells: _reproduce faster_

_____

2. Explain how the cell cycle is normally controlled, including reference to the role of **tumour-suppressor genes**:

_____

_____

_____

_____

3. With reference to the role of **oncogenes**, explain how the normal controls over the cell cycle can be lost:

_____

_____

_____

_____

Periodicals:
Rebels without a cause

Related activities: Mitosis and the Cell Cycle, Cell Growth and Cancer

Cell Growth and Cancer

Cancer is a term describing a large group of diseases characterised by the progressive and uncontrolled growth of abnormal cells. There is no single cause for all the forms of cancer; environmental, genetic, and biological factors are usually implicated. Certain risk factors increase a person's chance of getting cancer. Some risk factors, such as exposure to tobacco smoke, are controllable, while others, such as gender, are not. Because cancers arise as a result of damage to DNA, those factors that cause cellular damage, e.g. exposure to the chemicals in cigarette smoke, increase the risk of cancers developing.

Stages in the Formation of Cancer

The growth of a cancer begins when the genes controlling cell growth and multiplication (**oncogenes**) are transformed by agents known as **carcinogens**. Most well studied is the p53 gene which normally acts to prevent cell division in damaged cells. Scientists have found that the p53 gene is altered in 40% of all cancers. Once a cell is transformed into a tumour-forming type (**malignant**), the change in its oncogenes is passed on to all offspring cells:

Cancer cells ignore density-dependent inhibition and continue to multiply even after contacting one another, piling up until the nutrient supply becomes limiting.

1. Benign tumour cells
Defects (mutations) in one or two controlling genes cause the formation of a benign tumour. This is a localised population of proliferating cells where formation of new cells is matched by cell death.

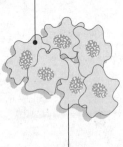

2. Malignant tumour cells
More mutations may cause the cells to become malignant. These cells stop producing a chemical that prevents blood vessels from forming. New capillaries grow into the tumour, providing it with nutrients.

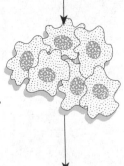

3. Metastasis
The new capillaries also provide a route for the malignant cells to break away from the tumour and travel to other parts of the body where they start new cancers.

Malignant cells break away from tumour mass and spread through the body through the **blood** or **lymphatic systems**.

Risk Factors for Cancer

The greatest **uncontrollable risk factor** for cancer is **ageing**; most cancers occur in people over the age of 65. Others include **genetic predisposition** (family history) and gender. **Controllable risk factors** include **lifestyle factors**, such as **tobacco use**. Not unexpectedly, different kinds of cancer are associated with different risk factors. For example, sunlight exposure increases the risk of skin cancers. Some major risk factors include the following:

Tobacco use is related to a wide range of cancers; smoking alone causes one third of all cancer deaths.

Excessive alcohol intake, especially when associated with tobacco use, is associated with oral cancers.

A highly processed, high fat diet is associated with higher risk of various cancers, including colon cancer.

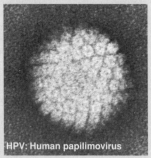

HPV: Human papilimovirus

Certain Infections are associated with the development of cancers. HPV is strongly linked to cervical cancer.

Ionising radiation and hazardous substances, such as asbestos and formaldehyde, cause cell damage that can lead to cancer.

Unprotected exposure to ultraviolet light causes early ageing of the skin and damage that can lead to the development of skin cancers.

1. Explain the mechanism by which the risk factors described above increase the chance of developing cancer:

_____

2. Explain why it can be difficult to determine the causative role of a single risk factor in the development of a cancer:

_____

_____

Differentiation of Human Cells

Stem cells are undifferentiated cells found in multicellular organisms. They are characterised by two features. The first, **self renewal**, is the ability to undergo numerous cycles of cell division while maintaining an unspecialised state. The second, <u>**potency**</u>, is the ability to differentiate into specialised cells. **Totipotent** cells, produced in the first few divisions of a fertilized egg, can differentiate into any cell type, embryonic or extra-embryonic. **Pluripotent cells** are descended from totipotent cells and can give rise to any of the cells derived from the three germ layers (endoderm, mesoderm, and ectoderm). Embryonic stem cells

at the blastocyst stage and foetal stem cells are pluripotent. Adult (somatic) stem cells are termed **multipotent**. They are undifferentiated cells found among differentiated cells in a tissue or organ. These cells can give rise to several other cell types, but those types are limited mainly to the cells of the blood, heart, muscle and nerves. The primary roles of adult stem cells are to maintain and repair the tissue in which they are found. A potential use of stem cells is making cells and tissues for medical therapies, such as **cell replacement therapy** and **tissue engineering** (for example, for bone and skin grafts).

Stem Cells and Blood Cell Production

New blood cells are produced in the red bone marrow, which becomes the main site of blood production after birth, taking over from the foetal liver. All types of blood cells develop from a single cell type: called a **multipotent stem cell** or haemocytoblast. These cells are capable of mitosis and of differentiation into 'committed' precursors of each of the main types of blood cell.

Each of the different cell lines is controlled by a specific **growth factor**. When a stem cell divides, one of its daughters remains a stem cell, while the other becomes a precursor cell, either a **lymphoid cell** or **myeloid cell**. These cells continue to mature into the various type of blood cells, developing their specialised features and characteristic roles as they do so.

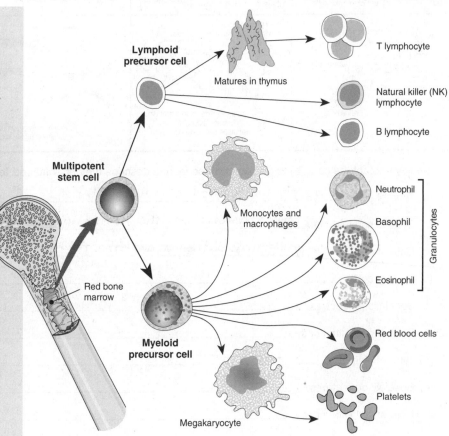

1. Describe the two defining features of stem cells:

   (a) _unspecialised_

   (b) _able to differentiate_

2. Distinguish between embryonic stem cells and adult stem cells with respect to their **potency** and their potential applications in medical technologies:

   s.cells - very potent, can differentiate into all diff types of cells. unlimited
   A. - less versatile, can only into some, limited.

3. Using an example, explain the purpose of stem cells in an adult: _in bone marrow to produce diff. types of blood cells._

4. Describe one potential advantage of using embryonic stem cells for tissue engineering technology: _- have a store of diff types of cells - easy to harvest unlike Adult._

Periodicals:
What is a stem cell?
Cell differentiation

Related activities: Differentiation of Human Cells
Web links: Stem Cells in the Spotlight, Stem Cell Resources

RA 3

Human Cell Specialisation

Animal cells are often specialised to perform particular functions. The eight specialised cell types shown below are representative of some 230 different cell types in humans. Each has specialised features that suit it to performing a specific role.

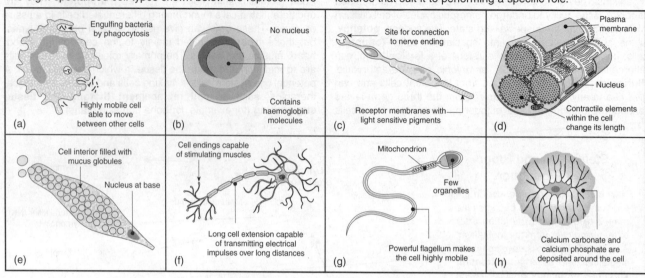

(a) Engulfing bacteria by phagocytosis. Highly mobile cell able to move between other cells.

(b) No nucleus. Contains haemoglobin molecules.

(c) Site for connection to nerve ending. Receptor membranes with light sensitive pigments.

(d) Plasma membrane. Nucleus. Contractile elements within the cell change its length.

(e) Cell interior filled with mucus globules. Nucleus at base.

(f) Cell endings capable of stimulating muscles. Long cell extension capable of transmitting electrical impulses over long distances.

(g) Mitochondrion. Few organelles. Powerful flagellum makes the cell highly mobile.

(h) Calcium carbonate and calcium phosphate are deposited around the cell.

1. Identify each of the cells (b) to (h) pictured above, and describe their **specialised features** and **role** in the body:

(a) Type of cell: _Phagocytic white blood cell (neutrophil)_

Specialised features: _Engulfs bacteria and other foreign material by phagocytosis_

Role of cell within body: _Destroys pathogens and other foreign material as well as cellular debris_

(b) Type of cell: _RBC_

Specialised features: _Biconcave, no contents, large SA:V_

Role of cell within body: _carry O2 + remove around body_

(c) Type of cell: _____

Specialised features: _____

Role of cell within body: _____

(d) Type of cell: _____

Specialised features: _____

Role of cell within body: _____

(e) Type of cell: _____

Specialised features: _____

Role of cell within body: _____

(f) Type of cell: _____

Specialised features: _____

Role of cell within body: _____

(g) Type of cell: _Sperm cell_

Specialised features: _Acrosome, flagella, N chr, mitochondria_

Role of cell within body: _reach egg, fertilise_

(h) Type of cell: _____

Specialised features: _____

Role of cell within body: _____

Related activities: Differentiation of Human Cells, Animal Cells

Differentiation of Plant Cells

The differentiation of plant cells occurs only at specific regions called **meristems**. Two types of growth can contribute to an increase in the size of a plant. **Primary growth**, which occurs in the **apical meristem** of the buds and root tips, increases the length (height) of a plant. **Secondary growth** (not discussed here) increases plant girth and occurs in the lateral meristem in the stem. All plants show primary growth but only some show secondary growth (the growth that produces woody tissues).

Primary Growth

Primary growth occurs at the **apical meristem** (root and shoot tips). Three types of **primary meristem** are produced from the apical meristem: procambium, protoderm, and ground meristem. In dicots, the **procambium** forms vascular bundles that are found in a ring near the epidermis and surrounded by cortex. As the procambium divides, the cells on the inside become primary **xylem** and those on the outside become primary **phloem**.

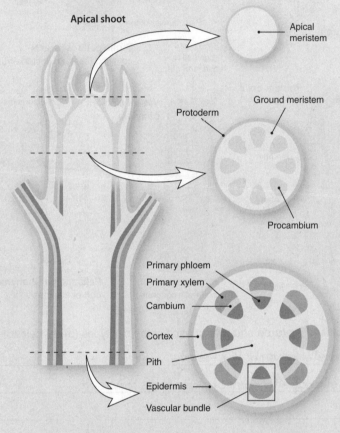

Primary Tissues Generated by the Meristem

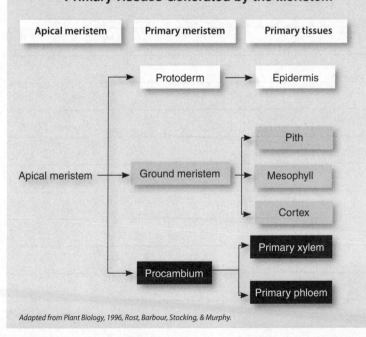

Adapted from Plant Biology, 1996, Rost, Barbour, Stocking, & Murphy.

1. Describe the role of the meristems in plants:

   Differentiate into other types.

2. Describe the location of the meristems and relate this to how plants grow:

   _____

   _____

   _____

   _____

   _____

3. Describe a distinguishing feature of meristematic tissue:

   _____

   _____

   _____

   _____

   _____

4. Discuss the structure and formation of the primary tissues in dicot plants:

   _____

   _____

   _____

   _____

   _____

   _____

   _____

   _____

   _____

   _____

   _____

   _____

   _____

   _____

   _____

Related activities: Root Cell Development

A 2

Plant Cell Specialisation

Plants show a wide variety of cell types. The vegetative plant body consists of three organs: stems, leaves, and roots. Flowers, fruits, and seeds comprise additional organs that are concerned with reproduction. The eight cell types illustrated below are representatives of these plant organ systems. Each has structural or physiological features that set it apart from the other cell types. The differentiation of cells enables each specialised type to fulfill a specific role in the plant.

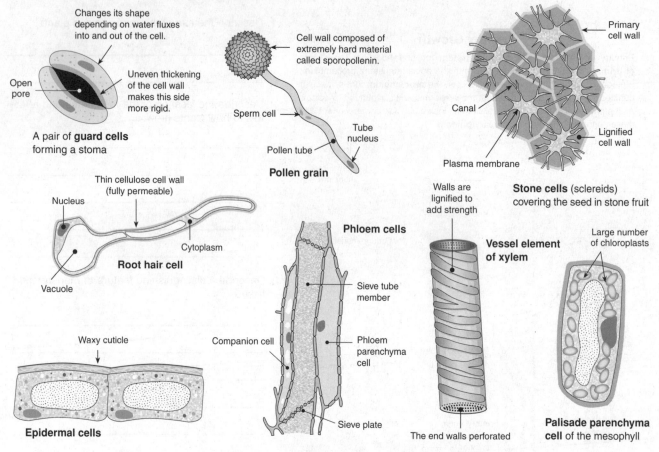

Changes its shape depending on water fluxes into and out of the cell.

Open pore

Uneven thickening of the cell wall makes this side more rigid.

A pair of **guard cells** forming a stoma

Cell wall composed of extremely hard material called sporopollenin.

Sperm cell

Tube nucleus

Pollen tube

Pollen grain

Primary cell wall

Canal

Plasma membrane

Lignified cell wall

Stone cells (sclereids) covering the seed in stone fruit

Thin cellulose cell wall (fully permeable)

Nucleus

Cytoplasm

Root hair cell

Vacuole

Walls are lignified to add strength

Vessel element of xylem

Large number of chloroplasts

Phloem cells

Sieve tube member

Waxy cuticle

Companion cell

Phloem parenchyma cell

Sieve plate

Epidermal cells

The end walls perforated

Palisade parenchyma cell of the mesophyll

1. Using the information given above, describe the **specialised features** and **role** of each of the cell types (b)-(h) below:

   (a) **Guard cell**: Features: Curved, sausage shaped cell, unevenly thickened.

      Role in plant: Turgor changes alter the cell shape to open or close the stoma.

   (b) **Pollen grain**: Features: _____

      Role in plant: _____

   (c) **Palisade parenchyma cell**: Features: _____

      Role in plant: _____

   (d) **Epidermal cell**: Features: _____

      Role in plant: _____

   (e) **Vessel element**: Features: _____

      Role in plant: _____

   (f) **Stone cell**: Features: _____

      Role in plant: _____

   (g) **Sieve tube member** (of phloem): Features: _____

      Role in plant: _____

   (h) **Root hair cell**: Features: _____

      Role in plant: _____

Related activities: Plant Cells, Differentiation of Plant Cells

Root Cell Development

In plants, cell division for growth (mitosis) is restricted to growing tips called **meristematic** tissue. These are located at the tips of every stem and root. This is unlike mitosis in a growing animal where cell divisions can occur all over the body. The diagram below illustrates the position and appearance of developing and growing cells in a plant root. Similar zones of development occur in the growing stem tips, which may give rise to specialised structures such as leaves and flowers.

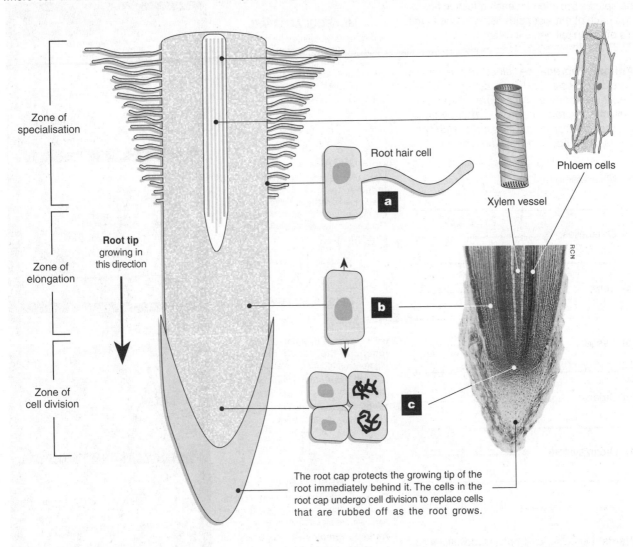

Zone of specialisation

Root hair cell

a

Phloem cells

Xylem vessel

RCN

Root tip growing in this direction

Zone of elongation

b

Zone of cell division

c

The root cap protects the growing tip of the root immediately behind it. The cells in the root cap undergo cell division to replace cells that are rubbed off as the root grows.

1. Briefly describe what is happening to the plant cells at each of the points labelled (**a**) to (**c**) in the diagram above:

   (a) _____

   (b) _____

   (c) _____

2. The light micrograph (below) shows a section of the cells of an onion root tip, stained to show up the chromosomes.

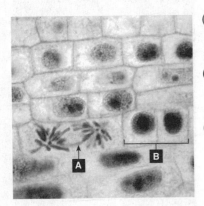

A

B

   (a) State the mitotic stage of the cell labelled A and explain your answer:

   Anaphase chr. separated to opp ends

   (b) State the mitotic stage just completed in the cells labelled B and explain:

   Telophase

   (c) If, in this example, 250 cells were examined and 25 were found to be in the process of mitosis, state the proportion of the cell cycle occupied by mitosis:

   _25/250_

3. Identify the cells that divide and specialise when a tree increases its girth (diameter): _meristem_

Related activities: Mitosis and the Cell Cycle, Plant Cell Specialisation

RL

Levels of Organisation

Organisation and the emergence of novel properties in complex systems are two of the defining features of living organisms. Organisms are organised according to a hierarchy of structural levels (below), each level building on the one below it. At each level, novel properties emerge that were not present at the simpler level. Hierarchical organisation allows specialised cells to group together into tissues and organs to perform a particular function. This improves efficiency of function in the organism.

In the spaces provided for each question below, assign each of the examples listed to one of the levels of organisation as indicated.

1. **Animals**: *adrenaline, blood, bone, brain, cardiac muscle, cartilage, collagen, DNA, heart, leucocyte, lysosome, mast cell, nervous system, neurone, phospholipid, reproductive system, ribosomes, Schwann cell, spleen, squamous epithelium.*

(a) Molecular level: _____

_____

(b) Organelles: _____

_____

(c) Cells: _____

_____

(d) Tissues: _Common_

function, articae

(e) Organs: _paticular_

function ung

(f) Organ system: _a paticular_

life function

endocrine

2. **Plants**: *cellulose, chloroplasts, collenchyma, companion cells, DNA, epidermal cell, fibres, flowers, leaf, mesophyll, parenchyma, pectin, phloem, phospholipid, ribosomes, roots, sclerenchyma, tracheid.*

(a) Molecular level: _____

_____

(b) Organelles: _____

_____

(c) Cells: _____

_____

(d) Tissues: _____

Xylem

(e) Organs: _____

_____

leaf

MOLECULAR LEVEL

Atoms and molecules form the most basic level of organisation. This level includes all the chemicals essential for maintaining life e.g. water, ions, fats, carbohydrates, amino acids, proteins, and nucleic acids.

ORGANELLE LEVEL

Many diverse molecules may associate together to form complex, highly specialised structures within cells called cellular organelles e.g. mitochondria, Golgi apparatus, endoplasmic reticulum, chloroplasts.

CELLULAR LEVEL

Cells are the basic structural and functional units of an organism. Each cell type has a different structure and function; the result of cellular differentiation during development.

Animal examples include: epithelial cells, osteoblasts, muscle fibres.

Plant examples include: sclereids, xylem vessels, sieve tubes.

TISSUE LEVEL

Tissues are composed of groups of cells of similar structure that perform a particular, related function.

Animal examples include: epithelial tissue, bone, muscle.

Plant examples include: phloem, chlorenchyma, endodermis, xylem.

ORGAN LEVEL

Organs are structures of definite form and structure, made up of two or more tissues.

Animal examples include: heart, lungs, brain, stomach, kidney.

Plant examples include: leaves, roots, storage organs, ovary.

ORGAN SYSTEM LEVEL

In animals, organs form parts of even larger units known as **organ systems**. An organ system is an association of organs with a common function, e.g. digestive system, cardiovascular system, and the urinary system. In all, eleven organ systems make up the **organism**.

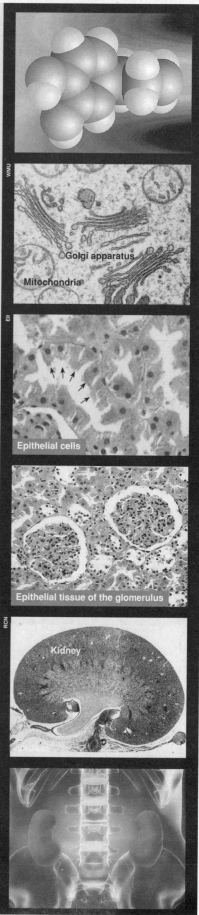

Golgi apparatus

Mitochondria

Epithelial cells

Epithelial tissue of the glomerulus

Kidney

Related activities: Animal Tissues, Plant Tissues

Animal Tissues

Tissues comprise cells and their associated extracellular substances grouped together to perform particular functions. Tissues improve the efficiency of operation because they enable tasks to be shared amongst various specialised cells. Animal tissues can be divided into four broad groups: **epithelial tissues**, **connective tissues**, **muscle**, and **nervous tissues**. Organs usually consist of several types of tissue. The heart mostly consists of cardiac muscle tissue, but also has epithelial tissue lining the heart chambers, connective tissue for strength and elasticity, and nervous tissue to direct the contractions of the cardiac muscle. Some features of tissue organisation are illustrated below using epithelial tissues as an example. Except for glandular epithelium, epithelial cells form fitted continuous sheets, held in place by desmosomes and tight junctions. Epithelia are classified as simple (single layered) or stratified (two or more layers), and the cells may be squamous (flat), cuboidal, or columnar (rectangular). Thus at least two adjectives describe any particular epithelium (e.g. stratified cuboidal).

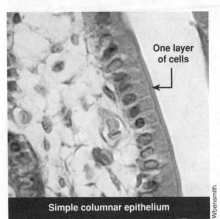

Simple columnar epithelium

The simple epithelium of the gastrointestinal tract is easily recognised by the regular column-like cells. it is specialised for secretion and absorption.

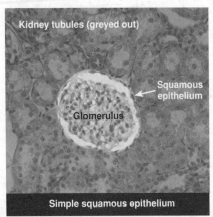

Kidney tubules (greyed out)

Squamous epithelium

Glomerulus

Simple squamous epithelium

Simple squamous epithelium lines surfaces where rapid diffusion is required such as in capillaries, alveoli, and lining the glomerular capsule (above).

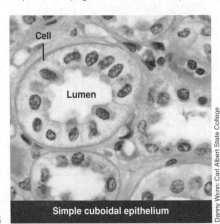

Cell

Lumen

Danny Wenn: Carl Albert State College

Simple cuboidal epithelium

Simple cuboidal epithelium is common in glands and their ducts and also lines the kidney tubules (above) and the surface of the ovaries.

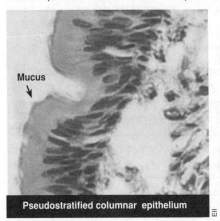

Mucus

Pseudostratified columnar epithelium

This epithelium lines most of the respiratory tract (e.g. trachea, above). The cells seem layered but all rest on the basement membrane. Goblet cells produce mucus which traps dust particles.

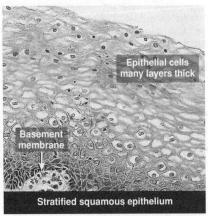

Epithelial cells many layers thick

Basement membrane

Stratified squamous epithelium

Stratified epithelium is more durable than simple epithelium because it has several layers. It has a protective role and can be ketratinised (e.g. in skin), or without keratin, e.g. in the vagina above.

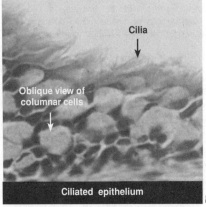

Cilia

Oblique view of columnar cells

Ciliated epithelium

Ciliated epithelial cells have hair-like outgrowths which produce rhythmic beating movements to move mucus along. It is found in passages of the airways, such as the bronchus (above).

1. Explain how the development of tissues improves functional efficiency: _____

_____

_____

2. Describe the primary function of each of the following epithelial tissues and give an example of where they occur:

(a) Stratified squamous epithelium: _____

(b) Simple squamous epithelium: _____

(c) Ciliated epithlelium: _____

(d) Glandular epithelium: _____

3. Identify the particular features of the following epithelial cells that contribute to their functional role in the tissue:

(a) Epithelial cells of the bronchus: _____

(b) Epithelial cells of the capillary endothelium: _____

Related activities: Levels of Organisation
Web links: Animal Tissues

R

Plant Tissues

Plant tissues are divided into two groups: simple and complex. **Simple tissues** contain only one cell type and form packing and support tissues. **Complex tissues** contain more than one cell type and form the conducting and support tissues of plants. Tissues are in turn grouped into tissue systems which make up the plant body. Vascular plants have three systems; the dermal, vascular, and ground tissue systems. The **dermal system** is the outer covering of the plant providing protection and reducing water loss. **Vascular tissue** provides the transport system by which water and nutrients are moved through the plant. The **ground tissue** system, which makes up the bulk of a plant, is made up mainly of simple tissues such as parenchyma, and carries out a wide variety of roles within the plant including photosynthesis, storage, and support.

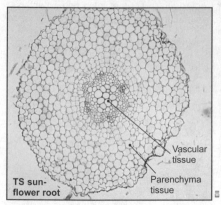

TS sun-flower root. Vascular tissue. Parenchyma tissue.

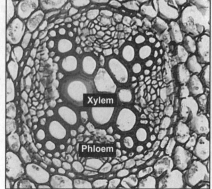

Xylem. Phloem.

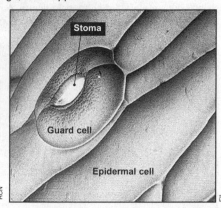

Stoma. Guard cell. Epidermal cell.

Simple Tissues

Simple tissues consists of only one or two cell types. **Parenchyma tissue** is the most common and involved in storage, photosynthesis, and secretion. **Collenchyma tissue** comprises thick-walled collenchyma cells alternating with layers of intracellular substances (pectin and cellulose) to provide flexible support. The cells of **sclerenchyma** tissue (fibres and sclereids) have rigid cell walls which provide support.

Complex Tissues

Xylem and phloem tissue (above left), which together make up the plant **vascular tissue** system, are complex tissues. Each comprises several tissue types including tracheids, vessel members, parenchyma and fibres in xylem, and sieve tube members, companion cells, parenchyma and sclerenchyma in phloem. **Dermal tissue** is also complex tissue and covers the outside of the plant. The composition of dermal tissue varies depending upon its location on the plant. Root epidermal tissue consist of epidermal cells which extend to root hairs (**trichomes**) for increasing surface area. In contrast, the epidermal tissue of leaves (above right) are covered by a waxy cuticle to reduce water loss, and specialised guard cells regulate water intake via the stomata (pores in the leaf through which gases enter and leave the leaf tissue).

1. The table below lists the major types of simple and complex plant tissue. Complete the table by filling in the role each of the tissue types plays within the plant. The first example has been completed for you.

Simple Tissue	Cell Type(s)	Role within the Plant
Parenchyma	Parenchyma cells	Involved in respiration, photosynthesis, storage and secretion.
Collenchyma		
Sclerenchyma		
Root endodermis	Endodermal cells	
Pericycle		
Complex Tissue		
Leaf mesophyll	Spongy mesophyll cells, palisade mesophyll cells	
Xylem	Tracheids vessels fibre parenchyma	water
Phloem	companion cells sieve tube fibre p	meagame, assimilates
Epidermis		

Related activities: Levels of Organisation, Xylem, Phloem
Web links: Photographic Atlas of Plant Anatomy

KEY TERMS: Word Find

Use the clues below to find the relevant key terms in the WORD FIND grid

```
K W F E Q B U C Q A S I M U X U G D I P L O I D W
V D T I F V V P R O P H A S E Y I P H L O E M B F
D M C E L L U L A R D I F F E R E N T I A T I O N
F A A W J W J B U Y E T E L O P H A S E B P I F Z
P W X H O M O L O G O U S C H R O M O S O M E S M
F F A C R O H A P L O I D C J A P I M B W G B S X
T Y F S Y C J N I N P W W N E U T R O P H I L Q V
C T H E R Y T H R O C Y T E S V A X H Y X I N K F
B Z B W K O T I S S U E M E T A P H A S E U R D B
C C E N T R I O L E S Z R O C Y T O K I N E S I S
D N A R E P L I C A T I O N B U X Y L E M N K E H
L U S O U M D Q G O R G A N Q G C D T C F E U I B
T G U F J P S V G N U C L E A R E N V E L O P E O
J U M I T O S I S R E D U C T I O N S F O F W B T
P Q N E Y F P S T E M C E L L U S X Y E S P J M B
K H U F W D W O F R K G B K U Q B S M W I S D T S
B N E W M I Y X E D A N A P H A S E B N V O M F E
```

The process occurring prior to cell division to produce a copy of all the DNA in the nucleus (2 words).

An undifferentiated cell, with the properties of self renewal and potency (2 words).

The process by which a less specialised cell becomes a more specialised cell type (2 words).

The phase of a cell cycle resulting in nuclear division.

The number of chromosomes in the gamete of an individual.

The number is the number of chromosomes in most cells except the gametes.

A stage in cell division when two daughter nuclei appear in the cell.

Meiosis is this type of division.

Plant tissue specialised for the transport of water and dissolved minerals.

Plant tissue specialised for the transport of dissolved sugar.

Paired microtubular structures in animal cells associated with cell division.

A collection of cells and tissues joined in structural unit to serve a common function.

Division of the cytoplasm of a eukaryotic cell to form two daughter cells.

Blood cells specialised to transport oxygen in the blood.

A stage of mitosis in which condensed chromosomes align in the middle of the cell.

A stage of mitosis in a eukaryotic cell in which chromosomes separate; each chromatid moves to opposite poles of the cell.

A type of white blood cell specialised to phagocytose foreign material.
A stage in mitosis when the chromatin condenses and becomes visible as chromosomes.

A collection of cells of the same origin and extracellular substances that carry out a specific function.

Chromosome pairs, one paternal and one maternal, of the same length, centromere position, and staining pattern with genes for the same characteristics at corresponding loci (2 words).

The double membrane around the nucleus, which breaks down during nuclear division and then reforms in the daughter cells (2 words).

Exchange Surfaces and Breathing

KEY CONCEPTS

▶ Cell size is limited by surface area to volume ratio.
▶ Multicellular organisms above a certain size need a exchange systems.
▶ Efficiency of gas exchange is a consequence of the structure of the gas exchange system.
▶ The distribution of the cells and tissues of the gas exchange system reflects their functional roles.
▶ Breathing ventilates the gas exchange surfaces.
▶ Lung function can be measured using spirometry.

KEY TERMS

alveolus (pl. alveoli)
breathing
breathing rate
bronchi
bronchioles
carbon dioxide
cartilage
cellular respiration
ciliated epithelium
diffusion
diaphragm
expiration (=exhalation)
Fick's law
gas exchange surface
gas exchange system
goblet cell
inspiration (=inhalation)
intercostal muscles
lung
oxygen
pulmonary ventilation
rib cage
smooth muscle
spirometer
surface area:volume ratio
surfactant
tidal volume
trachea
ventilation
vital capacity

OBJECTIVES

□ 1. Use the **KEY TERMS** to help you understand and complete these objectives.

Surface Area:Volume and Exchange Surfaces   pages 87-90, 94

□ 2. Distinguish between **cellular respiration** and **gas exchange** and explain why organisms need to exchange materials with their environment.

□ 3. Describe the relationship between an organism's size and its surface area (the surface area: volume ratio or SA:V). Explain why multicellular organisms above a certain size need specialised exchange surfaces.

□ 4. Describe how respiratory gases are exchanged across gas exchange surfaces. With reference to the **alveolus** of the mammalian **lung**, describe the essential features of an efficient gas exchange surface.

The Mammalian Gas Exchange System   pages 91, 93-94, 97

□ 5. Describe the structure and function of the gas exchange surfaces and related structures in humans (**trachea, bronchi, bronchioles, lungs**, and **alveoli**). Explain how these features contribute to efficient gas exchange.

□ 6. Describe the distribution of **cartilage, ciliated epithelium, goblet cells, smooth muscle**, and **elastic fibres** in the trachea, bronchi, bronchioles, and alveoli of the mammalian gas exchange system (e.g. a human).

□ 7. Describe the function of the cartilage, cilia, goblet cells, smooth muscle, and elastic fibres in the mammalian gas exchange system.

□ 8. Recognise the relationship between gas exchange surfaces (alveoli) and the blood vessels in the lung tissue (see *Transport in Animals*).

□ 9. Describe the mechanism of ventilation (**breathing**) in humans. Include reference to the following:

(a) The role of the **rib cage, diaphragm, intercostal muscles**, and **pleural membranes** in breathing.

(b) The role of **surfactants** in lung function.

(c) The distinction between **inspiration** (inhalation) as an active process and **expiration** (exhalation) as a passive process (during quiet breathing).

(d) How the demand for increased oxygen, e.g. during exercise, is met.

Measuring Gas Exchange   pages 92, 95-97

□ 10. Describe how a **spirometer** is used to measure **vital capacity, tidal volume, breathing rate**, and oxygen uptake in humans.

□ 11. Explain how the breathing (ventilation) rate and **pulmonary ventilation** (PV) rate are calculated and expressed. Interpret typical values for breathing rate, tidal volume, and PV obtained from a spirometer. Describe how each of these is affected by change in oxygen demand (e.g. as a result of exercise).

Periodicals:
Listings for this chapter are on page 338

Weblinks:
www.biozone.co.uk/
weblink/OCR-AS-2641.html

The Need for Gas Exchange

Living cells require energy for the activities of life. Energy is released in cells by the breakdown of sugars and other substances in the metabolic process called **cellular respiration**. As a consequence of this process, gases need to be exchanged between the respiring cells and the environment. In most organisms (with the exception of some bacterial groups) these gases are carbon dioxide (CO_2) and oxygen (O_2). The diagram below illustrates this process for an animal. Plant cells also respire, but their gas exchange budget is different because they also produce O_2 and consume CO_2 in photosynthesis.

The Need for Gas Exchange

Gas exchange is the process by which oxygen is acquired and carbon dioxide is removed. Cellular respiration creates a constant demand for oxygen (O_2) and a need to eliminate carbon dioxide gas (CO_2).

Gas exchange surfaces provide a means for gases to enter and leave the body. Some organisms use the body surface as the sole gas exchange surface, but many have specialised gas exchange structures (e.g. lungs, gills, or stomata). Amphibians use the body surface and simple lungs to provide for their gas exchange requirements.

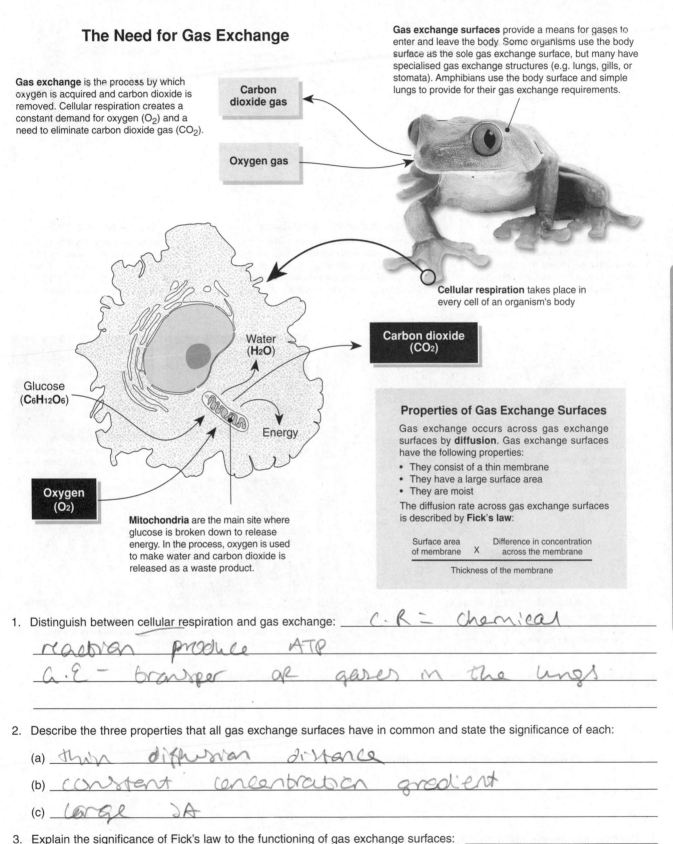

Carbon dioxide gas

Oxygen gas

Cellular respiration takes place in every cell of an organism's body

Water (H_2O)

Carbon dioxide (CO_2)

Glucose ($C_6H_{12}O_6$)

Energy

Oxygen (O_2)

Mitochondria are the main site where glucose is broken down to release energy. In the process, oxygen is used to make water and carbon dioxide is released as a waste product.

Properties of Gas Exchange Surfaces

Gas exchange occurs across gas exchange surfaces by **diffusion**. Gas exchange surfaces have the following properties:

- They consist of a thin membrane
- They have a large surface area
- They are moist

The diffusion rate across gas exchange surfaces is described by **Fick's law**:

$$\frac{\text{Surface area of membrane} \quad X \quad \text{Difference in concentration across the membrane}}{\text{Thickness of the membrane}}$$

1. Distinguish between cellular respiration and gas exchange: _C.R = chemical reaction produce ATP G.E - transfer of gases in the lungs_

2. Describe the three properties that all gas exchange surfaces have in common and state the significance of each:

   (a) _thin diffusion distance_

   (b) _constant concentration gradient_

   (c) _large SA_

3. Explain the significance of Fick's law to the functioning of gas exchange surfaces: _____

Periodicals:
Getting in and out

Related activities: Passive Transport Processes, Surface Area and Volume

A 2

Why a Gas Exchange System?

The way in which gas exchange is achieved is influenced by the animal's general body form and by the environment in which the animal lives. Small, aquatic organisms such as sponges, flatworms and cnidarians, require no specialised respiratory structures. Gases are exchanged between the surrounding water (or moist environment) and the body's cells by diffusion directly across the organism's surface. Larger animals require specialised gas exchange systems. The complexity of these is related to the efficiency of gas exchange required, which is determined by the oxygen demands of the organism.

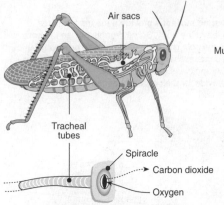

Gas exchange in insects

Insects, and sometimes spiders, transport gases via a system of branching tubes called **tracheae** or **tracheal tubes**. The gases move by diffusion across the moist lining directly to and from the tissues.

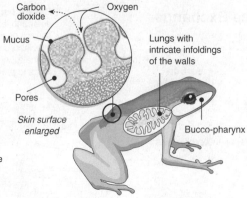

Gas exchange in amphibians

All amphibians make use of surface gas exchange. This is only possible if the surface is kept moist by secretions from mucous glands. At times of inactivity, the skin alone is often a sufficient surface with either water or air.

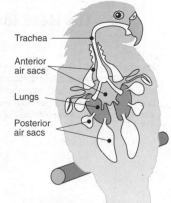

Gas exchange in birds

A bird has air sacs in addition to lungs. The air sacs act as bellows to ventilate the lungs but no gas exchange occurs in them. the suystem is very efficient as air flows in only one direction through the lung tissue.

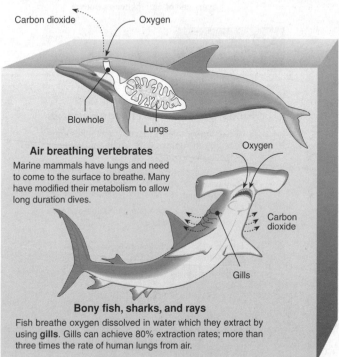

Air breathing vertebrates

Marine mammals have lungs and need to come to the surface to breathe. Many have modified their metabolism to allow long duration dives.

Bony fish, sharks, and rays

Fish breathe oxygen dissolved in water which they extract by using **gills**. Gills can achieve 80% extraction rates; more than three times the rate of human lungs from air.

Nudibranch snails have elaborate exposed gills to assist gas exchange.

Jellyfish increase their surface area for gas exchange by having ruffles.

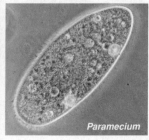

Gas exchange in protoctists such as *Paramecium* and *Amoeba* occurs by simple diffusion across the cell surface. Their high surface area: volume ratio means that no gas exchange system is needed.

1. Describe two reasons for the development of gas exchange structures and systems in animals:

   (a) _____

   (b) _____

2. Describe two ways in which air breathers manage to keep their gas exchange surfaces moist:

   (a) _____

   (b) _____

3. Describe two ways in which the efficiency of the gas exchange system is increased in vertebrates:

   (a) _____

   (b) _____

***Related activities**: Surface Area and Volume*
***Web links**: Avian Respiration*

Surface Area and Volume

When an object (e.g. a cell) is small it has a large surface area in comparison to its volume. In this case diffusion will be an effective way to transport materials (e.g. gases) into the cell. As an object becomes larger, its surface area compared to its volume is smaller. Diffusion is no longer an effective way to transport materials to the inside. For this reason, there is a physical limit for the size of a cell, with the effectiveness of diffusion being the controlling factor.

Diffusion in Organisms of Different Sizes

Respiratory gases and some other substances are exchanged with the surroundings by diffusion or active transport across the plasma membrane.

The **plasma membrane**, which surrounds every cell, functions as a selective barrier that regulates the cell's chemical composition. For each square micrometer of membrane, only so much of a particular substance can cross per second.

The surface area of an elephant is increased, for radiating body heat, by large flat ears.

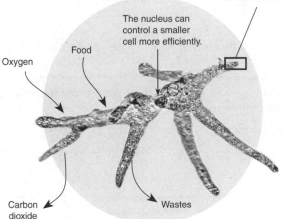

The nucleus can control a smaller cell more efficiently.

Food

Oxygen

Carbon dioxide

Wastes

A specialised gas exchange surface (lungs) and circulatory (blood) system are required to speed up the movement of substances through the body.

Respiratory gases cannot reach body tissues by diffusion alone.

Amoeba: The small size of single-celled protoctists, such as *Amoeba,* provides a large surface area relative to the cell's volume. This is adequate for many materials to be moved into and out of the cell by diffusion or active transport.

Multicellular organisms: To overcome the problems of small cell size, plants and animals became multicellular. They provide a small surface area compared to their volume but have evolved various adaptive features to improve their effective surface area.

Smaller is Better for Diffusion

One large cube

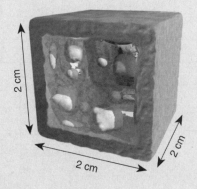

2 cm

2 cm

2 cm

Volume: = 8 cm³

Surface area: = 24 cm²

Eight small cubes

1 cm

1 cm

1 cm

Volume: = 8 cm³ for 8 cubes

Surface area: = 6 cm² for 1 cube

= 48 cm² for 8 cubes

The eight small cells and the single large cell have the same total volume, but their surface areas are different. The small cells together have twice the total surface area of the large cell, because there are more exposed (inner) surfaces. Real organisms have complex shapes, but the same principles apply.

The surface-area volume relationship has important implications for processes involving transport into and out of cells across membranes. For activities such as gas exchange, the surface area available for diffusion is a major factor limiting the rate at which oxygen can be supplied to tissues.

Exchange Surfaces and Breathing

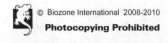

The diagram below shows four hypothetical cells of different sizes (cells do not actually grow to this size, their large size is for the sake of the exercise). They range from a small 2 cm cube to a larger 5 cm cube. This exercise investigates the effect of cell size on the efficiency of diffusion.

2 cm cube

3 cm cube

4 cm cube

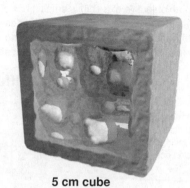
5 cm cube

1. Calculate the volume, surface area and the ratio of surface area to volume for each of the four cubes above (the first has been done for you). When completing the table below, show your calculations.

Cube size	Surface area	Volume	Surface area to volume ratio
2 cm cube	$2 \times 2 \times 6 = 24\ cm^2$ (2 cm x 2 cm x 6 sides)	$2 \times 2 \times 2 = 8\ cm^3$ (height x width x depth)	$24\ to\ 8 = 3{:}1$
3 cm cube			
4 cm cube			
5 cm cube			

2. Create a graph, plotting the surface area against the volume of each cube, on the grid on the right. Draw a line connecting the points and label axes and units.

3. State which increases the fastest with increasing size, the **volume** or **surface area**.

   Volume

4. Explain what happens to the ratio of surface area to volume with increasing size:

   _____

   _____

   _____

   _____

   _____

5. Diffusion of substances into and out of a cell occurs across the cell surface. Describe how increasing the size of a cell will affect the ability of diffusion to transport materials into and out of a cell:

   increasing the size will increase the rate

   of diffusion

   _____

   _____

   _____

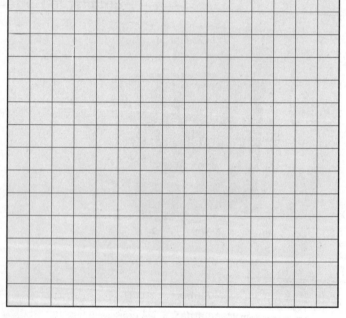

Breathing in Humans

In mammals, the mechanism of breathing (ventilation) provides a continual supply of fresh air to the lungs and helps to maintain a large diffusion gradient for respiratory gases across the gas exchange surface. Oxygen must be delivered regularly to supply the needs of respiring cells. Similarly, carbon dioxide, which is produced as a result of cellular metabolism, must be quickly eliminated from the body. Adequate lung ventilation is essential to these exchanges. The cardiovascular system participates by transporting respiratory gases to and from the cells of the body. The volume of gases exchanged during breathing varies according to the physiological demands placed on the body (e.g. by exercise). These changes can be measured using spirometry.

Inspiration (inhalation or breathing in)

During quiet breathing, inspiration is achieved by increasing the space (therefore decreasing the pressure) inside the lungs. Air then flows into the lungs in response to the decreased pressure inside the lung. Inspiration is always an active process involving muscle contraction.

1a External intercostal muscles contract causing the ribcage to expand and move up.

1b Diaphragm contracts and moves down.

2 Thoracic volume increases, lungs expand, and the pressure inside the lungs decreases.

3 Air flows into the lungs in response to the pressure gradient.

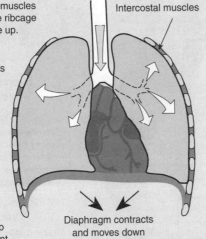

Intercostal muscles

Diaphragm contracts and moves down

Expiration (exhalation or breathing out)

During quiet breathing, expiration is achieved passively by decreasing the space (thus increasing the pressure) inside the lungs. Air then flows passively out of the lungs to equalise with the air pressure. In active breathing, muscle contraction is involved in bringing about both inspiration and expiration.

1 In **quiet breathing**, external intercostal muscles and diaphragm relax. Elasticity of the lung tissue causes recoil.

In **forced breathing**, the internal intercostals and abdominal muscles also contract to increase the force of the expiration.

2 Thoracic volume decreases and the pressure inside the lungs increases.

3 Air flows passively out of the lungs in response to the pressure gradient.

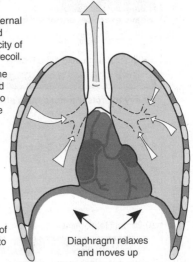

Diaphragm relaxes and moves up

Exchange Surfaces and Breathing

1. Explain the purpose of breathing: _to provide O₂ & remove CO₂_

2. (a) Describe the sequence of events involved in quiet breathing: _____

   (b) Explain the essential difference between this and the situation during heavy exercise or forced breathing: _____

3. Identify what other gas is lost from the body in addition to carbon dioxide: _____

4. Explain the role of the elasticity of the lung tissue in normal, quiet breathing: _____
   Elasticity of the lungs enables natural recoil of the lungs during expiration so it is a passive process not requiring energy.

5. Breathing rate is regulated through the medullary respiratory centre in response to demand for oxygen. The trigger for increased breathing rate is a drop in blood pH. Suggest why this is an appropriate trigger to increase breathing rate: _____

Related activities: Measuring Lung Function, Gas Transport in Humans
Web links: Respiratory Basics Learning Activity

RA 3

Responses to Exercise

Physical exercise places greater demands on the abilities of the body to maintain a steady state. Extra heat generated during exercise must be dissipated, oxygen demands increase, and there are more waste products produced. The body has an immediate response to exercise but also, over time, responds to the stress of repeated exercise (**training**) by adjusting its responses and improving its capacity for exercise and the efficiency with which it performs. The maintenance of homeostasis during exercise is principally the job of the circulatory and gas exchange systems, although the skin, kidneys, and liver are also important.

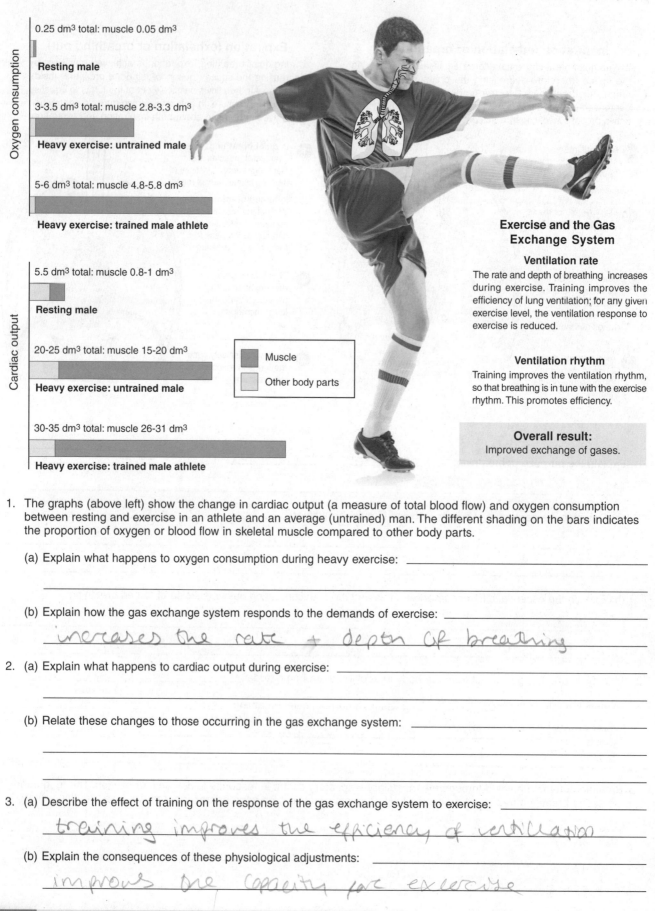

Oxygen consumption

0.25 dm³ total: muscle 0.05 dm³

Resting male

3-3.5 dm³ total: muscle 2.8-3.3 dm³

Heavy exercise: untrained male

5-6 dm³ total: muscle 4.8-5.8 dm³

Heavy exercise: trained male athlete

Cardiac output

5.5 dm³ total: muscle 0.8-1 dm³

Resting male

20-25 dm³ total: muscle 15-20 dm³

Heavy exercise: untrained male

30-35 dm³ total: muscle 26-31 dm³

Heavy exercise: trained male athlete

■ Muscle
□ Other body parts

Exercise and the Gas Exchange System

Ventilation rate

The rate and depth of breathing increases during exercise. Training improves the efficiency of lung ventilation; for any given exercise level, the ventilation response to exercise is reduced.

Ventilation rhythm

Training improves the ventilation rhythm, so that breathing is in tune with the exercise rhythm. This promotes efficiency.

Overall result:
Improved exchange of gases.

1. The graphs (above left) show the change in cardiac output (a measure of total blood flow) and oxygen consumption between resting and exercise in an athlete and an average (untrained) man. The different shading on the bars indicates the proportion of oxygen or blood flow in skeletal muscle compared to other body parts.

   (a) Explain what happens to oxygen consumption during heavy exercise: _____

   _____

   (b) Explain how the gas exchange system responds to the demands of exercise: _____
   increases the rate + depth of breathing

2. (a) Explain what happens to cardiac output during exercise: _____

   _____

   (b) Relate these changes to those occurring in the gas exchange system: _____

   _____

   _____

3. (a) Describe the effect of training on the response of the gas exchange system to exercise: _____
   training improves the efficiency of ventilation

   (b) Explain the consequences of these physiological adjustments: _____
   improves the capacity for exercise

The Human Gas Exchange System

Lungs are internal sac-like organs found in most amphibians, and all reptiles, birds, and mammals. The paired lungs of mammals are connected to the outside air by way of a system of tubular passageways: the trachea, bronchi, and bronchioles. Ciliated, mucus secreting epithelium lines this system of tubules, trapping and removing dust and pathogens before they reach the gas exchange surfaces. Each lung is divided into a number of lobes, each receiving its own bronchus. Each bronchus divides many times, terminating in the bronchioles from which arise 2-11 alveolar ducts and numerous **alveoli**. These provide a very large surface area (70 m²) for the exchange of respiratory gases by diffusion between the alveoli and the blood in the capillaries. The details of this exchange across the **aveolar-capillary membrane** are described on the next page.

Morphology of the Human Gas Exchange System

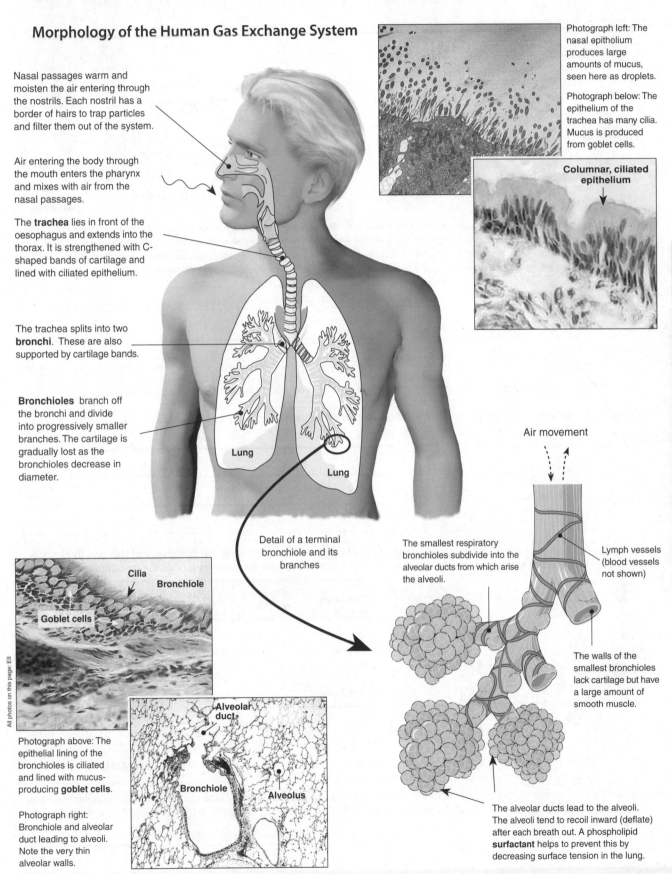

Nasal passages warm and moisten the air entering through the nostrils. Each nostril has a border of hairs to trap particles and filter them out of the system.

Air entering the body through the mouth enters the pharynx and mixes with air from the nasal passages.

The **trachea** lies in front of the oesophagus and extends into the thorax. It is strengthened with C-shaped bands of cartilage and lined with ciliated epithelium.

The trachea splits into two **bronchi**. These are also supported by cartilage bands.

Bronchioles branch off the bronchi and divide into progressively smaller branches. The cartilage is gradually lost as the bronchioles decrease in diameter.

Lung

Lung

Photograph left: The nasal epithelium produces large amounts of mucus, seen here as droplets.

Photograph below: The epithelium of the trachea has many cilia. Mucus is produced from goblet cells.

Columnar, ciliated epithelium

Detail of a terminal bronchiole and its branches

Air movement

The smallest respiratory bronchioles subdivide into the alveolar ducts from which arise the alveoli.

Lymph vessels (blood vessels not shown)

The walls of the smallest bronchioles lack cartilage but have a large amount of smooth muscle.

Cilia
Bronchiole
Goblet cells

All photos on this page: EII

Photograph above: The epithelial lining of the bronchioles is ciliated and lined with mucus-producing **goblet cells**.

Photograph right: Bronchiole and alveolar duct leading to alveoli. Note the very thin alveolar walls.

Alveolar duct
Bronchiole
Alveolus

The alveolar ducts lead to the alveoli. The alveoli tend to recoil inward (deflate) after each breath out. A phospholipid **surfactant** helps to prevent this by decreasing surface tension in the lung.

Exchange Surfaces and Breathing

Periodicals: Gas exchange in the lungs

Related activities: Breathing in Humans, Review of Lung Function
Web links: Vertebrate Lungs, Interactive Lungs

RA 2

An Alveolus

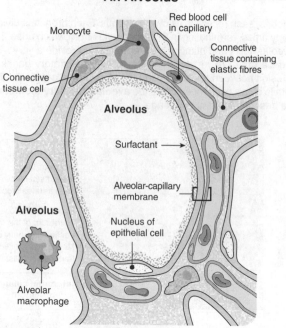

The Alveolar-Capillary Membrane

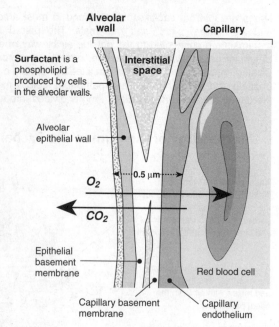

The diagram above illustrates the physical arrangement of the alveoli to the capillaries through which the blood moves. Phagocytic monocytes and macrophages are also present to protect the lung tissue. Elastic connective tissue gives the alveoli their ability to expand and recoil.

The **alveolar-capillary membrane** (gas exchange membrane) is the layered junction between the alveolar epithelial cells, the endothelial cells of the capillary, and their associated basement membranes (thin, collagenous layers underlying the epithelial tissues). Gases move freely across this membrane.

1. (a) Explain how the basic structure of the human gas exchange system provides such a large area for gas exchange:

   _____

   _____

   (b) Identify the general region of the lung where exchange of gases takes place: _____

2. Describe the structure and purpose of the alveolar-capillary membrane: _____

   _____

   _____

3. Describe the role of the surfactant in the alveoli: _____

   _____

4. Using the information above and opposite, complete the table below summarising the **histology of the gas exchange pathway**. Name each numbered region and use a tick or cross to indicate the presence or absence of particular tissues.

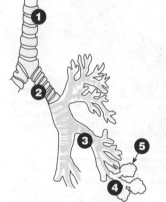

	Region	Cartilage	Ciliated epithelium	Goblet cells (mucus)	Smooth muscle	Connective tissue
1	Trachea	C shaped	✓	✓	✓	✓
2	bronchiole	C shaped	✓	✓	✓	✓
3	bronchus	gradually lost	✓	✓	✓	✓
4	Alveolar duct	none	✗	✗	✓	✓
5	Alveoli sac	none	✗	✗	very little	✓

5. Babies born prematurely are often deficient in surfactant. This causes respiratory distress syndrome; a condition where breathing is very difficult. From what you know about the role of surfactant, explain the symptoms of this syndrome:

   _____

   _____

Measuring Lung Function

Changes in lung volume can be measured using a technique called **spirometry**. Total adult lung capacity varies between 4 and 6 litres (L or dm³) and is greater in males. The **vital capacity**, which describes the volume exhaled after a maximum inspiration, is somewhat less than this because of the residual volume of air that remains in the lungs even after expiration. The exchange between fresh air and the residual volume is a slow process and the composition of gases in the lungs remains relatively constant. Once measured, the tidal volume can be used to calculate the **pulmonary ventilation rate** or PV, which describes the amount of air exchanged with the environment per minute. Measures of respiratory capacity provide one way in which a reduction in lung function can be assessed (for example, as might occur as result of disease or an obstructive lung disorder such asthma).

Measuring lung volume and oxygen consumption using spirometry

The apparatus used to measure the amount of air exchanged during breathing and the rate of breathing is a **spirometer** (also called a respirometer). A simple spirometer consists of a weighted drum, containing oxygen or air, inverted over a chamber of water (to provide a stable temperature). A tube connects the air-filled chamber with the subject's mouth, and soda lime in the system absorbs the carbon dioxide breathed out. Oxygen consumption is detected by the amount of fluid displacement in a manometer connected to the container. Breathing results in a trace called a spirogram, from which lung volumes can be measured directly.

During inspiration
Air is removed from the chamber, the drum sinks, and an upward deflection is recorded on the paper on the rotating drum.

During expiration
Air is added to the chamber, the drum rises, and a downward deflection is recorded.

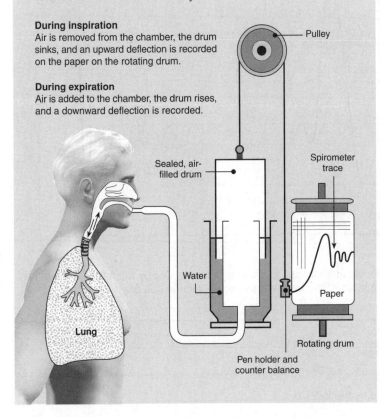

Pulley

Sealed, air-filled drum

Spirometer trace

Water

Paper

Lung

Rotating drum

Pen holder and counter balance

Lung Volumes and Capacities

The air in the lungs can be divided into volumes. Lung capacities are combinations of volumes.

DESCRIPTION OF VOLUME	Vol / dm³
Tidal volume (TV) Volume of air breathed in and out in a single breath	0.5
Inspiratory reserve volume (IRV) Volume breathed in by a maximum inspiration at the end of a normal inspiration	3.3
Expiratory reserve volume (ERV) Volume breathed out by a maximum effort at the end of a normal expiration	1.0
Residual volume (RV) Volume of air remaining in the lungs at the end of a maximum expiration	1.2

DESCRIPTION OF CAPACITY	
Inspiratory capacity (IC) = TV + IRV Volume breathed in by a maximum inspiration at the end of a normal expiration	3.8
Vital capacity (VC) = IRV + TV + ERV Volume that can be exhaled after a maximum inspiration.	4.8
Total lung capacity (TLC) = VC + RV The total volume of the lungs. Only a fraction of TLC is used in normal breathing	6.0

PRIMARY INDICATORS OF LUNG FUNCTION

Forced expiratory volume in 1 second (FEV_1)
The volume of air that is maximally exhaled in the first second of exhalation.

Forced vital capacity (FVC)
The total volume of air that can be forcibly exhaled after a maximum inspiration.

---

1. Describe how each of the following might be expected to influence values for lung volumes and capacities obtained using spirometry:

   (a) Height: _____

   (b) Gender: _____

   (c) Age: _____

2. A percentage decline in FEV_1 and FVC (to <80% of normal) are indicators of impaired lung function, e.g in asthma:

   (a) Explain why a forced volume is a more useful indicator of lung function than tidal volume:

   _____

   (b) Asthma is treated with drugs to relax the airways. Suggest how spirometry could be used during asthma treatment:

   _____

Respiratory gas	Approximate percentages of O_2 and CO_2		
	Inhaled air	Air in lungs	Exhaled air
O_2	21.0	13.8	16.4
CO_2	0.04	5.5	3.6

Above: The percentages of respiratory gases in air (by volume) during normal breathing. The percentage volume of oxygen in the alveolar air (in the lung) is lower than that in the exhaled air because of the influence of the **dead air volume** (the air in the spaces of the nose, throat, larynx, trachea and bronchi). This air (about 30% of the air inhaled) is unavailable for gas exchange.

Left: During exercise, the breathing rate, tidal volume, and PV increase up to a maximum (as indicated below).

Spirogram for a male during quiet and forced breathing, and during exercise

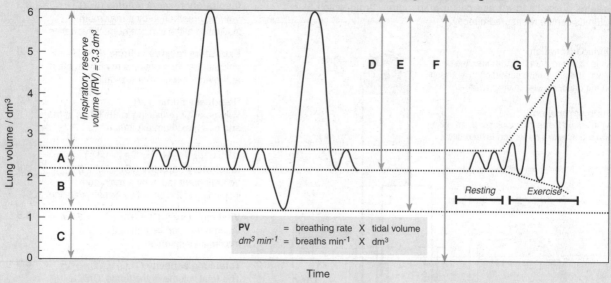

PV $= $ breathing rate X tidal volume
$dm^3 \ min^{-1} = $ breaths min^{-1} X dm^3

3. Using the definitions given on the previous page, identify the volumes and capacities indicated by the letters **A-F** on the spirogram above. For each, indicate the volume (vol) in dm^3. The inspiratory reserve volume has been identified:

(a) A: _____ Vol: _____ (d) D: _____ Vol: _____

(b) B: _____ Vol: _____ (e) E: _____ Vol: _____

(c) C: _____ Vol: _____ (f) F: _____ Vol: _____

4. Explain what is happening in the sequence indicated by the letter **G**: _____

_____

5. Calculate PV when breathing rate is 15 breaths per minute and tidal volume is 0.4 dm^3: _____

6. (a) Describe what would happen to PV during strenuous exercise: _____

(b) Explain how this is achieved: _____

_____

7. The table above gives approximate percentages for respiratory gases during breathing. Study the data and then:

(a) Calculate the difference in CO_2 between inhaled and exhaled air: _____

(b) Explain where this 'extra' CO_2 comes from: _____

(c) Explain why the dead air volume raises the oxygen content of exhaled air above that in the lungs: _____

_____

Review of Lung Function

The gas exchange system in humans includes the lungs and the system of tubes through which the air reaches them. Breathing (ventilation) provides a continual supply of fresh air to the lungs and helps to maintain a large diffusion gradient for respiratory gases across the gas exchange surface. The basic rhythm of breathing is controlled by the respiratory centre in the medulla of the hindbrain. The volume of gases exchanged during breathing varies according to the physiological demands placed on the body. These changes can be measured using spirometry. The following activity summarises the key features of gas exchange system structure and function. The stimulus material can be found in earlier exercises in this topic.

Components of the gas exchange system

Breathing

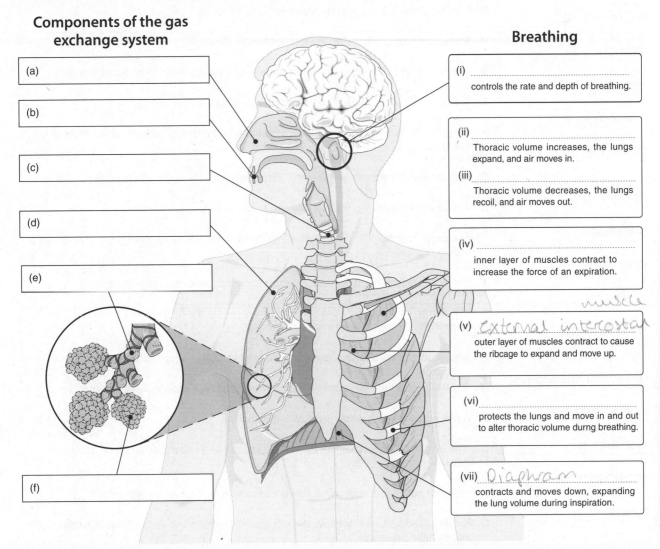

(a)

(b)

(c)

(d)

(e)

(f)

(i) _____ controls the rate and depth of breathing.

(ii) _____ Thoracic volume increases, the lungs expand, and air moves in.

(iii) _____ Thoracic volume decreases, the lungs recoil, and air moves out.

(iv) _____ inner layer of muscles contract to increase the force of an expiration.

(v) *external intercostal muscle* outer layer of muscles contract to cause the ribcage to expand and move up.

(vi) _____ protects the lungs and move in and out to alter thoracic volume durng breathing.

(vii) *Diaphram* contracts and moves down, expanding the lung volume during inspiration.

Exchange Surfaces and Breathing

1. On the diagram above, label the components of the gas exchange system (a-f) and the components and processes involved in breathing (i - vii).

2. Identify the volumes and capacities indicated by the letters **A - E** on the diagram of a spirogram below.

A = _____

B = _____

C = _____

D = _____

E = _____

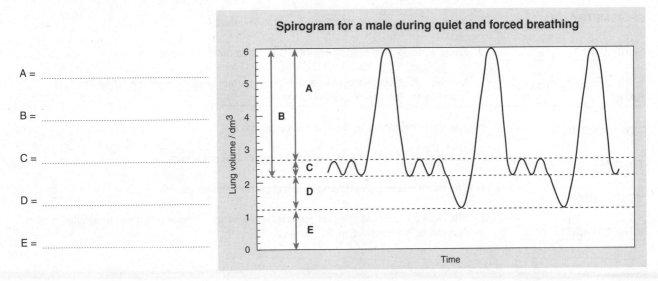

Spirogram for a male during quiet and forced breathing

Related activities: The Human Gas Exchange System, Breathing in Humans

RA 2

KEY TERMS: Mix and Match

INSTRUCTIONS: Test your vocab by matching each term to its correct definition, as identified by its preceding letter code.

ALVEOLI

BREATHING

BREATHING RATE

BRONCHI

BRONCHIOLES

CARBON DIOXIDE

CARTILAGE

CELLULAR RESPIRATION

DIAPHRAGM

EXPIRATION

FICK'S LAW

GAS EXCHANGE SURFACE

GAS EXCHANGE SYSTEM

GOBLET CELLS

INSPIRATION

INTERCOSTAL MUSCLES

LUNGS

OXYGEN

PULMONARY VENTILATION RATE (PV)

SPIROMETRY

SURFACTANT

SURFACE AREA:VOLUME RATIO

TIDAL VOLUME

TRACHEA

VENTILATION

VITAL CAPACITY

A An internal muscle separating the thorax and abdomen of mammals and involved in enabling thoracic volume changes during breathing.

B The act of breathing out or removing air from the lungs.

C Mucus-producing cells lining the gas exchange airways.

D Phospholipid substance responsible for reducing surface tension in the lung tissue.

E A thin membranous surface across which gas diffuse easily.

F A mathematical expression of the factors affecting the rate of diffusion of substances across a membrane.

G Gas required for aerobic respiration. Levels average 21% in the atmosphere.

H A method of measuring volume changes in the lung.

I Groups of muscles that run between the ribs and are mainly involved in the mechanical aspect of breathing.

J The active process of moving air or water across a gas exchange surface.

K The maximum amount of air a person can expel from the lungs after a maximum inspiration.

L The amount of surface area per unit volume of an object or collection of objects.

M Small air tubes that divide from the bronchi and become progressively smaller.

N A gas produced as a waste product of metabolism.

O The lung volume representing the normal volume of air displaced between normal inspiration and expiration.

P A strong but flexible connective tissue that reinforces the trachea and bronchi.

Q A specialised system to improve the efficiency of oxygen and carbon dioxide exchanges in multicellular organisms.

R The tube that connects the pharynx to the lungs. It is also called the windpipe.

S The rate at which a cycle of inspiration and expiration occurs (usually per minute).

T Internal gas exchange structures found in vertebrates.

U Large air tubes that branch from the trachea to enter the lungs.

V Microscopic structures in the lungs of most vertebrates that form the terminus of the bronchioles. The site of gas exchange.

W The act of breathing in or filling the lungs with air.

X The amount of air exchanged with the environment per minute.

Y The act of inhaling air into and exhaling air from the lungs.

Z The catabolic process in which the chemical energy in complex organic molecules is coupled to ATP production.

Transport in Animals

KEY CONCEPTS

▶ Most multicellular animals require an internal transport system.

▶ Circulatory systems may be open or closed.

▶ Closed systems can be single or double circuit.

▶ Heart structure is related to functional demand.

▶ The structure of blood vessels is related to their functional position in the circulatory system.

▶ Gas transport in mammals is a function of the blood.

KEY TERMS

arteries
arteriole
atrioventricular node (AVN)
atrioventricular valves
atrium (pl. atria)
blood
blood vessel
Bohr effect
capillaries
capillary network
cardiac cycle
closed circulatory systems
coronary circulation
diastole
double circulatory system
electrocardiogram (ECG)
haemoglobin
heart
intrinsic eart rate
lymph
lymphatic vessel
microcirculation
O2-HB dissociation curve
open circulatory system
plasma
pulmonary circulation
Purkyne tissue (*alt.* Purkinje)
semilunar valves
single circulatory system
sinoatrial node (SAN)
surface area:volume ratio
systemic circulation
systole
tissue fluid
veins
ventricle

Periodicals:
Listings for this chapter are on page 339

Weblinks:
www.biozone.co.uk/
weblink/OCR-AS-2641.html

OBJECTIVES

☐ 1. Use the **KEY TERMS** to help you understand and complete these objectives.

Diversity in Circulatory Systems
pages 100-104

☐ 2. Explain the requirement for a transport system in multicellular animals in terms of their size, **surface area: volume** ratio, and level of activity. .

☐ 3. Describe diversity in the structure and function of circulatory systems in animals, with reference to the **open circulatory system** of insects and the **closed circulatory system** of fish and other vertebrates.

☐ 4. Describe diversity in the structure and function of closed circulatory systems with reference to the **single circulatory system** in fish, and the **double circulatory system** in mammals.

The Mammalian Heart
pages 105-109

☐ 5. Describe the internal and external structure of the human heart in relation to its function. On diagrams identify **atria**, **ventricles**, **atrioventricular valves**, and **semilunar valves**, as well major vessels and the **coronary circulation**.

☐ 6. Relate differences in the thickness of the heart chambers to their functions.

☐ 7. Describe the changes in pressure and volume, and associated valve movements during the **cardiac cycle**.

☐ 8. Explain how the heart beat is initiated and maintained, including the role of the **sinoatrial node**, the **atrioventricular node**, the bundle of His, and the Purkyne tissue. Relate the activity of the SAN to the **intrinsic heart rate**. Explain how heart rate responds to increased metabolic requirements.

☐ 9. Interpret and explain an **electrocardiogram** (ECG) with respect to both normal and abnormal heart activity.

Blood and Gas Transport
pages 110-120

☐ 10. Describe the structure of **arteries**, **veins**, and **capillaries**. Relate the structure of each type of blood vessel to its specific function in the circulatory system.

☐ 11. Describe the organisation of blood vessels in a capillary network and their relationship to the lymphatic vessels. Distinguish between the composition of **blood, lymph, plasma**, and **tissue fluid**. Explain how tissue fluid is formed from the plasma.

☐ 12. Describe the role of haemoglobin (Hb) in transporting and delivering oxygen to the tissues. Explain how CO_2 is carried in the blood (including the role of Hb).

☐ 13. Describe the transport of oxygen in relation to the oxygen-haemoglobin dissociation curve. Compare the oxygen affinities and dissociation curves of adult and foetal haemoglobin and explain the significance of these differences.

☐ 14. CIE AND EXTENSION OCR: Describe short and long terms adjustments of the circulatory system, e.g. to altitude and in response to exercise.

Transport and Exchange Systems

Living cells require a constant supply of nutrients and oxygen, and continuous removal of wastes. Simple, small organisms can achieve this through **diffusion** across moist body surfaces without requiring specialized transport or exchange systems. Larger, more complex organisms require systems to facilitate exchanges as their surface area to volume ratio decreases. **Mass transport** (also known as mass flow or **bulk flow**)

describes the movement of materials at equal rates or as a single mass. Mass transport accounts for the long distance transport of fluids in living organisms. It includes the movement of blood in the circulatory systems of animals and the transport of water and solutes in the xylem and phloem of plants. In the diagram below, exchanges by diffusion are compared with mass transport to specific exchange sites.

Exchanges Across a Body Surface

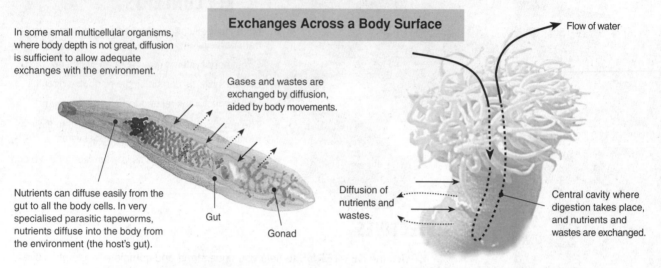

In some small multicellular organisms, where body depth is not great, diffusion is sufficient to allow adequate exchanges with the environment.

Gases and wastes are exchanged by diffusion, aided by body movements.

Nutrients can diffuse easily from the gut to all the body cells. In very specialised parasitic tapeworms, nutrients diffuse into the body from the environment (the host's gut).

Gut

Gonad

Platyhelminthes (liver fluke)

Flow of water

Diffusion of nutrients and wastes.

Central cavity where digestion takes place, and nutrients and wastes are exchanged.

Cnidarians (sea anemone)

Systems for Exchange and Transport

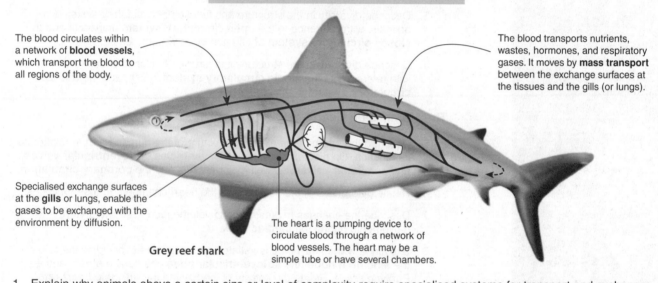

The blood circulates within a network of **blood vessels**, which transport the blood to all regions of the body.

The blood transports nutrients, wastes, hormones, and respiratory gases. It moves by **mass transport** between the exchange surfaces at the tissues and the gills (or lungs).

Specialised exchange surfaces at the **gills** or lungs, enable the gases to be exchanged with the environment by diffusion.

Grey reef shark

The heart is a pumping device to circulate blood through a network of blood vessels. The heart may be a simple tube or have several chambers.

1. Explain why animals above a certain size or level of complexity require specialised systems for transport and exchange:

_____

_____

_____

2. (a) Describe how materials move within the circulatory system of a vertebrate: _____

_____

(b) Contrast this with how materials are transported in a flatworm or single celled eukaryote:

_____

_____

(c) Identify two exchange sites in a vertebrate: _____

Related activities: Open Circulatory Systems, Closed Circulatory Systems

Open Circulatory Systems

Two basic types of circulatory systems have evolved in animals. Many invertebrates have an **open circulatory system**, while vertebrates have a **closed circulatory system**, consisting of a heart and a network of tube-like vessels. The circulatory systems of arthropods are open but varied in complexity. Insects, unlike most other arthropods, do not use a circulatory system to transport oxygen, which is delivered directly to the tissues via the system of tracheal tubes. In addition to its usual transport functions, the circulatory system may also be important in hydraulic movements of the whole body (as in many molluscs) or its component parts (e.g. newly emerged butterflies expand their wings through hydraulic pressure).

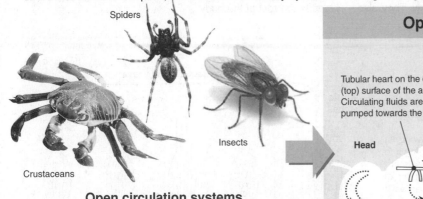

Spiders

Insects

Crustaceans

Open circulation systems

Arthropods and molluscs (except cephalopods) have open circulatory systems in which the blood is pumped by a tubular, or sac-like, heart through short vessels into large spaces in the body cavity. The blood bathes the cells before reentering the heart through holes (**ostia**). Muscle action may assist the circulation of the blood.

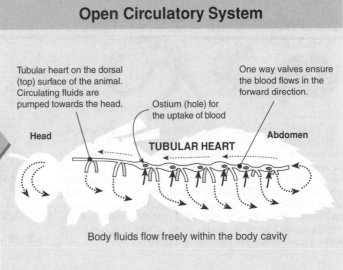

Open Circulatory System

Tubular heart on the dorsal (top) surface of the animal. Circulating fluids are pumped towards the head.

One way valves ensure the blood flows in the forward direction.

Ostium (hole) for the uptake of blood

Head

TUBULAR HEART

Abdomen

Body fluids flow freely within the body cavity

The circulatory system of crabs is best described as incompletely closed. The thoracic heart has three pairs of ostia a number of arteries, which leave the heart and branch extensively to supply various organs before draining into discrete channel-like sinuses.

In spiders, arteries from the dorsal heart empty the haemolymph into tissue spaces and then into a large ventral sinus that bathes the book lungs where gas exchange takes place. Venous channels conduct the haemolymph back to the heart.

The haemolymph occupies up to 40% of the body mass of an insect and is usually under low pressure due its lack of confinement in vessels. The circulation of the haemolymph is aided by body movements such as the ventilating movements of the abdomen.

1. Explain how an open circulatory system moves fluid (haemolymph) about the body: _____

   _____

   _____

2. Explain why arthropods do not bleed in a similar way to vertebrates: _____

   _____

3. Compare insects and decapod crustaceans (e.g. crabs) in the degree to which the circulatory system is closed:

   _____

   _____

4. (a) Explain why the crab's circulatory system is usually described as an open system: _____

   _____

   (b) Explain in what way this description is not entirely accurate: _____

   _____

© Biozone International 2008-2010
Photocopying Prohibited

Related activities: Closed Circulatory Systems
Web links: Animal Circulatory Systems

Transport in Animals

RA 3

Closed Circulatory Systems

Closed circulatory systems characteristic of vertebrates, annelids (earthworms) and cephalopods (octopus and squid). The blood is pumped by a heart through a series of arteries and veins. Oxygen is transported around the body by the blood and diffuses through capillary walls into the body cells. Closed circulatory systems are useful for large, active animals where oxygen can not easily be transported to the interior of the body. They also allow the animal more control over the distribution of blood flow by contracting or dilating blood vessels. Closed systems are the most developed in vertebrates where a chambered heart pumps the blood into blood vessels at high pressure. The system can also be divided into two separate regions, the pulmonary region taking up oxygen and the systematic region pumping oxygenated blood to the rest of the body.

INVERTEBRATE CLOSED SYSTEMS

Polychaete worm

Earthworm

Wiki: Hans Hillewaert

The closed systems of many annelids (e.g. earthworms) circulate blood through a series of vessels before returning it to the heart. In annelids, the dorsal and ventral blood vessels are connected by lateral vessels in every segment (right). The dorsal vessel receives blood from the lateral vessels and carries it towards the head. The ventral vessel carries blood posteriorly and distributes it to the segmental vessels. The dorsal vessel is contractile and is the main method of propelling the blood, but there are also several contractile aortic arches ('hearts') which act as accessory organs for blood propulsion.

VERTEBRATE CLOSED SYSTEMS

Rays

Bony fish

Sharks

Closed, single circuit systems

In closed circulation systems, the blood is contained within vessels and is returned to the heart after every circulation of the body. Exchanges between the blood and the fluids bathing the cells occurs by diffusion across capillaries. In single circuit systems, typical of fish, the blood goes directly from the gills to the body. The blood loses pressure at the gills and flows at low pressure around the body.

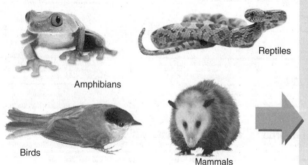

Reptiles

Amphibians

Birds

Mammals

Closed, double circuit systems

Double circulation systems occur in all vertebrates other than fish. The blood is pumped through a pulmonary circuit to the lungs, where it is oxygenated. The blood returns to the heart, which pumps the oxygenated blood, through a systemic circuit, to the body. In amphibians and most reptiles, the heart is not completely divided and there is some mixing of oxygenated and deoxygenated blood. In birds and mammals, the heart is fully divided and there is no mixing.

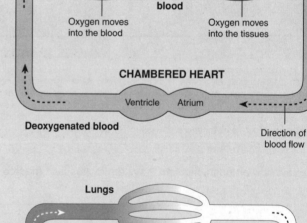

Contractle dorsal blood vessel

Aortic arches

Capillary networks

Tail

Head

Ventral blood vessel

Capillary bed   Gills   Systemic circulation

Oxygenated blood

Oxygen moves into the blood

Oxygen moves into the tissues

CHAMBERED HEART

Ventricle   Atrium

Deoxygenated blood

Direction of blood flow

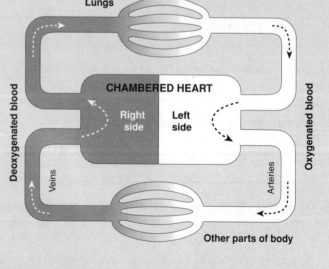

Lungs

Deoxygenated blood

CHAMBERED HEART

Right side

Left side

Oxygenated blood

Veins

Arteries

Other parts of body

RA 3

Related activities: *The Heart as a Pump, Blood Vessels*
Web links: *Animal Circulatory Systems*

Fish Heart

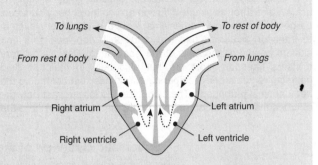

Conus arteriosus • Ventricle • Atrium • From rest of body • To gills • Ventral aorta • Atrioventricular valves • Sinoatrial valves • Sinus venosus • From rest of body

The fish heart is linear, with a sequence of three chambers in series (the conus may be included as a fourth chamber). Blood from the body first enters the heart through the sinus venosus, then passes into the atrium and the ventricle. A series of one-way valves between the chambers prevents reverse blood flow. Blood leaving the heart travels to the gills.

Mammalian Heart

To lungs • To rest of body • From rest of body • From lungs • Right atrium • Left atrium • Right ventricle • Left ventricle

In birds and mammals, the heart is fully partitioned into two halves, resulting in four chambers. Blood circulates through two circuits, with no mixing of the two. Oxygenated blood from the lungs is kept separated from the deoxygenated blood returning from the rest of the body.

1. Describe the main difference between closed and open systems of circulation: _open does not have a complete circuit of vessels for blood to pass through from heart to body and back._

2. (a) Describe where the blood flows to immediately after it has passed through the gills in a fish: _____

   (b) Relate this to the pressure at which the blood flows in the systemic circulation: _low pressure_

3. (a) Describe where the blood flows immediately after it has passed through the lungs in a mammal: _heart_

   (b) Relate this to the pressure at which the blood flows in the systemic circulation: _high pressure_

4. Explain the higher functional efficiency of a double circuit system, relative to a single circuit system: _____
   - maintain high BP
   - important for efficient transport

5. Hearts range from being simple contractile structures to complex chambered organs. Describe basic heart structure in:

   (a) Fish: _simple pump   A → V_

   (b) Mammals: _____

6. Explain how a closed circulatory system gives an animal finer control over the distribution of blood to tissues and organs:
   relax/constrict blood vessels enables animals to increase or restrict blood to certain areas.

Transport in Animals

The Human Circulatory System

Animal cells require a constant supply of nutrients and oxygen, and continuous removal of wastes. Simple, small organisms achieve this through simple diffusion across moist body surfaces. Larger, more complex organisms require a circulatory system to transport materials because diffusion is too inefficient and slow to supply all the cells of the body adequately. Circulatory systems transport materials, but also help to maintain fluid balance, regulate body temperature, and assist in defending the body against pathogens. The blood vessels form a vast network of tubes that carry blood away from the heart, transport it to the tissues, and then return it to the heart. The arteries, arterioles, capillaries, venules, and veins are organised into specific routes to circulate blood throughout the body. The figure below shows some of the **circulatory routes** through which the blood travels. The **pulmonary system** (or circulation) carries blood between the heart and lungs, and the **systemic system** (circulation) carries blood between the heart and the rest of the body. Two important subdivisions of the systemic circuit are the coronary (cardiac) circulation, which supplies the heart muscle, and the **hepatic portal circulation**, which runs from the gut to the liver.

Schematic Overview of the Human Circulatory System

Deoxygenated blood (coloured grey below) travels to the right side of the heart via the vena cavae. The heart pumps the deoxygenated blood to the lungs where it releases carbon dioxide and receives oxygen. The oxygenated blood (coloured white below) travels via the pulmonary vein back to the heart from where it is pumped to all parts of the body. The **venous system** (figure, left) returns blood from the capillaries to the heart. The **arterial system** (figure right) carries blood from the heart to the capillaries. **Portal systems** carry blood between two capillary beds.

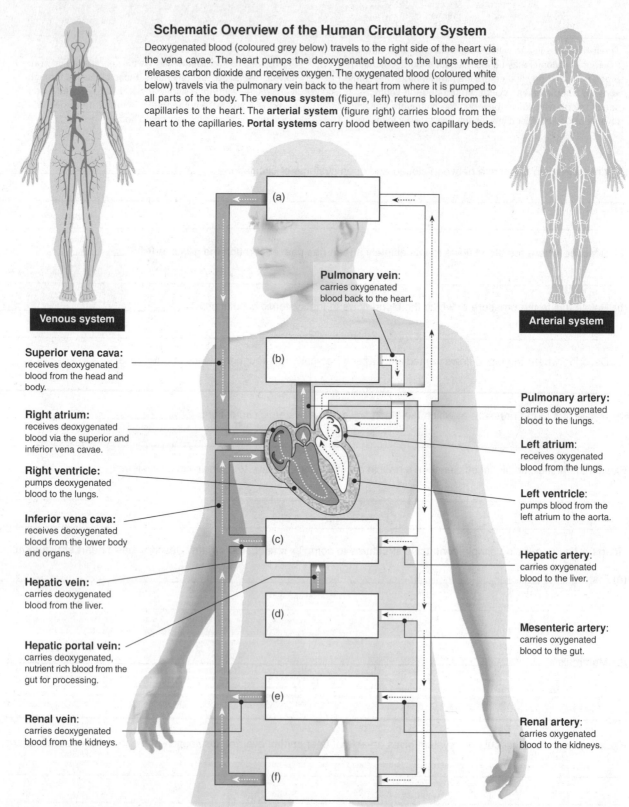

Venous system

Superior vena cava: receives deoxygenated blood from the head and body.

Right atrium: receives deoxygenated blood via the superior and inferior vena cavae.

Right ventricle: pumps deoxygenated blood to the lungs.

Inferior vena cava: receives deoxygenated blood from the lower body and organs.

Hepatic vein: carries deoxygenated blood from the liver.

Hepatic portal vein: carries deoxygenated, nutrient rich blood from the gut for processing.

Renal vein: carries deoxygenated blood from the kidneys.

Pulmonary vein: carries oxygenated blood back to the heart.

Arterial system

Pulmonary artery: carries deoxygenated blood to the lungs.

Left atrium: receives oxygenated blood from the lungs.

Left ventricle: pumps blood from the left atrium to the aorta.

Hepatic artery: carries oxygenated blood to the liver.

Mesenteric artery: carries oxygenated blood to the gut.

Renal artery: carries oxygenated blood to the kidneys.

1. Complete the diagram above by labelling the boxes with the organs or structures they represent.

Related activities: Closed Circulatory Systems, The Human Heart

Periodicals: *Venous disease*

The Human Heart

The heart is the centre of the human cardiovascular system. It is a hollow, muscular organ, weighing on average 342 grams. Each day it beats over 100 000 times to pump 3780 litres of blood through 100 000 kilometres of blood vessels. It comprises a system of four muscular chambers (two **atria** and two **ventricles**) that alternately fill and empty of blood, acting as a double pump.

The left side pumps blood to the body tissues and the right side pumps blood to the lungs. The heart lies between the lungs, to the left of the body's midline, and it is surrounded by a double layered **pericardium** of tough fibrous connective tissue. The pericardium prevents overdistension of the heart and anchors the heart within the **mediastinum**.

Human Heart Structure

(sectioned, anterior view)

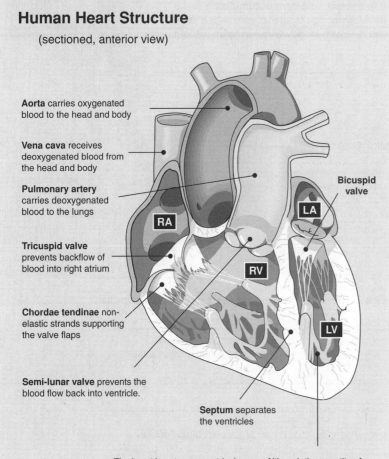

Aorta carries oxygenated blood to the head and body

Vena cava receives deoxygenated blood from the head and body

Pulmonary artery carries deoxygenated blood to the lungs

Tricuspid valve prevents backflow of blood into right atrium

Chordae tendinae non-elastic strands supporting the valve flaps

Semi-lunar valve prevents the blood flow back into ventricle.

Bicuspid valve

Septum separates the ventricles

The heart is not a symmetrical organ. Although the quantity of blood pumped by each side is the same, the walls of the left ventricle are thicker and more muscular than those of the right ventricle. The difference affects the shape of the ventricular cavities, so the right ventricle is twisted over the left.

Key to abbreviations

RA   Right atrium; receives deoxygenated blood via anterior and posterior vena cavae

RV   Right ventricle; pumps deoxygenated blood to the lungs via the pulmonary artery

LA   Left atrium; receives blood returning to the heart from the lungs via the pulmonary veins

LV   Left ventricle; pumps oxygenated blood to the head and body via the aorta

Top view of a heart in section, showing valves

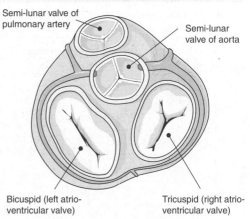

Semi-lunar valve of pulmonary artery

Semi-lunar valve of aorta

Bicuspid (left atrio-ventricular valve)

Tricuspid (right atrio-ventricular valve)

Posterior view of heart

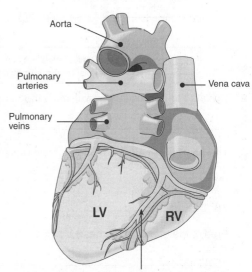

Aorta

Pulmonary arteries

Pulmonary veins

Vena cava

LV

RV

Coronary arteries: The high oxygen demands of the heart muscle are met by a dense capillary network. Coronary arteries arise from the aorta and spread over the surface of the heart supplying the cardiac muscle with oxygenated blood. Deoxygenated blood is collected by cardiac veins and returned to the right atrium via a large coronary sinus.

1. In the schematic diagram of the heart, below, label the four chambers and the main vessels entering and leaving them. The arrows indicate the direction of blood flow. Use large coloured circles to mark the position of each of the four valves.

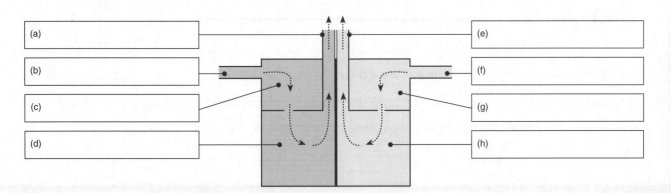

(a)

(b)

(c)

(d)

(e)

(f)

(g)

(h)

Transport in Animals

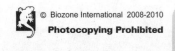

Periodicals: The heart

Related activities: Closed Circulatory Systems, Review of the Human Heart
Web links: Anatomy of the Heart, How the Heart Works

RA 2

Pressure Changes and the Asymmetry of the Heart

aorta, 100 mg Hg

The heart is not a symmetrical organ. The left ventricle and its associated arteries are thicker and more muscular than the corresponding structures on the right side. This asymmetry is related to the necessary pressure differences between the pulmonary (lung) and systemic (body) circulations (not to the distance over which the blood is pumped per se). The graph below shows changes blood pressure in each of the major blood vessel types in the systemic and pulmonary circuits (the horizontal distance not to scale). The pulmonary circuit must operate at a much lower pressure than the systemic circuit to prevent fluid from accumulating in the alveoli of the lungs. The left side of the heart must develop enough "spare" pressure to enable increased blood flow to the muscles of the body and maintain kidney filtration rates without decreasing the blood supply to the brain.

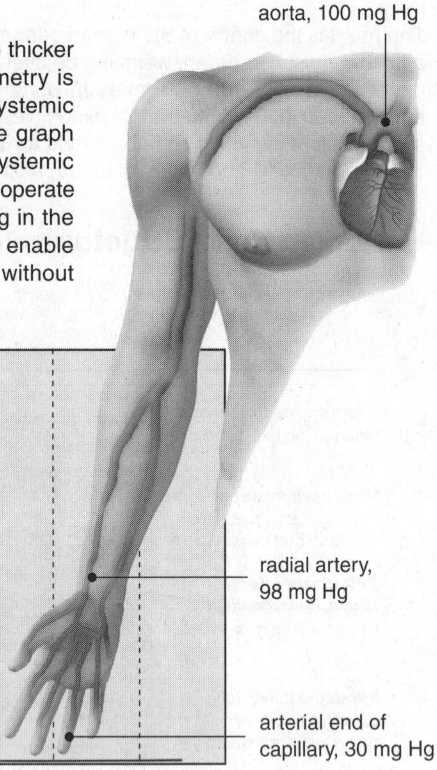

Blood pressure during contraction (systole)

Blood pressure during relaxation (diastole)

The greatest fall in pressure occurs when the blood moves into the capillaries, even though the distance through the capillaries represents only a tiny proportion of the total distance travelled.

radial artery, 98 mg Hg

arterial end of capillary, 30 mg Hg

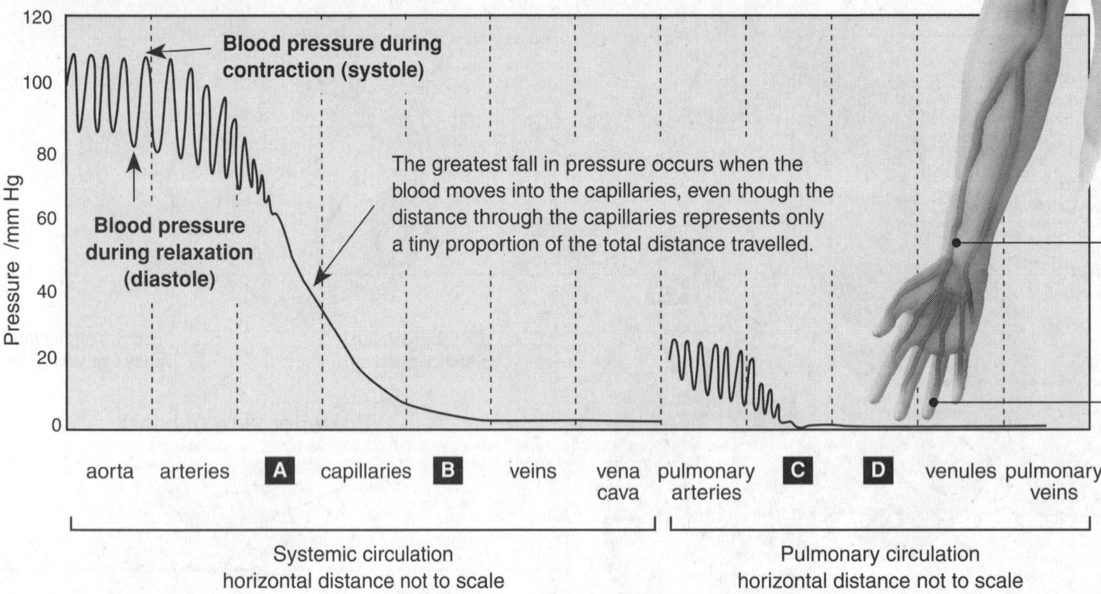

Pressure /mm Hg

aorta arteries **A** capillaries **B** veins vena cava pulmonary arteries **C** **D** venules pulmonary veins

Systemic circulation
horizontal distance not to scale

Pulmonary circulation
horizontal distance not to scale

2. Explain the purpose of the valves in the heart: *prevent back flow of blood – help fill the hearts chambers*

3. The heart is full of blood. Suggest two reasons why, despite this, it needs its own blood supply:

   (a) *to meet the high O_2 demands of the heart*

   (b) *must have its own waste system*

4. Predict the effect on the heart if blood flow through a coronary artery is restricted or blocked: *they won't get O_2 and cells will die leading to heart attack*

5. Identify the vessels corresponding to the letters **A-D** on the graph above:

   A: _____ B: _____ C: _____ D: _____

6. (a) Find out what is meant by the pulse pressure and explain how it is calculated: _____

   (b) Predict what happens to the pulse pressure between the aorta and the capillaries: _____

7. (a) Explain what you are recording when you take a pulse: *The expansion + recoil of the arteries that occur when the left ventricle contracts*

   (b) Name a place where pulse rate could best be taken and briefly explain why: _____

Control of Heart Activity

When removed from the body the cardiac muscle continues to beat. Therefore, the origin of the heartbeat is **myogenic**: the contractions arise as an intrinsic property of the cardiac muscle itself. The heartbeat is regulated by a special conduction system consisting of the pacemaker (**sinoatrial node**) and specialised conduction fibres called **Purkyne tissue**. The pacemaker sets a basic rhythm for the heart, but this rate is influenced by the cardiovascular control centre in the brainstem. Changing the rate and force of heart contraction is the main mechanism for controlling cardiac output in order to meet changing demands.

Generation of the Heartbeat

The basic rhythmic heartbeat is **myogenic**. The nodal cells (SAN and atrioventricular node) spontaneously generate rhythmic action potentials without neural stimulation. The normal resting rate of self-excitation of the SAN is about 50 beats per minute.

The amount of blood ejected from the left ventricle per minute is called the **cardiac output**. It is determined by the **stroke volume** (the volume of blood ejected with each contraction) and the **heart rate** (number of heart beats per minute).

> **Cardiac output**
> = **stroke volume** × **heart rate**

Cardiac muscle responds to stretching by contracting more strongly. The greater the blood volume entering the ventricle, the greater the force of contraction. This relationship is known as **Starling's Law.**

Intercalated discs

Mitochondrion

A TEM photo of cardiac muscle showing branched fibres (muscle cells). Each muscle fibre has one or two nuclei and many large mitochondria. **Intercalated discs** are specialised electrical junctions that separate the cells and allow the rapid spread of impulses through the heart muscle.

Sinoatrial node (SAN) is also called the **pacemaker**. It is a mass of specialised muscle cells near the opening of the superior vena cava. The pacemaker initiates the cardiac cycle, spontaneously generating action potentials that cause the atria to contract. The SAN sets the basic pace of the heart rate, although this rate is influenced by hormones and impulses from the autonomic nervous system.

Atrioventricular node (AVN) at the base of the atrium briefly delays the impulse to allow time for the atrial contraction to finish before the ventricles contract.

Bundle of His (atrioventricular bundle) containing Purkyne tissue – a tract of conducting fibres that distribute the action potentials over the ventricles causing ventricular contraction.

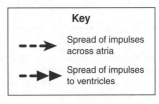

Key
- - -> Spread of impulses across atria
- - -►► Spread of impulses to ventricles

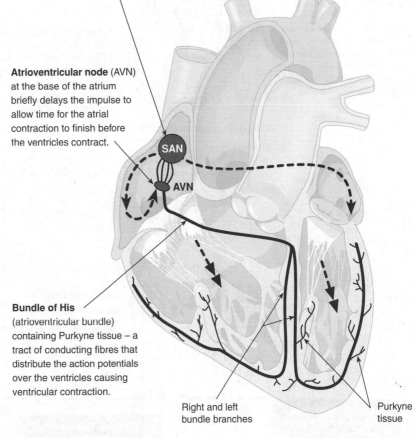

Right and left bundle branches

Purkyne tissue

1. Describe the role of each of the following in heart activity:

   (a) The sinoatrial node: _Initiates_

   (b) The atrioventricular node: _Delays impulse_

   (c) The bundle of His: _Distributes_

   (d) Intercalated discs: _____

2. (a) Explain what is meant by the heart beat being myogenic: _____

   independant of external stimulation

   (b) Describe the evidence for the myogenic nature of the heart beat: _____

   _____

3. Explain the significance of the delay in impulse conduction at the AVN: _____

   _____

 © Biozone International 2008-2010
Photocopying Prohibited

Periodicals:
Keeping pace: cardiac muscle and heartbeat

Related activities: The Human Heart, Review of the Human Heart

RDA 2

Transport in Animals

The Cardiac Cycle

The heart pumps with alternate contractions (**systole**) and relaxations (**diastole**). The **cardiac cycle** refers to the sequence of events of a heartbeat and involves three major stages: atrial systole, ventricular systole and complete cardiac diastole. Pressure changes within the heart's chambers generated by the cycle of contraction and relaxation are responsible for blood movement and cause the heart valves to open and close, preventing the backflow of blood. The noise of the blood when the valves open and close produces the heartbeat sound (**lubb-dupp**). The heart beat occurs in response to electrical impulses, which can be recorded as a trace, called an **electrocardiogram** or **ECG**. The ECG pattern is the result of the different impulses produced at each phase of the cardiac cycle, and each part is identified with a letter code. An ECG provides a useful method of monitoring changes in heart rate and activity and detection of heart disorders. The electrical trace is accompanied by volume and pressure changes (below).

The Cardiac Cycle

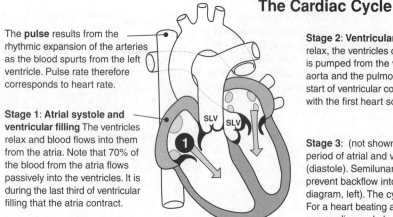

The **pulse** results from the rhythmic expansion of the arteries as the blood spurts from the left ventricle. Pulse rate therefore corresponds to heart rate.

Stage 1: Atrial systole and ventricular filling The ventricles relax and blood flows into them from the atria. Note that 70% of the blood from the atria flows passively into the ventricles. It is during the last third of ventricular filling that the atria contract.

Heart during ventricular filling

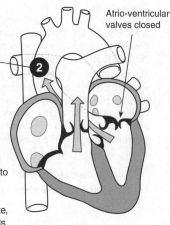

Atrio-ventricular valves closed

Stage 2: Ventricular systole The atria relax, the ventricles contract, and blood is pumped from the ventricles into the aorta and the pulmonary artery. The start of ventricular contraction coincides with the first heart sound.

Stage 3: (not shown) There is a short period of atrial and ventricular relaxation (diastole). Semilunar valves (**SLV**) close to prevent backflow into the ventricles (see diagram, left). The cycle begins again. For a heart beating at 75 beats per minute, one cardiac cycle lasts about 0.8 seconds.

Heart during ventricular contraction

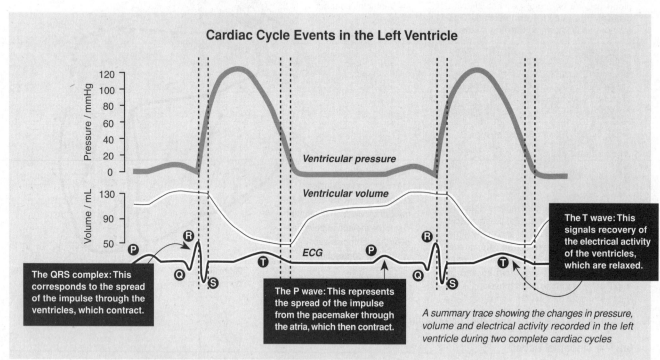

Cardiac Cycle Events in the Left Ventricle

Ventricular pressure

Ventricular volume

ECG

The QRS complex: This corresponds to the spread of the impulse through the ventricles, which contract.

The P wave: This represents the spread of the impulse from the pacemaker through the atria, which then contract.

The T wave: This signals recovery of the electrical activity of the ventricles, which are relaxed.

A summary trace showing the changes in pressure, volume and electrical activity recorded in the left ventricle during two complete cardiac cycles

1. Identify each of the following phases of an ECG by its international code:

    (a) Excitation of the ventricles and ventricular systole: _____

    (b) Electrical recovery of the ventricles and ventricular diastole: _____

    (c) Excitation of the atria and atrial systole: _____

2. Suggest the physiological reason for the period of electrical recovery experienced each cycle (the T wave):

    _____

3. Using the letters indicated, mark the points on trace above corresponding to each of the following:

    (a) E: Ejection of blood from the ventricle   (c) FV: Filling of the ventricle

    (b) AVC: Closing of the atrioventricular valve   (d) AVO: Opening of the atrioventricular valve

RA 2

Related activities: The Human Heart, Review of the Human Heart
Web links: Electrocardiogram, Cardiac Cycle Animation

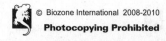
© Biozone International 2008-2010
Photocopying Prohibited

Review of the Human Heart

Large, complex organisms require a circulatory system to transport materials because diffusion is too inefficient and slow to supply all the cells of the body adequately. The circulatory system in humans transports nutrients, respiratory gases, wastes, and hormones, aids in regulating body temperature and maintaining fluid balance, and has a role in internal defence. All circulatory systems comprise a network of vessels, a circulatory fluid (blood), and a heart. This activity summarises key features of the structure and function of the human heart. The information for this activity can be found in the pages earlier in this topic.

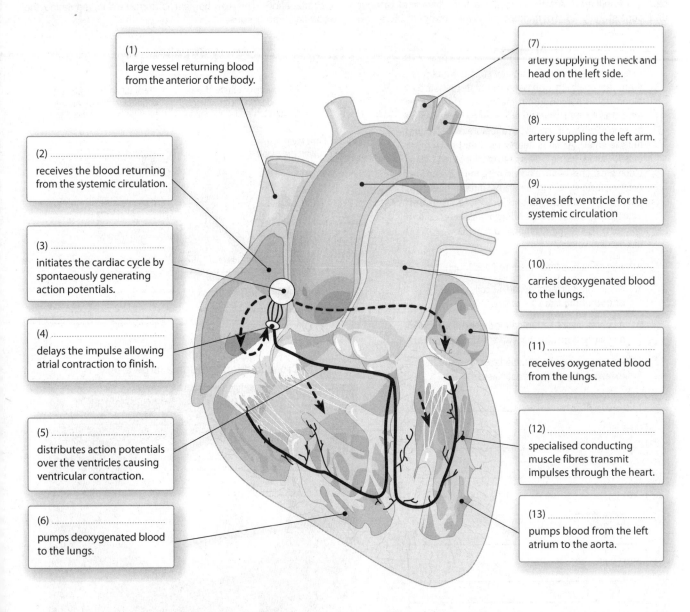

(1) .... large vessel returning blood from the anterior of the body.

(2) .... receives the blood returning from the systemic circulation.

(3) .... initiates the cardiac cycle by spontaeously generating action potentials.

(4) .... delays the impulse allowing atrial contraction to finish.

(5) .... distributes action potentials over the ventricles causing ventricular contraction.

(6) .... pumps deoxygenated blood to the lungs.

(7) .... artery supplying the neck and head on the left side.

(8) .... artery suppling the left arm.

(9) .... leaves left ventricle for the systemic circulation

(10) .... carries deoxygenated blood to the lungs.

(11) .... receives oxygenated blood from the lungs.

(12) .... specialised conducting muscle fibres transmit impulses through the heart.

(13) .... pumps blood from the left atrium to the aorta.

1. On the diagram above, label the identified components of heart structure and intrinsic control (**1-8**), and the components involved in extrinsic control of heart rate (**A-D**).

2. An **ECG** is the result of different impulses produced at each phase of the **cardiac cycle** (the sequence of events in a heartbeat). For each electrical event indicated in the ECG below, describe the corresponding event in the cardiac cycle:

A --
The spread of the impulse from the pacemaker (sinoatrial node) through the atria.

B --
The spread of the impulse through the ventricles.

C --
Recovery of the electrical activity of the ventricles.

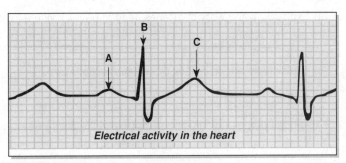

Electrical activity in the heart

3. Describe one treatment that may be indicated when heart rhythm is erratic or too slow: _____

   _____Artificial_____pacemaker_____to_____regulate_____

Related activities: The Human Heart, Control of Heart Activity, The Cardiac Cycle

RA 2

Transport in Animals

Arteries

In vertebrates, arteries are the blood vessels that carry blood away from the heart to the capillaries within the tissues. The large arteries that leave the heart divide into medium-sized (distributing) arteries. Within the tissues and organs, these distribution arteries branch to form very small vessels called **arterioles**, which deliver blood to capillaries. Arterioles lack the thick layers of arteries and consist only of an endothelial layer wrapped by a few smooth muscle fibres at intervals along their length. Resistance to blood flow is altered by contraction (**vasoconstriction**) or relaxation (**vasodilation**) of the blood vessel walls, especially in the arterioles. Vasoconstriction increases resistance and leads to an increase in blood pressure whereas vasodilation has the opposite effect. This mechanism is important in regulating the blood flow into tissues.

Arteries

Arteries have an elastic, stretchy structure that gives them the ability to withstand the high pressure of blood being pumped from the heart. At the same time, they help to maintain pressure by having some contractile ability themselves (a feature of the central muscle layer). Arteries nearer the heart have more elastic tissue, giving greater resistance to the higher blood pressures of the blood leaving the left ventricle. Arteries further from the heart have more muscle to help them maintain blood pressure. Between heartbeats, the arteries undergo elastic recoil and contract. This tends to smooth out the flow of blood through the vessel.

Arteries comprise three main regions (right):

1. A thin inner layer of epithelial cells called the **endothelium** lines the artery.

2. A central layer (the **tunica media**) of elastic tissue and smooth muscle that can stretch and contract.

3. An outer connective tissue layer (the **tunica externa**) has a lot of elastic tissue.

Artery Structure

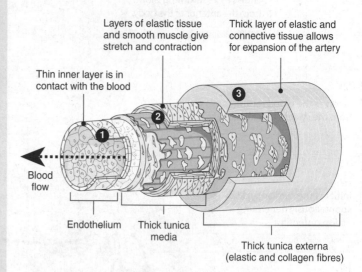

Layers of elastic tissue and smooth muscle give stretch and contraction

Thick layer of elastic and connective tissue allows for expansion of the artery

Thin inner layer is in contact with the blood

Blood flow

Endothelium

Thick tunica media

Thick tunica externa (elastic and collagen fibres)

Cross section through a large artery

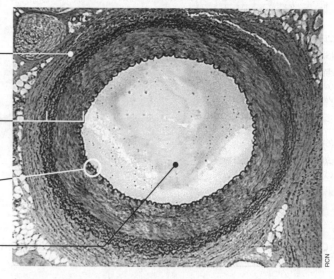

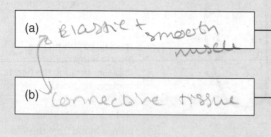

(a) *Elastic + smooth muscle*

(b) *Connective tissue*

(c) *Endothelium*

(d) *Lumen*

1. Using the diagram to help you, label the photograph (a)-(d) of the cross section through an artery (above).

2. (a) Explain why the walls of arteries need to be thick with a lot of elastic tissue: _*withstand the high BPressure*_

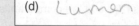

(b) Explain why arterioles lack this elastic tissue layer: _____

3. Explain the purpose of the smooth muscle in the artery walls: _*helps regulate blood flow + pressure*_

4. (a) Describe the effect of vasodilation on the diameter of an arteriole: _____

(b) Describe the effect of vasodilation on blood pressure: _____

Related activities: Veins, Capillaries
Web links: Arteries

Periodicals:
Blood pressure,
Cunning plumbing

© Biozone International 2008-2010
Photocopying Prohibited

Veins

Veins are the blood vessels that return blood to the heart from the tissues. The smallest veins (**venules**) return blood from the capillary beds to the larger veins. Veins and their branches contain about 59% of the blood in the body. The structural differences between veins and arteries are mainly associated with differences in the relative thickness of the vessel layers and the diameter of the lumen. These, in turn, are related to the vessel's functional role.

Veins

When several capillaries unite, they form small veins called **venules**. The venules collect the blood from capillaries and drain it into **veins**. Veins are made up of essentially the same three layers as arteries but they have less elastic and muscle tissue and a larger **lumen**. The venules closest to the capillaries consist of an **endothelium** and a tunica externa of connective tissue. As the venules approach the veins, they also contain the tunica media characteristic of veins (right). Although veins are less elastic than arteries, they can still expand enough to adapt to changes in the pressure and volume of the blood passing through them. Blood flowing in the veins has lost a lot of pressure because it has passed through the narrow capillary vessels. The low pressure in veins means that many veins, especially those in the limbs, need to have valves to prevent backflow of the blood as it returns to the heart.

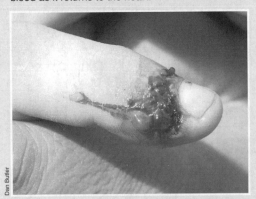

If a vein is cut, as is shown in this severe finger wound, the blood oozes out slowly in an even flow, and usually clots quickly as it leaves. In contrast, arterial blood spurts rapidly and requires pressure to staunch the flow.

Vein Structure

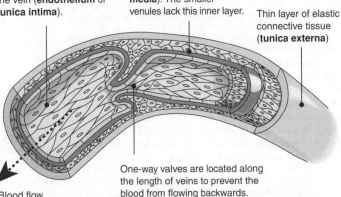

Inner thin layer of simple squamous epithelium lines the vein (**endothelium** or **tunica intima**).

Central thin layer of elastic and muscle tissue (**tunica media**). The smaller venules lack this inner layer.

Thin layer of elastic connective tissue (**tunica externa**)

One-way valves are located along the length of veins to prevent the blood from flowing backwards.

Blood flow

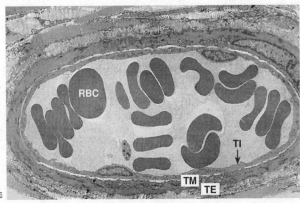

Above: TEM of a vein showing red blood cells (RBC) in the lumen, and the tunica intima (TI), tunica media (TM), and tunica externa (TE).

1. Contrast the structure of veins and arteries for each of the following properties:

   (a) Thickness of muscle and elastic tissue: _less than arteries_

   (b) Size of the lumen (inside of the vessel): _larger than arteries_

2. With respect to their functional roles, give a reason for the differences you have described above: _—no need for high BP_

3. Explain the role of the valves in assisting the veins to return blood back to the heart: _____

4. Blood oozes from a venous wound, rather than spurting as it does from an arterial wound. Account for this difference:

Related activities: Arteries, Capillaries
Web links: Veins

RA 2

Transport in Animals

Capillaries

In vertebrates, capillaries are very small vessels that connect arterial and venous circulation and allow efficient exchange of nutrients and wastes between the blood and tissues. Capillaries form networks or beds and are abundant where metabolic rates are high. Fluid that leaks out of the capillaries has an essential role in bathing the tissues.

Exchanges in Capillaries

Blood passes from the arterioles into the capillaries. Capillaries are small blood vessels with a diameter of just 4-10 μm. The only tissue present is an **endothelium** of squamous epithelial cells. Capillaries are so numerous that no cell is more than 25 μm from any capillary. It is in the capillaries that the exchange of materials between the body cells and the blood takes place.

Blood pressure causes fluid to leak from capillaries through small gaps where the endothelial cells join. This fluid bathes the tissues, supplying nutrients and oxygen, and removing wastes (right). The density of capillaries in a tissue is an indication of that tissue's metabolic activity. For example, cardiac muscle relies heavily on oxidative metabolism. It has a high demand for blood flow and is well supplied with capillaries. Smooth muscle is far less active than cardiac muscle, relies more on anaerobic metabolism, and does not require such an extensive blood supply.

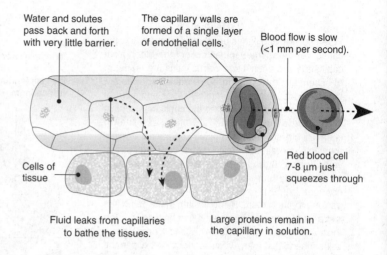

Water and solutes pass back and forth with very little barrier.

The capillary walls are formed of a single layer of endothelial cells.

Blood flow is slow (<1 mm per second).

Cells of tissue

Red blood cell 7-8 μm just squeezes through

Fluid leaks from capillaries to bathe the tissues.

Large proteins remain in the capillary in solution.

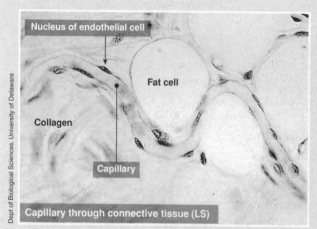

Nucleus of endothelial cell

Fat cell

Collagen

Capillary

Capillary through connective tissue (LS)

Dept of Biological Sciences. University of Delaware

Capillaries are found near almost every cell in the body. In many places, the capillaries form extensive branching networks. In most tissues, blood normally flows through only a small portion of a capillary network when the metabolic demands of the tissue are low. When the tissue becomes active, the entire capillary network fills with blood.

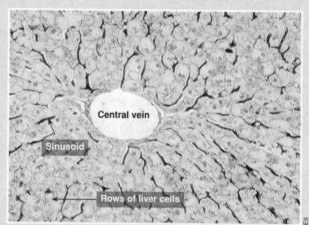

Central vein

Sinusoid

Rows of liver cells

Microscopic blood vessels in some dense organs, such as the liver (above), are called **sinusoids**. They are wider than capillaries and follow a more convoluted path through the tissue. Instead of the usual endothelial lining, they are lined with phagocytic cells. Like capillaries, sinusoids transport blood from arterioles to venules.

1. Describe the structure of a capillary, contrasting it with the structure of a vein and an artery:

_____

_____

_____

_____

_____

_____

2. Sinusoids provide a functional replacement for capillaries in some organs:

(a) Describe how sinusoids differ structurally from capillaries: _____

_____

(b) Describe in what way capillaries and sinusoids are similar: _____

_____

Related activities: Capillary networks, Formation of Tissue Fluid
Web links: Microcirculation

© Biozone International 2008-2010
Photocopying Prohibited

RA 2

Capillary Networks

Capillaries form branching networks where exchanges between the blood and tissues take place. The flow of blood through a capillary bed is called **microcirculation**. In most parts of the body, there are two types of vessels in a capillary bed: the **true capillaries**, where exchanges take place, and a vessel called a **vascular shunt**, which connects the arteriole and venule at either end of the bed. The shunt diverts blood past the true capillaries when the metabolic demands of the tissue are low (e.g. vasoconstriction in the skin when conserving body heat). When tissue activity increases, the entire network fills with blood.

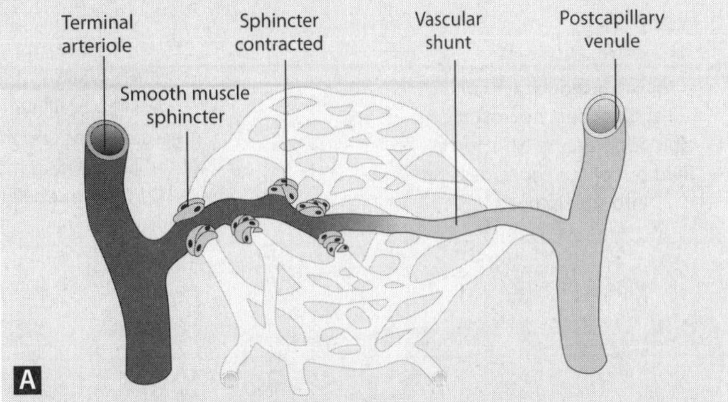

1. Describe the structure of a capillary network:

_____

_____

_____

_____

_____

_____

2. Explain the role of the smooth muscle sphincters and the vascular shunt in a capillary network:

_____

_____

_____

_____

_____

_____

3. (a) Describe a situation where the capillary bed would be in the condition labelled A:

_____

_____

_____

_____

(b) Describe a situation where the capillary bed would be in the condition labelled B:

_____

_____

_____

_____

4. Explain how a portal venous system differs from other capillary systems:

_____

_____

_____

_____

_____

_____

_____

A

When the sphincters contract (close), blood is diverted via the vascular shunt to the post-capillary venule, bypassing the exchange capillaries.

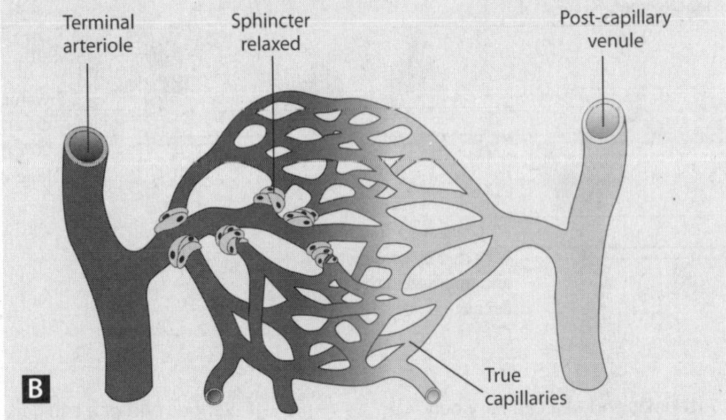

B

When the sphincters are relaxed (open), blood flows through the entire capillary bed allowing exchanges with the cells of the surrounding tissue.

Connecting Capillary Beds

The role of portal venous systems

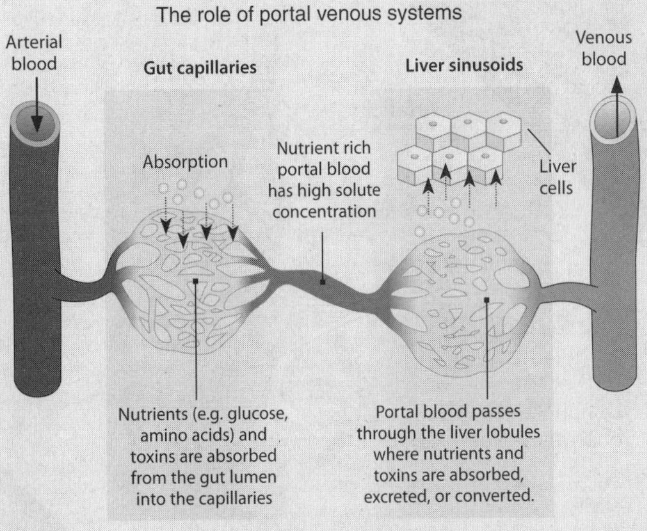

Nutrients (e.g. glucose, amino acids) and toxins are absorbed from the gut lumen into the capillaries

Portal blood passes through the liver lobules where nutrients and toxins are absorbed, excreted, or converted.

A portal venous system occurs when a capillary bed drains into another capillary bed through veins, without first going through the heart. Portal systems are relatively uncommon; most capillary beds drain into veins which then drain into the heart, not into another capillary bed. The diagram above depicts the hepatic portal system, which includes both capillary beds and the blood vessels connecting them.

Transport in Animals

Related activities: Capillaries

A 2

Formation of Tissue Fluid

The network of capillaries supplying the body's tissues ensures that no substance has to diffuse far to enter or leave a cell. Substances exchanged first diffuse through the interstitial fluid (or tissue fluid), which surrounds and bathes the cells. As with all cells, substances can move into and out of the endothelial cells of the capillary walls in several ways; by direct diffusion, by cytosis, and through gaps where the membranes are not joined by tight junctions. Some fenestrated capillaries are also more permeable than others. These specialised capillaries are important where absorption or filtration occurs (e.g. in the intestine or the kidney). Because capillaries are leaky, fluid flows across their plasma membranes. Whether fluid moves into or out of a capillary depends on the balance between the solute potential (ψs) of the blood and the blood pressure.

At the arteriolar end of a capillary bed, hydrostatic (blood) pressure (HP) forces fluid out of the capillaries and into the tissue fluid.

At the venous end of a capillary bed, hydrostatic pressure drops and most (90%) of the leaked fluid moves back into the capillaries.

Glucose, water, amino acids, ions, oxygen

As fluid leaks out through the capillary walls, it bathes the cells of the tissues

Water, CO_2 and other wastes. 90% of leaked fluid is reabsorbed

10% of leaked fluid is collected by lymph vessels and returned to the circulation near the heart

Arteriolar end of a capillary bed
Hydrostatic pressure plus solute potential is high at the arteriolar end

HP + ψs (inside) > HP + ψs (outside)

Venous end of a capillary bed
Hydrostatic pressure plus solute potential is low at the venous end

HP + ψs (inside) < HP + ψs (outside)

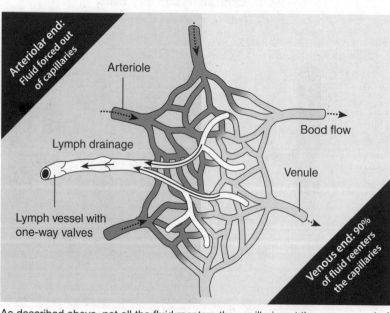

Arteriolar end: Fluid forced out of capillaries

Arteriole

Lymph drainage

Lymph vessel with one-way valves

Bood flow

Venule

Venous end: 90% of fluid reenters the capillaries

As described above, not all the fluid reenters the capillaries at the venous end of the capillary bed. This extra fluid is collected by the lymphatic vessels, a network of vessels alongside the blood vessels. Once the fluid enters the lymphatic vessels it is called lymph. The lymphatic vessels drain into the subclavian vein near the heart. Lymph is similar to tissue fluid but has more lymphocytes.

1. Describe the purpose of the tissue fluid:

   bathes the tissues providing O₂ + nutrients

2. Describe the features of capillaries that allow exchanges between the blood and other tissues:

3. Explain how hydrostatic (blood) pressure and solute potential operate to cause fluid movement at:

   (a) The arteriolar end of a capillary bed:

   hydrostatic fluid pushes f into capillaries

   (b) The venous end of a capillary bed:

   ↓ hydrostatic press ↑ solute potential tendency for water to reenter

3. Describe the two ways in which tissue fluid is returned to the general circulation:

   (a)

   (b) *lymph vessels drain back to the heart*

Related activities: Capillary Networks

Periodicals: A fair exchange

© Biozone International 2008-2010
Photocopying Prohibited

Blood

Blood makes up about 8% of body weight. Blood is a complex liquid tissue comprising cellular components suspended in plasma. If a blood sample is taken, the cells can be separated from the plasma by centrifugation. The cells (formed elements) settle as a dense red pellet below the transparent, straw-coloured plasma. Blood performs many functions: it transports nutrients, respiratory gases, hormones, and wastes; it has a role in thermoregulation through the distribution of heat; it defends against infection; and its ability to clot protects against blood loss. The examination of blood is also useful in diagnosing disease. The cellular components of blood are normally present in particular specified ratios. A change in the morphology, type, or proportion of different blood cells can therefore be used to indicate a specific disorder or infection (right).

Non-Cellular Blood Components

The non-cellular blood components form the plasma. Plasma is a watery matrix of ions and proteins and makes up 50-60% of the total blood volume.

Water
The main constituent of blood and lymph.
Role: Transports dissolved substances. Provides body cells with water. Distributes heat and has a central role in thermoregulation. Regulation of water content helps to regulate blood pressure and volume.

Mineral ions
Sodium, bicarbonate, magnesium, potassium, calcium, chloride.
Role: Osmotic balance, pH buffering, and regulation of membrane permeability. They also have a variety of other functions, e.g. Ca^{2+} is involved in blood clotting.

Plasma proteins
7-9% of the plasma volume.
Serum albumin
Role: Osmotic balance and pH buffering, Ca^{2+} transport.
Fibrinogen and prothrombin
Role: Take part in blood clotting.
Immunoglobulins
Role: Antibodies involved in the immune response.
α-globulins
Role: Bind/transport hormones, lipids, fat soluble vitamins.
β-globulins
Role: Bind/transport iron, cholesterol, fat soluble vitamins.
Enzymes
Role: Take part in and regulate metabolic activities.

Substances transported by non-cellular components
Products of digestion
Examples: sugars, fatty acids, glycerol, and amino acids.
Excretory products
Example: urea
Hormones and vitamins
Examples: insulin, sex hormones, vitamins A and B_{12}.
Importance: These substances occur at varying levels in the blood. They are transported to and from the cells dissolved in the plasma or bound to plasma proteins.

Cellular Blood Components

The cellular components of the blood (also called the formed elements) float in the plasma and make up 40-50% of the total blood volume.

Erythrocytes (red blood cells or RBCs)
5-6 million per mm^3 blood; 38-48% of total blood volume.
Role: RBCs transport oxygen (O_2) and a small amount of carbon dioxide (CO_2). The oxygen is carried bound to haemoglobin (Hb) in the cells. Each Hb molecule can bind four molecules of oxygen.

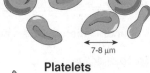

7-8 μm

Platelets

Small, membrane bound cell fragments derived from bone marrow cells; about 1/4 the size of RBCs.
0.25 million per mm^3 blood.
Role: To start the blood clotting process.

2 μm

Leucocytes (white blood cells)
5-10 000 per mm^3 blood
2-3% of total blood volume.
Role: Involved in internal defence. There are several types of white blood cells (see below).

Lymphocytes
T and B cells.
24% of the white cell count.
Role: Antibody production and cell mediated immunity.

Neutrophils
Phagocytes.
70% of the white cell count.
Role: Engulf foreign material.

Eosinophils
Rare leucocytes; normally 1.5% of the white cell count.
Role: Mediate allergic responses such as hayfever and asthma.

Basophils
Rare leucocytes; normally 0.5% of the white cell count.
Role: Produce heparin (an anti-clotting protein), and histamine. Involved in inflammation.

Transport in Animals

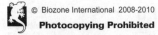

Periodicals:
Red blood cells

Related activities: Gas Transport in Humans, The Body's Defences

The Examination of Blood

Different types of microscopy give different information about blood. A SEM (right) shows the detailed external morphology of the blood cells. A fixed smear of a blood sample viewed with a light microscope (far right) can be used to identify the different blood cell types present, and their ratio to each other. Determining the types and proportions of different white blood cells in blood is called a **differential white blood cell count**. Elevated counts of particular cell types indicate allergy or infection.

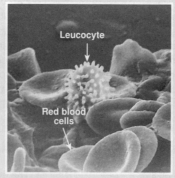

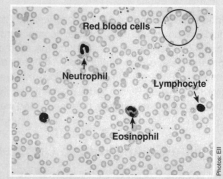

SEM of red blood cells and a leucocyte.   **Light microscope** view of a fixed blood smear.

1. For each of the following blood functions, identify the component(s) of the blood responsible and state how the function is carried out (the mode of action). The first one is done for you:

   (a) **Temperature regulation**. *Blood component involved*: _Water component of the plasma_

   Mode of action: _Water absorbs heat and dissipates it from sites of production (e.g. organs)_

   (b) **Protection against disease**. *Blood component*: _____

   Mode of action: _____

   (c) **Communication between cells, tissues, and organs**. *Blood component*: _____

   Mode of action: _____

   (d) **Oxygen transport**. *Blood component*: _____

   Mode of action: _____

   (e) **CO₂ transport**. *Blood components*: _____

   Mode of action: _____

   (f) **Buffer against pH changes**. *Blood components*: _____

   Mode of action: _____

   (g) **Nutrient supply**. *Blood component*: _____

   Mode of action: _____

   (h) **Tissue repair**. *Blood components*: _____

   Mode of action: _____

   (i) **Transport of hormones, lipids, and fat soluble vitamins**. *Blood component*: _____

   Mode of action: _____

2. Identify a feature that distinguishes red and white blood cells: _____

3. Explain two physiological advantages of red blood cell structure (lacking nucleus and mitochondria):

   (a) _____

   (b) _forces to anaerobicly !.. not using O₂ they carry_

4. Suggest what each of the following results from a differential white blood cell count would suggest:

   (a) Elevated levels of eosinophils (above the normal range): _____

   (b) Elevated levels of neutrophils (above the normal range): _____

   (c) Elevated levels of basophils (above the normal range): _____

   (d) Elevated levels of lymphocytes (above the normal range): _____

Gas Transport in Humans

The transport of respiratory gases around the body is the role of the blood and its respiratory pigments. Oxygen is transported throughout the body chemically bound to the respiratory pigment **haemoglobin** inside the red blood cells. In the muscles, oxygen from haemoglobin is transferred to and retained by **myoglobin**, a molecule that is chemically similar to haemoglobin except that it consists of only one haem-globin unit. Myoglobin has a greater affinity for oxygen than haemoglobin and acts as an oxygen store within muscles, releasing the oxygen during periods of prolonged or extreme muscular activity. If the myoglobin store is exhausted, the muscles are forced into oxygen debt and must respire anaerobically. The waste product of this, lactic acid, accumulates in the muscle and is transported (as lactate) to the liver where it is metabolised under aerobic conditions.

Gas Exchange and Transport

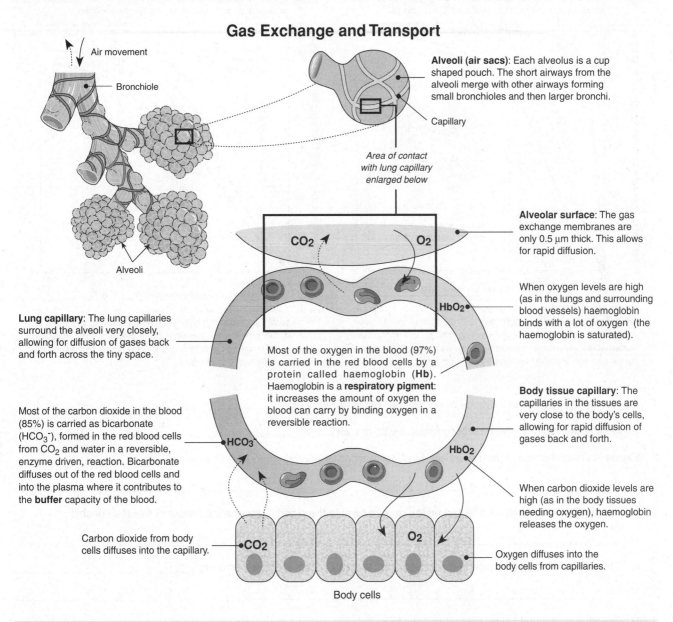

Air movement

Bronchiole

Alveoli

Alveoli (air sacs): Each alveolus is a cup shaped pouch. The short airways from the alveoli merge with other airways forming small bronchioles and then larger bronchi.

Capillary

Area of contact with lung capillary enlarged below

Alveolar surface: The gas exchange membranes are only 0.5 µm thick. This allows for rapid diffusion.

CO_2 O_2

HbO_2

When oxygen levels are high (as in the lungs and surrounding blood vessels) haemoglobin binds with a lot of oxygen (the haemoglobin is saturated).

Lung capillary: The lung capillaries surround the alveoli very closely, allowing for diffusion of gases back and forth across the tiny space.

Most of the oxygen in the blood (97%) is carried in the red blood cells by a protein called haemoglobin (**Hb**). Haemoglobin is a **respiratory pigment**: it increases the amount of oxygen the blood can carry by binding oxygen in a reversible reaction.

Body tissue capillary: The capillaries in the tissues are very close to the body's cells, allowing for rapid diffusion of gases back and forth.

Most of the carbon dioxide in the blood (85%) is carried as bicarbonate (HCO_3^-), formed in the red blood cells from CO_2 and water in a reversible, enzyme driven, reaction. Bicarbonate diffuses out of the red blood cells and into the plasma where it contributes to the **buffer** capacity of the blood.

HCO_3^-

HbO_2

When carbon dioxide levels are high (as in the body tissues needing oxygen), haemoglobin releases the oxygen.

Carbon dioxide from body cells diffuses into the capillary.

CO_2

O_2

Oxygen diffuses into the body cells from capillaries.

Body cells

Transport of carbon dioxide in the blood

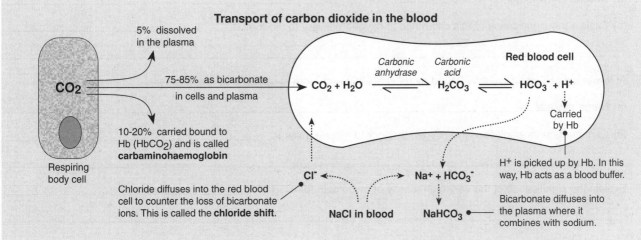

5% dissolved in the plasma

75-85% as bicarbonate in cells and plasma

10-20% carried bound to Hb ($HbCO_2$) and is called **carbaminohaemoglobin**

CO_2

Respiring body cell

Chloride diffuses into the red blood cell to counter the loss of bicarbonate ions. This is called the **chloride shift**.

Carbonic anhydrase   *Carbonic acid*   **Red blood cell**

$$CO_2 + H_2O \rightleftharpoons H_2CO_3 \rightleftharpoons HCO_3^- + H^+$$

Carried by Hb

Cl^- $Na^+ + HCO_3^-$

H^+ is picked up by Hb. In this way, Hb acts as a blood buffer.

NaCl in blood   **NaHCO_3**

Bicarbonate diffuses into the plasma where it combines with sodium.

Transport in Animals

Related activities: The Human Gas Exchange System

Oxygen does not easily dissolve in blood, but is carried in chemical combination with haemoglobin (Hb) in red blood cells. The most important factor determining how much oxygen is carried by Hb is the level of oxygen in the blood. The greater the oxygen tension, the more oxygen will combine with Hb. This relationship can be illustrated with an oxygen-haemoglobin dissociation curve as shown below (Fig. 1). In the lung capillaries, (high O_2), a lot of oxygen is picked up and bound by Hb. In the tissues, (low O_2), oxygen is released. In skeletal muscle, myoglobin picks up oxygen from haemoglobin and therefore serves as an oxygen store when oxygen tensions begin to fall. The release of oxygen is enhanced by the **Bohr effect** (Fig. 2).

Respiratory Pigments and the Transport of Oxygen

Fig. 1: Dissociation curves for haemoglobin and myoglobin at normal body temperature for foetal and adult human blood.

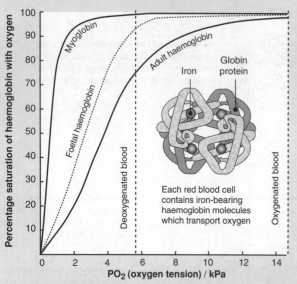

Fig. 2: Oxygen-haemoglobin dissociation curves for human blood at normal body temperature at different blood pH.

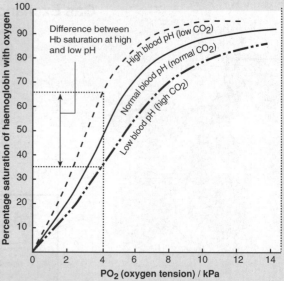

As oxygen level increases, more oxygen combines with haemoglobin (Hb). Hb saturation remains high, even at low oxygen tensions. Foetal Hb has a high affinity for oxygen and carries 20-30% more than maternal Hb. Myoglobin in skeletal muscle has a very high affinity for oxygen and will take up oxygen from haemoglobin in the blood.

As pH increases (lower CO_2), more oxygen combines with Hb. As the blood pH decreases (higher CO_2), Hb binds less oxygen and releases more to the tissues (**the Bohr effect**). The difference between Hb saturation at high and low pH represents the amount of oxygen released to the tissues.

1. (a) Identify two regions in the body where oxygen levels are very high: _____

   (b) Identify two regions where carbon dioxide levels are very high: _____

2. Explain the significance of the **reversible binding** reaction of haemoglobin (Hb) to oxygen: _____
   _____

3. (a) Haemoglobin saturation is affected by the oxygen level in the blood. Describe the nature of this relationship:
   _____

   (b) Comment on the significance of this relationship to oxygen delivery to the tissues: _____
   _____

4. (a) Describe how foetal Hb is different to adult Hb: _____

   (b) Explain the significance of this difference to oxygen delivery to the foetus: _____
   _____

5. At low blood pH, less oxygen is bound by haemoglobin and more is released to the tissues:

   (a) Name this effect: _Bohr effect_ _____

   (b) Comment on its significance to oxygen delivery to respiring tissue: _____
   _____

6. Explain the significance of the very high affinity of myoglobin for oxygen: _____
   _____

7. Identify the two main contributors to the buffer capacity of the blood: _____

The Effects of High Altitude

The air at high altitudes contains less oxygen than the air at sea level. Air pressure decreases with altitude so the pressure (therefore amount) of oxygen in the air also decreases. Sudden exposure to an altitude of 2000 m would make you breathless on exertion and above 7000 m most people would become unconscious. The effects of altitude on physiology are related to this lower oxygen availability. Humans and other animals can make some physiological adjustments to life at altitude; this is called acclimatisation. Some of the changes to the cardiovascular and respiratory systems to high altitude are outlined below.

Mountain Sickness

Altitude sickness or mountain sickness is usually a mild illness associated with trekking to altitudes of 5000 metres or so. Common symptoms include headache, insomnia, poor appetite and nausea, vomiting, dizziness, tiredness, coughing, and breathlessness. The best way to avoid mountain sickness is to ascend to altitude slowly (no more than 300 m per day above 3000 m). Continuing to ascend with mountain sickness can result in more serious illnesses: accumulation of fluid on the brain (cerebral oedema) and accumulation of fluid in the lungs (pulmonary oedema). These complications can be fatal if not treated with oxygen and a rapid descent to lower altitude.

Physiological Adjustment to Altitude

Effect	Minutes	Days	Weeks
Increased heart rate	←——————→		
Increased breathing		←——————→	
Concentration of blood		←——→	
Increased red blood cell production			←——————→
Increased capillary density			←——→

The human body can make adjustments to life at altitude. Some of these changes take place almost immediately: breathing and heart rates increase. Other adjustments may take weeks (see above). These responses are all aimed at improving the rate of supply of oxygen to the body's tissues. When more permanent adjustments to physiology are made (increased blood cells and capillary networks) heart and breathing rates can return to normal.

People who live permanently at high altitude, e.g. Tibetans, Nepalese, and Peruvian Indians, have physiologies adapted (genetically, through evolution) to high altitude. Their blood volumes and red blood cell counts are high, and they can carry heavy loads effortlessly despite a small build. In addition, their metabolism uses oxygen very efficiently.

Llamas, vicunas, and Bactrian camels are well suited to high altitude life. Vicunas and llamas, which live in the Andes, have high blood cell counts and their red blood cells live almost twice as long as those in humans. Their haemoglobin also picks up and offloads oxygen more efficiently than the haemoglobin of most mammals.

1. (a) Describe the general effects of high altitude on the body: _____

   _____

   (b) Name the general term given to describe these effects: _____

2. (a) Identify one short term physiological adaptation that humans make to high altitude: _____

   _____

   (b) Explain how this adaptation helps to increase the amount of oxygen the body receives: _____

   _____

3. (a) Describe one longer term adaptation that humans can make to living at high altitude: _____

   _____

   (b) Explain how this adaptation helps to increase the amount of oxygen the body receives: _____

   _____

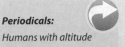
Periodicals:
Humans with altitude

Related activities: Blood, Gas Transport in Humans

(E)A 2

Transport in Animals

Exercise and Blood Flow

Exercise promotes health by improving the rate of blood flow back to the heart (called the venous return). This is achieved by strengthening all types of muscle and by increasing the efficiency of the heart. During exercise blood flow to different parts of the body changes in order to cope with the extra demands of the muscles, the heart, and the lungs.

1. The following table gives data for the **rate** of blood flow to various parts of the body at rest and during strenuous exercise. **Calculate** the **percentage** of the total blood flow that each organ or tissue receives under each regime of activity.

Organ or tissue	At rest		Strenuous exercise	
	$cm^3\ min^{-1}$	% of total	$cm^3\ min^{-1}$	% of total
Brain	700	14	750	4.2
Heart	200		750	
Lung tissue	100		200	
Kidneys	1100		600	
Liver	1350		600	
Skeletal muscles	750		12 500	
Bone	250		250	
Skin	300		1900	
Thyroid gland	50		50	
Adrenal glands	25		25	
Other tissue	175		175	
TOTAL	5000	**100**	17 800	**100**

2. Explain how the body increases the rate of blood flow during exercise: _____

_____

_____

3. (a) State approximately how many times the total rate of blood flow increases between rest and exercise: _____

   (b) Explain why the increase is necessary: _____

   _____

   _____

4. (a) Identify which organs or tissues show no change in the rate of blood flow with exercise:

   _____

   (b) Explain why this is the case: _____

   _____

   _____

5. (a) Identify which organs or tissues show the most change in the rate of blood flow with exercise:

   _____

   (b) Explain why this is the case: _____

   _____

   _____

Related activities: *Control of Heart Activity*

KEY TERMS: Memory Card Game

The cards below have a keyword or term printed on one side and its definition printed on the opposite side. The aim is to win as many cards as possible from the table. To play the game.....

1) Cut out the cards and lay them definition side down on the desk. You will need one set of cards between two students.

2) Taking turns, choose a card and, BEFORE you pick it up, state your own best definition of the keyword to your opponent.

3) Check the definition on the opposite side of the card. If both you and your opponent agree that your stated definition matches, then keep the card. If your definition does not match then return the card to the desk.

4) Once your turn is over, your opponent may choose a card.

Artery	Capillary	Vein
Diastole	Semilunar valves	Electro-cardiogram
Cardiac cycle	Sinoatrial node	Purkyne tissue
Tissue fluid	Atrio-ventricular node	Blood
Lymph	Haemoglobin	Ventricle
Atrium	Systole	Intrinsic heart rate

Transport in Animals

R 2

When you've finished the game keep these cutouts and use them as flash cards!

Large blood vessel that returns blood to the heart.

The smallest of a body's blood vessels (only one cell thick) and part of the microcirculation.

A large blood vessel with a thick, muscled wall which carries blood away from the heart.

An externally recorded interpretation of the electrical activity of the heart over time.

The heart valves located in the arteries leaving the heart.

The phase of the cardiac cycle when the heart fills with blood after contraction.

Specialised myocardial fibres that conduct electrical impulses to the ventricles to cause coordinated contractions of the heart muscle.

The impulse-generating (pacemaker) tissue located in the right atrium of the heart.

The sequence of events from the beginning of one heartbeat to the beginning of the next.

Circulatory fluid comprising numerous cell types, which moves respiratory gases and nutrients around the body.

Specialised tissue between the atria and the ventricles of the heart that delays the pacemaker impulse and thus coordinates contractions of the chambers.

A fluid derived from the blood plasma by leakage through capillaries. It bathes the tissues and is also called interstitial fluid.

A thick-walled chamber of the heart that pumps blood into arteries.

The iron-containing pigment in the blood responsible for binding and transporting oxygen within red blood cells.

A clear fluid contained within the lymphatic system. Similar in composition to the interstitial fluid.

A term describing the spontaneous rhythm of the heart.

The phase of the cardiac cycle where the heart muscle is contracting and pressure is being generated within the chambers of the heart.

A thin-walled chamber that collects blood returning to the heart.

Transport in Plants

KEY TERMS

apoplastic pathway

bulk (=mass) flow

capillary action

Casparian strip

cell turgor

cohesion-tension hypothesis

companion cell

cortex

endodermis

flaccid

guard cells

mass flow hypothesis

osmosis

phloem

root hair

root pressure

sieve tube members

sink

source

stomata

symplastic pathway

translocation

transpiration

transpiration pull

transpiration rate

turgor

vacuolar pathway

vessel (xylem vessel)

water potential

xylem

KEY CONCEPTS

▶ Specialised transport tissues are a feature of plants above a certain size.

▶ Gases enter and leave the plant mainly through stomata, but water loss is a consequence of this.

▶ Plants have adaptations to reduce water loss.

▶ Transpiration drives water uptake in plants.

▶ Translocation moves carbohydrate around the plant from sources to sinks.

OBJECTIVES

☐ 1. Use the **KEY TERMS** to help you understand and complete these objectives.

Transport Tissues — pages 124-129

☐ 2. Explain the need for transport systems in multicellular plants in relation to size and surface area to volume ratio.

☐ 3. Recognise that transport tissues in plants also have a role in support. Describe the roles of **cell turgor**, transport tissues (especially **xylem**), and cellulose in supporting the plant.

☐ 4. Use photographs and diagrams to describe the distribution of **xylem** and **phloem** in a dicotyledonous stem, leaf, and root.

☐ 5. Describe the composition and function of transport tissues in angiosperms: **xylem** (**vessels**, tracheids, fibres, xylem parenchyma) and **phloem** (**companion cells**, **sieve tube members**, fibres, phloem parenchyma).

☐ 6. Describe the adaptations of a dicot primary root, including the structure and role of **root hairs** and the **endodermis**, as well as xylem, and phloem.

Plant Transport Processes — pages 130-136

☐ 7. Define the term transpiration and recognise it as a necessary consequence of gas exchange in plants. Describe the role of **stomata** in the movement of gases into and out of the spongy mesophyll of the leaf.

☐ 8. Describe factors affecting the rate of transpiration. Describe how a potometer is used to Investigate the effect of physical factors (e.g. humidity, light, air movement, and temperature) on **transpiration rate**.

☐ 9. Describe the mechanism and pathways for water uptake in plant roots. Include reference to the role of **osmosis**, gradients in water potential, and the **symplastic**, **apoplastic**, and **vacuolar pathways** through the root. Explain the role of the **endodermis** and the **Casparian strip** in the movement of water into the symplast.

☐ 10. Use **water potential** to explain how water and dissolved minerals are moved up the plant from the roots to the leaves in the **transpiration stream**. Explain the roles of **cohesion-tension** (capillary action), **transpiration pull**, and **root pressure**, identifying the relative importance of each.

☐ 11. Describe the adaptations of **xerophytes** that help to reduce water loss.

☐ 12. Describe and explain **translocation** in the **phloem**, identifying **sources** and **sinks** in sucrose transport. Evaluate the evidence for and against the **mass flow** (pressure-flow) **hypothesis** for the mechanism of translocation.

Periodicals:

Listings for this chapter are on page 339

Weblinks:

www.biozone.co.uk/
weblink/OCR-AS-2641.html

Transport in Plants

The support and transport systems in plants are closely linked; many of the same tissues are involved in both systems. Primitive plants (e.g. mosses and liverworts) are small and low growing, and have no need for support and transport systems. If a plant is to grow to any size, it must have ways to hold itself up against gravity and to move materials around its body. The body of a flowering plant has three parts: **roots** anchor the plant and absorb nutrients from the soil, **leaves** produce sugars by photosynthesis, and **stems** link the roots to the leaves and provide support for the

leaves and reproductive structures. Stems have distinct points, called **nodes**, at which leaves and buds attach. The region of the stem between two nodes is called the **internode**. Regardless of their shape or location, all stems can be distinguished as such by the presence of nodes and internodes. Vascular tissues (xylem and phloem) link all plant parts so that water, minerals, and manufactured food can be transported between different regions. All plants rely on fluid pressure within their cells (turgor) to give some support to their structure.

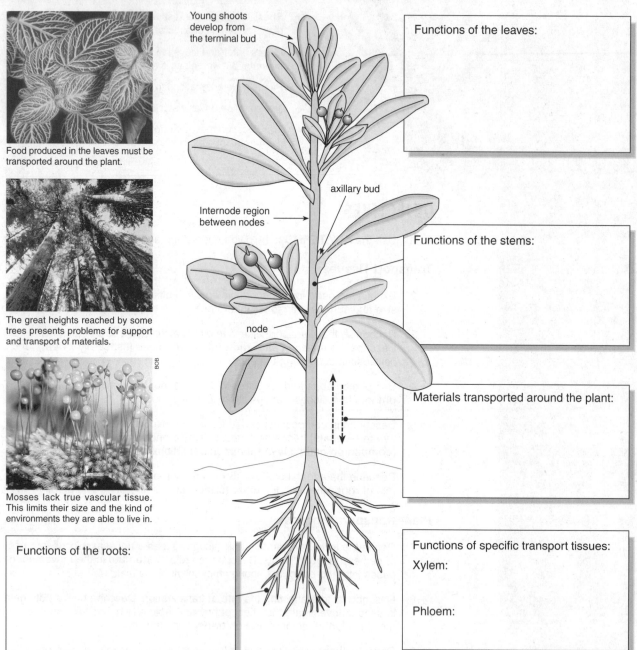

Food produced in the leaves must be transported around the plant.

The great heights reached by some trees presents problems for support and transport of materials.

Mosses lack true vascular tissue. This limits their size and the kind of environments they are able to live in.

Young shoots develop from the terminal bud

Internode region between nodes

axillary bud

node

Functions of the leaves:

Functions of the stems:

Materials transported around the plant:

Functions of the roots:

Functions of specific transport tissues:
Xylem:

Phloem:

1. In the boxes provided in the diagram above:
   (a) List the main functions of the leaves, roots and stems (remember that the leaves themselves have leaf veins).
   (b) List the materials that are transported around the plant body.
   (c) Describe the functions of the transport tissues: xylem and phloem.

2. Name the solvent for all the materials that are transported around the plant: _____

3. State what processes are involved in the transport of sap in the following tissues:
   (a) The xylem: _____
   (b) The phloem: _____

Related activities: Stems and Roots, Xylem, Phloem, Uptake at the Root, Translocation

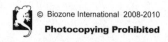 © Biozone International 2008-2010
Photocopying Prohibited

Stems and Roots

The stem and root systems of plants are closely linked. Stems are the primary organs for supporting the plant, whereas roots anchor the plant in the ground, absorb water and minerals from the soil, and transport these materials to other parts of the plant body. Roots may also act as storage organs, storing excess carbohydrate reserves until they are required by the plant. Like most parts of the plant, stems and roots contain vascular tissues. These take the form of bundles containing the xylem and phloem

and strengthening fibres. The entire plant body, including the roots and stems is covered in an epidermis but, unlike most of the plant, the root epidermis has only a thin cuticle that presents no barrier to water entry. Young roots are also covered with **root hairs**. Compared with stems, roots are relatively simple and uniform in structure, and their features are associated with aeration of the tissue and transport of water and minerals form the soil. Dicot stems and roots are described below.

Dicot Stem Structure

In dicots, the vascular bundles are arranged in an orderly fashion around the stem. Each vascular bundle contains **xylem** (to the inside) and **phloem** (to the outside). Between the phloem and the xylem is the **vascular cambium**; a layer of cells that divide to produce the thickening of the stem. The middle of the stem, called the **pith**, is filled with thin-walled parenchyma cells. The vascular bundles in dicots are arranged in an orderly way around the periphery of the stem (below).

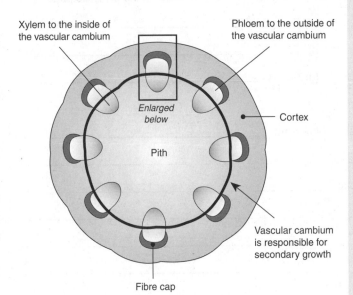

Xylem to the inside of the vascular cambium

Phloem to the outside of the vascular cambium

Enlarged below

Cortex

Pith

Vascular cambium is responsible for secondary growth

Fibre cap

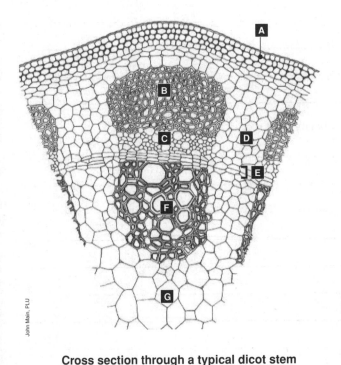

John Main, PLU

A
B
C
D
E
F
G

Cross section through a typical dicot stem

Dicot Root Structure

The primary tissues of a dicot root are simple in structure. The large cortex is made up of parenchyma (packing) cells, which store starch and other substances. The air spaces between the cells are essential for aeration of the root tissue, which is non-photosynthetic. The vascular tissue, xylem (X) and phloem (P) forms a central cylinder through the root and is surrounded by the **pericycle**, a ring of cells from which lateral roots arise.

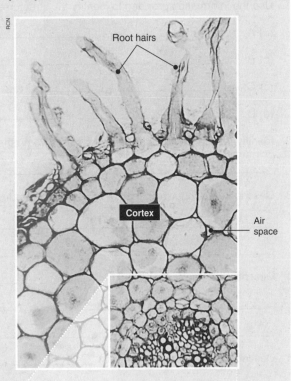

RCN

Root hairs

Cortex

Air space

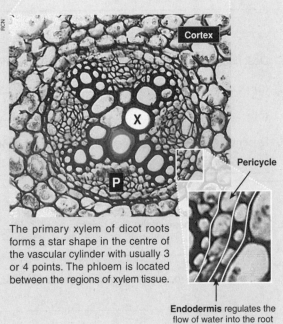

RCN

Cortex

X

P

Pericycle

The primary xylem of dicot roots forms a star shape in the centre of the vascular cylinder with usually 3 or 4 points. The phloem is located between the regions of xylem tissue.

Endodermis regulates the flow of water into the root

Related activities: Xylem, Phloem, Uptake at the Root
Web links: Photographic Atlas of Plant Anatomy, Types of Roots

RA 2

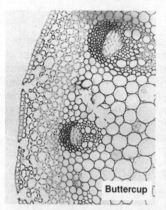

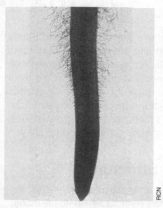

Buttercup

In plants with photosynthetic stems, CO_2 enters the stem through stomata in the epidermis. The air spaces in the cortex are more typical of leaf mesophyll than stem cortex.

Strawberry plants send out runners. These are above-ground, trailing stems that form roots at their nodes. The plant uses this mechanism to spread vegetatively over a wide area.

Root hairs are located just behind the region of cell elongation in the root tip. The root tip is covered by a slimy root cap. This protects the dividing cells of the tip and lubricates root movement.

The roots and their associated root hairs provide a very large surface area for the uptake of water and ions, as shown in this photograph of the roots of a hydroponically grown plant.

1. Use the information provided to identify the structures **A-G** in the photograph of the dicot stem on the previous page:

   (a) A: _____

   (b) B: _____

   (c) C: _____

   (d) D: _____

   (e) E: _____

   (f) F: _____

   (g) G: _____

2. Identify the feature that distinguishes stems from other parts of the plant: _____

   _____

3. Describe a distinguishing feature of stem structure in dicots: _____

   _____

   _____

4. Describe the role of the vascular cambium: _____

   _____

   _____

5. Describe three functions of roots: _____

   _____

   _____

6. Describe two distinguishing features of internal anatomy of a primary dicot root:

   (a) _____

   (b) _____

7. Describe the role of the parenchyma cells of the root cortex: _____

   _____

   _____

8. Explain the purpose of the root hairs: _____

   _____

9. Explain why the root tip is covered by a cap of cells: _____

   _____

   _____

Leaf Structure

The main function of leaves is as photosynthetic organs in which the sun's radiant energy is collected and used to drive the fixation of carbon dioxide. Regardless of their varying forms, foliage leaves comprise epidermal, mesophyll, and vascular (xylem and phloem) tissues. These tissues are organised in such a way as to maximise photosynthesis by maximising capture of sunlight energy and facilitating diffusion of gases into and out of the leaf

tissue. Gases enter and leave the leaf by way of **stomata**. Inside the leaf (as illustrated below), the large air spaces and loose arrangement of the spongy mesophyll provide a large surface area for gas exchange. The basic structure of a dicot leaf is described below. However, the mesophyll (the packing tissue of the leaf) may be variously arranged according to the particular photosynthetic adaptations of the leaf.

Dicot Leaf Structure

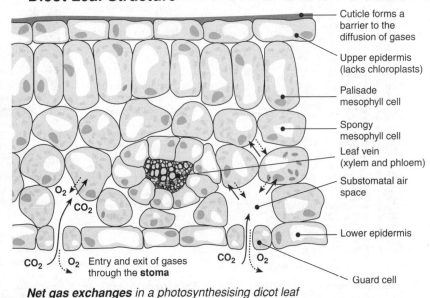

- Cuticle forms a barrier to the diffusion of gases
- Upper epidermis (lacks chloroplasts)
- Palisade mesophyll cell
- Spongy mesophyll cell
- Leaf vein (xylem and phloem)
- Substomatal air space
- Lower epidermis
- Guard cell

O_2 · CO_2

CO_2 / O_2 — Entry and exit of gases through the **stoma**

CO_2 / O_2

Net gas exchanges in a photosynthesising dicot leaf

Respiring plant cells use oxygen (O_2) and produce carbon dioxide (CO_2). These gases move in and out of the plant and through the air spaces by diffusion. Angiosperms have many air spaces between the cells of the stems, leaves, and roots. These air spaces are continuous and gases are able to move freely through them and into the plant's cells via the **stomata** (*sing*. stoma).

When the plant is photosynthesising, the situation is more complex. Overall there is a net consumption of CO_2 and a net production of oxygen. CO_2 fixation maintains a gradient in CO_2 concentration between the inside of the leaf and the atmosphere. Oxygen is produced in excess of respiratory needs and diffuses out of the leaf. These **net** exchanges are indicated by the arrows on the diagram.

Stoma

Nucleus of epidermal cell

The surface of the leaf epidermis of a dicot illustrating the density and scattered arrangement of the pores or **stomata**. In dicots, stomata are usually present only on the lower leaf surface.

The leaves of dicots can be distinguished from those of monocots by their netted pattern of leaf veins. Monocots, in contrast, generally have leaves with parallel venation.

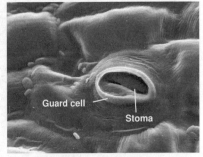

Guard cell

Stoma

Guard cells on each side of a stoma (pl. stomata) regulate the entry and exit of gases and water vapour. Stomata permit gas exchange but are also the major routes for water loss.

1. Describe two adaptive features of leaves:

   (a) _____

   (b) _____

2. Identify the region of a dicot leaf where most of the chloroplasts are found: _____

3. Explain the purpose of the air spaces in the leaf tissue: _____

   _____

   _____

4. (a) Describe how gases enter and leave the leaf tissue: _____

   _____

   (b) Explain how this movement is regulated: _____

   _____

   _____

Related activities: Transpiration, Adaptations of Xerophytes
Web links: Leaf Tissues, Leaf Structure

RA 2

Xylem

Xylem is the principal **water conducting tissue** in vascular plants. It is also involved in conducting dissolved minerals, in food storage, and in supporting the plant body. As in animals, tissues in plants are groupings of different cell types that work together for a common function. Xylem is a **complex tissue**. In angiosperms, it is composed of five cell types: tracheids, vessels, xylem parenchyma, sclereids (short sclerenchyma cells), and fibres. The tracheids and vessel elements form the bulk of the tissue. They are heavily strengthened and are the conducting cells of the xylem. Parenchyma cells are involved in storage, while fibres and sclereids provide support. When mature, xylem is dead.

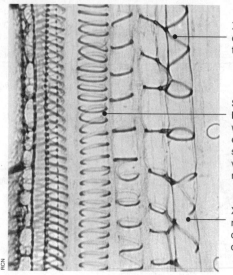

Xylem vessels form continuous tubes throughout the plant.

Spiral thickening of **lignin** around the walls of the vessel elements give extra strength allowing the vessels to remain rigid and upright.

Xylem is dead when mature. Note how the cells have lost their cytoplasm.

The Structure of Xylem Tissue

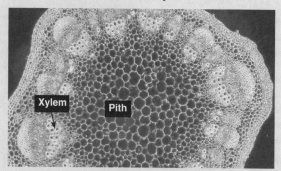

Xylem   Pith

This cross section through the stem, *Helianthus* (sunflower) shows the central pith, surrounded by a peripheral ring of vascular bundles. Note the xylem vessels with their thick walls.

Fibres are a type of sclerenchyma cell. They are associated with vascular tissues and usually occur in groups. The cells are very elongated and taper to a point and the cell walls are heavily thickened. Fibres give mechanical support to tissues, providing both strength and elasticity.

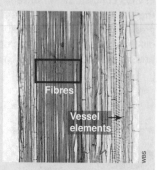

Fibres

Vessel elements

Vessel element

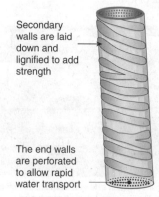

Secondary walls are laid down and lignified to add strength

The end walls are perforated to allow rapid water transport

Tip of tracheid cell

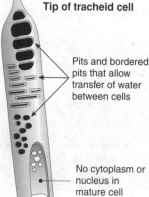

Pits and bordered pits that allow transfer of water between cells

No cytoplasm or nucleus in mature cell

Vessel elements and tracheids are the two conducting cells types in xylem. Tracheids are long, tapering hollow cells. Water passes from one tracheid to another through thin regions in the wall called **pits**. Vessel elements have pits, but the end walls are also perforated and water flows unimpeded through the stacked elements.

Vessel elements are found only in the xylem of angiosperms. They are large diameter cells that offer very low resistance to water flow. The possession of vessels (stacks of vessel elements) provides angiosperms with a major advantage over gymnosperms and ferns as they allow for very rapid water uptake and transport.

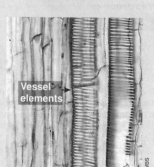

Vessel elements

1. Describe the function of **xylem**: _____

_____

2. Identify the four main cell types in xylem and explain their role in the tissue:

   (a) _____

   (b) _____

   (c) _____

   (d) _____

3. Describe one way in which xylem is strengthened in a mature plant: _____

4. Describe a feature of vessel elements that increases their efficiency of function: _____

_____

Related activities: *Plant Cell Specialisation, Plant Tissues, Uptake at the Root, Transpiration*   **Web links**: *Photographic Atlas of Plant Anatomy*

Phloem

Like xylem, **phloem** is a complex tissue, comprising a variable number of cell types. Phloem is the principal **food (sugar) conducting tissue** in vascular plants, transporting dissolved sugars around the plant. The bulk of phloem tissue comprises the **sieve tubes** (sieve tube members and sieve cells) and their companion cells. The sieve tubes are the principal conducting cells in phloem and are closely associated with the **companion cells** (modified parenchyma cells) with which they share a mutually dependent relationship. Other parenchyma cells, concerned with storage, occur in phloem, and strengthening fibres and sclereids (short sclerenchyma cells) may also be present. Unlike xylem, phloem is alive when mature.

LS through a sieve tube end plate

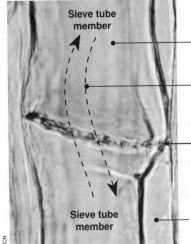

Sieve tube member

The sieve tube members lose most of their organelles but are still alive when mature

Sugar solution flows in both directions

Sieve tube end plate
Tiny holes (arrowed in the photograph below) perforate the sieve tube elements allowing the sugar solution to pass through.

Sieve tube member

Companion cell: a cell adjacent to the sieve tube member, responsible for keeping it alive

TS through a sieve tube end plate

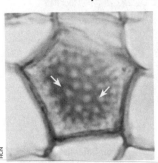

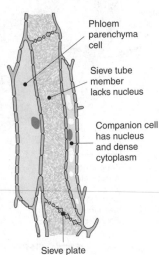

Phloem parenchyma cell

Sieve tube member lacks nucleus

Companion cell has nucleus and dense cytoplasm

Adjacent sieve tube members are connected via **sieve plates** through which the phloem sap flows.

Sieve plate

The Structure of Phloem Tissue

Phloem is alive at maturity and functions in the transport of sugars and minerals around the plant. Like xylem, it forms part of the structural vascular tissue of plants.

Fibres are associated with phloem as they are in xylem. Here they are seen in cross section where you can see the extremely thick cell walls and the way the fibres are clustered in groups. See the previous page for a view of fibres in longitudinal section.

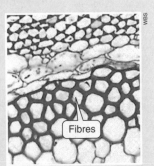

Fibres

In this cross section through a buttercup root, the smaller companion cells can be seen lying alongside the sieve tube members. It is the sieve tube members that, end on end, produce the **sieve tubes**. They are the conducting tissue of phloem.

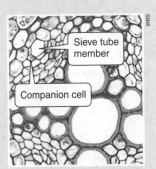

Sieve tube member

Companion cell

In this longitudinal section of a buttercup root, each sieve tube member has a thin **companion cell** associated with it. Companion cells retain their nucleus and control the metabolism of the sieve tube member next to them. They also have a role in the loading and unloading of sugar into the phloem.

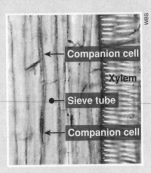

Companion cell

Xylem

Sieve tube

Companion cell

1. Describe the function of **phloem**: _____

2. Describe two differences between xylem and phloem: _____

   _____

3. Explain the purpose of the **sieve plate** at the ends of each sieve tube member: _____

   _____

4. (a) Name the conducting cell type in phloem: _____

   (b) Explain two roles of the companion cell in phloem: _____

   _____

5. State the purpose of the phloem parenchyma cells: _____

6. Identify a type of cell that provides strengthening in phloem: _____

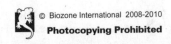
Related activities: Plant Cell Specialisation, Plant Tissues, Uptake at the Root, Translocatiom   *Web links:* Photographic Atlas of Plant Anatomy

RA 2

Uptake in the Root

Plants need to take up water and minerals constantly. They must compensate for the continuous loss of water from the leaves and provide the materials they need for the manufacture of food. The uptake of water and minerals is mostly restricted to the younger, most recently formed cells of the roots and the root hairs. Some water moves through the plant tissues via the plasmodesmata of the cells (the **symplastic route**), but most passes through the free spaces between cell walls (the **apoplast**). Water uptake is assisted by root pressure, which arises because the soil and root tissue has a higher water potential than other plant tissues. Two processes are involved in water and ion uptake: diffusion (osmosis in the case of water) and active transport.

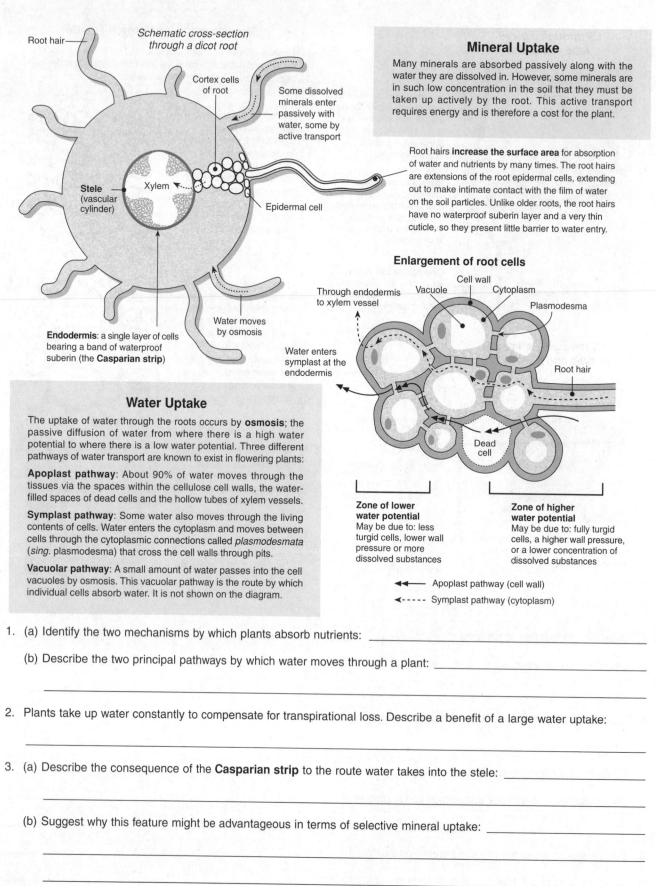

Mineral Uptake

Many minerals are absorbed passively along with the water they are dissolved in. However, some minerals are in such low concentration in the soil that they must be taken up actively by the root. This active transport requires energy and is therefore a cost for the plant.

Root hairs **increase the surface area** for absorption of water and nutrients by many times. The root hairs are extensions of the root epidermal cells, extending out to make intimate contact with the film of water on the soil particles. Unlike older roots, the root hairs have no waterproof suberin layer and a very thin cuticle, so they present little barrier to water entry.

Enlargement of root cells

Labels in schematic cross-section: Root hair, *Schematic cross-section through a dicot root*, Cortex cells of root, Some dissolved minerals enter passively with water, some by active transport, Stele (vascular cylinder), Xylem, Epidermal cell, Water moves by osmosis, **Endodermis**: a single layer of cells bearing a band of waterproof suberin (the **Casparian strip**).

Labels in enlargement: Through endodermis to xylem vessel, Vacuole, Cell wall, Cytoplasm, Plasmodesma, Water enters symplast at the endodermis, Root hair, Dead cell.

Zone of lower water potential
May be due to: less turgid cells, lower wall pressure or more dissolved substances

Zone of higher water potential
May be due to: fully turgid cells, a higher wall pressure, or a lower concentration of dissolved substances

◄◄──── Apoplast pathway (cell wall)
◄----- Symplast pathway (cytoplasm)

Water Uptake

The uptake of water through the roots occurs by **osmosis**; the passive diffusion of water from where there is a high water potential to where there is a low water potential. Three different pathways of water transport are known to exist in flowering plants:

Apoplast pathway: About 90% of water moves through the tissues via the spaces within the cellulose cell walls, the water-filled spaces of dead cells and the hollow tubes of xylem vessels.

Symplast pathway: Some water also moves through the living contents of cells. Water enters the cytoplasm and moves between cells through the cytoplasmic connections called *plasmodesmata* (*sing*. plasmodesma) that cross the cell walls through pits.

Vacuolar pathway: A small amount of water passes into the cell vacuoles by osmosis. This vacuolar pathway is the route by which individual cells absorb water. It is not shown on the diagram.

1. (a) Identify the two mechanisms by which plants absorb nutrients: _____

   (b) Describe the two principal pathways by which water moves through a plant: _____

   _____

2. Plants take up water constantly to compensate for transpirational loss. Describe a benefit of a large water uptake:

   _____

3. (a) Describe the consequence of the **Casparian strip** to the route water takes into the stele: _____

   _____

   (b) Suggest why this feature might be advantageous in terms of selective mineral uptake: _____

   _____

   _____

Transpiration

Plants lose water all the time, despite the adaptations they have to help prevent it (e.g. waxy leaf cuticle). Approximately 99% of the water a plant absorbs from the soil is lost by evaporation from the leaves and stem. This loss, mostly through stomata, is called **transpiration** and the flow of water through the plant is called the **transpiration stream**. Plants rely on a gradient in water potential (ψ) from the roots to the air to move water through their cells. Water flows passively from soil to air along a gradient of decreasing water potential. The gradient in water potential is the driving force in the ascent of water up a plant. A number of processes contribute to water movement up the plant: transpiration pull, cohesion, and root pressure. Transpiration may seem to be a wasteful process, but it has benefits. Evaporative water loss cools the plant and the transpiration stream helps the plant to maintain an adequate mineral uptake, as many essential minerals occur in low concentrations in the soil.

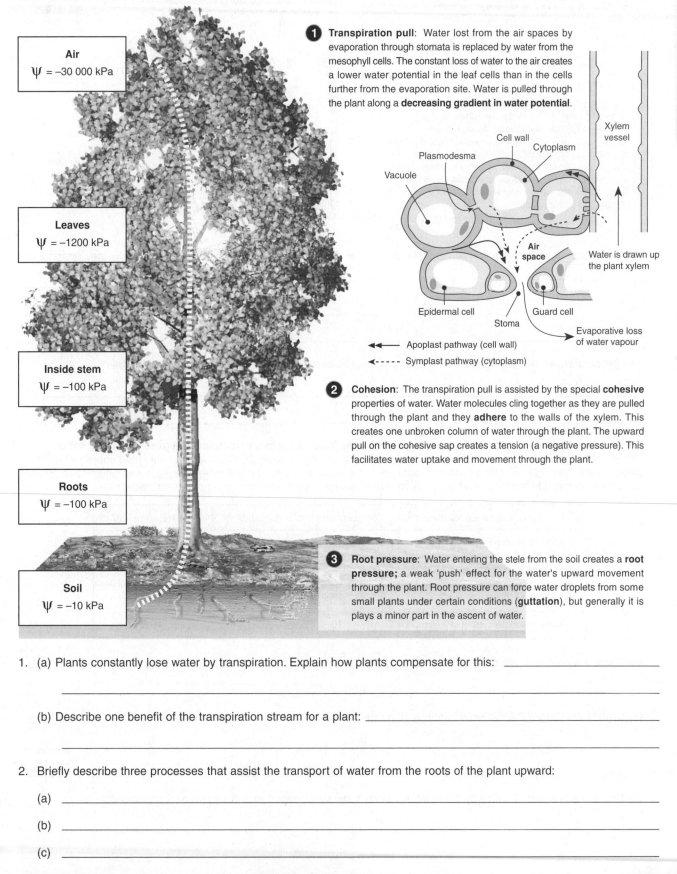

Air
ψ = −30 000 kPa

Leaves
ψ = −1200 kPa

Inside stem
ψ = −100 kPa

Roots
ψ = −100 kPa

Soil
ψ = −10 kPa

1 **Transpiration pull**: Water lost from the air spaces by evaporation through stomata is replaced by water from the mesophyll cells. The constant loss of water to the air creates a lower water potential in the leaf cells than in the cells further from the evaporation site. Water is pulled through the plant along a **decreasing gradient in water potential**.

Xylem vessel

Cell wall

Cytoplasm

Plasmodesma

Vacuole

Air space

Water is drawn up the plant xylem

Epidermal cell

Guard cell

Stoma

Evaporative loss of water vapour

◄◄——— Apoplast pathway (cell wall)
◄----- Symplast pathway (cytoplasm)

2 **Cohesion**: The transpiration pull is assisted by the special **cohesive** properties of water. Water molecules cling together as they are pulled through the plant and they **adhere** to the walls of the xylem. This creates one unbroken column of water through the plant. The upward pull on the cohesive sap creates a tension (a negative pressure). This facilitates water uptake and movement through the plant.

3 **Root pressure**: Water entering the stele from the soil creates a **root pressure**; a weak 'push' effect for the water's upward movement through the plant. Root pressure can force water droplets from some small plants under certain conditions (**guttation**), but generally it is plays a minor part in the ascent of water.

1. (a) Plants constantly lose water by transpiration. Explain how plants compensate for this: _____

_____

(b) Describe one benefit of the transpiration stream for a plant: _____

_____

2. Briefly describe three processes that assist the transport of water from the roots of the plant upward:

(a) _____

(b) _____

(c) _____

Periodicals:
How trees lift water

Related activities: *Passive Transport Processes*
Web links: *Transpiration Animation*

DA 3

The Potometer

A potometer is a simple instrument for investigating transpiration rate (water loss per unit time). The equipment is simple and easy to obtain. A basic potometer, such as the one shown right, can easily be moved around so that transpiration rate can be measured under different environmental conditions

Some of the physical conditions investigated are:

- Humidity or vapour pressure (high or low)

- Temperature (high or low)

- Air movement (still or windy)

- Light level (high or low)

- Water supply

It is also possible to compare the transpiration rates of plants with different adaptations e.g. comparing transpiration rates in plants with rolled leaves vs rates in plants with broad leaves. If possible, experiments like these should be conducted simultaneously using replicate equipment. If conducted sequentially, care should be taken to keep the environmental conditions the same for all plants used.

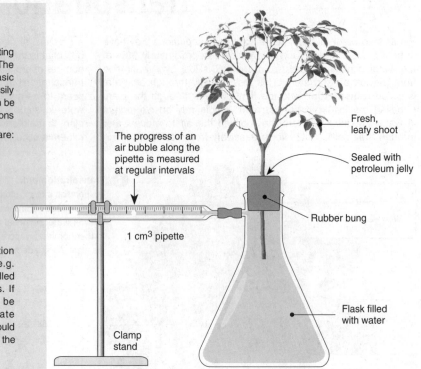

The progress of an air bubble along the pipette is measured at regular intervals

1 cm³ pipette

Clamp stand

Fresh, leafy shoot

Sealed with petroleum jelly

Rubber bung

Flask filled with water

3. Describe three environmental conditions that increase the rate of transpiration in plants, explaining how they operate:

(a) _____

(b) _____

(c) _____

4. The **potometer** (above) is an instrument used to measure transpiration rate. Briefly explain how it works:

_____

_____

5. An experiment was conducted on transpiration from a hydrangea shoot in a potometer. The experiment was set up and the plant left to stabilise (environmental conditions: still air, light shade, 20°C). The plant was then subjected to different environmental conditions and the water loss was measured each hour. Finally, the plant was returned to original conditions, allowed to stabilise and transpiration rate measured again. The data are presented below:

Experimental conditions	Temperature / °C	Humidity / %	Transpiration / gh⁻¹
(a) Still air, light shade, 20°C	18	70	1.20
(b) Moving air, light shade, 20°C	18	70	1.60
(c) Still air, bright sunlight, 23°C	18	70	3.75
(d) Still air and dark, moist chamber, 19.5°C	18	100	0.05

(a) Name the control in this experiment: _____

_____

(b) Identify the factors that increased transpiration rate, explaining how each has its effect: _____

_____

_____

(c) Suggest a possible reason why the plant had such a low transpiration rate in humid, dark conditions:

_____

_____

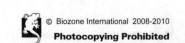

Adaptations of Xerophytes

Plants adapted to dry conditions are called **xerophytes** and they show structural (xeromorphic) and physiological adaptations for water conservation. These typically include small, hard leaves, and epidermis with a thick cuticle, sunken stomata, succulence, and permanent or temporary absence of leaves. Xerophytes may live in humid environments, provided that their roots are in dry microenvironments (e.g. the roots of epiphytic plants that grow on tree trunks or branches). The nature of the growing environment is important in many other situations too. **Halophytes** (salt tolerant plants) and alpine species may also show xeromorphic features in response to the scarcity of obtainable water and high transpirational losses in these environments.

Leaves modified into spines or hairs to reduce water loss. Light coloured spines reflect solar radiation.

Squat, rounded shape reduces surface area. The surface tissues of many cacti are tolerant of temperatures in excess of 50°C.

Shallow, but extensive fibrous root system.

Stem becomes the major photosynthetic organ, plus a reservoir for water storage.

Water table low

Seaweeds, which are protoctists, not plants, tolerate drying between tides even though they have no xeromorphic features.

A waxy coating of **suberin** on mangrove roots excludes 97% of salt from the water.

Dry Desert Plant

Desert plants, such as cacti, must cope with low or sporadic rainfall and high transpiration rates. A number of structural adaptations (diagram left) reduce water losses, and enable them to access and store available water. Adaptations such as waxy leaves also reduce water loss and, in many desert plants, germination is triggered only by a certain quantity of rainfall.

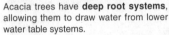

Acacia trees have **deep root systems**, allowing them to draw water from lower water table systems.

The outer surface of many succulents are coated in fine hairs, which traps air close to the surface reducing transpiration rate.

Ocean Margin Plant

Land plants that colonise the shoreline must have adaptations to obtain water from their saline environment while maintaining their osmotic balance. In addition, the shoreline is often a windy environment, so they frequently show xeromorphic adaptations that enable them to reduce transpirational water losses.

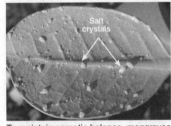

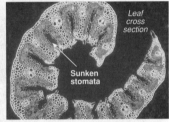

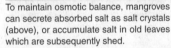

To maintain osmotic balance, mangroves can secrete absorbed salt as salt crystals (above), or accumulate salt in old leaves which are subsequently shed.

Grasses found on shoreline coasts (where it is often windy), curl their leaves and have sunken stomata to reduce water loss by transpiration.

Methods of water conservation in various plant species

Adaptation for water conservation	Effect of adaptation	Example
Thick, waxy cuticle to stems and leaves	Reduces water loss through the cuticle.	*Pinus* sp. ivy (*Hedera*), sea holly (*Eryngium*), prickly pear (*Opuntia*).
Reduced number of stomata	Reduces the number of pores through which water loss can occur.	Prickly pear (*Opuntia*), *Nerium* sp.
Stomata sunken in pits, grooves, or depressions Leaf surface covered with fine hairs Massing of leaves into a rosette at ground level	Moist air is trapped close to the area of water loss, reducing the diffusion gradient and therefore the rate of water loss.	**Sunken stomata**: *Pinus* sp., *Hakea* sp. **Hairy leaves**: lamb's ear. **Leaf rosettes**: dandelion (*Taraxacum*), daisy.
Stomata closed during the light, open at night	CAM metabolism: CO_2 is fixed during the night, water loss in the day is minimised.	**CAM plants**, e.g. American aloe, pineapple, *Kalanchoe*, *Yucca*.
Leaves reduced to scales, stem photosynthetic Leaves curled, rolled, or folded when flaccid	Reduction in surface area from which transpiration can occur.	**Leaf scales**: broom (*Cytisus*). **Rolled leaf**: marram grass (*Ammophila*), *Erica* sp.
Fleshy or succulent stems Fleshy or succulent leaves	When readily available, water is stored in the tissues for times of low availability.	**Fleshy stems**: *Opuntia*, candle plant (*Kleinia*). **Fleshy leaves**: *Bryophyllum*.
Deep root system below the water table	Roots tap into the lower water table.	Acacias, oleander.
Shallow root system absorbing surface moisture	Roots absorb overnight condensation.	Most cacti

Periodicals:
Cacti

Related activities: *Transpiration*
Web links: *Desert Plant Survival, Some Adaptations to Habitat*

Adaptations in halophytes and drought tolerant plants

Ice plant (*Carpobrotus*): The leaves of many desert and beach dwelling plants are fleshy or succulent. The leaves are triangular in cross section and crammed with water storage cells. The water is stored after rain for use in dry periods. The shallow root system is able to take up water from the soil surface, taking advantage of any overnight condensation.

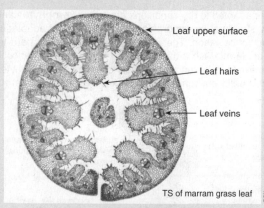

Leaf upper surface

Leaf hairs

Leaf veins

TS of marram grass leaf

Marram grass (*Ammophila*): The long, wiry leaf blades of this beach grass are curled downwards with the stomata on the inside. This protects them against drying out by providing a moist microclimate around the stomata. Plants adapted to high altitude often have similar adaptations.

Ball cactus (*Echinocactus grusonii*): In many cacti, the leaves are modified into long, thin spines which project outward from the thick fleshy stem. This reduces the surface area over which water loss can occur. The stem stores water and takes over as the photosynthetic organ. As in succulents, a shallow root system enables rapid uptake of surface water.

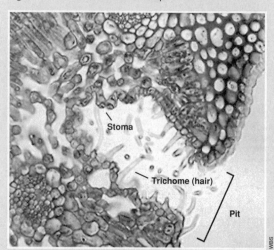

Stoma

Trichome (hair)

Pit

Oleander is a xerophyte from the mediterranean region with many water conserving features. It has a thick multi-layered epidermis and the stomata are sunken in trichome-filled pits on the leaf underside. The pits restrict water loss to a greater extent than they reduce uptake of carbon dioxide.

1. Explain the purpose of **xeromorphic** adaptations: _____

_____

2. Describe three xeromorphic adaptations of plants:

(a) _____

(b) _____

(c) _____

3. Describe a physiological mechanism by which plants can reduce water loss during the daylight hours:

_____

_____

4. Explain why creating a moist microenvironment around the areas of water loss reduces transpiration rate:

_____

_____

5. Explain why seashore plants (halophytes) exhibit many desert-dwelling adaptations: _____

_____

_____

Translocation

Phloem transports the organic products of photosynthesis (sugars) through the plant in a process called **translocation**. In angiosperms, the sugar moves through the sieve elements, which are arranged end-to-end and perforated with sieve plates. Apart from water, phloem sap comprises mainly sucrose (up to 30%). It may also contain minerals, hormones, and amino acids, in transit around the plant. Movement of sap in the phloem is from a **source** (a plant organ where sugar is made or mobilised) to a **sink** (a plant organ where sugar is stored or used). Loading sucrose into the phloem at a source involves energy expenditure; it is slowed or stopped by high temperatures or respiratory inhibitors. In some plants, unloading the sucrose at the sinks also requires energy, although in others, diffusion alone is sufficient to move sucrose from the phloem into the cells of the sink organ.

Transport in the Phloem by Pressure-Flow

Phloem sap moves from source (region where sugar is produced or mobilised) to sink (region where sugar is used or stored) at rates as great as 100 m h^{-1}: too fast to be accounted for by cytoplasmic streaming. The most acceptable model for phloem movement is the **pressure-flow** (bulk flow) hypothesis. Phloem sap moves by bulk flow, which creates a pressure (hence the term "pressure-flow"). The key elements in this model are outlined below and in steps 1-4 right. For simplicity, the cells that lie between the source or sink cells and the phloem sieve-tube have been omitted.

1 Loading sugar into the phloem from a source (e.g. leaf cell) increases the solute concentration (decreases the water potential, ψ) inside the sieve-tube cells. This causes the sieve-tubes to take up water from the surrounding tissues by osmosis.

2 The water absorption creates a hydrostatic pressure that forces the sap to move along the tube (bulk flow), just as pressure pushes water through a hose.

3 The gradient of pressure in the sieve tube is reinforced by the active unloading of sugar and consequent loss of water by osmosis at the sink (e.g. root cell).

4 Xylem recycles the water from sink to source.

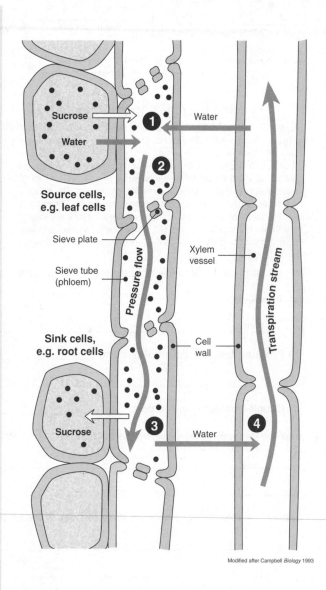

Modified after Campbell *Biology* 1993

Measuring Phloem Flow

Experiments investigating flow of phloem often use aphids. Aphids feed on phloem sap (left) and act as natural **phloem probes**. When the mouthparts (stylet) of an aphid penetrate a sieve-tube cell, the pressure in the sieve-tube force-feeds the aphid. While the aphid feeds, it can be severed from its stylet, which remains in place in the phloem. The stylet serves as a tiny tap that exudes sap. Using different aphids, the rate of flow of this sap can be measured at different locations on the plant.

1. (a) Explain what is meant by '**source to sink**' flow in phloem transport: _____

_____

(b) Name the usual **source** and **sink** in a growing plant:

Source: _____ Sink: _____

(c) Name another possible **source** region in the plant and state when it might be important: _____

_____

(d) Name another possible **sink** region in the plant and state when it might be important: _____

_____

2. Explain why energy is required for translocation and where it is used: _____

_____

Periodicals:
High tension

Related activities: Xylem, Phloem, Active and Passive Transport
Web links: Sucrose Transport

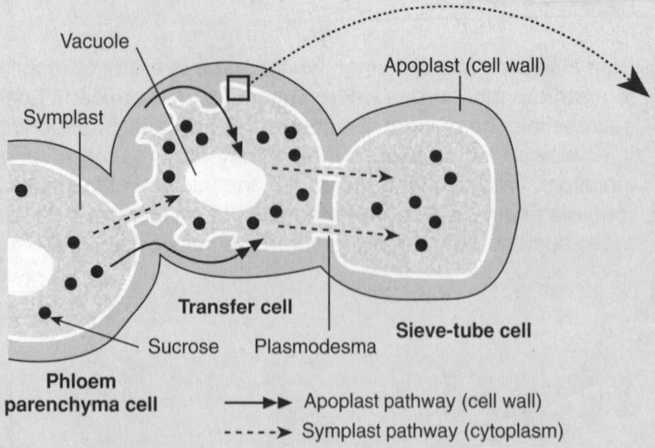

Loading Sucrose into the Phloem

Sugar (sucrose) can travel to the phloem sieve-tubes through both apoplastic and symplastic pathways. It is loaded into the phloem sieve-tube cells via modified companion cells, called **transfer cells** (above). Loading sucrose into the phloem requires active transport. Using a **coupled transport** (secondary pump) mechanism (right), transfer cells expend energy to accumulate the sucrose. The sucrose then passes into the sieve tube through plasmodesmata. The transfer cells have wall ingrowths that increase surface area for the transport of solutes. Using this mechanism, some plants can accumulate sucrose in the phloem to 2-3 times the concentration in the mesophyll.

Above: Proton pumps generate a hydrogen ion gradient across the membrane of the transfer cell. This process requires expenditure of energy. The gradient is then used to drive the transport of sucrose, by coupling the sucrose transport to the diffusion of H^+ back into the cell.

3. In your own words, describe what is meant by the following:

(a) Translocation: _____

_____

(b) Pressure-flow movement of phloem: _____

_____

(c) Coupled transport of sucrose: _____

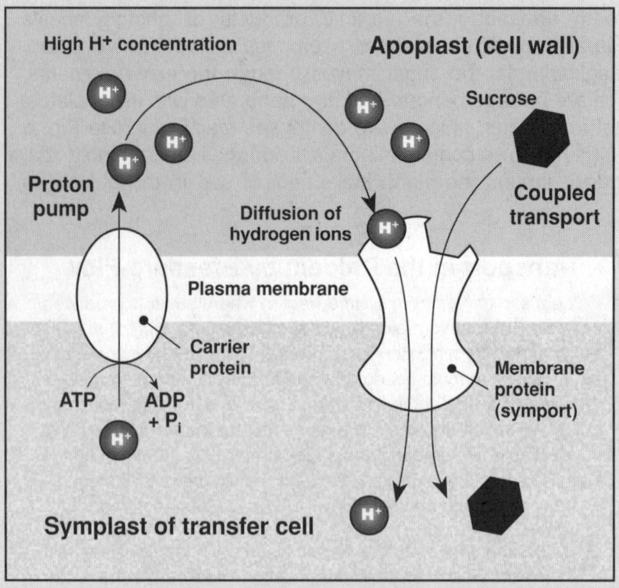

4. Briefly explain why water follows the sucrose as the sucrose is loaded into the phloem sieve-tube cell:

_____

_____

5. Explain the role of the companion (transfer) cell in the loading of sucrose into the phloem: _____

_____

_____

6. Contrast the composition of the fluid in the phloem and xylem (see the activities on xylem and phloem if you need help):

_____

_____

_____

7. Explain why it is necessary for phloem to be alive to be functional, whereas xylem can function as a dead tissue:

_____

_____

_____

8. The sieve plate represents a significant barrier to effective mass flow of phloem sap. Suggest why the presence of the sieve plate is often cited as evidence against the pressure-flow model for phloem transport:

_____

_____

_____

KEY TERMS: Mix and Match

INSTRUCTIONS: *Test your vocab by matching each term to its correct definition, as identified by its preceding letter code.*

APOPLASTIC PATHWAY

BULK FLOW

CAPILLARY ACTION

CASPARIAN STRIP

COHESION-TENSION HYPOTHESIS

COMPANION CELLS

CORTEX

ENDODERMIS

GUARD CELLS

MASS FLOW HYPOTHESIS

OSMOSIS

PHLOEM

ROOT HAIR

ROOT PRESSURE

SIEVE TUBE CELLS

SINK

SOURCE

STOMATA

SYMPLASTIC PATHWAY

TRANSLOCATION

TRANSPIRATION

TRANSPIRATION PULL

TURGOR

VESSEL

WATER POTENTIAL

XYLEM

A Any part of the plant that is using or storing sugar.

B Water conducting cells in the xylem of angiosperms but absent from most gymnosperms.

C The passive movement of water molecules across a partially permeable membrane down a gradient in water potential.

D The outer layer of a plant stem or root, bounded on the outside by the epidermis and on the inside by the endodermis.

E One proposed mechanism for the movement of sugars from source to sink in the phloem.

F A measure of the potential energy of water per unit volume relative to pure water. It quantifies the tendency of water to move from one area to another.

G Complex plant tissue specialised for the transport of water and dissolved mineral ions.

H Elongated cells in phloem for transporting carbohydrate (sugar).

I Living cells in close association with the sieve-tube members in phloem.

J Specialised cells, which occur in pairs and which regulate movement of gases and water vapour through the stomata.

K Pores in the leaf surface through which gases can move.

L The tendency of fluids in narrow tubes to move upwards, against the pull of gravity.

M The hypothesis for the movement of water through the plant based on transpiration pull and the cohesive properties of water.

N Pathway for water and mineral ions through the root via the spaces within the cellulose cell walls.

O Movement of water and solutes together as a single mass due to a pressure gradient.

P A thin layer of parenchyma tissue found just outside the vascular cylinder in roots, which helps to regulate the passive movements of water and ions,.

Q Vascular tissue that conducts sugars through the plant. Characterised by the presence of sieve tubes.

R The transport of materials within a plant by the phloem.

S Any part of the plant that is producing or releasing sugar.

T Pathway for water and mineral ions through the root via the living contents of the cells.

U A force resulting from the evaporation of water from the surfaces of cells in the interior of the leaves.

V The pressure of the cell contents that provides support to plant cells.

W Osmotic pressure within the cells of a root system that aids the movement of water through the plant. It occurs when the soil moisture level is high during the night or when transpiration is low during the day.

X A tubular outgrowth of an epidermal cell on a plant root.

Y A band of waterproof material in the radial and transverse walls of the endodermis, which blocks the passive movement of materials, such as water and solutes into the stele.

Z The loss of water vapour by plants, mainly from leaves via the stomata.

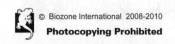

Biological Molecules

Water and inorganic ions	• The structure and properties of water • Water as the universal solvent • Inorganic ions essential for life
Amino acids and proteins	• The structure of amino acids • Protein structure: 1°, 2°, 3°, 4° • Enzymes as biological catalysts • Enzyme cofactors and inhibitors
Lipids and carbohydrates	• Structure and roles of carbohydrates • Structure and roles of lipids • Condensation and hydrolysis
Nucleic acids	• The structure and role of DNA • DNA replication • mRNA: genes to proteins

Important in this section...

• *Understand the structure and role of biological molecules*
• *Explain how pathogens and lifestyle can cause disease*
• *Understand the basis for classifying organisms*
• *Describe conservation strategies and their importance*

Food and Health

Diet and health	• Human global nutrition • Diet and malnutrition • Malnutrition and obesity • Cardiovascular disease and lifestyle
Increasing food production	• The green revolution • Selective breeding • Producing food with microbes • High input systems for food production

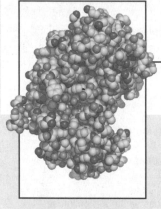

Four elements (H,C O, and N) make up 99% of the substance of all organisms. These form macromolecules of biological importance.

Malnutrition is associated with poor health. Increasing food production is a challenge facing modern societies.

Molecules, Biodiversity, Food and Health

The chemistry of life is based around a relatively small number of organic compounds.

Humans require a balanced diet but providing this on a global scale is a challenge. Pathogens causes disease but some diseases are the result of many interacting risk factors.

Identifying and conserving biodiversity will benefit humans both directly and indirectly.

The body's defences protect against diseases caused by pathogens. Many non-infectious diseases can be attributed to smoking.

Evolution is responsible for the biodiversity on Earth. Identifying and conserving biodiversity is a high priority for sound reasons.

Pathogens and human disease	• The nature of pathogens • Bacterial pathogens: TB and cholera • Viral pathogens: HIV and influenza • Protozoan pathogens: malaria
Defence and the immune system	• The body's non-specific defences • Specific defence: the immune system • Vaccines and vaccination • New medicines
The health effects of smoking	• Respiratory diseases • Smoking and lung disease • Smoking and cardiovascular disease • The epidemiological evidence

Biodiversity and the classification of organisms	• What is biodiversity? • Quantifying diversity: diversity indices • Classification systems new and old • Classification keys
Evolution: producing diversity on Earth	• What is variation? • Evolution by natural selection • Evidence for evolution: then and now • Speciation
Maintaining and restoring biodiversity	• Diversity and ecosystem stability • The potential impact of global warming • Agricultural benefits of higher diversity • Strategies for conserving biodiversity

Pathogens, Disease, & Defence

Biodiversity and Evolution

Biological Molecules

KEY CONCEPTS

▶ Organic molecules are central to living systems.

▶ Water's properties make it essential to life.

▶ Protein, carbohydrates, and lipids are three key groups of biological macromolecules.

▶ Enzymes are biological catalysts and regulate the metabolism of cells.

▶ Enzyme activity may depend on the presence of cofactors. Inhibition can regulate enzyme activity.

KEY TERMS

activation energy
active site
amino acid
amphipathic
anabolism
Benedict's test
biuret test
carbohydrate
catabolism
catalyst
cholesterol
coenzyme
cofactor
colorimetry
condensation
denaturation
disaccharide
emulsion test
endergonic reaction
enzyme
enzyme inhibition
exergonic reaction
fibrous protein
hydrolysis
globular protein
hydrogen bond
I2/KI test
inorganic ion
lipid
macromolecule
monomer
monosaccharide
optimum (for enzyme)
organic molecule
polar molecule
polymer
polysaccharide
protein

OBJECTIVES

☐ 1. Use the **KEY TERMS** to help you understand and complete these objectives.

Biological Molecules   pages 138-154, 161

☐ 2. Describe the importance of **inorganic ions** and **organic molecules** (e.g. **carbohydrates, proteins, lipids,** and **nucleic acids**) in biological systems.

☐ 3. Describe the structure of **water**. Explain the physical and chemical properties of water that are important in biological systems.

☐ 4. Describe the general structure and formula of an **amino acid**. Explain the basis for the different properties of amino acids.

☐ 5. Describe how **peptide bonds** are formed and broken in the synthesis (by **condensation**) and **hydrolysis** of **dipeptides** and **polypeptides**.

☐ 6. Distinguish between a protein's **primary structure** and **secondary structure**. Explain the role of **hydrogen bonding** in the secondary structure.

☐ 7. Explain how the **tertiary structure** of a protein arises. Describe how the tertiary structure of a globular protein (e.g. enzyme) is related to function.

☐ 8. Explain the **quaternary structure** of a protein with reference to haemoglobin.

☐ 9. Describe the structural and functional diversity of proteins, including reference to **globular** e.g. haemoglobin, and **fibrous** (e.g. collagen) **proteins**.

☐ 10. Describe **isomers** of **glucose**. Describe how **glycosidic bonds** are formed and broken in the synthesis (by **condensation**) and **hydrolysis** of a **disaccharide** (maltose) and a **polysaccharide** (amylose).

☐ 11. Compare and contrast the structure of glucose polymers: **starch, cellulose,** and **glycogen**. Relate the structure to their biological function in each case.

☐ 12. Describe the structure of **triglycerides** (fats and oils), **phospholipids,** and **cholesterol**. Relate the structure to their biological function in each case.

☐ 13. Describe simple tests for reducing and non-reducing sugars, starch, proteins, and lipids. Describe the use of **colorimetry** to determine the concentration of a reducing sugar (e.g. glucose) in a solution.

Enzymes   pages 155-160

☐ 14. Describe the properties and mode of action of **enzymes**, including the role of the **active site, specificity,** and **activation energy**. Explain models for enzyme function and outline role of enzymes in **anabolism** and **catabolism**.

☐ 15. Describe the effect of pH, temperature, substrate concentration, and enzyme concentration on enzyme activity. Recognise that enzymes can be **denatured**.

☐ 16. Using examples, explain the role of **cofactors** in enzyme activity.

☐ 17. Explain the effect of **competitive** and **non-competitive inhibitors** on enzyme activity. Distinguish between reversible and non-reversible inhibition.

Periodicals:

Listings for this chapter are on page 339

Weblinks:

www.biozone.co.uk/ weblink/OCR-AS-2641.html

Teacher Resource CD-ROM:

Applications of Enzymes

The Biochemical Nature of the Cell

Water is the main component of organisms, and provides an equable environment in which metabolic reactions can occur. Apart from water, most other substances in cells are compounds of carbon, hydrogen, oxygen, and nitrogen. The combination of carbon atoms with the atoms of other elements provides a huge variety of molecular structures, collectively called **organic** **molecules**. The organic molecules that make up living things can be grouped into four broad classes: carbohydrates, lipids, proteins, and nucleic acids. These are discussed in more detail in subsequent activities in this chapter. In addition, a small number of **inorganic ions** are also essential for life as components of larger molecules or extracellular fluids.

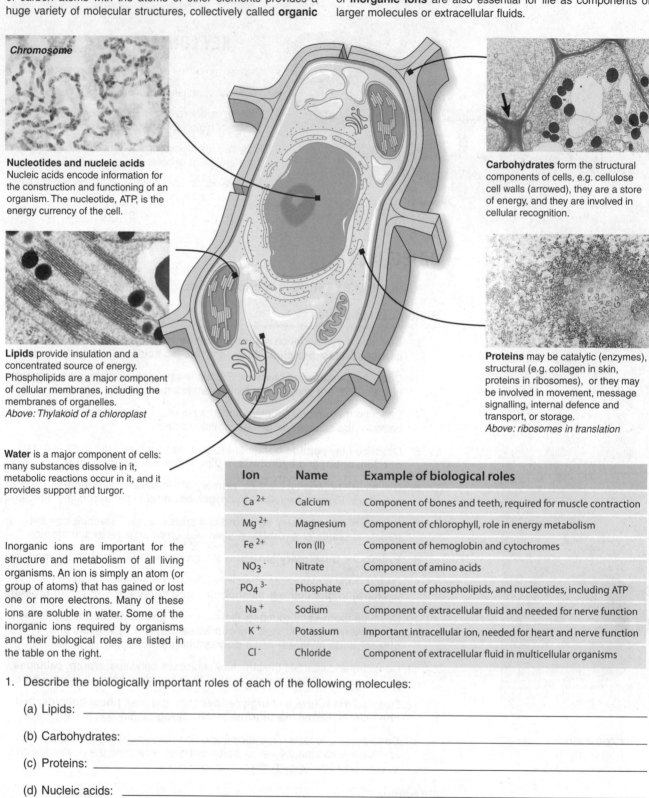

Chromosome

Nucleotides and nucleic acids
Nucleic acids encode information for the construction and functioning of an organism. The nucleotide, ATP, is the energy currency of the cell.

Lipids provide insulation and a concentrated source of energy. Phospholipids are a major component of cellular membranes, including the membranes of organelles.
Above: Thylakoid of a chloroplast

Water is a major component of cells: many substances dissolve in it, metabolic reactions occur in it, and it provides support and turgor.

Carbohydrates form the structural components of cells, e.g. cellulose cell walls (arrowed), they are a store of energy, and they are involved in cellular recognition.

Proteins may be catalytic (enzymes), structural (e.g. collagen in skin, proteins in ribosomes), or they may be involved in movement, message signalling, internal defence and transport, or storage.
Above: ribosomes in translation

Inorganic ions are important for the structure and metabolism of all living organisms. An ion is simply an atom (or group of atoms) that has gained or lost one or more electrons. Many of these ions are soluble in water. Some of the inorganic ions required by organisms and their biological roles are listed in the table on the right.

Ion	Name	Example of biological roles
Ca^{2+}	Calcium	Component of bones and teeth, required for muscle contraction
Mg^{2+}	Magnesium	Component of chlorophyll, role in energy metabolism
Fe^{2+}	Iron (II)	Component of hemoglobin and cytochromes
NO_3^-	Nitrate	Component of amino acids
PO_4^{3-}	Phosphate	Component of phospholipids, and nucleotides, including ATP
Na^+	Sodium	Component of extracellular fluid and needed for nerve function
K^+	Potassium	Important intracellular ion, needed for heart and nerve function
Cl^-	Chloride	Component of extracellular fluid in multicellular organisms

1. Describe the biologically important roles of each of the following molecules:

    (a) Lipids: _____

    (b) Carbohydrates: _____

    (c) Proteins: _____

    (d) Nucleic acids: _____

2. Using examples, distinguish between inorganic and organic molecules and their general roles in cells:

    _____

    _____

    _____

Related activities: *Organic Molecules, Water and Inorganic Ions*
Web links: *A Closer look at Water*

The Role of Water

Water is the most abundant of the smaller molecules making up living things, and typically makes up about two-thirds of any organism. Water is a liquid at room temperature and many substances dissolve in it. It is a medium inside cells and for aquatic life. Water takes part in, and is a common product of, many reactions. Water molecules are **polar** and have a weak attraction for each other and inorganic ions, forming large numbers weak hydrogen bonds. It is this feature that gives water many of its unique properties, including its low viscosity and its chemical behaviour as a **universal solvent**.

Important Properties of Water

A lot of energy is required before water will change state so aquatic environments are thermally stable and sweating and transpiration cause rapid cooling.

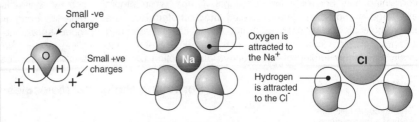

Water molecule
Formula: H_2O

Water surrounding a positive ion (Na^+)

Oxygen is attracted to the Na^+

Water surrounding a negative ion (Cl^-)

Hydrogen is attracted to the Cl^-

Small -ve charge

Small +ve charges

The most important feature of the chemical behaviour of water is its **dipole** nature. It has a small positive charge on each of the two hydrogens and a small negative charge on the oxygen.

Water is colourless, with a high transmission of visible light, so light penetrates tissue and aquatic environments.

Ice is less dense than water. Consequently ice floats, insulating the underlying water and providing valuable habitat.

Water has low viscosity, strong cohesive properties, and high surface tension. It can flow freely through small spaces.

Biological Molecules

1. On the diagram above, showing a positive and a negative ion surrounded by water molecules, draw the positive and negative charges on the water molecules (as shown in the example provided).

2. Explain the importance of the **dipole nature** of water molecules to the chemistry of life: _____
   _____
   _____

3. For (a)-(f), identify the important property of water, and describe an example of that property's biological significance:

   (a) Property important in the clarity of seawater: _____

   Biological significance: _____

   (b) Property important in the transport of water in xylem: _____

   Biological significance: _____

   (c) Property important in the relatively stable temperature of water bodies: _____

   Biological significance: _____

   (d) Property important in the transport of glucose around the body: _____

   Biological significance: _____

   (e) Property important in the cooling effect of evaporation: _____

   Biological significance: _____

   (f) Property important in ice floating: _____

   Biological significance: _____

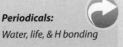
Periodicals:
Water, life, & H bonding

Related activities: Biochemical Nature of the Cell, Organic Molecules
Web links: Hydrogen Bonds and Water, Water and pH

RA 2

Organic Molecules

Organic molecules are those chemical compounds containing carbon that are found in living things. Specific groups of atoms, called **functional groups**, attach to a carbon-hydrogen core and confer specific chemical properties on the molecule. Some organic molecules in organisms are small and simple, containing only one or a few functional groups, while others are large complex assemblies called **macromolecules**. The macromolecules that make up living things can be grouped into four classes: carbohydrates, lipids, proteins, and nucleic acids. An understanding of the structure and function of these molecules is necessary to many branches of biology, especially biochemistry, physiology, and molecular genetics. The diagram below illustrates some of the common ways in which biological molecules are portrayed. Note that the **molecular formula** expresses the number of atoms in a molecule, but does not convey its structure; this is indicated by the **structural formula**. Molecules can also be represented as **models**. A ball and stick model shows the arrangement and type of bonds while a space filling model gives a more realistic appearance of a molecule, showing how close the atoms really are.

Portraying Biological Molecules

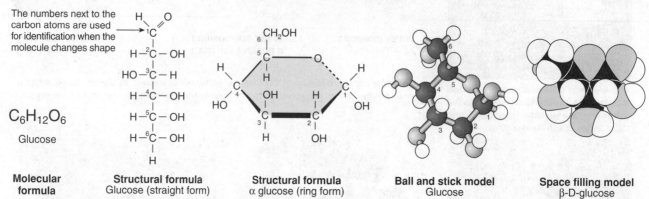

The numbers next to the carbon atoms are used for identification when the molecule changes shape

$C_6H_{12}O_6$
Glucose

Molecular formula

Structural formula Glucose (straight form)

Structural formula α glucose (ring form)

Ball and stick model Glucose

Space filling model β-D-glucose

Biological molecules may also include atoms other than carbon, oxygen, and hydrogen atoms. Nitrogen and sulfur are components of molecules such as amino acids and nucleotides. Some molecules contain the **C=O** (carbonyl) group. If this group is joined to at least one hydrogen atom it forms an **aldehyde**. If it is located between two carbon atoms, it forms a **ketone**.

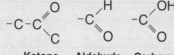

Ketone   **Aldehyde**   **Carboxyl**

Examples of Biological Molecules

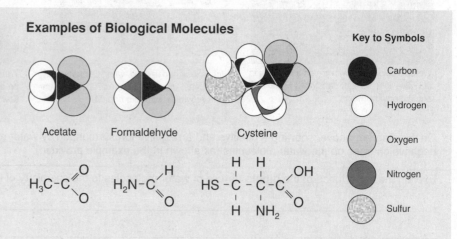

Acetate   Formaldehyde   Cysteine

Key to Symbols

● Carbon
○ Hydrogen
○ Oxygen
○ Nitrogen
○ Sulfur

1. Identify the three main elements comprising the structure of organic molecules: _____

2. Name two other elements that are also frequently part of organic molecules: _____

3. State how many covalent bonds a carbon atom can form with neighbouring atoms: _____

4. Distinguish between molecular and structural formulae for a given molecule: _____
   _____
   _____

5. Describe what is meant by a functional group: _____
   _____

6. Classify formaldehyde according to the position of the C=O group: _____

7. Identify a functional group always present in amino acids: _____

8. Identify the significance of cysteine in its formation of disulfide bonds: _____
   _____

Related activities: Biochemical Nature of the Cell, Amino Acids, Proteins

Amino Acids

Amino acids are the basic units from which proteins are made. Plants can manufacture all the amino acids they require from simpler molecules, but animals must obtain a certain number of ready-made amino acids (called **essential amino acids**) from their diet. The distinction between essential and non-essential amino acids is somewhat unclear though, as some amino acids can be produced from others and some are interconvertible by the urea cycle. Amino acids can combine to form peptide chains in a **condensation reaction**. The reverse reaction, the hydrolysis of peptide chains, releases free water and single amino acids.

Structure of Amino Acids

There are over 150 amino acids found in cells, but only 20 occur commonly in proteins. The remaining, non-protein amino acids have specialised roles as intermediates in metabolic reactions, or as neurotransmitters and hormones. All amino acids have a common structure (see right). The only difference between the different types lies with the 'R' group in the general formula. This group is variable, which means that it is different in each kind of amino acid.

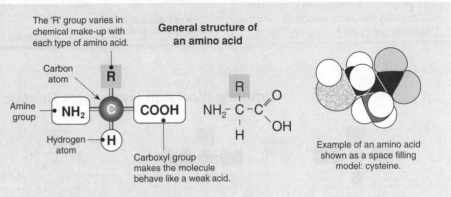

The 'R' group varies in chemical make-up with each type of amino acid.

General structure of an amino acid

Carbon atom — R

Amine group — NH_2 — C — COOH

Hydrogen atom — H

Carboxyl group makes the molecule behave like a weak acid.

Example of an amino acid shown as a space filling model: cysteine.

Properties of Amino Acids

Three examples of amino acids with different chemical properties are shown right, with their specific 'R' groups outlined. The 'R' groups can have quite diverse chemical properties.

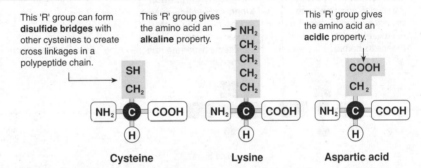

This 'R' group can form **disulfide bridges** with other cysteines to create cross linkages in a polypeptide chain.

This 'R' group gives the amino acid an **alkaline** property.

This 'R' group gives the amino acid an **acidic** property.

Cysteine   **Lysine**   **Aspartic acid**

A polypeptide chain

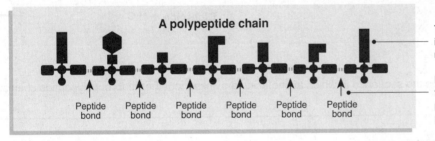

Peptide bond   Peptide bond   Peptide bond   Peptide bond   Peptide bond   Peptide bond

The order of amino acids in a protein is directed by the order of nucleotides in DNA and mRNA.

Peptide bonds link amino acids together in long polymers called polypeptide chains. These may form part or all of a protein.

The amino acids are linked together by peptide bonds to form long chains of up to several hundred amino acids (called polypeptide chains). These chains may be functional units (complete by themselves) or they may need to be joined to other polypeptide chains before they can carry out their function. In humans, not all amino acids can be manufactured by our body: ten must be taken in with our diet (eight in adults). These are the 'essential amino acids'. They are indicated by the symbol ◆ on the right. Those indicated with as asterisk are also required by infants.

Amino acids occurring in proteins

Alanine	Glycine	Proline
Arginine *	Histidine *	Serine
Asparagine	Isoleucine ◆	Threonine ◆
Aspartic acid	Leucine ◆	Tryptophan ◆
Cysteine	Lysine ◆	Tyrosine
Glutamine	Methionine ◆	Valine ◆
Glutamic acid	Phenylalanine ◆	

1. Describe the biological function of amino acids: _____

_____

_____

_____

2. Describe what makes each of the 20 amino acids found in proteins unique: _____

_____

_____

_____

Related activities: Organic Molecules, Proteins
Web links: Amino Acids and Proteins

A 2

Biological Molecules

Optical Isomers of Amino Acids

All amino acids, apart from the simplest one (glycine) show optical isomerism. The two forms that these optical isomers can take relate to the arrangement of the four bonding sites on the carbon atom. This can result in two different arrangements as shown on the diagrams on the right. With a very few minor exceptions, only the **L-forms** are found in living organisms.

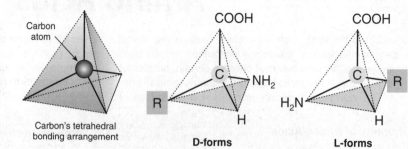

Carbon atom

Carbon's tetrahedral bonding arrangement

D-forms **L-forms**

Condensation and Hydrolysis Reactions

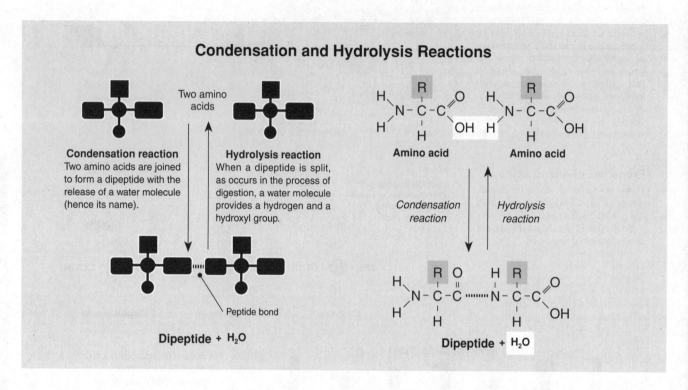

Condensation reaction
Two amino acids are joined to form a dipeptide with the release of a water molecule (hence its name).

Hydrolysis reaction
When a dipeptide is split, as occurs in the process of digestion, a water molecule provides a hydrogen and a hydroxyl group.

Peptide bond

Dipeptide + H₂O

Amino acid   Amino acid

Condensation reaction   *Hydrolysis reaction*

Dipeptide + H₂O

3. Describe the process that determines the sequence in which amino acids are linked together to form polypeptide chains:

_____

_____

4. Explain what is meant by **essential amino acids**: _____

_____

_____

5. Describe briefly the process of the **condensation** reaction for amino acids: _____

_____

_____

_____

6. Describe briefly the process of the **hydrolysis** reaction for amino acids: _____

_____

_____

_____

7. Name the optical isomeric form that occurs in nearly all amino acids in living things: _____

Proteins

The precise folding up of a protein into its **tertiary structure** creates a three dimensional arrangement of the active 'R' groups. The way each 'R' group faces with respect to the others gives the protein its unique chemical properties. If a protein loses this precise structure (denaturation), it is usually unable to carry out its biological function. Proteins are often classified on the basis of structure (globular vs fibrous). Some of the properties used for the basis of structural classification are outlined over the page.

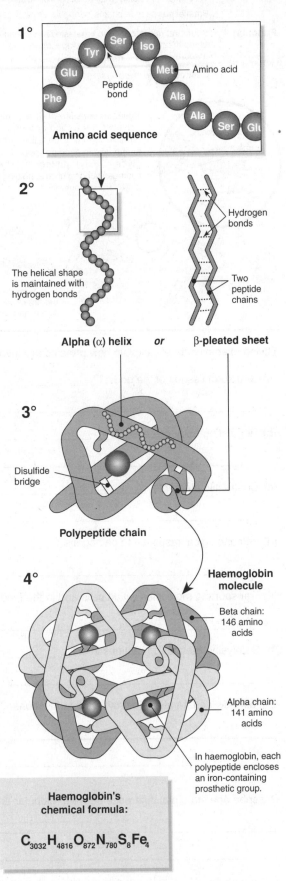

1° Amino acid

Peptide bond

Amino acid sequence

2° Hydrogen bonds

The helical shape is maintained with hydrogen bonds

Two peptide chains

Alpha (α) helix *or* **β-pleated sheet**

3°

Disulfide bridge

Polypeptide chain

4°

Haemoglobin molecule

Beta chain: 146 amino acids

Alpha chain: 141 amino acids

In haemoglobin, each polypeptide encloses an iron-containing prosthetic group.

Haemoglobin's chemical formula:

$$C_{3032}H_{4816}O_{872}N_{780}S_8Fe_4$$

Primary Structure - 1° *(amino acid sequence)*

Strings of hundreds of amino acids link together with peptide bonds to form molecules called polypeptide chains. There are 20 different kinds of amino acids that can be linked together in a vast number of different combinations. This sequence is called the **primary structure**. It is the arrangement of attraction and repulsion points in the amino acid chain that determines the higher levels of organisation in the protein and its biological function.

Secondary Structure - 2° *(α-helix or ß pleated sheet)*

Polypeptides become folded in various ways, referred to as the secondary (2°) structure. The most common types of 2° structures are a coiled α-**helix** and a β-**pleated sheet**. Secondary structures are maintained with hydrogen bonds between neighbouring CO and NH groups. H-bonds, although individually weak, provide considerable strength when there are a large number of them. The example, right, shows the two main types of secondary structure. In both, the **'R' side groups** (not shown) project out from the structure. Most globular proteins contain regions of α-helices together with β-sheets. Keratin (a fibrous protein) is composed almost entirely of α-helices. Fibroin (silk protein), is another fibrous protein, almost entirely in β-sheet form.

Tertiary Structure - 3° *(folding)*

Every protein has a precise structure formed by the folding of the secondary structure into a complex shape called the **tertiary structure**. The protein folds up because various points on the secondary structure are attracted to one another. The strongest links are caused by bonding between neighbouring **cysteine** amino acids which form disulfide bridges. Other interactions that are involved in folding include weak ionic and hydrogen bonds as well as hydrophobic interactions.

Quaternary Structure - 4°

Some proteins (such as enzymes) are complete and functional with a tertiary structure only. However, many complex proteins exist as aggregations of polypeptide chains. The arrangement of the polypeptide chains into a functional protein is termed the **quaternary structure**. The example (right) shows a molecule of haemoglobin, a globular protein composed of 4 polypeptide sub-units joined together; two identical **beta chains** and two identical **alpha chains**. Each has a haem (iron containing) group at the centre of the chain, which binds oxygen. Proteins containing non-protein material are **conjugated proteins**. The non-protein part is the **prosthetic group**.

Denaturation of Proteins

Denaturation refers to the loss of the three-dimensional structure (and usually also the biological function) of a protein. Denaturation is often, although not always, permanent. It results from an alteration of the bonds that maintain the secondary and tertiary structure of the protein, even though the sequence of amino acids remains unchanged. Agents that cause denaturation are:

- **Strong acids and alkalis**: Disrupt ionic bonds and result in coagulation of the protein. Long exposure also breaks down the primary structure of the protein.

- **Heavy metals**: May disrupt ionic bonds, form strong bonds with the carboxyl groups of the R groups, and reduce protein charge. The general effect is to cause the precipitation of the protein.

- **Heat and radiation** (e.g. UV): Cause disruption of the bonds in the protein through increased energy provided to the atoms.

- **Detergents and solvents**: Form bonds with the non-polar groups in the protein, thereby disrupting hydrogen bonding.

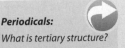
Biological Molecules

Periodicals: *What is tertiary structure?*

Related activities: *Enzymes, Modification of Proteins, Biochemical Tests*
Web links: *Amino Acids and Proteins*

RA 2

Structural Classification of Proteins

Fibrous Proteins

Properties
- Water insoluble
- Very tough physically; may be supple or stretchy
- Parallel polypeptide chains in long fibres or sheets

Function
- Structural role in cells and organisms *e.g. collagen found in connective tissue, cartilage, bones, tendons, and blood vessel walls.*
- Contractile *e.g. myosin, actin*

Collagen consists of three helical polypeptides wound around each other to form a 'rope'. Every third amino acid in each polypeptide is a glycine (Gly) molecule where hydrogen bonding occurs, holding the three strands together.

Hydrogen bond

Glycine

Fibres form due to cross links between collagen molecules.

Globular Proteins

Properties
- Easily water soluble
- Tertiary structure critical to function
- Polypeptide chains folded into a spherical shape

Function
- Catalytic *e.g. enzymes*
- Regulatory *e.g. hormones (insulin)*
- Transport *e.g. haemoglobin*
- Protective *e.g. antibodies*

α chain

Leu Tyr Gln Leu Gln Asn Tyr Cys Asn

disulfide bond

Ser Val Cys Gln

Cys Ser Val Leu Arg

β chain Tyr Gly

Gly Ile Val Glu Gln Cys Ala Leu Phe

Cys Ala Phe

Phe Val Asn Gln His Leu Cys Gly Ser His Leu Val Gln Tyr

Thr

Pro

Lys

Ala

Bovine insulin is a relatively small protein consisting of two polypeptide chains (an α chain and a β chain). These two chains are held together by disulfide bridges between neighbouring cysteine (Cys) molecules.

1. Giving examples, briefly explain how proteins are involved in the following functional roles:

   (a) Structural tissues of the body: _____

   _____

   (b) Regulating body processes: _____

   _____

   (c) Contractile elements: _____

   _____

   (d) Immunological response to pathogens: _____

   _____

   (e) Transporting molecules within cells and in the bloodstream: _____

   _____

   (f) Catalysing metabolic reactions in cells: _____

   _____

2. Explain how denaturation destroys protein function: _____

   _____

   _____

3. Describe one structural difference between globular and fibrous proteins: _____

   _____

   _____

4. Determine the total number of amino acids in the α and β chains of the insulin molecule illustrated above:

   (a) α chain: _____   (b) β chain: _____

Modification of Proteins

Proteins may be modified after they have been produced by ribosomes. After they pass into the interior of rough endoplasmic reticulum, some proteins may have carbohydrates added to them to form **glycoproteins**. Proteins may be further altered in the Golgi apparatus. The **Golgi apparatus** functions principally as a system for processing, sorting, and modifying proteins. Proteins that are to be secreted from the cell are synthesized by

ribosomes on the rough endoplasmic reticulum and transported to the Golgi apparatus. At this stage, carbohydrates may be removed or added in a step-wise process. Some of the possible functions of glycoproteins are illustrated below. Other proteins may have fatty acids added to them to form **lipoproteins**. These modified proteins transport lipids in the plasma between various organs in the body (e.g. gut, liver, and adipose tissue).

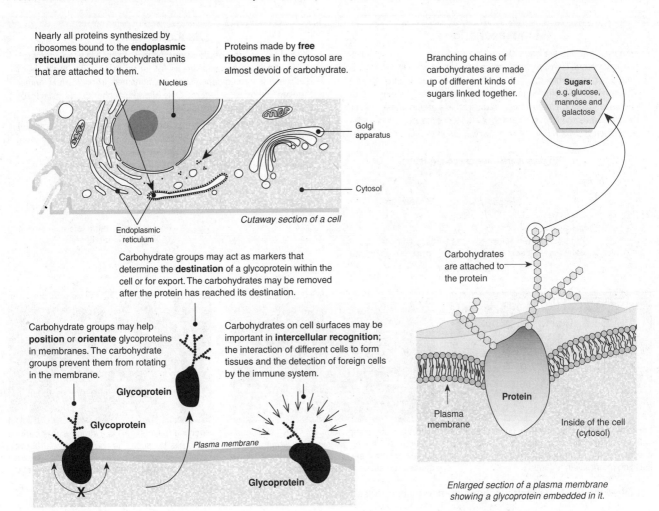

Nearly all proteins synthesized by ribosomes bound to the **endoplasmic reticulum** acquire carbohydrate units that are attached to them.

Proteins made by **free ribosomes** in the cytosol are almost devoid of carbohydrate.

Nucleus

Golgi apparatus

Cytosol

Cutaway section of a cell

Endoplasmic reticulum

Branching chains of carbohydrates are made up of different kinds of sugars linked together.

Sugars: e.g. glucose, mannose and galactose

Carbohydrates are attached to the protein

Carbohydrate groups may act as markers that determine the **destination** of a glycoprotein within the cell or for export. The carbohydrates may be removed after the protein has reached its destination.

Carbohydrate groups may help **position** or **orientate** glycoproteins in membranes. The carbohydrate groups prevent them from rotating in the membrane.

Glycoprotein

Glycoprotein

Carbohydrates on cell surfaces may be important in **intercellular recognition**; the interaction of different cells to form tissues and the detection of foreign cells by the immune system.

Plasma membrane

Glycoprotein

X

Plasma membrane

Protein

Inside of the cell (cytosol)

Enlarged section of a plasma membrane showing a glycoprotein embedded in it.

Biological Molecules

1. (a) Explain what a **glycoprotein** is: _____

    (b) Briefly describe three **roles** of glycoproteins: _____

    _____

    _____

2. (a) Explain what a **lipoprotein** is: _____

    (b) Briefly describe the **role** of lipoproteins: _____

    _____

    _____

3. Suggest why proteins made by free ribosomes in the cytosol are usually free of carbohydrate: _____

    _____

4. Explain why the orientation of a protein in the plasma membrane might be important: _____

    _____

    _____

Related activities: The Role of Membranes in Cells, Packaging Proteins

RA 2

Monosaccharides and Disaccharides

Carbohydrates are a family of organic molecules made up of carbon, hydrogen, and oxygen atoms with the general formula $(CH_2O)_x$. The most common arrangements found in sugars are hexose (6 sided) or pentose (5 sided) rings. Carbohydrates are versatile macromolecules. The way in which carbohydrate monomers are linked together provides a variety of structurally and functionally different molecules. Sugars play a central role in cells, providing energy and, in some cells, contributing to support. They are the major component of most plants (60-90% of the dry weight) and are used by humans as a cheap food source, and a source of fuel, housing, and clothing. Disaccharides important in human nutrition include lactose, sucrose and maltose.

Monosaccharides

Monosaccharides are used as a primary energy source for fuelling cell metabolism. They are **single-sugar** molecules and include glucose (grape sugar and blood sugar) and fructose (honey and fruit juices). The commonly occurring monosaccharides contain between three and seven carbon atoms in their carbon chains and, of these, the 6C hexose sugars occur most frequently. All monosaccharides are classified as **reducing** sugars (i.e. they can participate in reduction reactions).

Single sugars (monosaccharides)

Triose

C—C—C

e.g. glyceraldehyde

Pentose

e.g. ribose, deoxyribose

Hexose

e.g. glucose, fructose, galactose

Disaccharides

Disaccharides are **double-sugar** molecules and are used as energy sources and as building blocks for larger molecules. The type of disaccharide formed depends on the monomers involved and whether they are in their α- or β- form. Only a few disaccharides (e.g. lactose) are classified as reducing sugars.

Sucrose = α-glucose + β-fructose (simple sugar found in plant sap)
Maltose = α-glucose + α-glucose (a product of starch hydrolysis)
Lactose = β-glucose + β-galactose (milk sugar)
Cellobiose = β-glucose + β-glucose (from cellulose hydrolysis)

Double sugars (disaccharides)

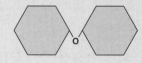

Examples sucrose, lactose, maltose, cellobiose

Lactose, a milk sugar, is made up of β-glucose + β-galactose. Milk contains 2-8% lactose by weight. It is the primary carbohydrate source for suckling mammalian infants.

Maltose is composed of two α-glucose molecules. These germinating wheat seeds contain maltose because the plant breaks down their starch stores to use it for food.

Sucrose (table sugar) is a simple sugar derived from plants such as sugar cane (above), sugar beet, or maple sap. It is composed of an α-glucose molecule and a β-fructose molecule.

1. Describe the two major functions of monosaccharides:

   (a) _____

   _____

   (b) _____

   _____

2. The breakdown of a disaccharide into its constituent monosaccharide units is an enzyme catalysed hydrolysis (see opposite). For each of the following common dissacharides, identify the enzyme responsible for the catalysis and the products of the hydrolysis, and describe an example of where this enzyme might naturally occur:

   (a) Lactose: Enzyme: _____ Products of hydrolysis: _____

   Found: _____

   (b) Maltose: Enzyme: _____ Products of hydrolysis: _____

   Found: _____

   (c) Sucrose: Enzyme: _____ Products of hydrolysis: _____

   Found: _____

3. Use your understanding of disaccharide chemistry to suggest how the digestive disorder lactose intolerance arises:

   _____

A 1

Related activities: Biochemical Tests, Carbohydrate Chemistry
Web links: Biomolecules: Carbohydrates

Periodicals:
Glucose & glucose
containing carbohydrates

© Biozone International 2008-2010
Photocopying Prohibited

Carbohydrate Chemistry

Monomers are linked together by **condensation reactions**, so called because linking two units together results in the production of a water molecule. The reverse reaction, in which compound sugars are broken down into their constituent monosaccharides, is called **hydrolysis**. It splits polymers into smaller units by breaking the bond between two monomers. Hydrolysis literally means breaking with water, and so requires the addition of a water molecule to occur. Carbohydrates also exist as **isomers**. Isomers are compounds with the same molecular formula, but they have a different structural formula. Because of this they have different properties. For example, when α–glucose polymers are linked together they form starch, but β–glucose polymers form cellulose. In all carbohydrates, the structure is closely related to their functional properties.

Isomerism

Compounds with the same chemical formula (same types and numbers of atoms) may differ in the arrangement of their atoms. Such variations in the arrangement of atoms in molecules are called **isomers**. In **structural isomers** (such as fructose and glucose, and the α and β glucose, right), the atoms are linked in different sequences. **Optical isomers** are identical in every way but are mirror images of each other.

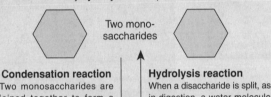

α glucose   β glucose

Condensation and Hydrolysis Reactions

Monosaccharides can combine to form compound sugars in what is called a **condensation** reaction. Compound sugars can be broken down by **hydrolysis** to simple monosaccharides.

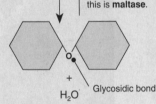

Two mono-saccharides

Condensation reaction

Two monosaccharides are joined together to form a disaccharide with the release of a water molecule (hence its name). Energy is supplied by a nucleotide sugar (e.g. ADP-glucose).

Hydrolysis reaction

When a disaccharide is split, as in digestion, a water molecule is used as a source of hydrogen and a hydroxyl group. The reaction is catalysed by enzymes. For maltose (right), this is **maltase**.

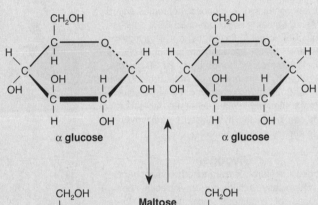

α glucose   α glucose

Maltose

+
H₂O   Glycosidic bond

Disaccharide + water

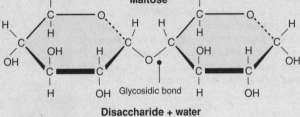

Glycosidic bond

Disaccharide + water

1. Distinguish between structural and optical isomers in carbohydrates, describing examples of each:

_____

_____

_____

2. Explain briefly how compound sugars are formed and broken down: _____

_____

_____

3. Using examples, explain how the isomeric structure of a carbohydrate may affect its chemical behaviour:

_____

_____

_____

_____

_____

Polysaccharides

Polysaccharides or complex carbohydrates are straight or branched chains of many monosaccharides (sometimes many thousands) of the same or different types. The most common polysaccharides, cellulose, starch, and glycogen, contain only glucose, but their properties are very different. These differences are a function of the glucose isomer involved and the types of glycosidic linkages joining the glucose monomers. Different polysaccharides, based on the same sugar monomer, can thus be highly soluble and a source of readily available energy or a strong structural material that resists being digested.

Cellulose

Cellulose is a structural material in plants and is made up of unbranched chains of β-glucose molecules held together by 1,4 glycosidic links. As many as 10,000 glucose molecules may be linked together to form a straight chain. Parallel chains become cross-linked with hydrogen bonds and form bundles of 60-70 molecules called **microfibrils**. Cellulose microfibrils are very strong and are a major component of the structural components of plants, such as the cell wall (photo, right).

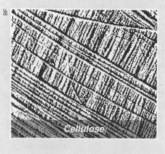
Cellulose

Starch

Starch is also a polymer of glucose, but it is made up of long chains of α-glucose molecules linked together. It contains a mixture of 25-30% amylose (unbranched chains linked by α-1,4 glycosidic bonds) and 70-75% amylopectin (branched chains with α-1, 6 glycosidic bonds every 24-30 glucose units). Starch is an energy storage molecule in plants and is found concentrated in insoluble starch granules within plant cells (see photo, right). Starch can be easily hydrolysed by enzymes to soluble sugars when required.

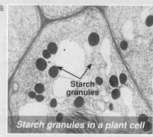
Starch granules in a plant cell

Glycogen

Glycogen, like starch, is a branched polysaccharide. It is chemically similar to amylopectin, being composed of α-glucose molecules, but there are more α-1,6 glycosidic links mixed with α-1,4 links. This makes it more highly branched and water-soluble than starch. Glycogen is a storage compound in animal tissues and is found mainly in liver and muscle cells (photo, right). It is readily hydrolysed by enzymes to form glucose.

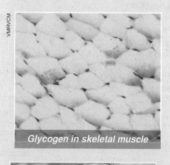
Glycogen in skeletal muscle

Chitin

Chitin is a tough modified polysaccharide made up of chains of β-glucose molecules. It is chemically similar to cellulose but each glucose has an amine group (–NH2) attached. After cellulose, chitin is the second most abundant carbohydrate. It is found in the cell walls of fungi and is the main component of the exoskeleton of insects (right) and other arthropods.

Chitinous insect exoskeleton

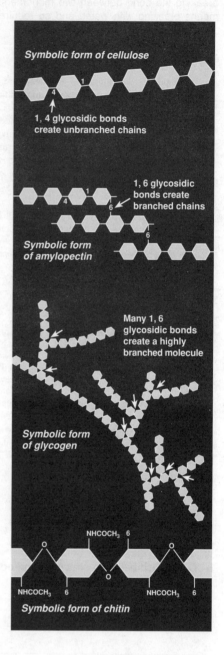

Symbolic form of cellulose

1, 4 glycosidic bonds create unbranched chains

1, 6 glycosidic bonds create branched chains

Symbolic form of amylopectin

Many 1, 6 glycosidic bonds create a highly branched molecule

Symbolic form of glycogen

Symbolic form of chitin

1. Explain why polysaccharides are such a good source of energy: _____

_____

2. Discuss the structural differences between the polysaccharides starch and glycogen, explaining how the differences in structure contribute to the functional properties of the molecule:

_____

_____

_____

_____

_____

Related activities: Carbohydrate Chemistry

Periodicals:
Designer starches

Cellulose and Starch

Cellulose is the most common molecule on Earth, making up one third of the volume of plants and one half of the volume of wood. Cellulose is made of thousands of β-glucose monomers arranged in a single, unbranched chain. In plants, cellulose makes up the bulk of the cell wall, providing both the strength required to keep the cell's shape and the support for the plant stem or trunk. As a material that can be exploited by humans, cellulose provides fibres which can be made into thread for textiles, and wood used for framing and cladding buildings. In contrast, **starch** is made of α-glucose monomers, and is a compact, branching molecule. It has no structural function, but can be hydrolysed to release soluble sugars for energy.

Cellulose Structure and Function

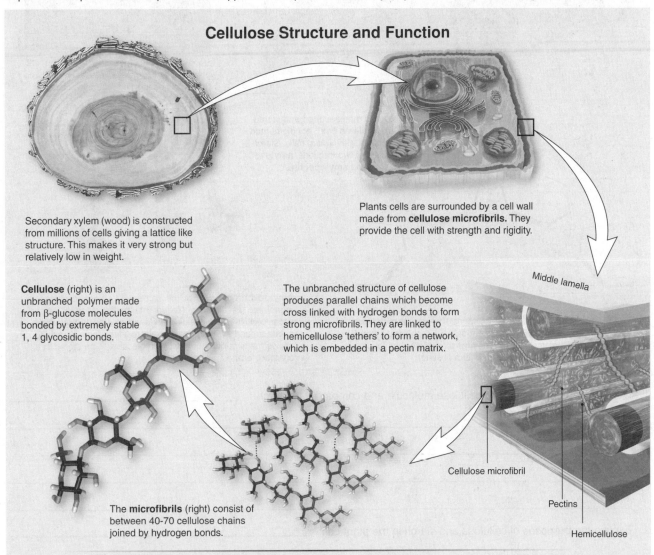

Secondary xylem (wood) is constructed from millions of cells giving a lattice like structure. This makes it very strong but relatively low in weight.

Plants cells are surrounded by a cell wall made from **cellulose microfibrils.** They provide the cell with strength and rigidity.

Middle lamella

Cellulose (right) is an unbranched polymer made from β-glucose molecules bonded by extremely stable 1, 4 glycosidic bonds.

The unbranched structure of cellulose produces parallel chains which become cross linked with hydrogen bonds to form strong microfibrils. They are linked to hemicellulose 'tethers' to form a network, which is embedded in a pectin matrix.

The **microfibrils** (right) consist of between 40-70 cellulose chains joined by hydrogen bonds.

Cellulose microfibril

Pectins

Hemicellulose

Biological Molecules

Cellulose and Tensile Strength

The **tensile strength** of a material refers to its ability to resist the strain of a pulling force. The simplest way to test tensile strength is to hang the material from a support and increase the tension by adding weight at the other end until the material separates. The tensile strength is then expressed as the force applied per the area of material in cross section. Typically this is written as pascals (Pa) or Newtons per square metre (Nm^{-2}), but is also often expressed as pounds per square inch (PSI). Cellulose fibrils show extremely high tensile strength (table below).

Material	Tensile strength (GPa)
Cellulose	17.8
Silk	1
Human hair	0.38
Rubber	0.15
Steel wire	2.2
Nylon (6/6)	0.75

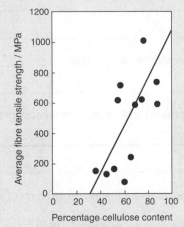

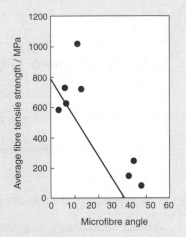

Plants produce fibres of various strengths that can be used for textiles, ropes, and construction. In general, the greater the amount of cellulose in the fibre the greater its tensile strength (above left). The tensile strength of the plant fibre is also influenced by the orientation of the cellulose microfibrils. Microfibrils orientated parallel to the length of the fibre (0°) provide the greatest fibre strength (above right).

Periodicals:
Designer starches

Related activities: Polysaccharides

RA 3

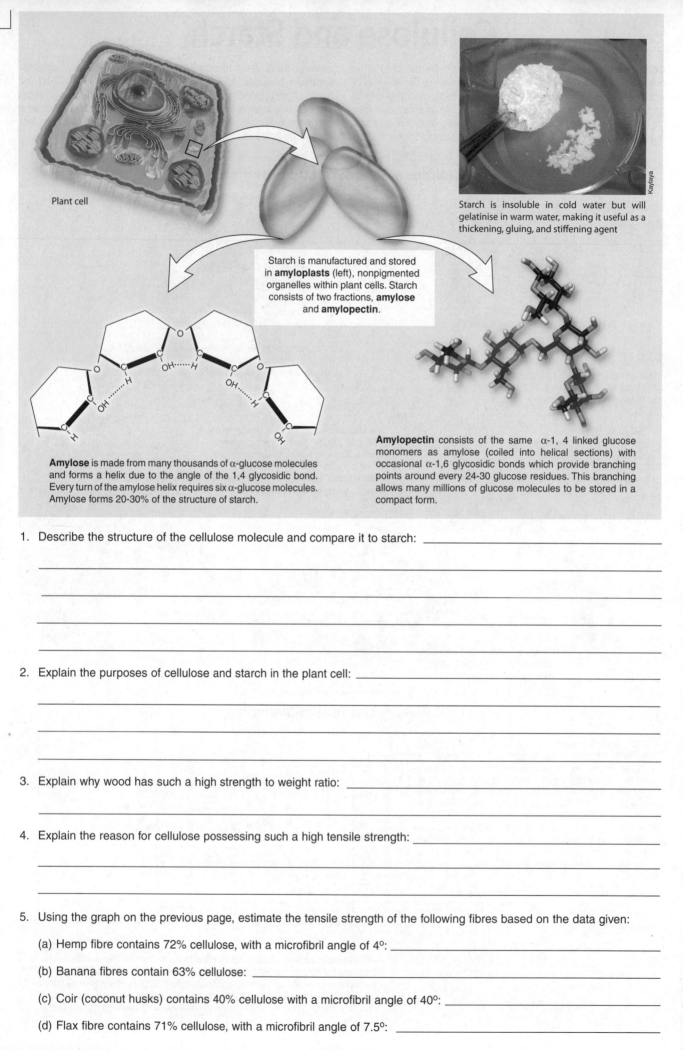

Plant cell

Starch is insoluble in cold water but will gelatinise in warm water, making it useful as a thickening, gluing, and stiffening agent

Starch is manufactured and stored in **amyloplasts** (left), nonpigmented organelles within plant cells. Starch consists of two fractions, **amylose** and **amylopectin**.

Amylose is made from many thousands of α-glucose molecules and forms a helix due to the angle of the 1,4 glycosidic bond. Every turn of the amylose helix requires six α-glucose molecules. Amylose forms 20-30% of the structure of starch.

Amylopectin consists of the same α-1, 4 linked glucose monomers as amylose (coiled into helical sections) with occasional α-1,6 glycosidic bonds which provide branching points around every 24-30 glucose residues. This branching allows many millions of glucose molecules to be stored in a compact form.

1. Describe the structure of the cellulose molecule and compare it to starch: _____

_____

_____

_____

_____

2. Explain the purposes of cellulose and starch in the plant cell: _____

_____

_____

_____

3. Explain why wood has such a high strength to weight ratio: _____

_____

4. Explain the reason for cellulose possessing such a high tensile strength: _____

_____

_____

5. Using the graph on the previous page, estimate the tensile strength of the following fibres based on the data given:

(a) Hemp fibre contains 72% cellulose, with a microfibril angle of 4°: _____

(b) Banana fibres contain 63% cellulose: _____

(c) Coir (coconut husks) contains 40% cellulose with a microfibril angle of 40°: _____

(d) Flax fibre contains 71% cellulose, with a microfibril angle of 7.5°: _____

Lipids

Lipids are a group of organic compounds with an oily, greasy, or waxy consistency. They are relatively insoluble in water and tend to be water-repelling (e.g. cuticle on leaf surfaces). Lipids are important biological fuels, some are hormones, and some serve as structural components in plasma membranes. Proteins and carbohydrates may be converted into fats by enzymes and stored within cells of adipose tissue. During times of plenty, this store is increased, to be used during times of food shortage.

Neutral Fats and Oils

The most abundant lipids in living things are **neutral fats**. They make up the fats and oils found in plants and animals. Fats are an economical way to store fuel reserves, since they yield more than twice as much energy as the same quantity of carbohydrate. Neutral fats are composed of a glycerol molecule attached to one (monoglyceride), two (diglyceride) or three (triglyceride) fatty acids. The fatty acid chains may be saturated or unsaturated (see below). **Waxes** are similar in structure to fats and oils, but they are formed with a complex alcohol instead of glycerol.

Glycerol | Fatty acids

9   12   15

Triglyceride: an example of a neutral fat

Condensation

Glycerol | Fatty acids

↓

Triglyceride | Water

Triglycerides form when glycerol bonds with three fatty acids. Glycerol is an alcohol containing three carbons. Each of these carbons is bonded to a hydroxyl (-OH) group.

When glycerol bonds with the fatty acid, an **ester bond** is formed and water is released. Three separate condensation reactions are involved in producing a triglyceride.

Saturated and Unsaturated Fatty Acids

Fatty acids are a major component of neutral fats and phospholipids. About 30 different kinds are found in animal lipids. **Saturated fatty acids** contain the maximum number of hydrogen atoms. **Unsaturated fatty acids** contain some carbon atoms that are double-bonded with each other and are not fully saturated with hydrogens. Lipids containing a high proportion of saturated fatty acids tend to be solids at room temperature (e.g. butter). Lipids with a high proportion of unsaturated fatty acids are oils and tend to be liquid at room temperature. This is because the unsaturation causes kinks in the straight chains so that the fatty acids do not pack closely together. Regardless of their degree of saturation, fatty acids yield a large amount of energy when oxidised.

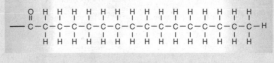

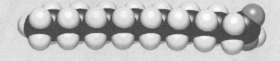

Formula (above) and molecular model (below) for **palmitic acid** (a saturated fatty acid)

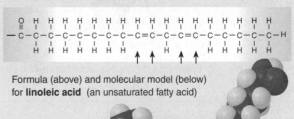

Formula (above) and molecular model (below) for **linoleic acid** (an unsaturated fatty acid)

Biological Molecules

1. (a) Distinguish between saturated and unsaturated fatty acids: _____

_____

(b) Explain how the type of fatty acid present in a neutral fat or phospholipid is related to that molecule's properties:

_____

_____

_____

2. Explain how neutral fats can provide an animal with:

(a) Energy: _____

(b) Water: _____

(c) Insulation: _____

Phospholipids

Phospholipids are the main component of cellular membranes. They consist of a glycerol attached to two fatty acid chains and a phosphate (PO_4^{3-}) group. The phosphate end of the molecule is attracted to water (it is hydrophilic) while the fatty acid end is repelled (hydrophobic). The hydrophobic ends turn inwards in the membrane to form a **phospholipid bilayer**.

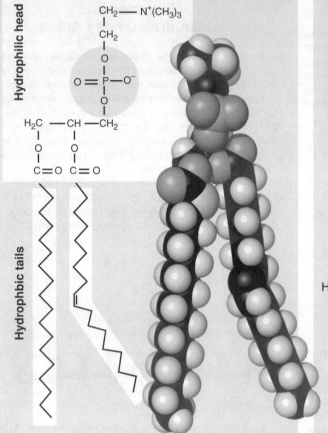

Hydrophilic head

Hydrophbic tails

Steroids and Cholesterol

Although steroids are classified as lipids, their structure is quite different to that of other lipids. Steroids have a basic structure of three rings made of 6 carbon atoms each and a fourth ring containing 5 carbon atoms. Examples of steroids include the male and female sex hormones (testosterone and oestrogen), and the hormones cortisol and aldosterone.

Cholesterol, while not a steroid itself, is a sterol lipid and is a precursor to several steroid hormones. It is present in the plasma membrane, where it regulates membrane fluidity by preventing the phospholipids packing too closely together.

Like phospholipids, cholesterol is amphipathic. The hydroxyl (-OH) group on cholesterol interacts with the polar head groups of the membrane phospholipids, while the steroid ring and hydrocarbon chain tuck into the hydrophobic portion of the membrane. This helps to stabilise the outer surface of the membrane and reduce its permeability to small water-soluble molecules.

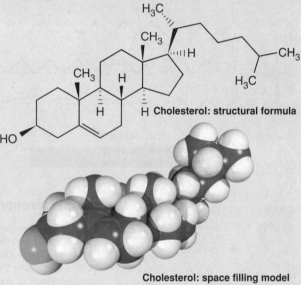

Cholesterol: structural formula

Cholesterol: space filling model

3. Outline the key **chemical** difference between a phospholipid and a triglyceride: _____

_____

_____

_____

4. Explain why saturated fats (e.g. lard) are solid at room temperature: _____

_____

5. (a) Relate the structure of phospholipids to their chemical properties and their functional role in cellular membranes:

_____

_____

_____

(b) Suggest how the cell membrane structure of an Arctic fish might differ from that of tropical fish species:

_____

_____

6. Explain how the structure of cholesterol enables it to perform structural and functional roles within membranes:

_____

_____

Enzymes

Most enzymes are proteins. They are capable of catalysing (speeding up) biochemical reactions and are therefore called biological **catalysts**. Enzymes act on one or more compounds (called the **substrate**). They may break down a single substrate molecule into simpler substances, or join two or more substrate molecules together. The enzyme itself is unchanged in the reaction; its presence merely allows the reaction to take place more rapidly. The part of the enzyme into which the substrate binds and undergoes reaction is the **active site**. It is a function of the polypeptide's complex tertiary structure.

Enzyme Structure

The model on the right illustrates the enzyme *Ribonuclease S*, which breaks up RNA molecules. It is a typical enzyme, being a globular protein and composed of up to several hundred atoms. The darkly shaded areas are part of the **active site** and make up the **cleft**; the region into which the substrate molecule(s) are drawn.

The correct positioning of these sites is critical for the catalytic reaction to occur. The substrate (RNA in this case) is drawn into the cleft by the active sites. By doing so, it puts the substrate molecule under stress, causing the reaction to proceed more readily.

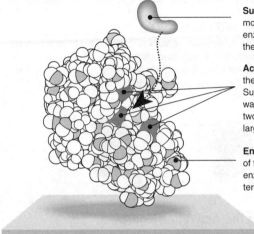

Substrate molecule: Substrate molecules are the chemicals that an enzyme acts on. They are drawn into the cleft of the enzyme.

Active site: These attraction points draw the substrate to the enzyme's surface. Substrate molecule(s) are positioned in a way to promote a reaction: either joining two molecules together or splitting up a larger one (as in this case).

Enzyme molecule: The complexity of the active site is what makes each enzyme so specific (i.e. precise in terms of the substrate it acts on).

Source: After *Biochemistry*, (1981) by Lubert Stryer

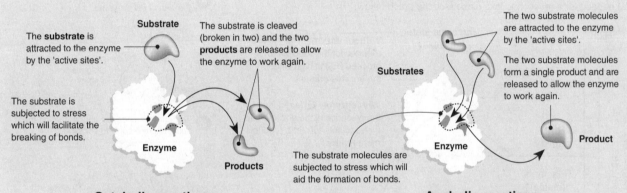

The **substrate** is attracted to the enzyme by the 'active sites'.

Substrate

The substrate is cleaved (broken in two) and the two **products** are released to allow the enzyme to work again.

The substrate is subjected to stress which will facilitate the breaking of bonds.

Enzyme

Products

The two substrate molecules are attracted to the enzyme by the 'active sites'.

The two substrate molecules form a single product and are released to allow the enzyme to work again.

Substrates

The substrate molecules are subjected to stress which will aid the formation of bonds.

Enzyme

Product

Catabolic reactions

Some enzymes can cause a single substrate molecule to be drawn into the active site. Chemical bonds are broken, causing the substrate molecule to break apart to become two separate molecules. Catabolic reactions break down complex molecules into simpler ones and involve a net release of energy, so they are called exergonic. **Examples**: *hydrolysis, cellular respiration*.

Anabolic reactions

Some enzymes can cause two substrate molecules to be drawn into the active site. Chemical bonds are formed, causing the two substrate molecules to form bonds and become a single molecule. Anabolic reactions involve the net use of energy (they are endergonic) and build more complex molecules and structures from simpler ones. **Examples**: *protein synthesis, photosynthesis*.

Biological Molecules

1. Explain what is meant by the active site of an enzyme and relate it to the enzyme's tertiary structure:

_____

_____

2. Explain what might happen to an enzyme's activity if the gene encoding its production was altered by a mutation:

_____

_____

3. Distinguish between **catabolism** and **anabolism**, giving an example of each and identifying each reaction as **endergonic** or **exergonic**:

_____

_____

_____

_____

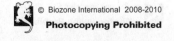

How Enzymes Work

Chemical reactions in cells are accompanied by energy changes. The amount of energy released ($-\Delta G$) or taken up ($+\Delta G$) is directly related to the tendency of a reaction to run to completion (for all the reactants to form products). Any reaction, even an exergonic reaction, needs to raise the energy of the substrate to an unstable transition state before the reaction will proceed (below left). The amount of energy required to do this is the activation energy (Ea). Enzymes work by lowering the Ea for any given reaction. They do this by orienting the substrate, or by adding charges or otherwise inducing strain in the substrate so that bonds are destabilised and the substrate is more reactive. Our current 'induced-fit' model of enzyme function is supported by studies of enzyme inhibitors, which show that enzymes are flexible and change shape when interacting with the substrate.

How Enzymes Work

The **lock and key** model proposed earlier last century suggested that the (perfectly fitting) substrate was simply drawn into a matching cleft on the enzyme molecule (below). This model was supported by early X-ray crystallography but has since been modified to recognise the flexibility of enzymes (the induced fit model, described right).

Substrate

1 **Enzyme** **2** **3**

Products

Lowering the Activation Energy

The presence of an enzyme simply makes it easier for a reaction to take place. All **catalysts** speed up reactions by influencing the stability of bonds in the reactants. They may also provide an alternative reaction pathway, thus lowering the activation energy (E_a) needed for a reaction to take place (see the graph below).

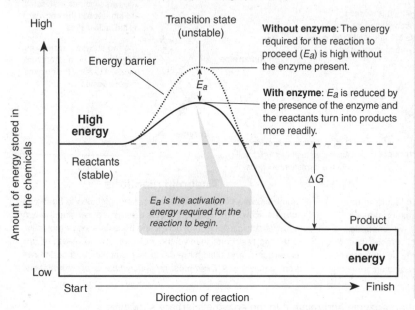

High

Transition state (unstable)

Energy barrier

E_a

High energy

Reactants (stable)

Amount of energy stored in the chemicals

E_a is the activation energy required for the reaction to begin.

ΔG

Product

Low energy

Low

Start → Finish

Direction of reaction

Without enzyme: The energy required for the reaction to proceed (E_a) is high without the enzyme present.

With enzyme: E_a is reduced by the presence of the enzyme and the reactants turn into products more readily.

The Current Model: Induced Fit

An enzyme's interaction with its substrate is best regarded as an **induced fit** (below). The shape of the enzyme changes when the substrate fits into the cleft. The reactants become bound to the enzyme by weak chemical bonds. This binding can weaken bonds within the reactants themselves, allowing the reaction to proceed more readily.

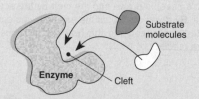

Substrate molecules

Enzyme Cleft

1 Two substrate molecules are drawn into the cleft of the enzyme.

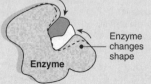

Enzyme changes shape

Enzyme

2 The enzyme changes shape, forcing the substrate molecules to combine.

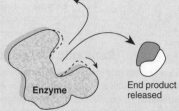

Enzyme

End product released

3 The resulting end product is released by the enzyme which returns to its normal shape, ready to receive more.

1. Explain the mechanism by which enzymes act as **biological catalysts**: _____

_____

_____

_____

2. Describe the key features of the '**lock and key**' model of enzyme action and explain its deficiencies as a working model:

_____

_____

3. Describe the current '**induced fit**' model of enzyme action, explaining how it differs from the lock and key model:

_____

_____

_____

_____

A 2

Related activities: Enzyme Reaction Rates
Web links: How Enzymes Work

Periodicals:
Enzymes: fast and flexible

© Biozone International 2008-2010
Photocopying Prohibited

Enzyme Reaction Rates

Enzymes are sensitive molecules. They often have a narrow range of conditions under which they operate properly. For most of the enzymes associated with plant and animal metabolism, there is little activity at low temperatures. As the temperature increases, so too does the enzyme activity, until the point is reached where the temperature is high enough to damage the enzyme's structure. At this point, the enzyme ceases to function; a phenomenon called enzyme or protein **denaturation**.

Extremes in acidity (pH) can also cause the protein structure of enzymes to denature. Poisons often work by denaturing enzymes or occupying the enzyme's active site so that it does not function. In some cases, enzymes will not function without cofactors, such as vitamins or trace elements. In the four graphs below, the rate of reaction or degree of enzyme activity is plotted against each of four factors that affect enzyme performance. Answer the questions relating to each graph:

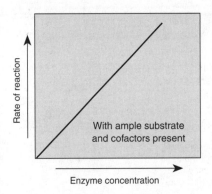

1. **Enzyme concentration**
   (a) Describe the change in the rate of reaction when the enzyme concentration is increased (assuming there is plenty of the substrate present):

   _____

   _____

   (b) Suggest how a cell may vary the amount of enzyme present in a cell:

   _____

   _____

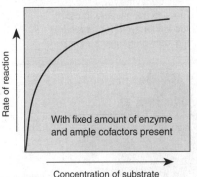

2. **Substrate concentration**
   (a) Describe the change in the rate of reaction when the substrate concentration is **increased** (assuming a fixed amount of enzyme and ample cofactors):

   _____

   _____

   (b) Explain why the rate changes the way it does: _____

   _____

   _____

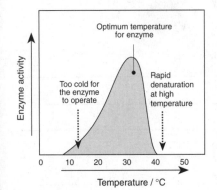

3. **Temperature**
   Higher temperatures speed up all reactions, but few enzymes can tolerate temperatures higher than 50–60°C. The rate at which enzymes are **denatured** (change their shape and become inactive) increases with higher temperatures.

   (a) Describe what is meant by an optimum temperature for enzyme activity:

   _____

   _____

   (b) Explain why most enzymes perform poorly at low temperatures:

   _____

   _____

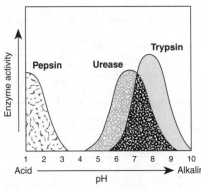

4. **pH (acidity/alkalinity)**
   Like all proteins, enzymes are **denatured** by extremes of **pH** (very acid or alkaline). Within these extremes, most enzymes are still influenced by pH. Each enzyme has a preferred pH range for optimum activity.

   (a) State the optimum pH for each of the enzymes:

   Pepsin: _____ Trypsin: _____ Urease: _____

   (b) Pepsin acts on proteins in the stomach. Explain how its optimum pH is suited to its working environment:

   _____

   _____

Related activities: Enzymes, Enzyme Cofactors, Enzyme Inhibitors

RDA 2

Biological Molecules

Enzyme Cofactors

Nearly all enzymes are made of protein, although RNA has been demonstrated to have enzymatic properties. Some enzymes (e.g. pepsin) consist of only protein. Other enzymes require the addition of extra non-protein components to complete their catalytic properties. In these cases, the protein portion is called the **apoenzyme**, and the additional chemical component is called a **cofactor**. Neither the apoenzyme nor the cofactor has catalytic activity on its own. Cofactors may be organic molecules (e.g. vitamin C and the coenzymes in the respiratory chain) or inorganic ions (e.g. Ca^{2+}, Zn^{2+}). They also may be tightly or loosely bound to the enzyme. Permanently bound cofactors are called **prosthetic groups**, whereas temporarily attached molecules, which detach after a reaction are called **coenzymes**. Some cofactors include both an organic and a non-organic component. Examples include the haem prosthetic groups, which consist of an iron atom in the centre of a porphyrin ring.

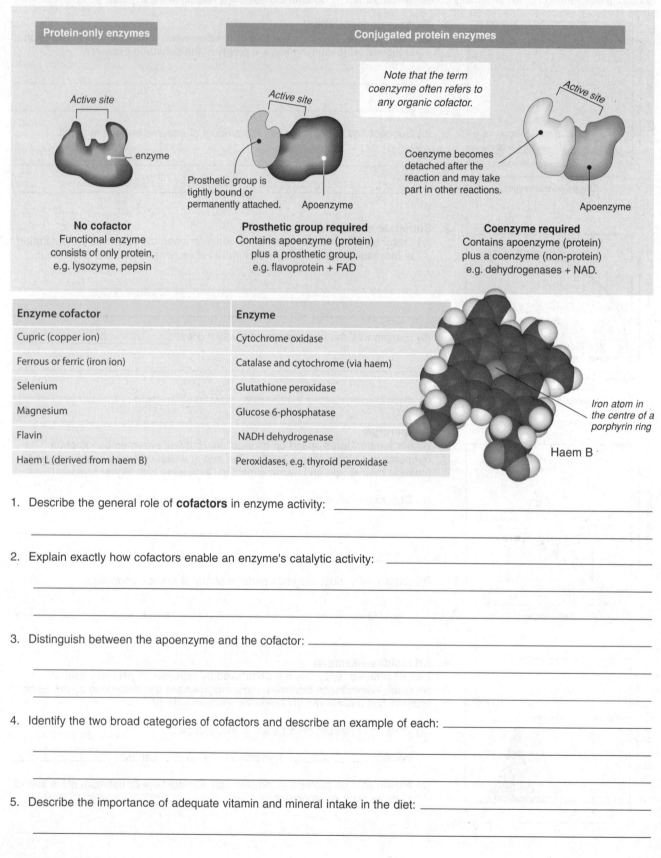

Protein-only enzymes

Active site

enzyme

No cofactor
Functional enzyme consists of only protein, e.g. lysozyme, pepsin

Conjugated protein enzymes

Note that the term coenzyme often refers to any organic cofactor.

Active site

Prosthetic group is tightly bound or permanently attached.

Apoenzyme

Prosthetic group required
Contains apoenzyme (protein) plus a prosthetic group, e.g. flavoprotein + FAD

Active site

Coenzyme becomes detached after the reaction and may take part in other reactions.

Apoenzyme

Coenzyme required
Contains apoenzyme (protein) plus a coenzyme (non-protein) e.g. dehydrogenases + NAD.

Enzyme cofactor	Enzyme
Cupric (copper ion)	Cytochrome oxidase
Ferrous or ferric (iron ion)	Catalase and cytochrome (via haem)
Selenium	Glutathione peroxidase
Magnesium	Glucose 6-phosphatase
Flavin	NADH dehydrogenase
Haem L (derived from haem B)	Peroxidases, e.g. thyroid peroxidase

Iron atom in the centre of a porphyrin ring

Haem B

1. Describe the general role of **cofactors** in enzyme activity: _____

_____

2. Explain exactly how cofactors enable an enzyme's catalytic activity: _____

_____

_____

3. Distinguish between the apoenzyme and the cofactor: _____

_____

_____

4. Identify the two broad categories of cofactors and describe an example of each: _____

_____

_____

5. Describe the importance of adequate vitamin and mineral intake in the diet: _____

_____

_____

Related activities: Enzymes, Enzyme Reaction Rates

Enzyme Inhibitors

Enzymes may be deactivated, temporarily or permanently, by chemicals called enzyme inhibitors. **Irreversible inhibitors** bind tightly to the enzyme, either at the active site or remotely from it, and are not easily displaced. **Reversible inhibitors** can be displaced from the enzyme and have a role as enzyme regulators in metabolic pathways. **Competitive inhibitors** compete directly with the substrate for the active site, and their effect can be overcome by increasing the concentration of available substrate. A **non-competitive inhibitor** does not occupy the active site, but distorts it so that the substrate and enzyme can no longer interact. Both competitive and non-competitive inhibition may be irreversible, in which case the inhibitors involved act as poisons. In contrast, the action of allosteric regulators, including allosteric inhibitors (below), is always reversible.

Allosteric Enzyme Regulation

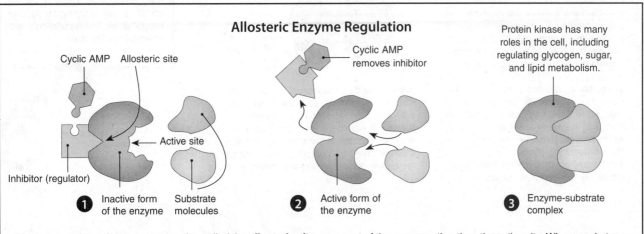

Cyclic AMP Allosteric site

Inhibitor (regulator)

1 Inactive form of the enzyme Substrate molecules

Cyclic AMP removes inhibitor

2 Active form of the enzyme

Protein kinase has many roles in the cell, including regulating glycogen, sugar, and lipid metabolism.

3 Enzyme-substrate complex

Active site

Allosteric regulators have a receptor site, called the **allosteric site**, on a part of the enzyme other than the active site. When a substance binds to the allosteric site, it regulates the activity of the enzyme. Often the action is inhibitory (as shown above for protein kinase), but allosteric regulators can also switch an enzyme from its inactive to its active form. Thus, they can serve as regulators of metabolic pathways. The activity of the enzyme protein kinase is regulated by the level of cyclic AMP in the cell. When a regulatory inhibitor protein binds reversibly to its allosteric site, the enzyme is inactive. Cyclic AMP removes the allosteric inhibitor and activates the enzyme.

Biological Molecules

Competitive Inhibition

Competitive inhibitors compete with the normal substrate for the enzyme's active site.

A competitive inhibitor occupies the active site only temporarily and so the inhibition is reversible.

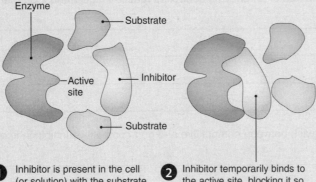

Enzyme Substrate Active site Inhibitor Substrate

1 Inhibitor is present in the cell (or solution) with the substrate

2 Inhibitor temporarily binds to the active site, blocking it so that the substrate cannot bind

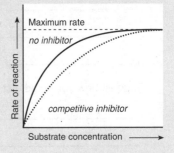

Fig.1 Effect of competitive inhibition on enzyme reaction rate at different substrate concentration

Maximum rate
no inhibitor
competitive inhibitor
Rate of reaction
Substrate concentration

Non-competitive Inhibition

Non-competitive inhibitors bind with the enzyme at a site other than the active site. They inactivate the enzyme by altering its shape.

This is very like allosteric inhibition, but allosteric inhibition is always reversible, whereas non-competitive inhibition may be irreversible.

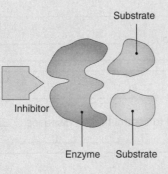

Substrate Inhibitor Enzyme Substrate

Active site cannot bind the substrates

1 Without the inhibitor bound, the enzyme can bind the substrate

2 When the inhibitor binds, the enzyme changes shape.

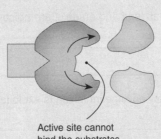

Fig.2 Effect of non-competitive inhibition on enzyme reaction rate at different substrate concentration

Maximum rate
no inhibitor
non-competitive inhibitor
Rate of reaction
Substrate concentration

Poisons are Irreversible Inhibitors

Some enzyme inhibitors are poisons because the enzyme-inhibitor binding is irreversible. Irreversible inhibitors form strong covalent bonds with an enzyme. These inhibitors may act at, near, or remote from the active site and modify the enzyme's structure to such an extent that it ceases to work. For example, the poison **cyanide** is an irreversible enzyme inhibitor that combines with the copper and iron in the active site of **cytochrome c oxidase** and blocks cellular respiration.

Since many enzymes contain sulfhydryl (-SH), alcohol, or acidic groups as part of their active sites, any chemical that can react with them may act as an irreversible inhibitor. Heavy metals, Ag^+, Hg^{2+}, or Pb^{2+}, have strong affinities for -SH groups and destroy catalytic activity. Most heavy metals are non-competitive inhibitors.

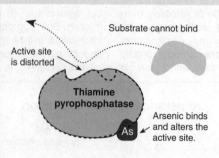

Substrate cannot bind

Active site is distorted

Thiamine pyrophosphatase

Arsenic binds and alters the active site.

Arsenic and phosphorus share some structural similarities so arsenic will often substitute for phosphorus in biological systems. It therefore targets widely dispersed enzyme reactions. Arsenic can act as either a competitive or a non-competitive inhibitor (as above) depending on the enzyme.

Drugs

Many drugs work by irreversible inhibition of a pathogen's enzymes. Penicillin and related antibiotics inhibit a bacterial enzyme (transpeptidase) which forms some of the linkages in the bacterial cell wall. Susceptible bacteria cannot complete cell wall synthesis and cannot divide but human cells are unaffected by the drug.

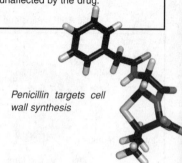

Penicillin targets cell wall synthesis

1. Distinguish between **competitive** and **non-competitive** inhibition: _____

_____

_____

_____

2. (a) Compare and contrast the effect of competitive and non-competitive inhibition on the relationship between the substrate concentration and the rate of an enzyme controlled reaction (figs 1 and 2 on the previous page):

_____

_____

_____

_____

(b) Suggest how you could distinguish between competitive and non-competitive inhibition in an isolated system:

_____

_____

3. Compare and contrast non-competitive inhibition and inhibition by an **allosteric regulator**: _____

_____

_____

4. Explain why heavy metals, such as lead and arsenic, are poisonous: _____

_____

_____

5. (a) Using an example, explain how enzyme inhibition is exploited to control human diseases: _____

_____

_____

(b) Explain why the drug is poisonous to the target organism, but not to humans: _____

_____

Biochemical Tests

Biochemical tests are used to detect the presence of molecules such as lipids, proteins, and carbohydrates (sugars and starch). For tests where the presence of a substance is indicated by a colour change, **colormetric analysis** of a known dilution series (e.g. five prepared dilutions of glucose) can be used to produce a **calibration curve**. This curve can then be used to quantify that substance in samples of unknown concentration. Simple biochemical tests are thus useful, but they are crude. To analyse a mix of substances, a technique such as chromatography is required. For example, a positive Benedict's test indicates the presence of reducing sugar(s), but chromatography will distinguish different sugars (e.g. fructose and glucose).

Simple Food Tests

Proteins: The Biuret Test

Reagent:	Biuret solution.
Procedure:	A sample is added to biuret solution and gently heated.
Positive result:	Solution turns from blue to lilac.

Starch: The Iodine Test

Reagent:	Iodine.
Procedure:	Iodine solution is added to the sample.
Positive result:	Blue-black staining occurs.

Lipids: The Emulsion Test

Reagent:	Ethanol.
Procedure:	The sample is shaken with ethanol. After settling, the liquid portion is distilled and mixed with water.
Positive result:	The solution turns into a cloudy-white emulsion of suspended lipid molecules.

Sugars: The Benedict's Test

Reagent:	Benedict's solution.
Procedure:	Non reducing sugars: The sample is boiled with dilute hydrochloric acid (acid hydrolysis), then cooled and neutralised. A test for reducing sugars is then performed.
	Reducing sugar: Benedict's solution is added, and the sample is placed in a water bath.
Positive result:	Solution turns from blue to orange to red-brown.

Colorimetric Analysis of Glucose

Prepare glucose standards

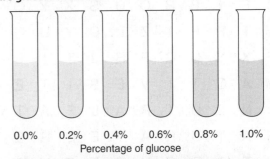

0.0% 0.2% 0.4% 0.6% 0.8% 1.0%

Percentage of glucose

Benedict's reagent in boiling water bath 4-10 minutes

Produce the calibration curve

Cool and filter samples as required. Using a red filter, measure the absorbance (at 735 nm) for each of the known dilutions and use these values to produce a calibration curve for glucose.

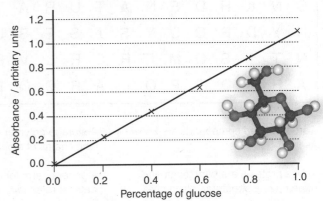

Biological Molecules

1. Explain why lipids must be mixed in ethanol before they will form an emulsion in water: _____

_____

2. (a) Explain how you could quantify the amount of glucose in a range of commercially available glucose drinks:

_____

_____

(b) Explain how you would proceed if the absorbance values you obtained for most of your 'unknowns' were outside the range of your calibration curve:

_____

_____

3. Explain the purpose of acid hydrolysis of a non-reducing sugar before testing with Benedict's reagent: _____

_____

4. Explain why the emulsion of lipids, ethanol, and water appears cloudy: _____

_____

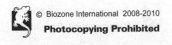

Related activities: Proteins, Carbohydrates, Lipids
Web links: Paper Chromatography

A 2

KEY TERMS: Word Find

Use the clues below to find the relevant key terms in the WORD FIND grid

```
O U X I N D U C E D F I T D C O L O R I M E T R Y
N E X E R G O N I C W J E G L Y C O G E N O J S V
J G L O B U L A R C V H R W K W D I P E P T I D E
C F I B R O U S X Y O L T A M I N O A C I D S P Q
M T P F I Z L Q Y U O F I W D E L X I Y V A H H L
F X Y U C F H J T I L Z A A N A B O L I C Y A O Y
M A C T I V E S I T E Q R C H R F Q E O O P H S M
I N E U T R A L F A T S Y A T P U Y N X N Y C P O
S A T U R A T E D R E N S Z S O M M Z I D C E H L
S T A R C H B Z K M W A T E R M R D Y M E D L O E
J Z O X D R C H I T I N R B E F Z S M I N X L L C
I V E L O R G A N I C Q U A B V W S E H S M U I U
A P R I M A R Y S T R U C T U R E B S I A A L P L
Q N K H D E N A T U R A T I O N M B R R T B O I A
H Y D R O L Y S I S F B U Y H L R R O O I G S D R
N T I S O M E R S R L V R F R U L Q O N O Q E S R
A H D D I P O L A R C Q E G W B R X W K N Z K I Z
```

Carbon-based compounds are known as this.

These proteins are very tough and often have a structural role in cells.

The loss of a protein's three dimensional functional structure is called this.

The formula that describes the number of atoms in a molecule.

This inorganic ion is a component of haemoglobin.

A molecule, like water, in which the opposite ends are oppositely charged.

The most abundant lipids in living things (2 words).

The emulsion tests detects these.

A carbohydrate storage molecule found in muscle and liver tissue.

A polysaccharide found in the exoskeleton of arthropods.

An important structural polysaccharide in plants.

A storage polymer in plants made up of long chains of alpha-glucose.

A general term for a reaction in which water is released.

These lipid molecules naturally form bilayers.

The building blocks of proteins (2 words).

The sequence of amino acids in a protein is called this (2 words).

The region of an enzyme responsible for substrate binding and reaction catalysis (2 words).

Substances required by an enzyme to enable its catalytic function.

Known as the universal solvent.

Reactions that build larger molecules from smaller ones are called this.

The splitting of a molecule into smaller components by addition of a water molecule.

Reactions that release energy are called this.

The structure of a protein maintained by disulfide bonds and hydrophilic and hydrophobic interactions.

A method for determining concentration of a solution on the basis of the density of a colour change.

Currently accepted model for enzyme function (2 words).

A fatty acid containing the maximum number of hydrogen atoms is called this.

These proteins are water soluble and have catalytic and regulatory roles in cells.

The product of a condensation reaction between two amino acids
The forms a molecule can take are called this.

These are biological catalysts.

Nucleic Acids

KEY TERMS

adenine
amino acids
antiparallel
base-pairing rule
codon
cytosine
DNA
DNA ligase
DNA polymerase
DNA replication
gene expression
genetic code
genome
guanine
helicase
histone
hydrogen bonds
lagging strand
leading strand
nucleic acids
nucleotides
Okazaki fragments
protein
purine
pyrimidine
replication fork
RNA (mRNA, rRNA, tRNA)
semi-conservative
        replication
thymine
uracil

KEY CONCEPTS

▶ Nucleic acids have a central role in storing genetic information and controlling the behaviour of cells.

▶ DNA is a self-replicating polynucleotide constructed according to strict base-pairing rules.

▶ DNA replication is a semi-conservative process controlled by enzymes.

▶ The universal genetic code, through gene expression, contains the information to construct proteins.

OBJECTIVES

☐ 1. Use the **KEY TERMS** to help you understand and complete these objectives.

Chromosome Structure   pages 164-166

☐ 2. EXTENSION: Describe the structure of eukaryote **chromosomes**, explaining how DNA is packaged and organised in the nucleus by **histones**.

Nucleic Acids   pages 164, 167-175

☐ 3. Recognise examples of **nucleic acids** and describe their role in biological systems. Understand the role of condensation reactions in the formation of **polynucleotides** (nucleic acids).

☐ 4. Describe the components of a (mono)**nucleotide**: a 5C sugar (**ribose** or **deoxyribose**), a nitrogenous base (**purine** or **pyrimidine**), and a phosphate. Identify the purines and pyrimidines that form nucleotides in **DNA** and **RNA**.

☐ 5. Describe the Watson-Crick **double-helix** model of DNA. Include reference to the base-pairing rule, the anti-parallel strands, and the role of **hydrogen bonding** between **purines** and **pyrimidines**.

☐ 6. Contrast the structure and function of **RNA** and **DNA**. Outline and functional roles of **mRNA**, **tRNA**, and **rRNA**.

☐ 7. Explain the **semi-conservative replication** of DNA during interphase, including the role of **DNA polymerase**, **helicase**, and **DNA ligase**.

☐ 8. Explain the significance of DNA replication in the 5' to 3' direction. Relate this to the formation of the **leading strand** and the **lagging strand**.

☐ 9. Demonstrate an understanding of the **base-pairing rule** for creating a complementary strand from a single strand of DNA.

Gene Expression   pages 176-178

☐ 10. Describe a gene as a sequence of DNA nucleotides coding for a mRNA product (e.g. a polypeptide). Recall the role of DNA and RNA in living organisms (see #3).

☐ 11. CIE only: Describe how the nucleotide sequence (the **genetic code**) codes for the amino acid sequence in a polypeptide.

☐ 12. CIE only: Describe how the information in DNA is used to construct polypeptides, including the role of **messenger RNA**, **transfer RNA**, and **ribosomes**.

Periodicals:
Listings for this chapter are on page 339

Weblinks:
www.biozone.co.uk/
weblink/OCRAS-2160.html

Teacher Resource CD-ROM:
The Meselson–Stahl Experiment

DNA Molecules

Even the smallest DNA molecules are extremely long. The DNA from the small *Polyoma* virus, for example, is 1.7 µm long; about three times longer than the longest proteins. The DNA comprising a bacterial chromosome is 1000 times longer than the cell into which it has to fit. The amount of DNA present in the nucleus of the cells of eukaryotic organisms varies widely from one species to another. In vertebrate sex cells, the quantity of DNA ranges from 40 000 **kb** to 80 000 000 **kb**, with humans about in the middle of the range. The traditional focus of DNA research has been on those DNA sequences that code for proteins, yet protein-coding DNA accounts for less than 2% of the DNA in

human chromosomes. The rest of the DNA, once dismissed as non-coding 'evolutionary junk', is now recognised as giving rise to functional RNA molecules, many of which have already been identified as having important regulatory functions. While there is no clear correspondence between the complexity of an organism and the number of protein-coding genes in its genome, this is not the case for non-protein-coding DNA. The genomes of more complex organisms contain much more of this so-called "non-coding" DNA. These RNA-only 'hidden' genes tend to be short and difficult to identify, but the sequences are highly conserved and clearly have a role in inheritance, development, and health.

Total length of DNA in viruses, bacteria, and eukayotes

Taxon	Organism	Base pairs (in 1000s, or kb)	Length
Viruses	Polyoma or SV40	5.1	1.7 µm
	Lambda phage	48.6	17 µm
	T2 phage	166	56 µm
	Vaccinia	190	65 µm
Bacteria	Mycoplasma	760	260 µm
	E. coli (from human gut)	4600	1.56 mm
Eukaryotes	Yeast	13 500	4.6 mm
	Drosophila (fruit fly)	165 000	5.6 cm
	Human	2 900 000	99 cm

Kilobase (kb)

A kilobase is unit of length equal to 1000 base pairs of a double-stranded nucleic acid molecule (or 1000 bases of a single-stranded molecule). One kb of double stranded DNA has a length of 0.34 µm. (1 µm = 1/1000 mm)

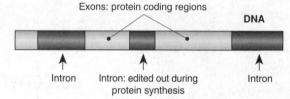

Exons: protein coding regions

DNA

Intron

Intron: edited out during protein synthesis

Intron

Most protein-coding genes in eukaryotic DNA are not continuous and may be interrupted by 'intrusions' of other pieces of DNA. Protein-coding regions (**exons**) are interrupted by non-protein-coding regions called **introns**. Introns range in frequency from 1 to over 30 in a single 'gene' and also in size (100 to more than 10 000 bases). Introns are edited out of the protein-coding sequence during protein synthesis, but probably, after processing, go on to serve a regulatory function.

Giant lampbrush chromosomes

Lampbrush chromosomes are large chromosomes found in amphibian eggs, with lateral loops of DNA that produce a brushlike appearance under the microscope. The two scanning electron micrographs (below and right) show minute strands of DNA giving a fuzzy appearance in the high power view.

Loops of DNA

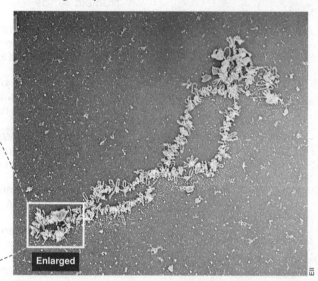

Enlarged

1. Consult the table above and make the following comparisons. Determine how much more DNA is present in:

   (a) The bacterium *E. coli* compared to the Lambda Phage virus: _____

   (b) Human cells compared to the bacteria *E. coli:* _____

2. State what proportion of DNA in a eukaryotic cell is used to code for proteins or structural RNA: _____

3. Describe two reasons why geneticists have reevaluated their traditional view that one gene codes for one polypeptide:

   (a) _____

   _____

   (b) _____

   _____

Related activities: The Simplest Case: Genes to Proteins

Periodicals:
DNA: 50 years of the double helix

Eukaryote Chromosome Structure

The chromosomes of eukaryote cells (such as those from plants and animals) are complex in their structure compared to those of prokaryotes. The illustration below shows a chromosome during the early stage of meiosis. Here it exists as a chromosome consisting of two chromatids. A non-dividing cell would have chromosomes with the 'equivalent' of a single chromatid only. The chromosome consists of a protein coated strand which coils in three ways during the time when the cell prepares to divide.

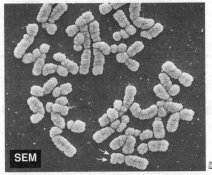

A cluster of human chromosomes seen during metaphase of cell division. Individual chromatids (arrowed) are difficult to discern on these double chromatid chromosomes.

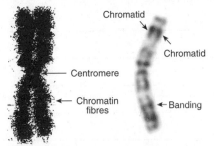

Chromatid
Chromatid
Centromere
Chromatin fibres
Banding

Chromosome TEM   Human chromosome 3

A human chromosome from a dividing white blood cell (above left). Note the compact organisation of the chromatin in the two chromatids. The LM photograph (above right) shows the banding visible on human chromosome 3.

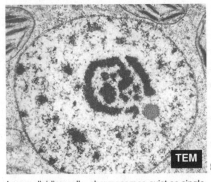

In non-dividing cells, chromosomes exist as single-armed structures. They are not visible as coiled structures, but are 'unwound' to make the genes accessible for transcription (above).

Looped domains

The evidence for the existence of looped domains comes from the study of giant lampbrush chromosomes in amphibian oocytes (above). Under electron microscopy, the lateral loops of the DNA-protein complex have a brushlike appearance.

The Packaging of Chromatin

Chromatin structure is based on successive levels of DNA packing. **Histone proteins** are responsible for packing the DNA into a compact form. Without them, the DNA could not fit into the nucleus. Five types of histone proteins form a complex with DNA, in a way that resembles "beads on a string". These beads, or **nucleosomes**, form the basic unit of DNA packing.

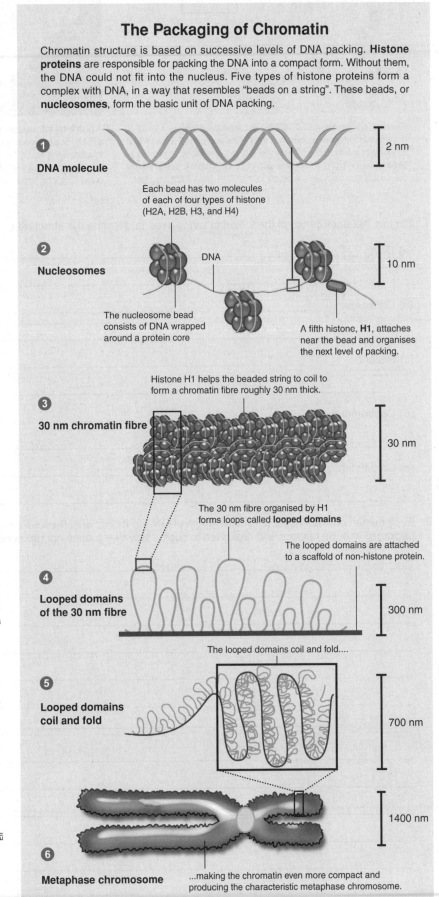

1 DNA molecule — 2 nm

Each bead has two molecules of each of four types of histone (H2A, H2B, H3, and H4)

2 Nucleosomes — DNA — 10 nm

The nucleosome bead consists of DNA wrapped around a protein core

A fifth histone, **H1**, attaches near the bead and organises the next level of packing.

Histone H1 helps the beaded string to coil to form a chromatin fibre roughly 30 nm thick.

3 30 nm chromatin fibre — 30 nm

The 30 nm fibre organised by H1 forms loops called **looped domains**

The looped domains are attached to a scaffold of non-histone protein.

4 Looped domains of the 30 nm fibre — 300 nm

The looped domains coil and fold....

5 Looped domains coil and fold — 700 nm

1400 nm

6 Metaphase chromosome — ...making the chromatin even more compact and producing the characteristic metaphase chromosome.

Nucleic Acids

Related activities: DNA Molecules
Web links: Chromosome Structure

EDA 2

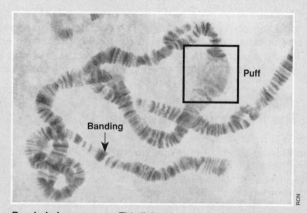

Banded chromosome: This light microscope photo is a view of the polytene chromosomes in a salivary gland cell of a sandfly. It shows a banding pattern that is thought to correspond to groups of genes. Regions of chromosome **puffing** are thought to occur where the genes are being transcribed into mRNA (see SEM on right).

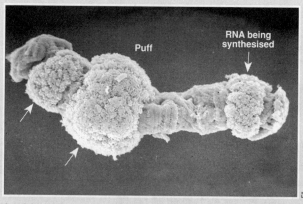

A **polytene chromosome** viewed with a scanning electron microscope (SEM). The arrows indicate localised regions of the chromosome that are uncoiling to expose their genes (puffing) to allow transcription of those regions. Polytene chromosomes are a special type of chromosome consisting of a large bundle of chromatids bound tightly together.

1. Explain the significance of the following terms used to describe the structure of chromosomes:

(a) DNA: _____

_____

(b) Chromatin: _____

_____

(c) Histone: _____

_____

(d) Centromere: _____

_____

(e) Chromatid: _____

_____

2. Each human cell has about a 1 metre length of DNA in its nucleus. Discuss the mechanisms by which this DNA is packaged into the nucleus and organised in such a way that it does not get ripped apart during cell division:

_____

_____

_____

_____

_____

_____

_____

_____

_____

_____

_____

_____

_____

Nucleic Acids

Nucleic acids are a special group of chemicals in cells concerned with the transmission of inherited information. They have the capacity to store the information that controls cellular activity. The central nucleic acid is called **deoxyribonucleic acid** (DNA). DNA is a major component of chromosomes and is found primarily in the nucleus, although a small amount is found in mitochondria and chloroplasts. Other **ribonucleic acids** (RNA) are involved in the 'reading' of the DNA information. All nucleic acids are made up of simple repeating units called **nucleotides**, linked together to form chains or strands, often of great length. The strands vary in the sequence of the bases found on each nucleotide. It is this sequence which provides the 'genetic code' for the cell. In addition to nucleic acids, certain nucleotides and their derivatives are also important as suppliers of energy (**ATP**) or as hydrogen ion and electron carriers in respiration and photosynthesis (NAD, NADP, and FAD).

Chemical Structure of a Nucleotide

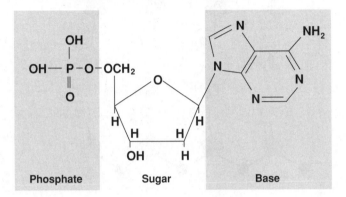

Phosphate   Sugar   Base

Symbolic Form of a Nucleotide

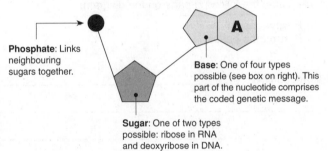

Phosphate: Links neighbouring sugars together.

Base: One of four types possible (see box on right). This part of the nucleotide comprises the coded genetic message.

Sugar: One of two types possible: ribose in RNA and deoxyribose in DNA.

Nucleotides are the building blocks of DNA. Their precise sequence in a DNA molecule provides the genetic instructions for the organism to which it governs. Accidental changes in nucleotide sequences are a cause of mutations, usually harming the organism, but occasionally providing benefits.

Bases

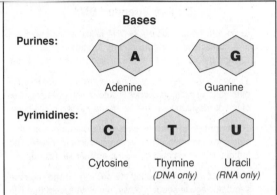

Purines:
Adenine   Guanine

Pyrimidines:
Cytosine   Thymine   Uracil
            (DNA only) *(RNA only)*

The two-ringed bases above are **purines** and make up the longer bases. The single-ringed bases are **pyrimidines**. Although only one of four kinds of base can be used in a nucleotide, **uracil** is found only in RNA, replacing **thymine**. DNA contains: A, T, G, and C, while RNA contains A, U, G, and C.

Sugars

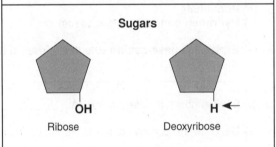

OH   H
Ribose   Deoxyribose

Deoxyribose sugar is found only in DNA. It differs from **ribose** sugar, found in RNA, by the lack of a single oxygen atom (arrowed).

RNA Molecule

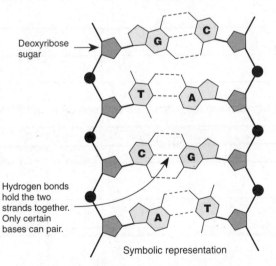

In RNA, uracil replaces thymine in the code.

Ribose sugar

Ribonucleic acid (RNA) comprises a *single strand* of nucleotides linked together.

DNA Molecule

Deoxyribose sugar

Hydrogen bonds hold the two strands together. Only certain bases can pair.

Symbolic representation

DNA Molecule

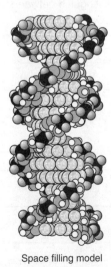

Space filling model

Deoxyribonucleic acid (DNA) comprises a *double strand* of nucleotides linked together. It is shown unwound in the symbolic representation (left). The DNA molecule takes on a twisted, double helix shape as shown in the space filling model on the right.

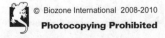
Related activities: DNA Molecules, Creating a DNA Molecule

Nucleic Acids

A 2

Formation of a nucleotide

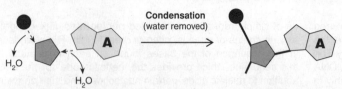

Condensation
(water removed)

A nucleotide is formed when phosphoric acid and a base are chemically bonded to a sugar molecule. In both cases, water is given off, and they are therefore condensation reactions. In the reverse reaction, a nucleotide is broken apart by the addition of water (**hydrolysis**).

Formation of a dinucleotide

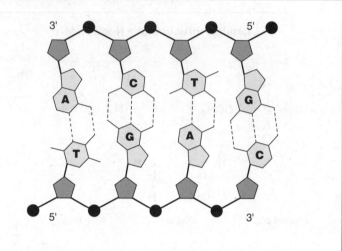

Two nucleotides are linked together by a condensation reaction between the phosphate of one nucleotide and the sugar of another.

Double-Stranded DNA

The **double-helix** structure of DNA is like a ladder twisted into a corkscrew shape around its longitudinal axis. It is 'unwound' here to show the relationships between the bases.

- The way the correct pairs of bases are attracted to each other to form hydrogen bonds is determined by the number of bonds they can form and the shape (length) of the base.

- The **template strand** the side of the DNA molecule that stores the information that is transcribed into mRNA. The template strand is also called the **antisense strand**.

- The other side (often called the **coding strand**) has the same nucleotide sequence as the mRNA except that T in DNA substitutes for U in mRNA. The coding strand is also called the **sense strand**.

1. The diagram above depicts a double-stranded DNA molecule. Label the following parts on the diagram:

   (a) **Sugar** (deoxyribose)   (d) **Purine** bases
   (b) **Phosphate**   (e) **Pyrimidine** bases
   (c) **Hydrogen bonds** (between bases)

2. (a) Explain the **base-pairing rule** that applies in double-stranded DNA: _____

     _____

   (b) Explain how this differs in mRNA: _____

   (c) Describe the purpose of the hydrogen bonds in double-stranded DNA: _____

     _____

3. Describe the functional role of nucleotides: _____

   _____

4. Distinguish between the **template strand** and **coding strand** of DNA, identifying the functional role of each:

   _____

   _____

   _____

5. Complete the following table summarising the differences between DNA and RNA molecules:

	DNA	RNA
Sugar present		
Bases present		
Number of strands		
Relative length		

Creating a DNA Model

Although DNA molecules can be enormous in terms of their molecular size, they are made up of simple repeating units called **nucleotides**. A number of factors control the way in which these nucleotide building blocks are linked together. These factors cause the nucleotides to join together in a predictable way. This is referred to as the **base pairing rule** and can be used to construct a complementary DNA strand from a template strand, as illustrated in the exercise below:

DNA Base Pairing Rule			
Adenine	is always attracted to	**Thymine**	A ⟷ T
Thymine	is always attracted to	**Adenine**	T ⟷ A
Cytosine	is always attracted to	**Guanine**	C ⟷ G
Guanine	is always attracted to	**Cytosine**	G ⟷ C

1. Cut around the nucleotides on page 171 and separate each of the 24 nucleotides by cutting along the columns and rows (see arrows indicating two such cutting points). Although drawn as geometric shapes, these symbols represent chemical structures.

2. Place one of each of the four kinds of nucleotide on their correct spaces below:

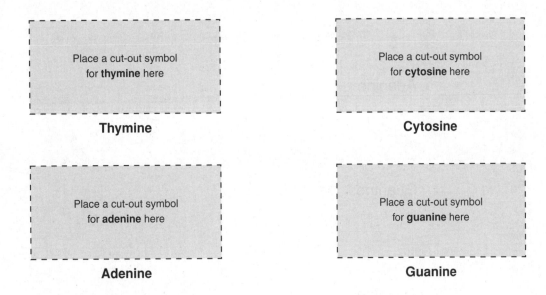

Place a cut-out symbol for **thymine** here

Thymine

Place a cut-out symbol for **cytosine** here

Cytosine

Place a cut-out symbol for **adenine** here

Adenine

Place a cut-out symbol for **guanine** here

Guanine

3. Identify and **label** each of the following features on the *adenine* nucleotide immediately above:
   phosphate, sugar, base, hydrogen bonds

4. Create one strand of the DNA molecule by placing the 9 correct 'cut out' nucleotides in the labelled spaces on the following page (DNA molecule). Make sure these are the right way up (with the **P** on the left) and are aligned with the left hand edge of each box. Begin with thymine and end with guanine.

5. Create the complementary strand of DNA by using the base pairing rule above. Note that the nucleotides have to be arranged upside down.

6. Under normal circumstances, it is not possible for adenine to pair up with guanine or cytosine, nor for any other mismatches to occur. Describe the two factors that prevent a mismatch from occurring:

   (a) Factor 1: _____

   _____

   (b) Factor 2: _____

   _____

7. Once you have checked that the arrangement is correct, you may glue, paste or tape these nucleotides in place.

NOTE:	There may be some value in keeping these pieces loose in order to practise the base pairing rule. For this purpose, *removable tape* would be best.

Related activities: Nucleic Acids, The DNA Molecule

PA 2

Nucleic Acids

DNA Molecule

Put the named nucleotides on the left hand side to create the template strand

Put the matching **complementary** nucleotides opposite the template strand

Thymine

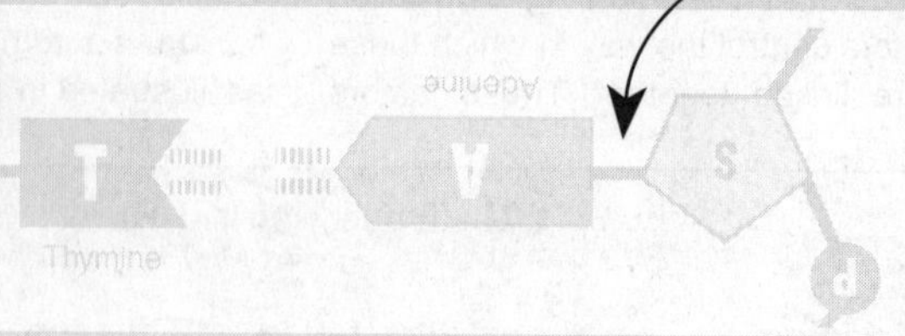

Cytosine

Adenine

Adenine

Guanine

Thymine

Thymine

Cytosine

Guanine

Nucleotides

Tear out this page along the perforation and separate each of the 24 nucleotides by cutting along the columns and rows (see arrows indicating the cutting points).

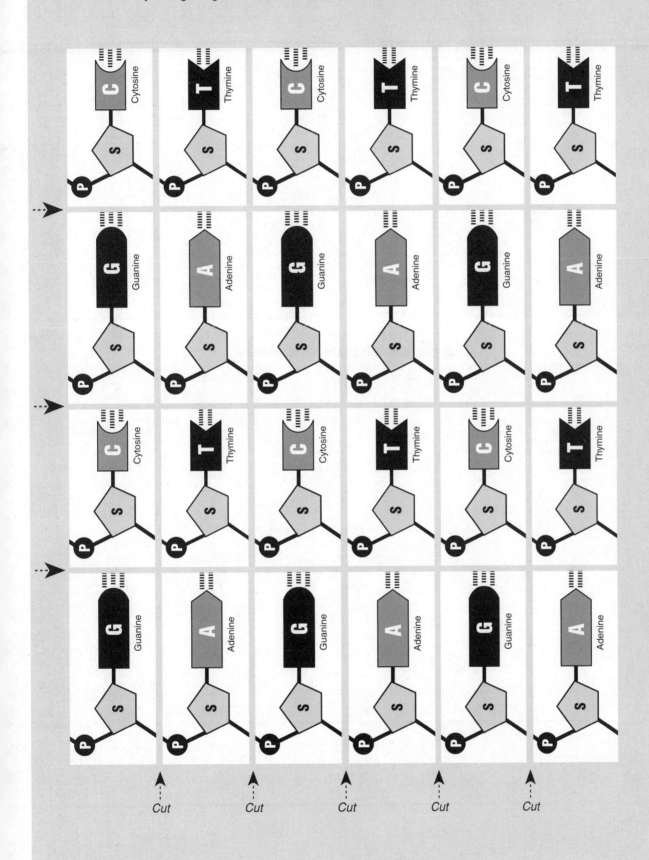

Nucleic Acids

Cut   Cut   Cut   Cut   Cut

This page is left blank deliberately

DNA Replication

DNA replication is a necessary preliminary step for cell division (both mitosis and meiosis). This process ensures that each resulting cell receives a complete set of genes from the original cell. After DNA replication, each chromosome is made up of two chromatids, joined at the **centromere**. Each **chromatid** contains half original (parent) DNA and half new (daughter) DNA. The two chromatids will become separated during cell division to form two separate chromosomes. During DNA replication, nucleotides are added at a region called the **replication fork**. The position of the replication fork moves along the chromosome as replication progresses. This whole process occurs simultaneously for each chromosome of a cell and the entire process is tightly controlled by enzymes. The diagram below describes essential steps in the process, while that on the next page identifies the role of the enzymes at each stage.

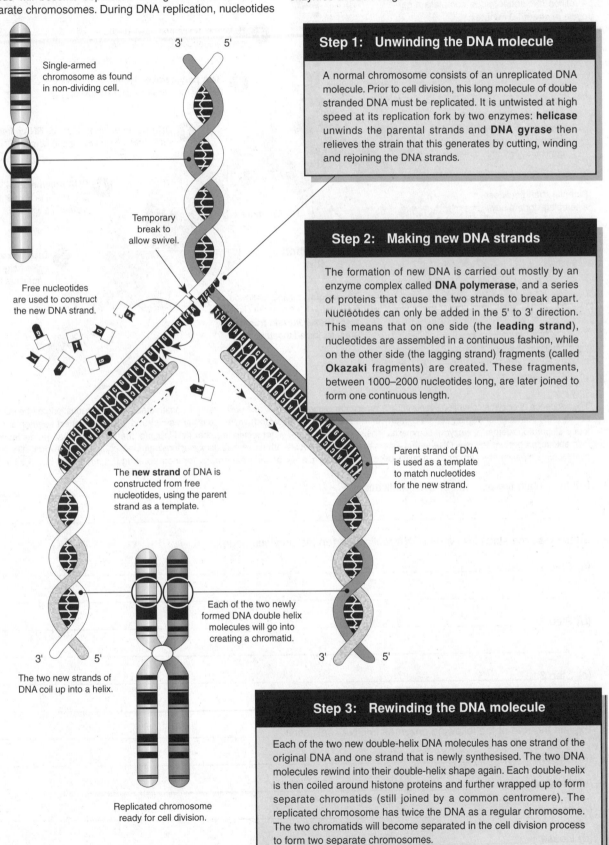

Single-armed chromosome as found in non-dividing cell.

Temporary break to allow swivel.

Free nucleotides are used to construct the new DNA strand.

The **new strand** of DNA is constructed from free nucleotides, using the parent strand as a template.

Parent strand of DNA is used as a template to match nucleotides for the new strand.

Each of the two newly formed DNA double helix molecules will go into creating a chromatid.

The two new strands of DNA coil up into a helix.

Replicated chromosome ready for cell division.

Step 1: Unwinding the DNA molecule

A normal chromosome consists of an unreplicated DNA molecule. Prior to cell division, this long molecule of double stranded DNA must be replicated. It is untwisted at high speed at its replication fork by two enzymes: **helicase** unwinds the parental strands and **DNA gyrase** then relieves the strain that this generates by cutting, winding and rejoining the DNA strands.

Step 2: Making new DNA strands

The formation of new DNA is carried out mostly by an enzyme complex called **DNA polymerase**, and a series of proteins that cause the two strands to break apart. Nucleotides can only be added in the 5' to 3' direction. This means that on one side (the **leading strand**), nucleotides are assembled in a continuous fashion, while on the other side (the lagging strand) fragments (called **Okazaki** fragments) are created. These fragments, between 1000–2000 nucleotides long, are later joined to form one continuous length.

Step 3: Rewinding the DNA molecule

Each of the two new double-helix DNA molecules has one strand of the original DNA and one strand that is newly synthesised. The two DNA molecules rewind into their double-helix shape again. Each double-helix is then coiled around histone proteins and further wrapped up to form separate chromatids (still joined by a common centromere). The replicated chromosome has twice the DNA as a regular chromosome. The two chromatids will become separated in the cell division process to form two separate chromosomes.

Periodicals:
DNA polymerase

Related activities: Mitosis and the Cell Cycle, Review of DNA Replication
Web links: DNA Replication

DA 3

Nucleic Acids

Enzyme Control of DNA Replication

DNA replication occurs during interphase of the cell cycle at an astounding rate. As many as 4000 nucleotides per second are replicated. This explains how under ideal conditions, bacterial cells with as many as 4 million nucleotides, can complete a cell cycle in about 20 minutes.

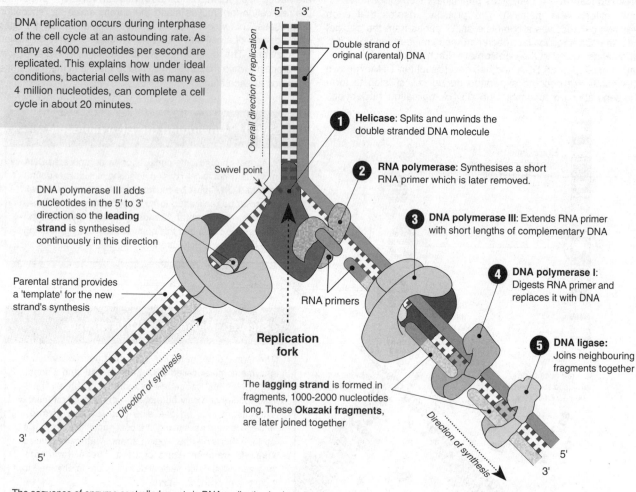

Overall direction of replication

5'   3'

Double strand of original (parental) DNA

Swivel point

1 **Helicase**: Splits and unwinds the double stranded DNA molecule

2 **RNA polymerase**: Synthesises a short RNA primer which is later removed.

3 **DNA polymerase III**: Extends RNA primer with short lengths of complementary DNA

4 **DNA polymerase I**: Digests RNA primer and replaces it with DNA

5 **DNA ligase**: Joins neighbouring fragments together

DNA polymerase III adds nucleotides in the 5' to 3' direction so the **leading strand** is synthesised continuously in this direction

Parental strand provides a 'template' for the new strand's synthesis

RNA primers

Replication fork

The **lagging strand** is formed in fragments, 1000-2000 nucleotides long. These **Okazaki fragments**, are later joined together

Direction of synthesis

3'
5'

5'
3'

Direction of synthesis

The sequence of enzyme controlled events in DNA replication is shown above (1-5). Although shown as separate, many of the enzymes are found clustered together as enzyme complexes. These enzymes are also able to 'proof-read' the new DNA strand as it is made and correct mistakes. The polymerase enzyme can only work in one direction, so that one new strand is constructed as a continuous length (the leading strand) while the other new strand is made in short segments to be later joined together (the lagging strand). **NOTE** that the nucleotides are present as deoxynucleoside triphosphates. When hydrolysed, these provide the energy for incorporating the nucleotide into the strand.

1. Briefly explain the purpose of DNA replication: _____

    _____

2. Summarise the steps involved in DNA replication (on the previous page):

    (a) Step 1: _____

    _____

    (b) Step 2: _____

    _____

    (c) Step 3: _____

    _____

3. Explain the role of the following enzymes in DNA replication: _____

    (a) Helicase: _____

    (b) DNA polymerase I: _____

    (c) DNA polymerase III: _____

    (d) Ligase: _____

4. Determine the time it would take for a bacteria to replicate its DNA (see note in diagram above): _____

Review of DNA Replication

The diagram below summarises the main steps in DNA replication. You should use this activity to test your understanding of the main features of DNA replication, using the knowledge gained in the previous activity to fill in the missing information. You should attempt this from what you have learned, but refer to the previous activity if you require help.

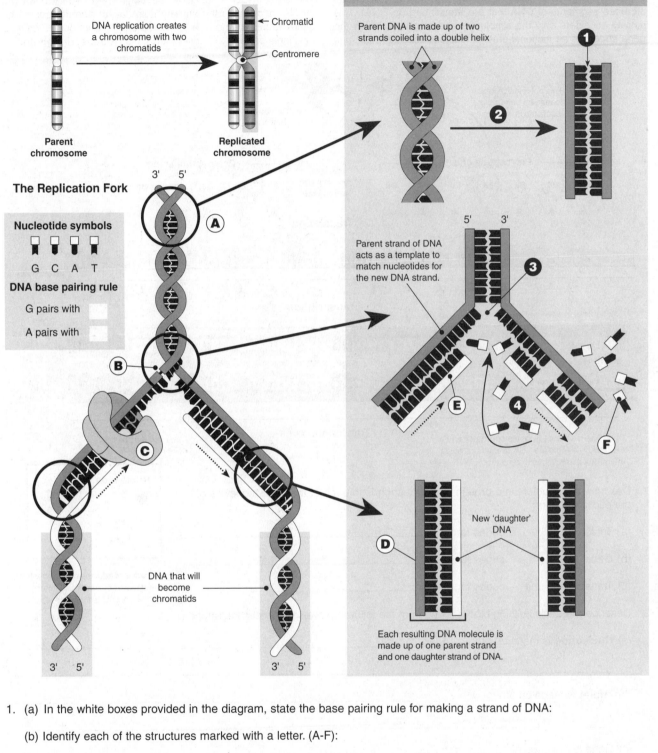

Purpose of DNA Replication

DNA replication creates a chromosome with two chromatids

← Chromatid

← Centromere

Parent chromosome

Replicated chromosome

The Replication Fork

Nucleotide symbols

G C A T

DNA base pairing rule

G pairs with

A pairs with

3' 5'

(A)

(B)

(C)

DNA that will become chromatids

3' 5'   3' 5'

Detail of Stages in DNA Replication

Parent DNA is made up of two strands coiled into a double helix

1

2

Parent strand of DNA acts as a template to match nucleotides for the new DNA strand.

5' 3'

3

(E) 4

(F)

New 'daughter' DNA

(D)

Each resulting DNA molecule is made up of one parent strand and one daughter strand of DNA.

1. (a) In the white boxes provided in the diagram, state the base pairing rule for making a strand of DNA:

(b) Identify each of the structures marked with a letter. (A-F):

A: _____   C: _____   E: _____

B: _____   D: _____   F: _____

3. Match each of the processes (1-4) to the correct summary of the process provided below:

☐ Unwinding of parent DNA double helix   ☐ Unzipping of parent DNA

☐ Free nucleotides occupy spaces alongside exposed bases   ☐ DNA strands are joined by base pairing

Related activities: DNA Replication
Web links: DNA Replication

RA 1

The Simplest Case: Genes to Proteins

The traditionally held view of genes was as sections of DNA coding only for protein. This view has been revised in recent years with the discovery that much of the nonprotein-coding DNA encodes functional RNAs; it is not all non-coding "junk" DNA as was previously assumed. In fact, our concept of what constitutes a gene is changing rapidly and now encompasses all those segments of DNA that are transcribed (to RNA). This activity considers only the simplest scenario: one in which the gene codes for a functional protein. **Nucleotides**, the basic unit

of genetic information, are read in groups of three (**triplets**). Some triplets have a special controlling function in the making of a polypeptide chain. The equivalent of the triplet on the mRNA molecule is the **codon**. Three codons can signify termination of the amino acid chain (UAG, UAA and UGA in the mRNA code). The codon AUG is found at the beginning of every gene (on mRNA) and marks the starting point for reading the gene. The genes required to form a functional end-product (in this case, a functional protein) are collectively called a **transcription unit**.

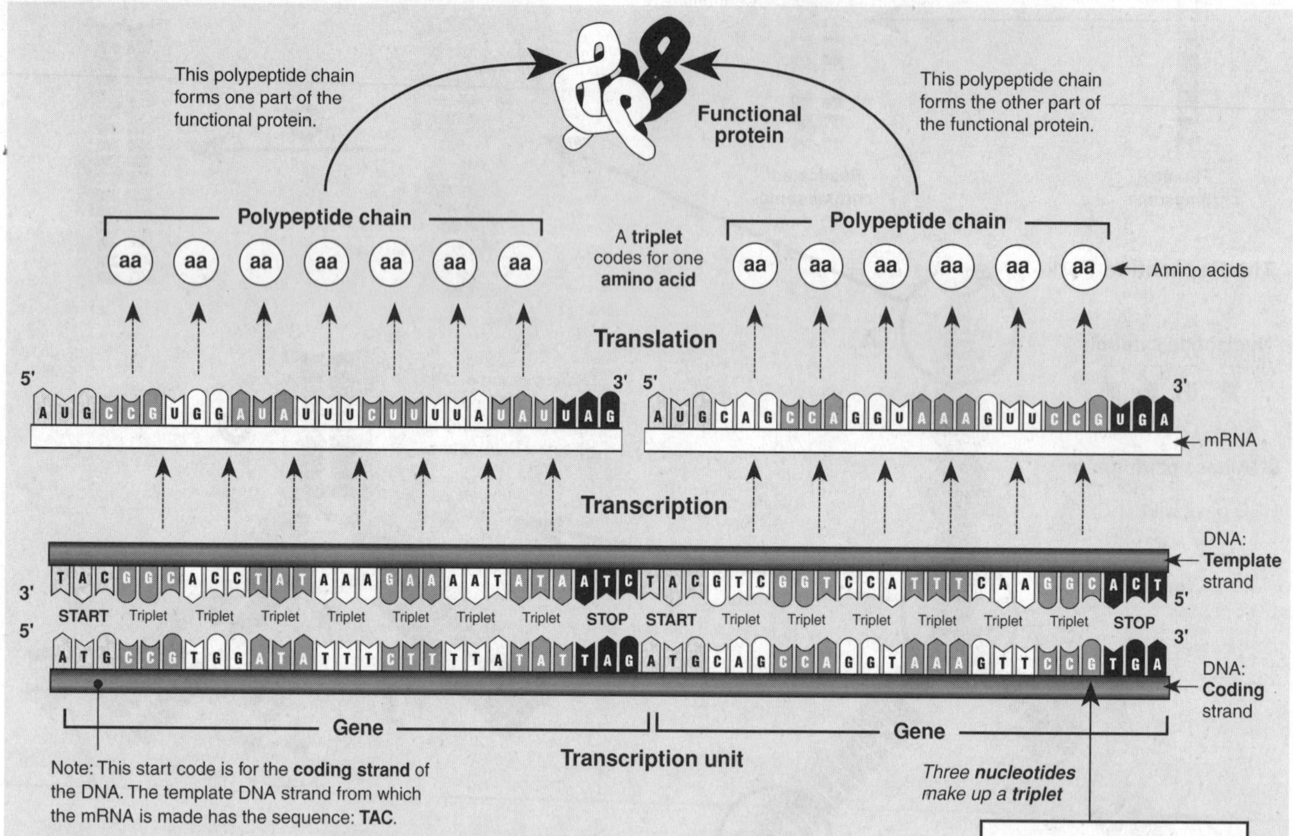

This polypeptide chain forms one part of the functional protein.

This polypeptide chain forms the other part of the functional protein.

Functional protein

Polypeptide chain

A triplet codes for one amino acid

Polypeptide chain

← Amino acids

Translation

← mRNA

Transcription

DNA: **Template** strand

START   Triplet   Triplet   Triplet   Triplet   Triplet   Triplet   Triplet   STOP   START   Triplet   Triplet   Triplet   Triplet   Triplet   Triplet   STOP

DNA: **Coding** strand

Gene   **Transcription unit**   **Gene**

Note: This start code is for the **coding strand** of the DNA. The template DNA strand from which the mRNA is made has the sequence: **TAC**.

*Three **nucleotides** make up a **triplet***

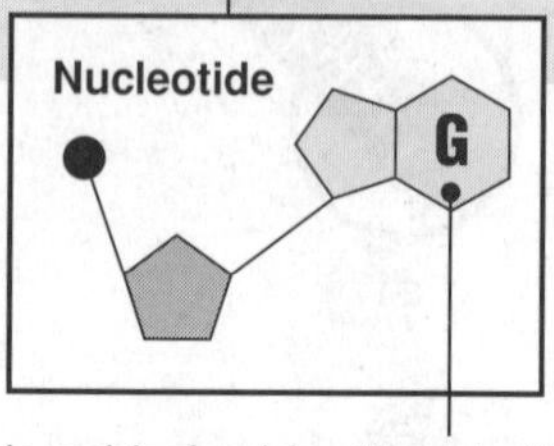

Nucleotide

In models of nucleic acids, nucleotides are denoted by their base letter. (In this case: **G** is for guanine)

1. Describe the structure in a protein that corresponds to each of the following levels of genetic information:

   (a) Triplet   codes for: _____

   (b) Gene   codes for: _____

   (c) Transcription unit   codes for: _____

2. Describe the basic building blocks for each of the following levels of genetic information:

   (a) **Nucleotide** is made up of: _____

   _____

   (b) **Triplet** is made up of: _____

   _____

   (c) **Gene** is made up of: _____

   _____

   (d) **Transcription unit** is made up of: _____

   _____

3. Describe the steps involved in forming a functional protein: _____

   _____

Related activities: The Genetic Code

Periodicals:

What is a gene?

The Genetic Code

The genetic code consists of the sequence of bases arranged along the DNA molecule. It consists of a four-letter alphabet (from the four kinds of bases) and is read as three-letter words (called the triplet code on the DNA, or the codon on the mRNA). Each of the different kinds of triplet codes for a specific amino acid. This code is represented at the bottom of the page in the **mRNA-amino acid table**. There are 64 possible combinations of the four kinds of bases making up three-letter words. As a result, there is some **degeneracy** in the code; a specific amino acid may have several triplet codes.

1. (a) Use the base-pairing rule for DNA replication to create the complementary strand for the template strand below.

    (b) For the same DNA strand, determine the mRNA sequence and then use the mRNA–amino acid table to determine the corresponding amino acid sequence. Note that in mRNA, uracil (U) replaces thymine (T) and pairs with adenine.

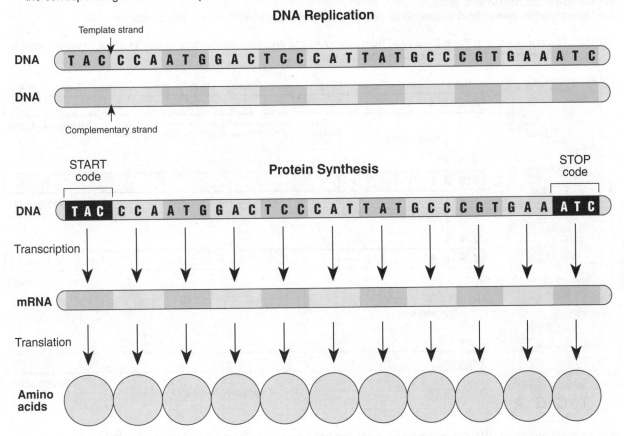

mRNA - Amino Acid Table

How to read the table

The table on the right is used to 'decode' the genetic code as a sequence of amino acids in a polypeptide chain, from a given mRNA sequence. The amino acid names are shown as three letter abbreviations (e.g. Ser = serine). To work out which amino acid is coded for by a codon (3 bases in the mRNA), carry out the following steps:

i   Look for the first letter of the codon in the row on the left hand side of the table.

ii Look for the column that intersects the same row from above that matches the second base.

iii Locate the third base in the codon by looking along the row on the right hand side that matches your codon.

Example: **GAU** codes for Asp (asparagine)

2. (a) State the mRNA START and STOP codons: _____

    (b) Describe the function of the START and STOP codons in a mRNA sequence: _____

    _____

Related activities: Amino Acids, The Simplest Case: Genes to Proteins

A 1

Nucleic Acids

Analysing a DNA Sample

The nucleotide (base sequence) of a section of DNA can be determined using DNA sequencing techniques. The base sequence determines the amino acid sequence of the resultant protein therefore the DNA tells us what type of protein that gene encodes. This exercise reviews the areas of DNA replication, transcription, and translation using an analysis of a gel electrophoresis column. **Attempt it after you have completed the rest of this topic.** Remember that the gel pattern represents the sequence in the synthesised strand.

1. Determine the amino acid sequence of a protein from the nucleotide sequence of its DNA, with the following steps:

   (a) Determine the sequence of **synthesised DNA** in the gel
   (b) Convert it to the complementary sequence of the **sample DNA**
   (c) Complete the **mRNA** sequence
   (d) Determine the **amino acid** sequence by using the *mRNA - amino acid table* in this workbook.

   NOTE: The nucleotides in the gel are read from bottom to top and the sequence is written in the spaces provided from left to right (the first four have been done for you).

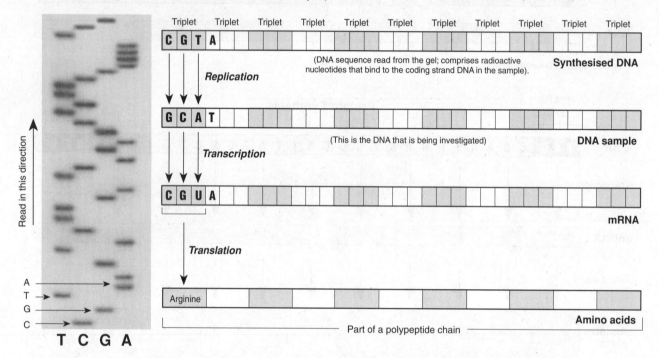

2. For each single strand DNA sequence below, write the base sequence for the **complementary DNA** strand:

   (a) DNA: T A C   T A G   C C G   C G A   T T T   A C A   A T T

   DNA: _____

   (b) DNA: T A C   G C C   T T A   A A G   G G C   C G A   A T C

   DNA: _____

   (c) Identify the cell process that this exercise represents: _____

3. For each single strand DNA sequence below, write the base sequence for the **mRNA** strand and the **amino acid** that it codes for (refer to the mRNA-amino acid table to determine the amino acid sequence):

   (a) DNA: T A C   T A G   C C G   C G A   T T T   A C A   A T T

   mRNA: _____

   Amino
   acids: _____

   (b) DNA: T A C   G C C   T T A   A A G   G G C   C G A   A T C

   mRNA: _____

   Amino
   acids: _____

   (c) Identify the cell process that this exercise represents: _____

KEY TERMS: What Am I?

THE OBJECT OF THIS GAME is to guess the unknown term from clues given to you by your team. Teams can be two or more, or you can play against individuals.
1) Cut out the cards below. You will need one set per team.
2) Shuffle the cards and deal them, face down, to each person in your team.

3) Affix tape to the back of the card so it can be stuck to your forehead. At no stage look at the word on the card!
4) One team starts. The members of your team give you a clue, one at a time, up to a maximum of **three** clues about what your term is. Do not use the word(s) on your card!

5) The clue should be a single point e.g. *"You feed on grass."*
6) If you guess correctly, your team receives another turn and the score is recorded. If you cannot guess, then the turn passes to the other team.
7) The game can be ended after one round or many.

Adenine	DNA replication	Helicase
Lagging strand	Leading strand	Nucleic acid
Nuceotide	Hyrogen bond	Protein
Gene expression	RNA	Okazaki fragments
Thymine	DNA	Purine
DNA polymerase	Pyrimidine	Uracil

Nucleic Acids

R 2

These cards have been deliberately left blank

Food and Health

KEY CONCEPTS

▶ A balanced diet is essential for good health.

▶ Imbalances in nutrient and energy intake can lead to obesity and malnutrition.

▶ Cardiovascular disease is associated with certain risk factors, including high LDL:HDL ratios.

▶ Selective breeding has been used to produce crops and animals for food production.

▶ Humans use a variety of methods to increase food production and reduce food spoilage.

KEY TERMS

antibiotic
atherosclerosis
balanced diet
carbohydrate
cardiovascular disease
cholesterol
coronary heart disease
deficiency (of nutrient)
domestication
fats
fertiliser
freezing
high density lipoprotein
(HDL)
hybrid vigour
inbreeding
irradiation
LDL:HDL ratio
low density lipoprotein
(LDL)
malnutrition
micronutrient
microorganism
mineral
obesity
outcrossing
pesticide
pickling
protein
salting
selective breeding
trans-fats
vitamin

OBJECTIVES

☐ 1. Use the **KEY TERMS** to help you understand and complete these objectives.

Diet and Health

pages 183-192

☐ 2. Explain what is meant by a **balanced diet**, recognising its components in terms of adequate nutrient intake and the functions of **protein**, **fats**, **carbohydrates**, **vitamins**, and **minerals**. Recognise that dietary guidelines help people to make informed food choices.

☐ 3. Explain how an unbalanced diet can lead to **malnutrition** and **obesity**. Distinguish between deficiencies in energy and protein, and **micronutrient** deficiencies. Recognise the link between poor nutrition, obesity, micronutrient deficiencies, and poor health. Explain the link between poverty and poor nutrition in both developed and developing countries.

☐ 4. Describe the health risks associated with obesity, including its role as a risk factor in **cardiovascular disease**.

☐ 5. Discuss the possible links between diet and **coronary heart disease** and critically evaluate the evidence for such links.

☐ 6. Describe the functional differences between **high-density lipoproteins** (HDL) and **low-density lipoproteins** (LDL). Discuss the possible effects of high blood **cholesterol** levels on the cardiovascular system. In particular, discuss the role of **LDL:HDL ratio** on risk of cardiovascular disease.

Food Production and Preservation

pages 182, 193-203

☐ 7. Recognise that plants are the basis of all food chains, and that as such humans are dependent on them for all food. Understand the role that the **green revolution** has played in increasing global food production rates.

☐ 8. Discuss how **selective breeding** is used to breed high yield crop plants, or crops with properties such as increased resistance to disease and pests.

☐ 9. Describe how selective breeding is used to breed highly productive domestic animals. Explain how the domestic animals bred for food today have been obtained by selective breeding from their wild ancestors. Appreciate the value in maintaining the biodiversity of these ancient lines.

☐ 10. Discuss how **fertilisers**, **pesticides**, and **antibiotics** are used to increase food production. Include reference to the implications of these practices.

☐ 11. Describe how **microorganisms** are used in food technology industries to make food for humans. Include reference to both the advantages and disadvantages of using such techniques.

☐ 12. Understand that **food preservation** techniques act by reducing the growth of microbial contaminants. Explain how common food preservation techniques prevent the spoilage of food by microorganisms. Include reference to: **salting**, adding sugar, **pickling**, **freezing**, **heat treatment**, and **irradiation**.

Periodicals:

Listings for this chapter are on page 339

Weblinks:

www.biozone.co.uk/
weblink/OCR-AS-2641.html

Global Human Nutrition

Globally, 854 million people are undernourished and, despite advances in agricultural practices and technologies, the number of hungry people in the world continues to rise. The majority of these people live in developing nations, but 9 million live in industrialised countries. Over 6 million people die annually from starvation, while millions of others suffer debilitating diseases as a result of malnutrition. Protein deficiencies (such as kwashiorkor), are common amongst the world's malnourished, because the world's poorest nations consume only a fraction of the world's protein resources, surviving primarily on cereal crops. Political and environmental factors contribute significantly to the world's hunger problem. In some countries, food production is sufficient to meet needs, but inadequate distribution methods cause food shortages in some regions.

Human Nutritional Requirements

A **balanced diet,** taken from the components below, is essential for human growth, development, metabolism, and good health. In many developing countries, deficiency diseases and starvation are prevalent either because of an absolute scarcity of food or because of inadequate nutrition. In many developed Western nations, an oversupply of cheap, nutritionally poor and highly processed food is contributing to an increase of diet-related diseases such as obesity, diabetes, and heart disease. **Malnutrition** (a lack of specific nutrients), once commonly associated with undernutrition, is now rising in developed nations over consuming on poor quality processed foods.

beans

lentils

Proteins (supplied by beans and pulses, and animal products such as meat and fish) are essential to growth and repair of muscle and other tissues. Unlike animal protein, plant protein is incomplete and sources must be chosen to complement one another nutritionally. Deficiencies result in kwashiorkor or marasmus.

Carbohydrates (right) are supplied in breads, starchy vegetables, cereals, and grains. They form the staple of most diets and provide the main energy source for the body.

Fats (left) and oils provide an energy source and are important for absorption of fat soluble vitamins.

Minerals (inorganic elements) and **vitamins** (essential organic compounds) are both required for numerous normal body functions. They are abundant in fruit and vegetables (right)

Humans and Agriculture

Via photosynthesis, plants are the ultimate source of food and metabolic energy for nearly all animals. Besides foods (e.g. grains, fruits, and vegetables), plants also provide humans with shelter, clothing, medicines, fuels, and the raw materials from which many other products are made.

Plant tissues provide the energy for almost all heterotrophic life. Low technology, low-input subsistence agriculture can provide enough food to supply a family unit, but the diet may not always be balanced because of limited access to seed and fertilisers and limitations on the crops that can be grown regionally.

Industrialised agriculture produces high yields per unit of land at cheaper prices, but has a large environmental impact because of high inputs of energy, fertilisers, and pesticides. **Wheat production** (left) and animal "factory farming" are examples. Rice production is also an example of intensive agriculture, but remains largely traditional (not mechanised) in many parts of the world.

Plantation agriculture is practised mainly in tropical countries solely to produce a high value cash crop for sale in developed countries. Typical crops include bananas (left), cotton, coffee, sugarcane, tobacco, and cocoa. Cash crops can deprive subsistence farmers of the land they need to grow their own food.

1. Contrast the primary causes of malnutrition in developed and developing countries: _____

_____

_____

_____

_____

_____

2. *"One of the likely effects of a global fuel crisis would be food shortage"*. Explain this statement:_____

_____

_____

_____

_____

A Balanced Diet

Nutrients are required for metabolism, tissue growth and repair, and as an energy source. Good nutrition (provided by a **balanced diet**) is recognised as a key factor in good health. Conversely poor nutrition (malnutrition) may cause ill-health or **deficiency diseases**. A diet refers to the quantity and nature of the food eaten. While not all foods contain all the representative nutrients, we can obtain the required balance of different nutrients by eating a wide variety of foods. In a recent overhaul of previous dietary recommendations, the health benefits of monounsaturated fats (such as olive and canola oils), fish oils, and whole grains have

been recognised, and people are being urged to reduce their consumption of highly processed foods and saturated (rather than total) fat. Those on diets that restrict certain food groups (e.g. vegans) must take care to balance their intake of foods to ensure an adequate supply of protein and other nutrients (e.g. iron and B vitamins). **Reference Nutrient Intakes** (RNIs) (see the next page) provide nutritional guidelines for different sectors of the population in the UK. RNIs help to define the upper and lower limits of adequate nutrient intake for most people, but they are not recommendations for intakes by individuals.

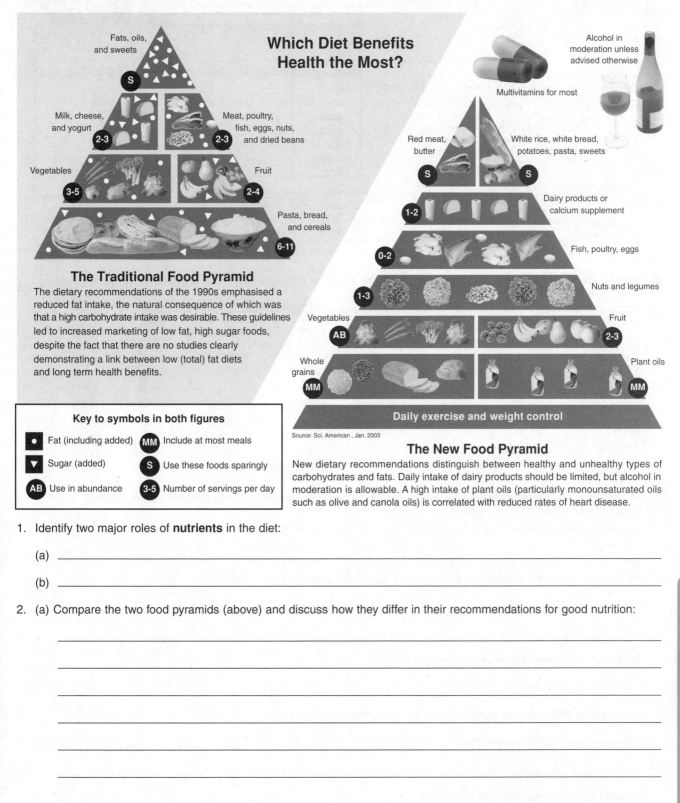

Which Diet Benefits Health the Most?

The Traditional Food Pyramid

The dietary recommendations of the 1990s emphasised a reduced fat intake, the natural consequence of which was that a high carbohydrate intake was desirable. These guidelines led to increased marketing of low fat, high sugar foods, despite the fact that there are no studies clearly demonstrating a link between low (total) fat diets and long term health benefits.

Source: Sci. American , Jan. 2003

The New Food Pyramid

New dietary recommendations distinguish between healthy and unhealthy types of carbohydrates and fats. Daily intake of dairy products should be limited, but alcohol in moderation is allowable. A high intake of plant oils (particularly monounsaturated oils such as olive and canola oils) is correlated with reduced rates of heart disease.

Key to symbols in both figures

- ● Fat (including added)
- ▼ Sugar (added)
- **AB** Use in abundance
- **MM** Include at most meals
- **S** Use these foods sparingly
- **3-5** Number of servings per day

1. Identify two major roles of **nutrients** in the diet:

   (a) _____

   (b) _____

2. (a) Compare the two food pyramids (above) and discuss how they differ in their recommendations for good nutrition:

   _____
   _____
   _____
   _____
   _____
   _____
   _____
   _____

Periodicals:
The good, the fad, and the unhealthy

Related activities: Deficiency Diseases, Dietary Disorders

DA 2

Food and Health

Nutritional Guidelines in the UK

In the UK, Dietary Reference Values (DRVs) provide guidelines for nutrient and energy intake for particular groups of the population. In a population, it is assumed that the nutritional requirements of the population as a whole are represented by a normal, bell-shaped, curve (below). DRVs collectively encompass RNIs, LRNIs, and EARs, and replace the earlier Recommended Daily Amounts (RDAs), which *recommended* nutrient intakes for particular groups in the population, including those with very high needs.

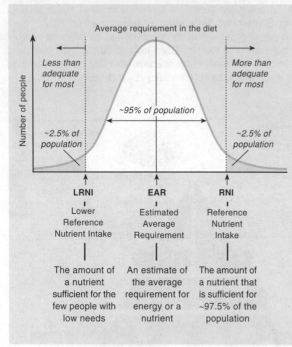

Average requirement in the diet

Number of people

Less than adequate for most

More than adequate for most

~95% of population

~2.5% of population

~2.5% of population

LRNI
Lower Reference Nutrient Intake

EAR
Estimated Average Requirement

RNI
Reference Nutrient Intake

The amount of a nutrient sufficient for the few people with low needs

An estimate of the average requirement for energy or a nutrient

The amount of a nutrient that is sufficient for ~97.5% of the population

Table 1 (below): Estimated Average Requirements (EAR) for energy, and Reference Nutrient Intakes (RNIs) for selected nutrients, for UK males and females aged 19-50 years (per day).

Source: Dept of Health. Dietary Reference Values for Food Energy and Nutrients for the UK, 1991.

Age range	Reference Nutrient Intakes (RNIs)					EARs	
	Protein (g)	Calcium (mg)	Iron (mg)	Folate (μg)	Vit.C (mg)	EAR (MJ) Males	Females
Males							
19-50 years	55.5	700	8.7	700	40	10.60	
Females							
19-50 years	45.0	700	14.8	600	40		8.10
Pregnant	51.0	1250	14.8	700	50		8.90
Lactating	56.0	1250	14.8	950	70		10.20

DRVs have been set for population groups within the UK, taking into account age and gender. Only a portion of the table is shown here.

RNIs are provided for each constituent of a balanced diet.

EARs for energy are based on the present lifestyles and activity levels of the UK population.

(b) Based on the information on the graph (right), state the evidence that might support the revised recommendations:

_____

_____

_____

_____

_____

_____

_____

■ Percentage of calories from fat in the traditional diet

□ Incidence of coronary heart disease per 10 000 men (over a 10 year period)

| | 38% | 40% |
| | 3000 | |

Country	Japan	Eastern Finland	Crete
	10%, 500		200
Type of fat consumed	Low fat	Saturated fat	Monounsaturated fat (olive oil)

3. With reference to the table above, contrast the nutritional requirements of non-pregnant and lactating women:

_____

_____

4. (a) Suggest in which way the older RDAs could have been misleading for many people: _____

_____

_____

(b) Explain how the DRVs differ from the older RDAs: _____

_____

5. Suggest how **DRVs** can be applied in each of the following situations:

(a) Dietary planning and assessment: _____

_____

(b) Food labelling and consumer information: _____

_____

Deficiency Diseases

Malnutrition is the general term for nutritional disorders resulting from not having enough food (starvation), not enough of the right food (deficiency), or too much food (obesity). Children under 5 are the most at risk from starvation and deficiency diseases because they are growing rapidly and are more susceptible to disease. Malnutrition is a key factor in the deaths of 6 million children each year and, in developing countries, dietary deficiencies are a major problem. In these countries, malnutrition usually presents as **marasmus** or **kwashiorkor** (energy and protein deficiencies).

Specific vitamin and mineral deficiencies (below and following page) in adults are associated with specific disorders, e.g. **beriberi** (vitamin B₁), **scurvy** (vitamin C), **rickets** (vitamin D), **pellagra** (niacin), or **anaemia** (iron). Vitamin deficiencies in childhood result in chronic, lifelong disorders. Deficiency diseases are rare in developed countries. People who do suffer from some form of dietary deficiency are either alcoholics, people with intestinal disorders that prevent proper nutrient uptake, or people with very restricted diets (e.g. vegans).

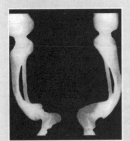

Vitamin D Deficiency

Lack of vitamin D in children produces the disease rickets. In adults a similar disease is called osteomalacia. Suffers typically show skeletal deformities (e.g. bowed legs, left) because inadequate amounts of calcium and phosphorus are incorporated into the bones. Vitamin D is produced by the skin when exposed to sunlight and it is vital for the absorption of calcium from the diet.

Vitamin A Deficiency

Vitamin A (found in animal livers, eggs, and dairy products) is essential for the production of light-absorbing pigments in the eye and for the formation of cell structures. Symptoms of deficiency include loss of night vision, inflammation of the eye, **keratomalacia** (damage to the cornea), and the appearance of **Bitots spots**, evident as foamy, opaque patches on the white of the eye (refer to photo).

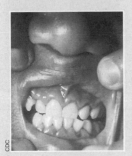

Vitamin C Deficiency

Vitamin C deficiency causes a disease known as scurvy. It is now rare in developed countries because of increased consumption of fresh fruit and vegetables. Inadequate vitamin C intake disturbs the body's normal production of collagen, a protein in connective tissue that holds body structures together. This results in poor wound healing, rupture of small blood vessels (visible bleeding in the skin), swollen gums, and loose teeth.

Vitamin B₁ Deficiency

Vitamin B₁ (thiamine) is required for respiratory metabolism, and nerve and muscle function. Lack of thiamine causes the metabolic disorder, **beriberi**, which occurs predominantly in underfed populations, or in breast fed babies whose mother is on a restrictive diet. Symptoms of beriberi include nerve degeneration, heart failure, and oedema (swelling caused by fluid accumulation). Without medical treatment, sufferers will die.

Kwashiorkor

A severe type of protein-energy deficiency in young children (1-3 years old), occurring mainly in poor rural areas in the tropics. Kwashiorkor occurs when a child is suddenly weaned on to a diet that is low in calories, protein, and certain essential micronutrients. The problem is often made worse by a poor appetite due to illnesses such as measles. Children have stunted growth, oedema (accumulation of fluid in the tissues), and are inactive, apathetic and weak. Resistance against infection is lost, which may be fatal.

Marasmus

Marasmus is the most common form of deficiency disease. It is a severe form of protein and energy malnutrition that usually occurs in famine or starvation conditions. Children suffering from marasmus are stunted and extremely emaciated. They have loose folds of skin on the limbs and buttocks, due to the loss of fat and muscle tissue. Unlike kwashiorkor sufferers, marasmus does not cause the bloated and elongated abdomen. However sufferers have no resistance to disease and common infections are typically fatal.

1. Distinguish between **malnutrition** and **starvation**:

_____

_____

_____

_____

2. For each of the following vitamins, identify the natural sources of the vitamin, its function, and effect of deficiency:

(a) Vitamin A: _____

Function: _____

Deficiency: _____

(b) Vitamin B₁: _____

Function: _____

Deficiency: _____

Periodicals:
The alphabet soup of vitamins

Related activities: *Dietary Disorders, A Balanced Diet*

RA 2

Food and Health

Common Mineral Deficiencies

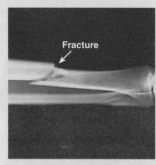
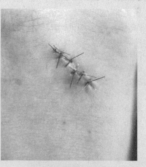
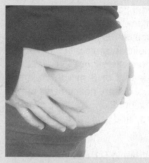
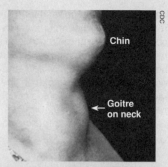

Calcium Deficiency

Calcium is required for enzyme function, formation of bones and teeth, blood clotting, and muscular contraction. Calcium deficiency causes poor bone growth and structure, increasing the tendency of bones to fracture and break. It also results in muscular spasms and poor blood clotting ability.

Zinc Deficiency

Zinc is found in red meat, poultry, fish, whole grain cereals and breads, legumes, and nuts. It is important for enzyme activity, production of insulin, making of sperm, and perception of taste. A deficiency in zinc causes growth retardation, a delay in puberty, muscular weakness, dry skin, and a delay in wound healing.

Iron Deficiency

Anaemia results from lower than normal levels of haemoglobin in red blood cells. Iron from the diet is required to produce haemoglobin. People most at risk include women during **pregnancy** and those with an inadequate dietary intake. Symptoms include fatigue, fainting, breathlessness, and heart palpitations.

Iodine Deficiency

Iodine is essential for the production of thyroid hormones. These hormones control the rate of metabolism, growth, and development. Shortage of iodine in the diet may lead to **goitre** (thyroid enlargement as shown above). Iodine deficiency is also responsible for some cases of thyroid underactivity (**hypothyroidism**).

(c) Vitamin C: _____

　　　Function: _____

　　　Deficiency: _____

(d) Vitamin D: _____

　　　Function: _____

　　　Deficiency: _____

3. Suggest why young children, pregnant women, and athletes are among the most susceptible to dietary deficiencies:

_____

_____

_____

4. Explain why a lack of iron leads to the symptoms of anaemia (fatigue and breathlessness): _____

_____

_____

5. Suggest why a zinc deficiency is associated with muscular weakness and a delay in puberty: _____

_____

_____

6. Using the example of **iodine**, explain how artificial dietary supplementation can be achieved and discuss its benefits:

_____

_____

7. Explain why people suffering from nutritional deficiencies have a poor resistance to disease: _____

_____

_____

_____

_____

Malnutrition and Obesity

Malnutrition describes an imbalance between what someone eats and what is needed to stay healthy. In economically developed areas of the world, most (but not all) forms of malnutrition are the result of poorly balanced nutrient intakes rather than a lack of food *per se*. Amongst the most common of these is **obesity**, as indicated by **Body Mass Index** (BMI) values in excess of 30. In Britain, 20-30% of all adults are obese. Obesity is commonly the result of excessive energy intake, usually associated with a highly processed diet. Obesity is a risk factor in a number of chronic diseases. Poor nutritional choices amongst the obese are often associated with low intakes of **micronutrients** (minerals and vitamins) and can lead to deficiency diseases such as anaemia. Somewhat paradoxically, obesity in developed countries is more common in poorly educated, lower socio-economic groups than amongst the wealthy, who usually have more options in terms of food choices. Obesity is also increasing in incidence in developing nations, such as China, even where hunger exists. While the simple explanation for excessive body fat is simple (energy in exceeds energy out), a complex of biological and socio-economic factors are implicated in creating the problems of modern obesity.

Obesity and Malnutrition

Micronutrients are needed for proper growth, development, and bodily function. Deficiencies result in stunted growth, poor neural development, and immune insufficiency. Women of reproductive age, children, the elderly, and people with compromised immune systems are most at risk. Obesity is increasingly a problem when income is restricted (below and overleaf). This is often related to the consumption of a highly processed, energy dense diet lacking in the essential nutrients found in fresh and unprocessed foods. Obese people are at risk of deficiencies in iron, vitamin A, B group vitamins, and trace minerals such as selenium.

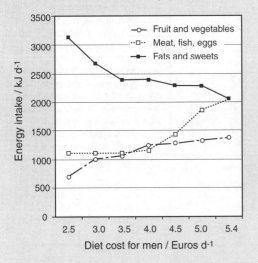

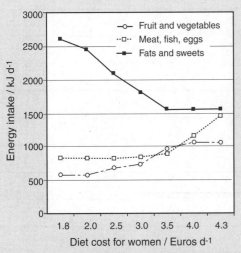

*ABOVE: **Change in energy intakes from different food sources**. A study of men and women in the Val-de-Marne region, France showed that their energy intakes from different food sources changed after imposing cost constraints. The less people had to spend on food, the greater the decline in their intake of fruits vegetables, and sources of protein.*
Data: American Journal of Clinical Nutrition (2004)

Possible Root Causes of Obesity

▶ Eating too many calories for energy needs.
▶ Increased consumption of calorie-dense foods and low fibre intake.
▶ Overconsumption of trans-fatty acids and refined carbohydrates.
▶ Reduced energy expenditure; a decline in physical activity.

Obesity more than doubles the risk of hypertension and stroke.

Obesity is a major independent risk factor for cardiovascular disease because it is associated with increased prevalence of cardiovascular risk factors, including **type 2 diabetes**, **high LDL cholesterol**, **low HDL cholesterol**, and high levels of triglycerides.

The heaviness of the chest wall and a higher-than-normal oxygen requirement in obese people restricts normal physical activity and increases respiratory problems.

Insulin resistance is prevalent in obese patients and is associated with non-alcoholic fatty liver disease. Obesity is associated with high bile cholesterol levels, gallstones and gall bladder disease.

Obesity is clearly associated with higher risk of certain types of cancers, including rectal, colon and breast cancer. Cancer survival rates are also lower among obese patients.

Obesity in premenopausal women is associated with irregular menstrual cycles and infertility.

Obesity carries a high risk of osteoarthritis of weight-bearing joints.

Body Mass Index

A common method of assessing obesity is the **body mass index** (BMI).

Body mass index (BMI) =

$$\frac{\text{weight of body (in kg)}}{\text{height (in metres) squared}}$$

A BMI of: 17 to 20 = underweight
20 to 25 = normal weight
25 to 30 = overweight
over 30 = obesity

$$BMI = \frac{90 \text{ kg}}{(1.68)^2} = 32$$

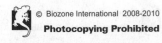
Food and Health

The Problem of the Malnourished Obese

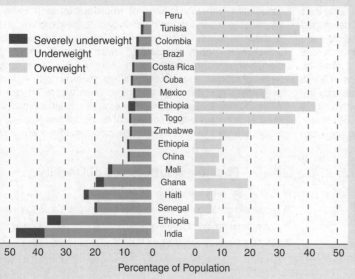

Severely underweight
Underweight
Overweight

Peru
Tunisia
Colombia
Brazil
Costa Rica
Cuba
Mexico
Ethiopia
Togo
Zimbabwe
Ethiopia
China
Mali
Ghana
Haiti
Senegal
Ethiopia
India

Percentage of Population

Increasingly, obesity is a problem for the world's poor, who often rely on nutritionally poor carbohydrate staples and are deficient in the micronutrients required for good health. A 1999 United Nations study found obesity to be growing rapidly in all developing countries, even in countries where hunger exists, such as China. The well-intentioned policies of some countries can compound the problem. For example, the Egyptian Food Subsidy Programme, which reduced the relative prices of energy-dense, nutrient-poor food items, is one of the major factors contributing to the emergence of obese and micronutrient deficient mothers in the country.

Left: Weight profiles in selected developing nations. WHO 1997, via FAO

Below: Percentage obesity in relation to income and education. American Journal of Clinical Nutrition, 2004

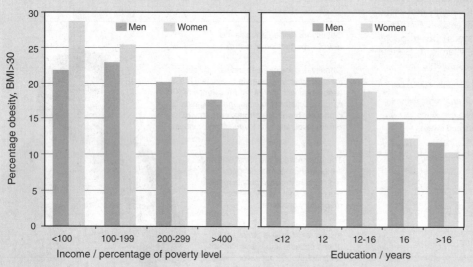

Percentage obesity, BMI>30

Men   Women

Income / percentage of poverty level

Men   Women

Education / years

In both developed and developing countries, many disparities in health are linked to inequalities in education and income. The highest rates of obesity occur amongst the most poorly educated, impoverished sectors of the population. In addition, there is an inverse relation between energy density (MJ per kg) and energy cost ($ per MJ), such that energy-dense refined foods may represent the lowest-cost option to consumers. Poverty and food insecurity are associated with lower food expenditures, low fruit and vegetable consumption, and lower-quality diets.

1. Explain why obesity is regarded as a form of malnutrition: _____

_____

2. Describe the two basic energy factors that determine how a person's weight will change: _____

_____

3. (a) Describe the evidence linking obesity to economic hardship: _____

_____

_____

_____

(b) Explain how this could also explain the rising incidence of obesity in developing nations: _____

_____

_____

4. Discuss the factors that are likely contributors to micronutrient deficiencies among the obese: _____

_____

_____

_____

Cardiovascular Disease

Cardiovascular disease (CVD) is a term describing all diseases involving the heart and blood vessels. It includes coronary heart disease (CHD), atherosclerosis, hypertension (high blood pressure), peripheral vascular disease, stroke, and congenital heart disorders. The British Heart Foundation estimates CVD is responsible for 34% of deaths in the UK, making it the single most common cause of death in the UK. The economic costs of the disease are high. In 2006, it was estimated that £14.4 billion was spent on treating CVD, and related lost productivity cost an additional £16 billion. Despite high mortality, CVD related deaths have been in decline since the late 1970s, mainly due to improvements in education about the disease and its risk factors, and advances in screening and treatment. Its continued prevalence is of considerable public health concern, particularly as many of the **risk factors** involved, such as cigarette smoking, obesity, and high blood cholesterol, are controllable.

Types of Cardiovascular Disease

Atherosclerosis (hardening of the arteries) is caused by deposits of fats and cholesterol on the inner walls of the arteries. Blood flow becomes restricted and increases the risk of blood clots (**thrombosis**). Complications arising as a result of atherosclerosis include heart attack (**infarction**), gangrene, and **stroke**. A stroke is the rapid loss of brain function due to a disturbance in the blood supply to the brain, and may result in death if the damage is severe. Speech, or vision and movement on one side of the body is often affected.

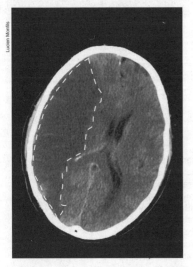

The CT scan (above) shows a brain affected by a severe cerebral infarction or ischaemic stroke. The loss of blood supply results in tissue death (outlined area). Blood clots resulting from atherosclerosis are a common cause of ischaemic stroke.

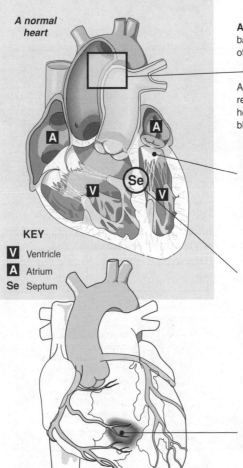

A normal heart

KEY

V	Ventricle
A	Atrium
Se	Septum

Restricted supply of blood to heart muscle resulting in myocardial infarction

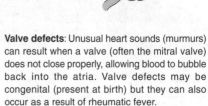

Aortic aneurysm: A ballooning and weakening of the wall of the aorta.

Aneurysms usually result from generalised heart disease and high blood pressure.

Valve defects: Unusual heart sounds (murmurs) can result when a valve (often the mitral valve) does not close properly, allowing blood to bubble back into the atria. Valve defects may be congenital (present at birth) but they can also occur as a result of rheumatic fever.

Septal defects: These hole-in-the-heart congenital defects occur where the dividing wall (**septum**) between the left and right sides of the heart is not closed. These defects may occur between the atria or the ventricles, and are sometimes combined with valve problems.

Myocardial infarction (*heart attack*): Occurs when an area of the heart is deprived of blood supply resulting in tissue damage or death. It is the major cause of death in developed countries. Symptoms of infarction include a sudden onset of chest pain, breathlessness, nausea, and cold clammy skin. Damage to the heart may be so severe that it leads to heart failure and even death (myocardial infarction is fatal within 20 days in 40 to 50% of all cases).

1. Explain what is meant by cardiovascular disease (CVD): _____

_____

2. Distinguish between congenital and acquired CVD, including reference to risk factors in disease development:

_____

_____

_____

3. Explain why the high rates of cardiovascular disease are such a major public health concern: _____

_____

_____

_____

Periodicals:
Coronary heart disease

Related activities: Malnutrition and Obesity, Smoking and the Cardiovascular System

RDA 3

Food and Health

Atherosclerosis

Atherosclerosis is a disease of the arteries caused by **atheromas** (fatty deposits) on the inner arterial walls. An atheroma is made up of cells (mostly macrophages) or cell debris, with associated fatty acids, cholesterol, calcium, and varying amounts of fibrous connective tissue. The accumulation of fat and plaques causes the lining of the arteries to degenerate. Atheromas weaken the arterial walls and eventually restrict blood flow through the arteries, increasing the risk of **aneurysm** (swelling of the artery wall) and **thrombosis** (blood clots). Complications arising as a result of atherosclerosis include heart attacks, strokes, and gangrene. A typical progression for the formation of an atheroma is illustrated below.

Initial lesion	Fatty streak	Intermediate lesion	Atheroma	Fibroatheroma	Complicated plaque
Atherosclerosis is triggered by damage to an artery wall caused by blood borne chemicals or persistent **hypertension**.	Low density lipoproteins (LDLs) accumulate beneath the endothelial cells. Macrophages follow and absorb them, forming foam cells.	Foam cells accumulate forming greasy yellow lesions called atherosclerotic plaques.	A core of extracellular lipids under a cap of fibrous tissue forms.	Lipid core and fibrous layers. Accumulated smooth muscle cells die. Fibres deteriorate and are replaced with scar tissue.	Calcification of plaque. Arterial wall may ulcerate. Hypertension may worsen. Plaque may break away causing a clot.

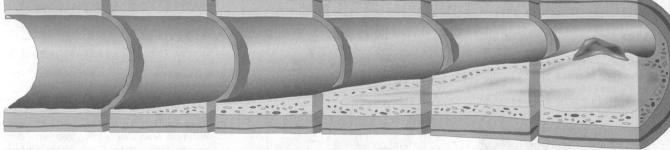

Earliest onset	From first decade	From third decade	From fourth decade
Growth mechanism	Growth mainly by lipid accumulation	Smooth muscle/ collagen increase	Thrombosis, haematoma
Clinical correlation	Clinically silent	Clinically silent or overt	

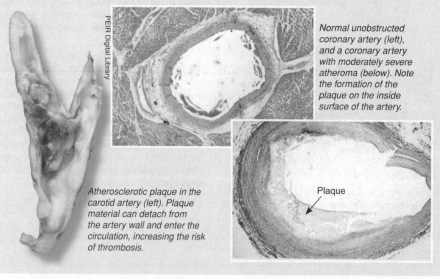

PEIR Digital Library

Normal unobstructed coronary artery (left), and a coronary artery with moderately severe atheroma (below). Note the formation of the plaque on the inside surface of the artery.

Plaque

Atherosclerotic plaque in the carotid artery (left). Plaque material can detach from the artery wall and enter the circulation, increasing the risk of thrombosis.

Recent studies indicate that most heart attacks are caused by the body's **inflammatory response** to a plaque. The inflammatory process causes young, soft, cholesterol-rich plaques to rupture and break into pieces. If these block blood vessels they can cause lethal heart attacks, even in previously healthy people.

Aorta opened lengthwise (above), with extensive atherosclerotic lesions (arrowed).

1. Explain why most people are unlikely to realise they are developing atherosclerosis until serious complications arise:

_____

_____

_____

2. Explain how an atherosclerostic plaque changes over time: _____

_____

_____

3. Describe some of the consequences of developing atherosclerosis: _____

_____

Cholesterol, Diet, and Heart Health

Cholesterol is a sterol lipid found in all animal tissues as part of cellular membranes. It regulates membrane fluidity and is an important precursor molecule in vitamin D synthesis, as well as many steroid hormones (e.g. testosterone and oestrogen). Cholesterol is not soluble in the blood and is transported within spherical particles called **lipoproteins**. There are various compositions of lipoproteins (e.g. HDL and LDL, below left) and the composition determines how cholesterol will be metabolised. Cholesterol has an essential role in body chemistry and is made by the body even when it is not taken in as part of the diet. However, high levels of cholesterol (particularly the ratio of LDL to HDL) are associated with cardiovascular disease.

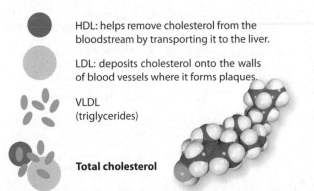

HDL: helps remove cholesterol from the bloodstream by transporting it to the liver.

LDL: deposits cholesterol onto the walls of blood vessels where it forms plaques.

VLDL (triglycerides)

Total cholesterol

Total blood cholesterol comprises low density lipoprotein (LDL), high density lipoprotein (HDL), and very low density lipoprotein (VLDL), which is the triglyceride carrying component in the blood.

Cholesterol and Risk of CVD

Abnormally high concentrations of LDL and lower concentrations of functional HDL are strongly associated with the development of atheroma. It is the **LDL:HDL ratio**, rather than total cholesterol itself, that provides the best indicator of risk for developing cardiovascular disease, and the risk profile is different for men and women (below). The LDL:HDL ratio is mostly genetically determined but can be influenced by body composition, diet, and exercise. Aerobic exercise and weight reduction both increase HDL and reduce LDL levels.

Ratio of LDL to HDL

Risk	Men	Women
Very low (half mean risk)	1.0	1.5
Mean risk	3.6	3.2
Moderate risk (2X mean risk)	6.3	5.0
High (3X mean risk)	8.0	6.1

Effects of HDL and LDL on CVD

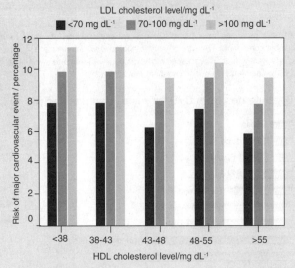

Above: A higher LDL increases CVD risk at any given level of HDL.

What About Diet?

The link between diet and heart disease was hypothesised more than 50 years ago when researchers found that high blood cholesterol was associated with the development of atherosclerotic plaques. They hypothesised that the corollary of this was that a diet high in cholesterol must contribute to high blood cholesterol.

Consumers were encouraged to switch to polyunsaturated oils and spreads (such as margarine) and reduce their intake of saturated fat. This was especially the case post-WWII, when butter was expensive. The dietary fat-heart disease theory is persuasive. It seems to make sense; if you eat more cholesterol, you must have more in your blood. But is it true?

1. (a) Describe the essential role of cholesterol in the body: _____

(b) The body has a certain minimum requirement for cholesterol. Explain how this is obtained: _____

2. (a) Explain the link between high LDL:HDL ratio and the risk of cardiovascular disease: _____

(b) Explain why this ratio is more important to medical practitioners than total blood cholesterol *per se*: _____

(c) Suggest how this ratio could be lowered in at-risk individuals: _____

Periodicals: Heart disease and cholesterol

Related activities: Cardiovascular Disease, Atherosclerosis

A 2

Food and Health

The Debate Over Dietary Cholesterol

Despite decades of research, notably in the United States (the Framingham Heart Study) and Finland (the North Karelia Project), the data do not support of a link between dietary intake of cholesterol and cholesterol in the blood. The situation is complex but what does seem to be agreed is that consumption of *trans-* fatty acids (TFAs) is associated with increased risk of heart disease. TFAs raise the level of LDL, but also decrease levels of HDL. The net increase in LDL/HDL ratio with TFAs is approximately double that due to saturated fat. Recommendations against TFA consumption accord with current dietary guidelines to decrease the amount of processed (highly refined) food that is consumed, while increasing the proportion of whole grains, fruits, vegetables, and unrefined vegetable oils.

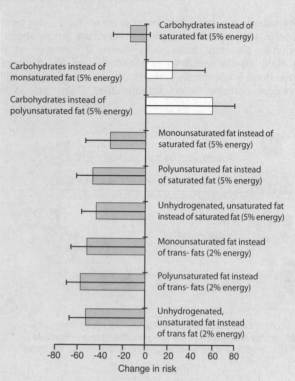

TFAs are produced when vegetable oils are hydrogenated to make them solid but spreadable. They are present in natural foods in only trace amounts but are widely used in processed foods because of their long shelf life and stability. They have no known benefit to human health.

Population Studies

High intakes of *trans-* fats are linked with increased heart disease risk in large-scale population studies. For example, the **Nurses' Health Study**, a cohort study involving more than 120,000 female nurses, found that CHD risk roughly doubled for each 2% increase in *trans-* fat calories consumed. They also found that replacing saturated and *trans-*fats with unsaturated fats produced a greater risk reduction than replacement with carbohydrates.

Deaths rates from CVD for males and females per 100,000 population

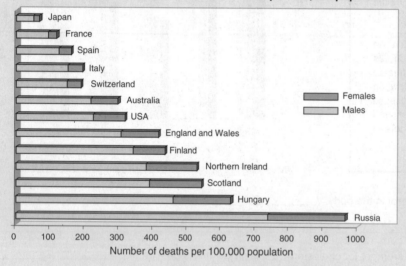

The bar graph (left) shows the mortality (per 100,000 population) attributable to cardiovascular disease (CVD) for men and women of selected countries. The rate of CVD is lowest in Japan and France and high in Northern Europe.

These differences between countries have been primarily attributed to diet (i.e. saturated fat, salt, vitamin and antioxidant content). Studies have found strong North - South gradients in both fruit and vegetable consumption between these countries. For example, a study in 1998 found that people in England consumed twice as much fruit (1.3 kg per person per week versus 0.7 kg) and one third more vegetables (1.3 kg per person per week versus 0.8 kg) than those living in Northern Ireland.

3. (a) Describe the evidence for the current recommendations that people reduce their intake of highly processed foods:

_____

_____

_____

(b) Describe the evidence supporting the link between fruit and vegetable intake and lower rates of death from CVD:

_____

_____

_____

The Green Revolution

Since the 1950s, most increases in global food production have come from increased yields per unit area of cropland rather than farming more land. The initial **green revolution** increased the intensity and frequency of cropping, using high inputs of fertilisers, pesticides, and water to increase yields in improved varieties. The **second green revolution** began in the 1960s and improved production by further developing high yielding crop varieties. The countries whose crop yields per unit of land area have increased during the two green revolutions are illustrated below. Several

agricultural research centres and **seed** or **gene banks** also play a key role in developing high yielding crop varieties. Most of the world's gene banks store the seeds of the hundred or so plant species that collectively provide approximately 90% of the food consumed by humans. However, some banks are also storing the seeds of species threatened with extinction or a loss of genetic diversity. Producing more food on less land is an important way of protecting biodiversity by saving large areas of natural habitat from being used to grow food.

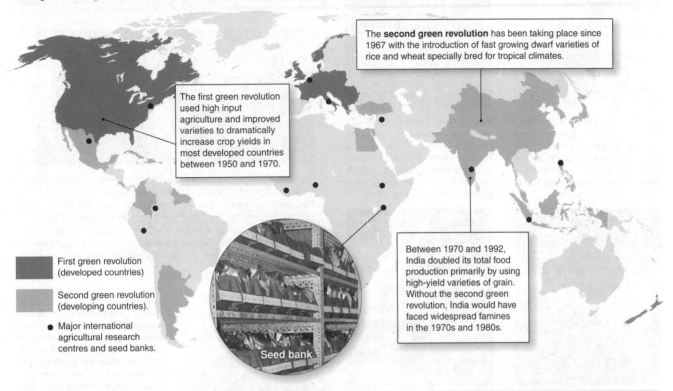

The **second green revolution** has been taking place since 1967 with the introduction of fast growing dwarf varieties of rice and wheat specially bred for tropical climates.

The first green revolution used high input agriculture and improved varieties to dramatically increase crop yields in most developed countries between 1950 and 1970.

Between 1970 and 1992, India doubled its total food production primarily by using high-yield varieties of grain. Without the second green revolution, India would have faced widespread famines in the 1970s and 1980s.

Seed bank

■ First green revolution (developed countries)

■ Second green revolution (developing countries).

● Major international agricultural research centres and seed banks.

High-input, intensive agriculture uses large amounts of fossil fuel energy, water, commercial inorganic fertilisers, and pesticides to produce large quantities of single crops (monocultures) from relatively small areas of land. At some point though, outputs diminish or even decline.

There are approximately 30 000 plant species with parts suitable for human consumption, but just three grain crops (wheat, rice, and corn) provide more than half the calories the world's population consumes. These crops have been the focus of the second green revolution.

Increased yields from industrialised agriculture depend on the extensive use of fossil fuels to run machinery, produce and apply fertilisers and pesticides, and pump water for irrigation. Since 1950, the use of fossil fuels in agriculture has increased four-fold.

1. Describe how the technologies of the first and second green revolutions differ: _____

_____

_____

_____

_____

_____

Food and Health

Periodicals: *What price more food?*

Related activities: *Global Human Nutrition, Selective Breeding in Crop Plants*
Web links: *Green Revolution, FAO: Crop and Grasslands Service*

A 2

The second green revolution (also called the **gene revolution**) is based on further developments in **selective breeding** and **genetic engineering**. It has grown rapidly in scope and importance since it began in 1967. Initially, it involved the development of fast growing, high yielding varieties of rice, corn, and wheat, which were specially bred for tropical and subtropical climates to meet global food demand. More recently, genetically modified seeds have been used to create plants with higher yields and specific tolerances (e.g. pest resistance, herbicide tolerance, or drought tolerance). GM seed is also used to improve the nutritional quality of crops (e.g. by increasing protein or vitamin levels), or to produce plants for edible vaccine delivery.

Recent Crop Developments

Winged bean

Upland rice

A new potential crop plant is the tropical winged bean (*Psophocarpus*). All parts of the plant are edible, it grows well in hot climates, and it is resistant to many of the diseases common to other bean species.

Most green revolution breeds are "high-responders", requiring optimum levels of water and fertiliser before they realise their yield potential. Under sub-optimal conditions they may not perform as well as traditional varieties.

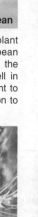

Wheat

Maize

Improvements in crop production have come from the modification of a few, well known species. Future research aims to maintain genetic diversity in high-yielding, disease resistant varieties.

A century ago, yields of maize (corn) in the USA were around 25 bushels per acre. In 1999, yields from hybrid maize were five to ten times this, depending on the growing conditions.

Improving Rice Crops

Rice is the world's second most important cereal crop, providing both a food and an income source to millions of people worldwide. Traditional rice strains lack many of the essential vitamins and minerals required by humans for good health and are susceptible to crop failure and low yields if not tended carefully. Advances in plant breeding, biotechnology, and genetic engineering (below) have helped to overcome these problems.

IR-8 rice

The second green revolution produced a high-yielding, semi-dwarf variety of rice called **IR-8** (above) in response to food shortages. IR-8 was developed by cross breeding two parental strains of rice and has shorter and stiffer stalks than either parent, allowing the plant to support heavier grain heads without falling over. More recently, a new improved variety, '**super rice**' has been developed to replace IR-8. Yields are expected to be 20% higher.

Genetic modification is being used to alter rice for a wide variety of purposes. **Golden rice** is genetically engineered to contain high levels of beta-carotene, which is converted in the body to vitamin A. This allows better nutrient delivery to people in poor underdeveloped countries where rice is the food staple. Other companies are focusing on improving the resistance of rice crops to insects, bacteria, and herbicides, improving yields, or delivering edible hepatitis and cholera vaccines in the rice.

2. Using examples, explain how the technologies of the second green revolution are being used to:

(a) Improve crop yields: _____

_____

_____

(b) Improve the nutritional quality of crops: _____

_____

_____

3. (a) Explain how countries currently suffering from food shortages might benefit from recent crop developments:

_____

_____

_____

(b) Describe the constraints that might exist on developing countries taking advantage of these potential benefits:

_____

_____

Selective Breeding in Crop Plants

Most agricultural plants and animals have undergone **selective breeding** (artificial selection), resulting in an astounding range of phenotypic variation over a relatively short period of time. Selective breeding involves breeding from the individuals with the most desirable phenotypes (e.g. high yield) to alter the average phenotype in the species. Wheat has been cultivated for more than 9000 years and has undergone many changes during the process of its domestication (below). Wheat's evolution involved two natural hybridisation events, accompanied by **polyploidy**. **Hybrids** are the offspring of genetically dissimilar parents and are important because they recombine the genetic characteristics of parental lines and show increased **heterozygosity**. This is associated with greater adaptability, survival, growth, and fertility in the offspring; a phenomenon known as **hybrid vigour**. Selective breeding is used to increase yield, and to produce disease and pest resistant crops (bottom).

Polyploidy Events in the Evolution of Wheat

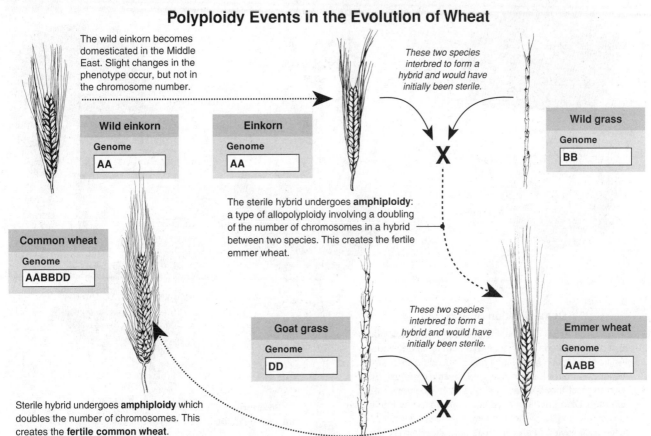

The wild einkorn becomes domesticated in the Middle East. Slight changes in the phenotype occur, but not in the chromosome number.

Wild einkorn Genome **AA**

Einkorn Genome **AA**

These two species interbred to form a hybrid and would have initially been sterile.

Wild grass Genome **BB**

The sterile hybrid undergoes **amphiploidy**: a type of allopolyploidy involving a doubling of the number of chromosomes in a hybrid between two species. This creates the fertile emmer wheat.

Common wheat Genome **AABBDD**

These two species interbred to form a hybrid and would have initially been sterile.

Goat grass Genome **DD**

Emmer wheat Genome **AABB**

Sterile hybrid undergoes **amphiploidy** which doubles the number of chromosomes. This creates the **fertile common wheat**.

Breeding programmes around the world are developing apples resistant to the bacterial disease that causes fireblight (above).

Modern wheat (above) has been selected for its non shattering heads, high yield, and high gluten (protein) content.

Hybrid corn varieties have been bred to minimise harm inflicted by insect pests such as the corn rootworm (above).

Food and Health

1. Explain how a farmer thousands of years ago was able to improve the phenotypic character of a cereal crop:

_____

_____

_____

_____

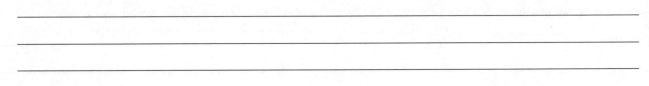

Periodicals: The adaptations of cereals

Related activities: The Green Revolution, Selective Breeding in Animals

RA 2

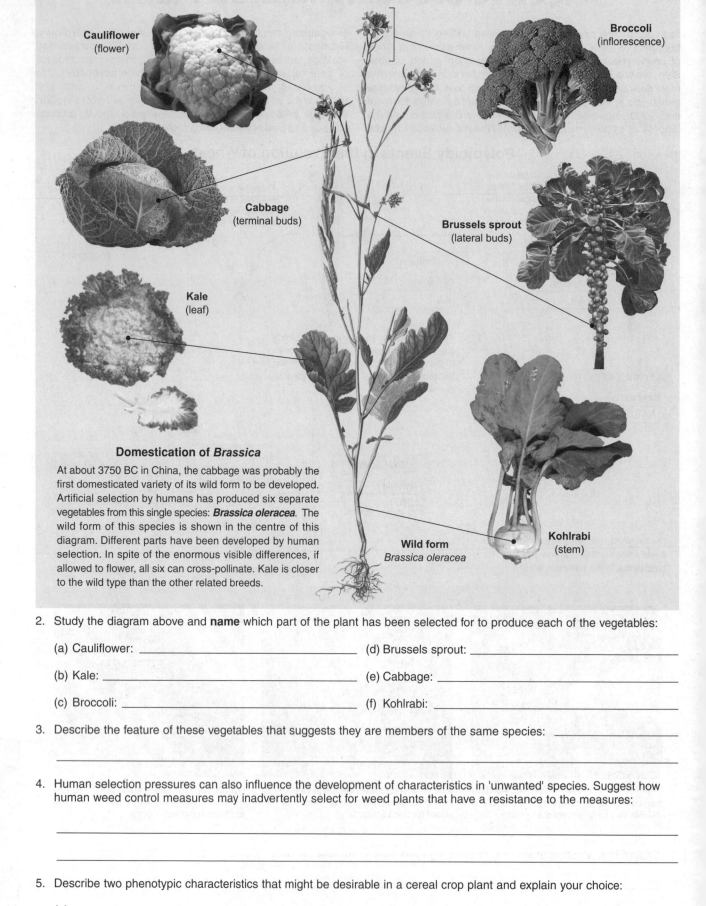

Domestication of *Brassica*

At about 3750 BC in China, the cabbage was probably the first domesticated variety of its wild form to be developed. Artificial selection by humans has produced six separate vegetables from this single species: ***Brassica oleracea***. The wild form of this species is shown in the centre of this diagram. Different parts have been developed by human selection. In spite of the enormous visible differences, if allowed to flower, all six can cross-pollinate. Kale is closer to the wild type than the other related breeds.

2. Study the diagram above and **name** which part of the plant has been selected for to produce each of the vegetables:

(a) Cauliflower: _____ (d) Brussels sprout: _____

(b) Kale: _____ (e) Cabbage: _____

(c) Broccoli: _____ (f) Kohlrabi: _____

3. Describe the feature of these vegetables that suggests they are members of the same species: _____

_____

4. Human selection pressures can also influence the development of characteristics in 'unwanted' species. Suggest how human weed control measures may inadvertently select for weed plants that have a resistance to the measures:

_____

_____

5. Describe two phenotypic characteristics that might be desirable in a cereal crop plant and explain your choice:

(a) _____

_____

(b) _____

_____

Selective Breeding in Animals

The domestication of livestock has a long history dating back at least 8000 years. Today's important stock breeds were all derived from wild ancestors that were domesticated by humans who then used **selective breeding** to produce livestock to meet specific requirements. Selective breeding of domesticated animals involves identifying desirable qualities (e.g. high wool production or meat yield), and breeding together individuals with those qualities so the trait is reliably passed on. Practices such as **inbreeding**, **line-breeding,** and **outcrossing** are used to select and 'fix' desirable traits in varieties. Today, modern breeding techniques often employ reproductive technologies, such as artificial insemination, so that the desirable characteristics of one male can be passed on to many females. These new technologies refine the selection process and increase the rate at which stock improvements are made. Rates are predicted to accelerate further as new technologies, such as genomic selection, become more widely available and less costly. Producing highly inbred lines of animals with specific traits can have disadvantages however. Homozygosity for a number of desirable traits can cause physiological or physical problems to the animal itself. For example, animals bred specifically for rapid weight gain often grow so fast that they have skeletal and muscular difficulties.

The Origin of Domestic Animals

PIG
Wild ancestor: Boar
Origin: Anatolia, 9000 years BP
Now: More than 12 distinct modern breeds, including the Berkshire (meat) and Tamworth (hardiness).

DOMESTIC FOWL
Wild ancestor: Red jungle fowl
Origin: Indus Valley, 4000 BP
Now: More than 60 breeds including Rhode Island Red (meat) and Leghorn (egg production).

Each domesticated breed has been bred from the wild ancestor (pictured). The date indicates the earliest record of the domesticated form (years before present or BP). Different countries have different criteria for selection, based on their local environments and consumer preferences.

GOAT
Wild ancestor: Bezoar goat
Origin: Iraq, 10 000 years BP
Now: approx. 35 breeds including Spanish (meat), Angora (fibre) and Nubian (dairy).

SHEEP
Wild ancestor: Asiatic mouflon
Origin: Iran, Iraq, Levant, 10 000 years BP
Now: More than 200 breeds including Merino (wool), Suffolk (meat), Friesian (milk), and dual purpose (Romney).

CATTLE
Wild ancestor: Auroch (extinct)
Origin: SW Asia, 10 000 years BP
Now: 800 modern breeds including the Aberdeen angus (meat), Friesian and Jersey (milk), and Zebu (draught).

1. Distinguish between inbreeding and out-crossing, explaining the significance of each technique in selective breeding:

_____

_____

_____

_____

2. Describe the contribution that new reproductive technologies are making to selective breeding:

_____

_____

Food and Health

Related activities: Selective Breeding in Crop Plants, Variation in Species

RA 2

Beef breeds: Simmental, Aberdeen-Angus, Hereford (above), Galloway, Charolais. Consumer demand has led to the shift towards continental breeds such as Charolais because they are large, with a high proportion of lean muscle. **Desirable traits**: high muscle to bone ratio, rapid growth and weight gain, hardy, easy calving, docile temperament.

Dairy breeds: Jersey, Friesian (above), Holstein, Aryshire. **Desirable traits**: high yield of milk with high butterfat, milking speed, docile temperament, and udder characteristics such as teat placement.

Special breeds: Some cattle are bred for their suitability for climate or terrain. Scottish highland cattle (above) are a hardy, long coated breed and produce well where other breeds cannot thrive.

Artificial Selection and Genetic Gain in Cattle

Cattle are selected on the basis of particular desirable traits (e.g. milk solids or muscle mass). Most of the genetic improvement in dairy cattle has relied on selection of high quality progeny from proven stock and extensive use of superior sires through artificial insemination (AI). In beef cattle, AI is useful for introducing new breeds.

Improved breeding techniques accelerate the **genetic gain**, i.e. the gain toward the desirable phenotype of a **breed**. The graph (below) illustrates the predicted gains based on artificial insemination and standard selection techniques (based on criteria such as production or temperament). These are compared with the predicted gains using breeding values and various reproductive technologies such as embryo multiplication and transfer (EMT) of standard and transgenic stock, marker (gene) assisted selection, and sib-selection (selecting bulls on the basis of their sisters' performance).

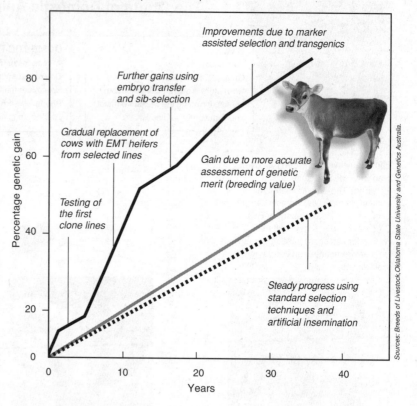

Sources: Breeds of Livestock,Oklahoma State University and Genetics Australia.

3. Describe some of the positive and negative outcomes of selective breeding in domestic animals: _____

_____

_____

_____

_____

_____

4. Identify the two methods by which most of the genetic progress in dairy cattle has been achieved:

(a) _____ (b) _____

5. Explain what is meant by the term **genetic gain** as it applies to livestock breeding: _____

_____

_____

6. Suggest why mixed breeds, such as Hereford-Friesian crosses, are popular for mixed beef/milk production:

_____

_____

Producing Food With Microorganisms

Bacteria and fungi are used extensively in many aspects of food technology. The microorganisms mainly act as production agents. They are used to turn ingredients into food, or to modify food ingredients, rather than as a raw ingredient itself. Microorganisms have traditionally been used in the production of fermented foods: alcoholic beverages, bread, and fermented dairy products. The control and efficiency of these uses have been greatly refined in recent times. The advent of genetic engineering has increased the range of microbial products available and has provided alternative sources for products that were once available only through expensive or wasteful means (e.g. production of the enzyme chymosin). Some microorganisms, such as the yeast *Saccharomyces cerevisiae*, are used as food additives because they contain several vitamins essential to good health. Future applications include the wider use of genetically engineered microbes in crop improvement.

Microorganisms in the Food Industry

Cheese production uses cultures of lactic acid bacteria (e.g. *Streptococcus* spp.) and genetically engineered microbial rennin, which is added to curd the milk protein. Microbial activity occurs at several stages to produce characteristic flavours and textures.

Yoghurt is produced from milk by the action of lactic acid bacteria particularly *Lactobacillus bulgaricus* and *Streptococcus thermophilus*. These bacteria break down milk proteins into peptides.

Soy sauce (shoyu): Filamentous fungi (*Aspergillus soyae* and *A.oryzeae*) digest soy proteins and solids. The culture is fermented in the presence of lactic acid bacteria (*Lactobacillus* spp.) and acid tolerant yeast (e.g. *Torulopsis*) over a year or more.

Bread (leavened): The sugars in the dough are fermented by the yeast, *Saccharomyces cerevisiae*, producing alcohol and CO_2. The gas causes the dough to rise, while the alcohol is converted to flavour compounds during baking.

Beer and wine: The sugars in fruits (wine) or grains (beers) are fermented by yeast (e.g. *Saccharomyces carlsbergensis, S. cerevisiae*) to alcohol. Beer production first requires a malting process to convert starches in the grain to fermentable sugars.

Sauerkraut production involves the fermentation of cabbage. The initial fermentation involves lactic acid bacteria (*Leuconostoc mesenteroides* and *Enterobacter cloacae*), followed by acid production with *Lactobacillus plantarum*.

Vinegar production uses cultures of acetic acid bacteria (e.g. *Acetobacter* and *Gluconobacter*). When you leave wine exposed to oxygen, these bacteria convert the alcoholic brew (ethanol) to ethanoic acid (vinegar).

Vitamins and amino acids are dietary supplements produced as by-products of faulty or altered microbial metabolism. Examples include lysine and vitamin B_{12}. Microbial species used include: *Corynebacterium glutamicum, Pseudomonas, Propionibacterium*.

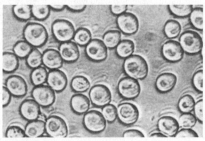

Commercial production of microorganisms: Baker's yeast (*Saccharomyces cerevisiae*, above) is sold for both industrial use and home brewing and baking. *Bacillus thuringiensis* is a widely used biological pest control agent. Nitrogen fixing bacteria (e.g. *Rhizobium*) are used to enhance plant nutrition.

1. Explain what is meant by the term **industrial microbiology**: _____

_____

2. Briefly describe two examples of how microorganisms are used for the production of alcoholic beverages:

_____

_____

_____

_____

Periodicals:
The microbiology of cheese

Related activities: New Medicines
Web links: Industrial Microbiology

Food and Health

A 2

Problems with Microorganisms in Food Technology

Many microorganisms used in food production are grown in **biofermenters** (below). There are a number of problems associated with these large scale fermentations, including providing adequate supply of oxygen and nutrients, maintaining a constant temperature and pH, and removing wastes as they build up so growth is not inhibited. Rigorous testing using scale models must be carried out to ensure that the problems associated with scaling up have been eliminated.

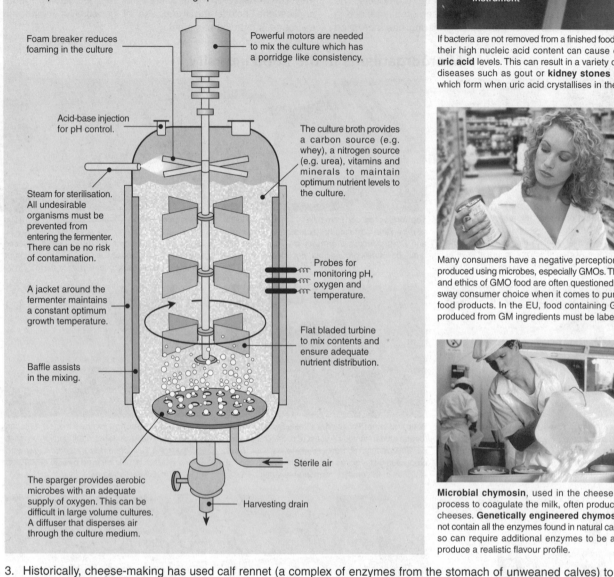

Foam breaker reduces foaming in the culture

Powerful motors are needed to mix the culture which has a porridge like consistency.

Acid-base injection for pH control.

The culture broth provides a carbon source (e.g. whey), a nitrogen source (e.g. urea), vitamins and minerals to maintain optimum nutrient levels to the culture.

Steam for sterilisation. All undesirable organisms must be prevented from entering the fermenter. There can be no risk of contamination.

Probes for monitoring pH, oxygen and temperature.

A jacket around the fermenter maintains a constant optimum growth temperature.

Flat bladed turbine to mix contents and ensure adequate nutrient distribution.

Baffle assists in the mixing.

Sterile air

The sparger provides aerobic microbes with an adequate supply of oxygen. This can be difficult in large volume cultures. A diffuser that disperses air through the culture medium.

Harvesting drain

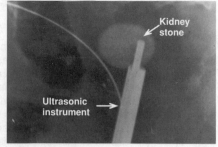

Kidney stone

Ultrasonic instrument

If bacteria are not removed from a finished food product, their high nucleic acid content can cause elevated **uric acid** levels. This can result in a variety of painful diseases such as gout or **kidney stones** (above), which form when uric acid crystallises in the kidney.

Many consumers have a negative perception of food produced using microbes, especially GMOs. The safety and ethics of GMO food are often questioned and can sway consumer choice when it comes to purchasing food products. In the EU, food containing GMOs or produced from GM ingredients must be labelled.

Microbial chymosin, used in the cheese making process to coagulate the milk, often produces bitter cheeses. **Genetically engineered chymosin** does not contain all the enzymes found in natural calf rennet, so can require additional enzymes to be added to produce a realistic flavour profile.

3. Historically, cheese-making has used calf rennet (a complex of enzymes from the stomach of unweaned calves) to coagulate the milk. Genetically engineered chymosin (rennin) is now most often used. Bearing this in mind, discuss:

   (a) Some advantages of using genetically modified chymosin: _____

   _____

   _____

   (b) Some disadvantages of using genetically modified chymosin: _____

   _____

   _____

4. Suggest why consumers may be more accepting of foods produced using yeast, rather than those made using bacteria:

   _____

   _____

   _____

   _____

Increasing Food Production

The accelerating demands of the world's population on food resources creates a need to produce more food, more quickly, at minimal cost. For the most part, this has involved intensive, industrialised agricultural systems, where high inputs of energy are used to obtain high yields per unit of land farmed. Such systems apply not just to crop plants, but to animals too, which are raised to slaughter weight at high densities in confined areas (a technique called **factory farming**). Producing food from a limited amount of land presents several challenges: to maximise yield while minimising losses to disease and pests, to ensure sustainability of the practice, and (in the case of animals) to meet certain standards of welfare and safety. Intensive agriculture makes use of chemical pesticides and fertilisers, as well as antibiotics and hormones to achieve these aims, often with deleterious effects on the environment and on crop and animal health. An alternative approach to meeting global food demands is to develop sustainable farming systems based on sound crop and animal management practices. Such systems make use of organic fertilisers and natural pest controls to achieve profitable, efficient food production that is sustainable in the long term.

One View: Intensive Farming Practices

Antibiotics are used in the intensive farming of **poultry** for egg and meat production. Proponents regard antibiotics as an important management tool to prevent, control, and treat disease, allowing farmers to raise healthy animals and produce safe food.

The application of inorganic fertilisers has been a major factor in the increased yields of industrialised agriculture. Excessive application can be detrimental however, leading to enrichment of water bodies and contamination of groundwater.

Fertilisers can be sprayed using aerial topdressing in inaccessible areas.

Clearing land of trees for agriculture can lead to slope instability, soil erosion, and land degradation.

Pesticides and fungicides are used extensively to control crop pests and diseases in industrialised agriculture. Indiscriminate use of these leads to increased resistance to commonly used chemicals and contamination of land and water.

Antibiotics are used to treat diseases such as mastitis in dairy cattle. Milk must be withheld until all antibiotic residues have disappeared.

Feedlots are a type of confined animal feeding operation which is used for rapidly feeding livestock, notably cattle (above left), up to slaughter weight. Diet for stock in feedlots are very dense in energy to encourage rapid growth and deposition of fat in the meat (marbling). As in many forms of factory farming, antibiotics are used to combat disease in the crowded environment.

1. Discuss the use of chemical pesticides and fertilisers to increase crop yields in intensive systems. In your discussion, include reference to the advantages and disadvantages of such applications:

_____

_____

_____

2. Animals raised for food production in intensive systems (e.g. feedlot cattle, and factory farmed pigs and poultry) are often supplied with antibiotics at low doses in the feed:

(a) Explain the benefits of this practice to production: _____

_____

(b) Discuss the environmental, health, and animal welfare issues associated with this practice: _____

_____

_____

_____

_____

Food and Health

Periodicals:
Poultry farming

Related activities: The Green Revolution
Web links: Factory Farming

A 2

An Alternative View: Sustainability and Diversity

Sustainable agricultural practices are economically viable and environmentally sound. Increasingly, farmers are investigating methods by which they can earn a reasonable living from the land, while remaining less reliant on government subsidies, and petroleum and chemical inputs. As in intensive farming, there are many approaches to sustainable agriculture. That pictured here just one.

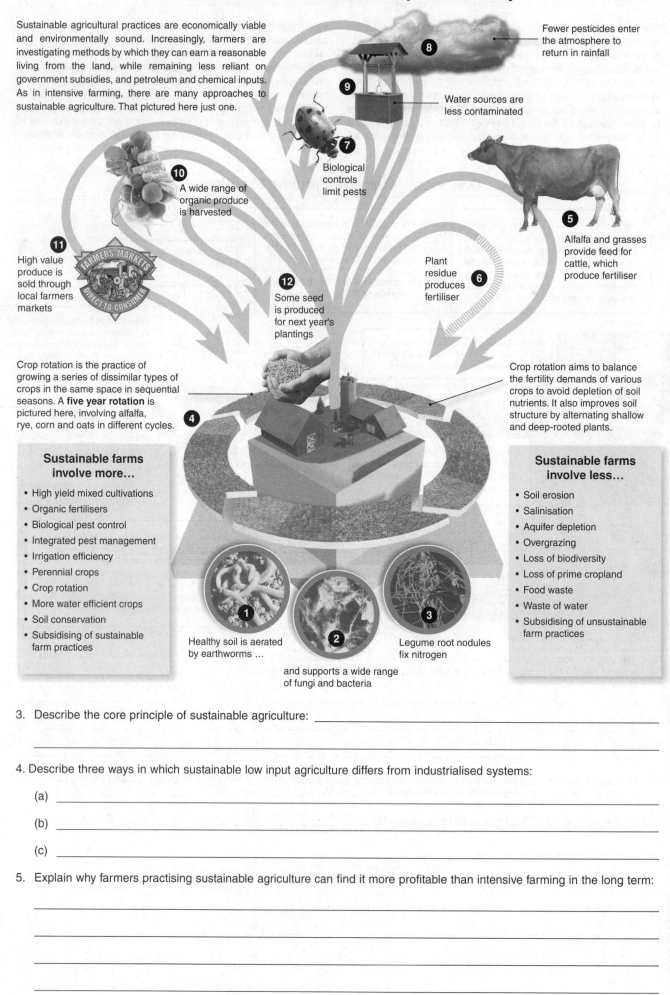

8 Fewer pesticides enter the atmosphere to return in rainfall

9 Water sources are less contaminated

7 Biological controls limit pests

10 A wide range of organic produce is harvested

5 Alfalfa and grasses provide feed for cattle, which produce fertiliser

11 High value produce is sold through local farmers markets

6 Plant residue produces fertiliser

12 Some seed is produced for next year's plantings

Crop rotation is the practice of growing a series of dissimilar types of crops in the same space in sequential seasons. A **five year rotation** is pictured here, involving alfalfa, rye, corn and oats in different cycles.

4

Crop rotation aims to balance the fertility demands of various crops to avoid depletion of soil nutrients. It also improves soil structure by alternating shallow and deep-rooted plants.

Sustainable farms involve more...

- High yield mixed cultivations
- Organic fertilisers
- Biological pest control
- Integrated pest management
- Irrigation efficiency
- Perennial crops
- Crop rotation
- More water efficient crops
- Soil conservation
- Subsidising of sustainable farm practices

Sustainable farms involve less...

- Soil erosion
- Salinisation
- Aquifer depletion
- Overgrazing
- Loss of biodiversity
- Loss of prime cropland
- Food waste
- Waste of water
- Subsidising of unsustainable farm practices

1 Healthy soil is aerated by earthworms ...

2 and supports a wide range of fungi and bacteria

3 Legume root nodules fix nitrogen

3. Describe the core principle of sustainable agriculture: _____

_____

4. Describe three ways in which sustainable low input agriculture differs from industrialised systems:

(a) _____

(b) _____

(c) _____

5. Explain why farmers practising sustainable agriculture can find it more profitable than intensive farming in the long term:

_____

_____

_____

_____

Food Preservation

Without intervention, **food spoilage** begins as soon as an item is picked, slaughtered, or manufactured. While many food spoilage factors (bruising, oxidation, humidity) alter the nutritional quality and appearance (texture, colour, flavour) of food, other spoilage factors can have more serious consequences. The consumption of food spoiled by the presence of microbes or their toxins, could cause illness, or even death, from **food poisoning**. For this reason, it is vital that perishable foods be preserved to prevent microbial growth and extend their shelf life. **Food preservation** describes any process that prevents or slows food spoilage. Basic storage and handling techniques often safeguard the flavour and appearance of food, but more advanced techniques (below) are required to ensure that food stored for a prolonged period remains safe for human consumption.

Preparing Korean kimchi (pickled vegetables)

Pickling preserves food by lowering the pH to below 4.6 and killing the microbes that cannot survive the acidic environment. There are two pickling methods. The first involves **adding salt** and allowing an anaerobic fermentation by the native bacteria (e.g. *Lactobacillus*). This fermentation produces lactic acid, which lowers the pH and inhibits the growth of harmful microbes. The second method involves storing the food in a prepared **acid solution**, often vinegar (acetic acid) to achieve the lower pH.

Preserving food in a **sugar syrup** produces a high osmotic pressure environment. Microbial cells have a lower osmotic pressure than the medium, so water leaves their cells causing cell dehydration and death. Fruits are commonly preserved in this manner.

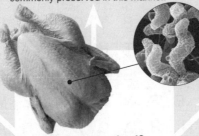

Why preserve food?
Food may harbour pathogenic organisms such as *Salmonella* and *Campylobacter* (above). Food preservation techniques aim to reduce or destroy the number of harmful microbes, increasing the food's shelf life and lowering the risk of food poisoning on consumption.

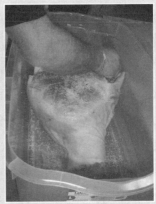

Curing ham by adding salt

Adding **salt** to food draws moisture out from the food by osmosis. The food environment becomes drier, reducing the available water and thereby inhibiting microbial growth. Salt concentrations of up to 20% are required for effective preservation. Meat, such as pork products (above), are often preserved in this way.

HTST equipment for pasteurising

Heat treatments are used with liquid food products such as milk and fruit juice to logarithmically reduce (but not eliminate) the numbers of microorganisms present. High temperature short time (HTST) treatments involve heating the product to 71.7°C for 15-20 seconds. The more extreme ultra-heat treatment (UHT) process, which also kills bacterial spores, requires heating to 135°C for 1-2 seconds before packaging into sterile containers. UHT products have a shelf life of 6-9 months, compared with up to two weeks for a HTST treated product.

Food irradiation uses ionising radiation (electron beams, X-rays, or gamma rays) to destroy microbes on food. Irradiation kills microorganisms by damaging their DNA, but it does not destroy prions or toxins. irradiation is commonly used to preserve fruit, vegetables, and spices but this must be clearly indicated (above).

Freezing turns the water contained within food into ice, reducing its water activity. With no available water present, microbial growth is inhibited. Freezing is not a method of sterilisation, and does not necessarily kill microbes, but the very low temperature (-18°C) slows down most chemical reactions, prolonging the shelf life of frozen food to several months. Fruit, vegetables (above), meat, fish, and shellfish are commonly frozen.

1. Describe a benefit and a disadvantage of using heat treatments to prolong the shelf life of a food product:

   (a) Benefit: _____

   (b) Disadvantage: _____

2. Explain how high salt and sugar act to prolong the shelf life of perishable foods: _____

   _____

   _____

   _____

Related activities: The Control of Disease

A 2

Food and Health

KEY TERMS: Word Find

Use the clues below to find the relevant key terms in the WORD FIND grid

```
O G C A R D I O V A S C U L A R D I S E A S E L V
O P V I T A M I N I X C T C H O L E S T E R O L T
R B Z X L M A L N U T R I T I O N D K O H V O X S
M J Y F E R T I L I S E R S S D B S L N G R B K B
V A L R I L W D N I E J I O J G E I K X Q N E H C
G B J Z Q A P C J Q B A L A N C E D D I E T S Z J
A T H E R O S C L E R O S I S E H L M B Z N I H N
U X S C O V O W H Y B R I D V I G O U R Q F T W Q
Y Z R A U V V L K M P Y Q G C B J I V C W M Y D B
N S D F L W A J C S T H R R A L P I C K L I N G D
I V S F M T R S E L E C T I V E B R E E D I N G X
P E S T I C I D E S N P O N Z M A A S L H D L C Y
R H S J Q I D N L H A G W T R A N S F A T S D L F
C J L O V X B X G Z L R Q G E X L F J V Q Q A W I
N B A D O M E S T I C A T I O N K L U H W D V N F
J E H K P F N R K A V R A N T I B I O T I C Z C P
Z I Q M C O R O N A R Y H E A R T D I S E A S E Q
M M G J K X W Y W F Q N B O D Y M A S S I N D E X
```

A substance or compound that kills bacteria or inhibits their growth.

A diet containing the correct proportions of the nutrients required for good health (2 words: 8, 4).

Sterol lipid found in the cell membranes and transported in the blood plasma of all animals.

The accumulation of fatty plaques within the walls of the coronary arteries.

Disease of the coronary circulation (3 words: 8, 5, 7).

Nutrients (usually) applied to the soil to promote growth.

Preservation of food with dry salt.

Preservation of food with a salt solution (brine).

This type of lipoprotein transports cholesterol to the liver for metabolism (abbrev).

The process of breeding plants and animals for particular genetic traits (2 words: 9, 8).

Improvement in any biological function as a result of outbreeding.

This type of lipoprotein promotes deposition of cholesterol in the artery wall.

Disease of the heart and blood vessels (2 words: 14, 7).

A term describing excessive body mass (as defined by a BMI>30).

The process whereby a population of animals or plants, through a process of selection, becomes accustomed to human provision and control.

Organic compound required as a nutrient in trace amounts by an organism.

Insufficient, excessive or imbalanced consumption of nutrients.

A commonly used measure of body fat based on height and weight (3 words: 4, 4, 5).

Fats produced in the hydrogenation of vegetable oils and implicated in dietary disorders.

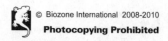

Pathogens and Human Disease

KEY CONCEPTS

KEY TERMS

aetiology
AIDS
antibiotic
disease
health
HIV
incidence
infectious disease
Influenzavirus
Koch's postulates
life expectancy
malaria
mortality
*Mycobacterium
        tuberculosis*
parasite
pathogen
Plasmodium
prevalence
protoctist
protozoan
retrovirus
T cell
tuberculosis
vector (for disease)
virus

OBJECTIVES

☐ 1. Use the **KEY TERMS** to help you understand and complete these objectives.

The Nature of Disease
pages 206-209, 211-212, 214, 221-226

☐ 2. Discuss what is meant by **health** and **disease**, and distinguish between infectious and non-infectious disease.

☐ 3. Define the terms **pathogen** and **parasite**, and discuss the meanings and use of these terms. Identify and describe pathogens in different taxa including the bacteria, viruses, and protoctists.

☐ 4. Recognise different methods for controlling infectious disease. Use examples to explain the significance of drug resistance to the effective control of diseases, including newly emerging diseases (e.g. new influenza strains).

☐ 5. CIE only: Describe the cause of **cholera**. Discuss the role of biological, social, and economic factors in the treatment, control, and prevention of cholera.

Case Study: Malaria
pages 212-213, 226

☐ 6. Describe **malaria** as an example of a disease caused by a protozoan. Understand that many pathogenic protozoans are highly specialised parasites with part of their life cycle occurring within a human.

☐ 7. Describe the cause and modes of transmission of malaria. Discuss the global importance of malaria and describe factors in its distribution.

☐ 8. Describe the roles of social, economic, and biological factors in the treatment, control, and prevention of malaria.

Case Study: HIV/AIDS
pages 215-221, 226

☐ 9. Describe the agent involved and the modes of transmission of **HIV/AIDS**. Assess the global (including economic) impact of the disease and describe factors in its distribution (including its regional distribution).

☐ 10. Identify stages in the development of an HIV infection, including the effect of HIV on the immune system. Explain why AIDS is termed a syndrome.

☐ 11. Describe social, economic, and biological factors in the treatment, control, and prevention of HIV/AIDS.

Case Study: Tuberculosis
pages 210, 226

☐ 12. Describe the causes and modes of transmission of **tuberculosis** (TB). Discuss the global importance of TB. Describe the roles of social, economic, and biological factors in the control and prevention of TB.

☐ 13. Explain the role of antibiotics in the treatment of TB. Discuss the difficulties associated with the treatment of TB (including the importance of bacterial resistance to antibiotics).

Periodicals:

Listings for this chapter are on page 340

Health vs Disease

Disease is more difficult to define than **health**, which is described as a state of complete physical, mental, and social well-being. A disease is usually associated with particular **symptoms** that help to define and diagnose it. The term **disease** is used to describe a condition whereby part or all of an organism's normal physiological function is upset. All diseases, with the exception of some mental diseases, can be classified as **physical diseases** (i.e. diseases that cause permanent or temporary damage to

the body). Physical diseases can be subdivided into two major groups: **infectious diseases** caused by an infectious agent (**pathogen**) and **non-infectious diseases** (most of which are better described as disorders). Non-infectious diseases are often not clearly the result of any single factor, but they can be further categorised into major subgroups according to their principal cause (outlined below). However, many diseases fall into more than one category, e.g. Alzheimer's disease and some cancers.

The Nature of Disease

Infectious Diseases

Infectious diseases are diseases that are caused by pathogens and which can be transmitted from one person to another. Most, although not all, pathogens are microorganisms, and they fall into five main categories: viruses, bacteria, fungi, protozoans, and multicellular parasites.

Deficiency Diseases

Deficiency diseases are non-infectious diseases caused by an inadequate or unbalanced diet, or by over eating. Examples include obesity, rickets, scurvy, marasmus, and kwashiorkor.

Social Diseases

Social diseases include a wide range of disorders that are influenced by living conditions and personal behaviour. They may or may not be caused by an infectious agent. Examples include obesity, sexually transmitted diseases, and lung cancer and emphysema due to smoking.

Mental Disorders

The term mental disorder encompasses a range of diseases that affect a person's thoughts, memory, emotions, and personal behaviour. Examples include Alzheimer's, schizophrenia, and depression.

Degenerative Diseases

Degenerative diseases are non-infectious diseases caused by ageing and the inability of the body to carry out effective repairs and regeneration. Examples include osteoarthritis, Alzheimer's disease, and many cancers.

Inherited Diseases

Some diseases result from inherited malfunctions in a body system and have no external cause. Defective genes may cause the failure of a body system throughout a person's life, or the onset of disease may occur later in life. Examples include cystic fibrosis, multiple sclerosis, Alzheimer's, and Huntington's disease.

Down syndrome is a congenital disease caused by having three copies of chromosome 21.

Smoking is a common social behaviour that causes lung cancer, chronic bronchitis, and emphysema.

Mental diseases encompass a range of often unrelated disorders involving disturbances to personality.

Asthma is a common, non-infectious respiratory disease with a number of underlying causes.

1. Discuss the differences between health and disease: _____

_____

_____

_____

2. Using illustrative examples, suggest why many diseases fall into more than one disease category:

_____

_____

Related activities: Infection and Disease

Infection and Disease

Infectious disease refers to disease caused by a **pathogen** (an infectious agent). Many pathogens, including all viruses, are also parasites in that they rely on a host organism for at least part of their life cycle. Pasteur's work on microorganisms in the late 1860s formed the basis of modern-day **aseptic techniques** and, in 1876-1877, **Robert Koch** established a sequence of experimental steps (**Koch's postulates**) for directly relating a specific microbe to a specific disease. During the past 100 years, the postulates have been invaluable in determining the specific agents of many diseases. As our knowledge base expands, infectious agents are being increasingly implicated in what were thought to be non-infectious diseases, including many cancers.

Koch's Postulates

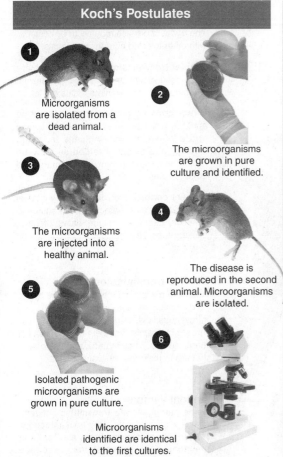

1 Microorganisms are isolated from a dead animal.

2 The microorganisms are grown in pure culture and identified.

3 The microorganisms are injected into a healthy animal.

4 The disease is reproduced in the second animal. Microorganisms are isolated.

5 Isolated pathogenic microorganisms are grown in pure culture.

6 Microorganisms identified are identical to the first cultures.

Koch isolated bacteria from a diseased animal, then injected them into a healthy animal, causing it to exhibit identical symptoms to the first. This demonstrated that a specific infectious disease (e.g. anthrax) was caused by a specific microorganism (*Bacillus anthracis*). Koch used the procedure to identify the bacteria that caused anthrax and tuberculosis.

Koch's findings are summarised as **Koch's postulates**:
1. The same pathogen must be present in every case of the disease.
2. The pathogen must be isolated from the diseased host and grown in pure culture.
3. The pathogen from the pure culture must then cause the disease when it is inoculated into a healthy, susceptible animal.
4. The pathogen must be isolated from the inoculated animal and shown to be the original organism.

Types of Pathogens

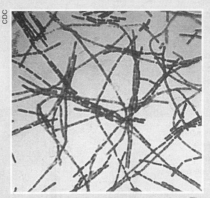

Bacillus anthracis bacterium causes anthrax. The anthrax bacillum can form long-lived spores.

Bacteria: All bacteria are prokaryotes, and are categorised according to the properties of their cell walls and features such as cell shape and arrangement, oxygen requirement, and motility. Many bacteria are useful, but the relatively few species that are pathogenic are responsible for enormous social and economic cost. This is especially so since the rise in incidence of antibiotic resistance.

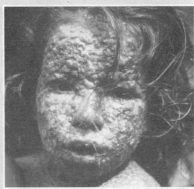

Photo: Bangladeshi girl with smallpox (1973). Smallpox was eradicated from the country in 1977.

Viral pathogens: Viruses are specialised intacelular **parasites**. They are responsible for many everyday diseases (e.g. the common cold), as well as more dangerous diseases, such as Ebola, and some diseases that have since been eradicated as a result of vaccination programmes (e.g. the *Variola* virus, which causes smallpox, above). Viruses are obligate intracellular parasites and need living host cells in order to multiply.

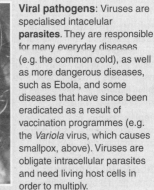

Malaria sporozoite moving through gut epithelia. The parasite is carried by a mosquito vector.

Eukaryotic pathogens: Eukaryotic pathogens (fungi, algae, protozoa, and parasitic worms) include those responsible for malaria and schistosomiasis. Many are highly specialised **parasites** with a number of hosts on which they are dependent. The malaria parasite for example has a mosquito and a human host. Like many other pathogens, this parasite has become resistant to the drugs used to treat it.

1. Explain the contribution of Robert Koch to the **aetiology** of disease: _____

2. Explain what is meant by a parasite and identify a pathogenic parasitic organism: _____

3. Explain why diseases caused by **intracellular parasites** are difficult to control and treat: _____

Transmission of Disease

The human body, like that of other large animals, is under constant attack by a wide range of potential parasites and pathogens. Once inside us, these organisms seek to reproduce and exploit us for food. Pathogens may be transferred from one individual to another by a number of methods (below). The transmission of infectious diseases can be virtually eliminated by observing appropriate personal hygiene procedures, and by chlorinating drinking water and providing adequate sanitation.

Portals of Entry

Respiratory tract
The mouth and nose are major entry points for pathogens, particularly airborne viruses, which are inhaled from other people's expelled mucus.

Examples: diphtheria, meningococcal meningitis, tuberculosis, whooping cough, influenza, measles, German measles (rubella), chickenpox.

Gastrointestinal tract
The mouth is one of the few openings where we deliberately place foreign substances into our body. Food is often contaminated with microorganisms, but most of these are destroyed in the stomach.

Examples: cholera, typhoid fever, mumps, hepatitis A, poliomyelitis, bacillary dysentery, salmonellosis.

Breaking the skin surface
The skin provides an effective barrier to the entry of most pathogens. However, a cut or abrasion will allow easy entry for pathogens. Some parasites and pathogens have adaptive features that allow them to penetrate the skin surface.

Examples: tetanus, gas gangrene, bubonic plague, hepatitis B, rabies, malaria, leptospirosis, and HIV.

Urinogenital openings
The urinogenital openings provide entry points for the pathogens responsible for sexually transmitted infections (STIs) and other opportunistic infections (i.e. thrush).

Examples: gonorrhoea, syphilis, HIV, and *E. coli* (a cause of urinary tract infections).

The Body Under Assault

Modes of Transmission

Contact transmission
The agent of disease may occur by contact with other infected humans or animals:

Droplet transmission: Mucus droplets are discharged into the air by coughing, sneezing, laughing, or talking within a radius of 1 m.

Direct contact: Direct transmission of an agent by physical contact between its source and a potential host. Includes touching, kissing, and sexual intercourse. May be person to person, or between humans and other animals.

Indirect contact: Includes touching objects that have been in contact with the source of infection. Examples include: eating utensils, drinking cups, bedding, toys, money, and used syringes.

Vehicle transmission
Agents of disease may be transmitted by a medium such as food, blood, water, intravenous fluids (e.g. drugs), and air. Airborne transmission refers to the spread of fungal spores, some viruses, and bacteria that are transported on dust particles.

Animal Vectors
Some pathogens are transmitted between hosts by other animals. Bites from arthropods (e.g. mosquitoes, ticks, fleas, and lice) and mammals (e.g. rodents) may introduce pathogens, while flies can carry pathogens on their feet. In 1897, **Ronald Ross** identified the *Anopheles* mosquito as the vector for malaria. He was the first to implicate insects in the transmission of disease.

1. State how pathogens benefit from invading a host: _____

   _____

2. Describe two personal hygiene practices that would minimise the risk of transmitting an infectious disease:

   (a) _____

   (b) _____

3. Identify the common **mode of transmission** and the **portal of entry** for the following pathogens:

   (a) Protozoan causing malaria: _____

   (b) Tetanus bacteria: _____

   (c) Cholera bacteria: _____

   (d) Common cold virus: _____

   (e) Tuberculosis bacteria: _____

   (f) HIV (AIDS) virus: _____

   (g) Gonorrhoea bacteria: _____

Related activities: Bacterial Diseases, Viral Diseases, The Control of Disease

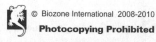

© Biozone International 2008-2010
Photocopying Prohibited

Bacterial Pathogens

Relatively few of the world's bacterial species cause disease. Those that do, the so-called pathogenic bacteria, have and range of adaptations that enable them to penetrate the defences of a host and cause an infection (diagram below). Bacterial diseases are commonly transmitted through food, water, air, or by direct contact. The **natural reservoir** (source of infection) of a disease varies from species to species, ranging from humans and other organisms, to sewage or contaminated water. Much of our control of bacterial disease is achieved through identifying reservoirs of infection and limiting the routes of transmission.

How Bacteria Invade a Host's Tissues

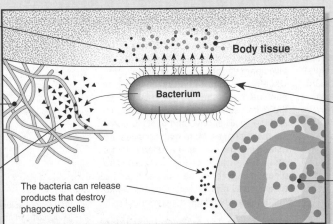

Toxins: Bacterial toxins can act locally to promote bacterial invasion (e.g. the enzymes that degrade collagen), or they may have cytotoxic activity and destroy cells directly.

Fibrin: Fibrous threads of protein are deposited when blood clots. This action by the host effectively limits the movement of pathogens in infected areas.

Enzymes are released that break down fibrin, allowing the bacteria greater freedom of movement.

The bacteria can release products that destroy phagocytic cells

Body tissue

Bacterium

The bacterium releases enzymes that degrade the connective tissue of the host, allowing the spread of infection.

Fimbriae: Fine, threadlike extensions from the bacterial cell, called fimbriae, enable the bacteria to attach to the mucous membranes and directly attack the nearby host tissues.

Phagocyte: These white blood cells are very effective in identifying and destroying foreign cells such as pathogens.

Methods of Bacterial Transmission

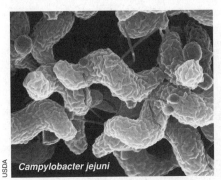

Campylobacter jejuni

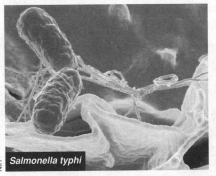

Salmonella typhi

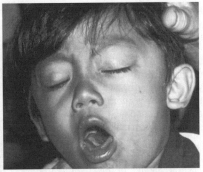

Foodborne bacterial diseases

Bacterial foodborne illnesses are caused by consuming food or beverages contaminated with bacteria or their toxins. Examples include *Salmonella* food poisoning, listeriosis, *E.coli* gastroenteritis and *Campylobacter* infection. Symptoms of bacterial food poisoning include fever, abdominal cramps, and diarrhoea. Some, including campylobacteriosis and salmonellosis, are associated with consuming raw or undercooked poultry.

Waterborne bacterial diseases

Waterborne bacterial pathogens are responsible for a number of serious diarrhoeal illnesses, including typhoid (*Salmonella typhi*), cholera (*Vibrio cholerae*), and shigellosis (*Shigella*). Transmission of these diseases is usually through faecal contamination of drinking water. The fever and diarrhoea associated with such diseases is not trivial; it is responsible for hundreds of thousands of deaths annually in countries where inadequate sanitation is a problem.

Airborne bacterial diseases

Airborne pathogens are transmitted on dust particles or droplets when people cough, sneeze, or exhale. Vaccination against certain airborne bacterial pathogens has been highly successful. **Whooping cough** (above) is a potentially fatal respiratory disease caused by *Bordetella pertussis*. The prevalence of this disease, whose symptoms are caused by the bacterial pertussis toxin, has declined dramatically following the introduction of immunisation programmes.

1. Describe two adaptations that enable bacteria to penetrate a host's defences and cause disease:

    (a) _____

    (b) _____

2. Explain how each of the following bacterial diseases can be controlled by targeting its mode of transmission:

    (a) Cholera: _____

    _____

    (b) Salmonellosis: _____

    _____

3. Explain why immunisation is often a good option for controlling airborne bacterial diseases: _____

_____

Related activities: Tuberculosis, Cholera, Antibiotics
Web links: Microbiology in Motion

RA 2

Tuberculosis

Tuberculosis (TB) is a contagious disease caused by the bacterium *Mycobacterium tuberculosis* (MTB). TB can affect many areas of the body, but the most common form, **pulmonary TB**, affects the lungs. Symptoms of pulmonary TB include a productive cough, and presence of the disease is indicated on a chest X-ray by opaque areas and large cavities in the lungs. TB is widespread, and increasing drug resistance is contributing to the impact of the disease. About 8% of TB patients also have HIV, so TB rates increase as HIV incidence rises. Effective treatment requires an aggressive and prolonged regime of antibiotics, and patients often do not comply with the full treatment. However, there are some positive trends. Globally, the incidence (new infection) rate is falling slowly and prevalence and death rates per capita are falling faster than TB incidence. If these trends are sustained, the WHO could achieve its goal of halving prevalence and death rates in some regions by 2015.

Infection and Transmission

TB is a contagious disease, and is spread through the air when infectious people cough, sneeze, talk, or spit (below). Transmission of TB does not require a large inoculum; a person needs to inhale only a small number of *Mycobacterium tuberculosis* (MTB) to be infected.

Left untreated, each person with active TB will infect on average between 10 and 15 people every year. People infected with MTB will not necessarily become ill with the disease. The immune system can 'wall off' the MTB which can lie dormant for years, protected by a thick waxy coat. When the immune system is weakened, the chance of becoming ill (showing active symptoms) is much greater.

Global Importance

TB is a disease of poverty and it has a disproportionate representation in poor countries. 95% of all TB cases and 98% of all TB deaths occur in developing countries, with most deaths occurring in sub-Saharan Africa and Asia. The drugs to treat TB are outdated and, to make things worse, the cases of multidrug-resistant TB (MDR-TB) are rising (0.5 million in 2006). However TB is treatable as long as patients follow the prolonged course of treatment to completion.

Effect on Lung Function

When MTB is inhaled, bacilli reach the lungs, where they are ingested by an alveolar macrophages (by phagocytosis). Usually the macrophages destroy the bacteria, but if they do not, they protect the microbes from the body's defences, and the bacilli survive and multiply within the macrophages. More macrophages are attracted to the area and a tubercle forms (hence the name of the disease). The disease may become dormant or the tubercle may rupture, releasing bacilli into the bronchioles (diagram, lower right).

Affected tissue is replaced by scarring (fibrosis) and cavities filled with cheese-like white necrotic material. During active disease, some of these cavities are joined to the air passages and this material can be coughed up. It contains living bacteria and can therefore pass on infection.

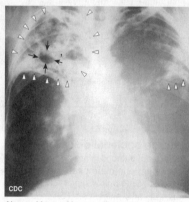

Above: X-ray of lungs affected by pulmonary TB. The white triangles indicate areas where the airspaces of the lung are congested with fluid, and the dark arrows indicate a cavity, from which infective material is coughed up. Surface area for gas exchange is reduced and lung function is adversely affected.

Stages in TB Infection

The series below illustrates stages in MTB infection.

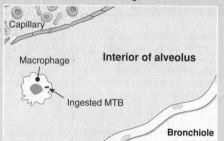

MTB enter the lung and are ingested by macrophages (phagocytic white blood cells).

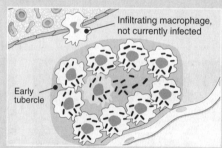

The multiplying bacteria cause the macrophages to swell and rupture. The newly released bacilli infect other macrophages. At this stage a tubercle may form and the disease may lie dormant.

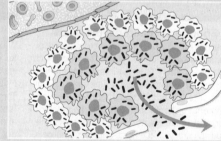

Eventually the tubercle ruptures, allowing bacilli to spill into the bronchiole. The bacilli can now be transmitted when the infected person coughs.

1. (a) Describe how TB infection affects the structure and function of the lung: _____

_____

_____

   (b) Relate this to the way in which TB is transmitted: _____

_____

2. Explain why the prevalence of TB is of global concern: _____

_____

_____

_____

RA 2

Related activities: *Bacterial Pathogens, Respiratory Diseases*
Web links: *Microbiology in Motion*

Periodicals:
Tuberculosis,
The white plague

© Biozone International 2008-2010
Photocopying Prohibited

Cholera

Cholera is an acute intestinal infection caused by the bacterium *Vibrio cholerae*. The disease has a short incubation period, from one to five days. The bacterium produces an enterotoxin that causes a copious, painless, watery diarrhoea that can quickly lead to severe dehydration and death if treatment is not promptly given. Most people infected with *V. cholerae* do not become ill, although the bacterium is present in their faeces for 7-14 days. When cholera appears in a community it is essential to take measures against its spread. These include: **hygienic disposal of human faeces,** provision of an adequate supply of **safe drinking water, safe food handling and preparation** (e.g. preventing contamination of food and cooking food thoroughly), and **effective general hygiene** (e.g. hand washing with soap). Cholera has reemerged as a global health threat after virtually disappearing from the Americas and most of Africa and Europe for more than a century. Originally restricted to the Indian subcontinent, cholera spread to Europe in 1817 in the first of seven pandemics. The current pandemic (below) shows signs of slowly abating, although under-reporting is a problem.

Symptoms

More than 90% of cases are of mild or moderate severity and are difficult to distinguish from other types of acute diarrhoea. Less than 10% of ill people develop typical cholera with signs of moderate or severe dehydration.

Treatment

Most cases of diarrhoea can be treated by giving a solution of oral rehydration salts. During an epidemic, 80-90% of diarrhoea patients can be treated by oral rehydration alone, but patients who become severely dehydrated must be given intravenous fluids. In severe cases, antibiotics can reduce the volume and duration of diarrhoea and reduce the presence of *V. cholerae* in the faeces.

Transmission

Cholera is spread by contaminated water and food. Sudden large outbreaks are usually caused by a contaminated water supply. *Vibrio cholerae* is often found in the aquatic environment and is part of the normal flora of brackish water and estuaries. Human beings are also one of the reservoirs of the pathogenic form of *Vibrio cholerae*.

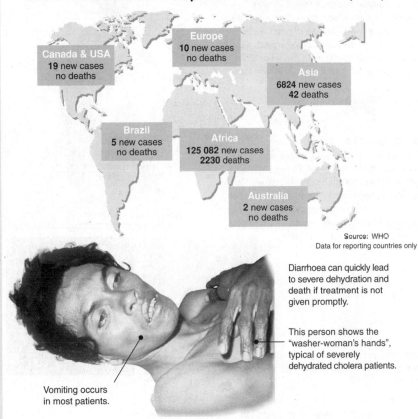

The Cholera Pandemic: Reported Cases and Deaths (2005)

Europe
10 new cases
no deaths

Canada & USA
19 new cases
no deaths

Asia
6824 new cases
42 deaths

Brazil
5 new cases
no deaths

Africa
125 082 new cases
2230 deaths

Australia
2 new cases
no deaths

Source: WHO
Data for reporting countries only

Diarrhoea can quickly lead to severe dehydration and death if treatment is not given promptly.

This person shows the "washer-woman's hands", typical of severely dehydrated cholera patients.

Vomiting occurs in most patients.

1. Identify the pathogen that causes cholera: _____

2. Describe the symptoms of cholera and explain why these symptoms are so dangerous if not treated quickly:

   _____

   _____

   _____

3. State how cholera is transmitted between people: _____

   _____

4. Describe the effective treatment of cholera at the following stages in the progression of the disease:

   (a) Mild onset of dehydration: _____

   (b) Severe symptoms: _____

5. Identify the risk factors associated with the incidence of cholera and relate these to social and economic conditions:

   _____

   _____

   _____

Related activities: Bacterial Pathogens, Control of Disease

A 2

Protozoan Diseases

Protozoa are one-celled, eukaryotic organisms that belong to the Kingdom Protoctista. Among the protozoans, there are many variations on cell structure. While most inhabit water and soil habitats, some are part of the natural microbiota of animals (i.e. they are microorganisms that live on or in animals). Relatively few of the nearly 20 000 species of protozoans cause disease; those that do are often highly specialised, intracellular parasites with complex life cycles involving one or more hosts. Under certain adverse conditions, some protozoans produce a protective capsule called a **cyst**. A cyst allows the protozoan to survive conditions unsuitable for survival. For specialised parasitic species, this includes survival for periods outside a host.

AMOEBAE

Amoebae move by extending projections of their cytoplasm. Several pathogenic amoebae infect humans and feed mainly on red blood cells. People become infected with the pathogen for amoebic microencephalitis while swimming in warm bodies of fresh water or hot springs, when the waterborne cysts pass across mucous membranes and infect blood, brain, and spinal cord. It's almost always fatal.

APICOMPLEXA

Plasmodium vivax

These protozoans are not mobile and tend to be intracellular parasites. They use special enzymes to penetrate the host's tissues. They have complex life cycles involving transmission between several host species. Apicomplexans include **Plasmodium**, which is spread by mosquitoes and causes **malaria**.

Plasmodium sporozoite moving through the cytoplasm of the intestinal epithelia

Ute Frevert, Plos

Giardia trophozoite, SEM

CDC

FLAGELLATES

Flagellates are usually spindle-shaped, with flagella projecting from the front end. The whiplike motion of the flagella pulls the cells through their environment. *Giardia* (inset) is found in the small intestine of mammals. It is passed in the faeces and its life cycle alternates between an actively swimming trophozoite (left) and an infective, resistant cyst. Those infected by *Giardia* develop the disease of the same name.

1. Some protozoans form cysts under certain conditions.

   (a) Explain what a **cyst** is: _____

   _____

   (b) Explain how the ability to form a cyst helps a parasitic protozoan to survive: _____

   _____

   _____

2. Several parasitic protozoans causing diseases in humans use other animal species as hosts for part of their life cycle. Identify the host (including class and genus) that is involved in part of the life cycle for malaria:

   _____

3. (a) The disease known as **giardia** is an increasingly common problem for campers. In seemingly remote areas, campers may contract this disease by drinking water from streams and lakes. Briefly explain the likely reason for this:

   _____

   (b) Suggest how *Giardia* infection could be avoided in areas where *Giardia* is present: _____

   _____

   (c) Suggest how local authorities could prevent *Giardia* becoming a problem in areas set aside for public use:

   _____

Related activities: Malaria
Web links: Animations for Parasitism

Malaria

Malaria is a serious parasitic disease. It is spread by bites of **Anopheles mosquitoes** and affects up to 300 million people in the tropics each year. Most cases of malaria occur in sub-Saharan Africa where the disease accounts for 20% of all childhood deaths. The parasites responsible for malaria are protozoa known as **plasmodia**. Four species can cause malaria in humans. Each spends part of its life cycle in humans and part in *Anopheles* mosquitoes. In many regions, transmission is seasonal, with the peak during and just after the rainy season. Epidemics often occur when conditions suddenly favour transmission in areas where people have little or no immunity. Children and non-immune pregnant women are most at risk, with young children accounting for most of the malaria deaths worldwide. Malaria, especially *falciparum* malaria, is often a medical emergency requiring hospitalisation. Treatment involves the use of antimalarial drugs and, in severe cases, blood transfusion. Symptoms, which appear one to two weeks after being bitten, include headache, shaking, chills, and fever. *Falciparum* malaria is more severe and it can be fatal within a few days of the first symptoms.

Malaria

Malaria occurs in over 100 countries and territories. More than 40% of the people in the world are at risk. Large areas of Central and South America, Hispaniola (Haiti and the Dominican Republic), Africa, the Indian subcontinent, Southeast Asia, the Middle East, and Oceania are considered malaria-risk areas (an area of the world that has malaria).

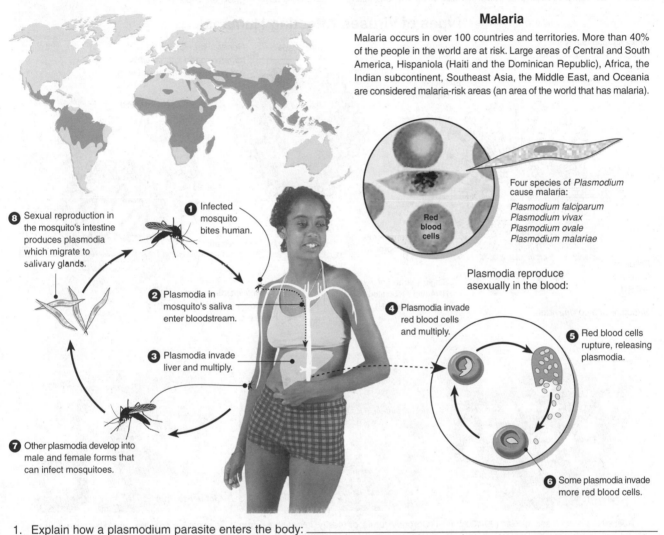

Four species of *Plasmodium* cause malaria:

Plasmodium falciparum
Plasmodium vivax
Plasmodium ovale
Plasmodium malariae

Red blood cells

Plasmodia reproduce asexually in the blood:

① Infected mosquito bites human.

② Plasmodia in mosquito's saliva enter bloodstream.

③ Plasmodia invade liver and multiply.

④ Plasmodia invade red blood cells and multiply.

⑤ Red blood cells rupture, releasing plasmodia.

⑥ Some plasmodia invade more red blood cells.

⑦ Other plasmodia develop into male and female forms that can infect mosquitoes.

⑧ Sexual reproduction in the mosquito's intestine produces plasmodia which migrate to salivary glands.

1. Explain how a plasmodium parasite enters the body: _____

_____

2. Suggest a way in which villagers could reduce the occurrence of malaria carrying mosquitoes in their immediate area:

_____

3. (a) Describe the biological factors important in the global occurrence of malaria: _____

_____

   (b) Suggest measures that could be the most cost effective in controlling the number of new malaria infections:

_____

_____

4. Global warming is expected to increase the geographical area of malaria infection. Explain why this is expected:

_____

_____

© Biozone International 2008-2010
Photocopying Prohibited

Periodicals:
Malaria,
Beating the bloodsuckers

Related activities: Resistance in Pathogens, The Control of Disease
Web links: Malaria Animation

RA 2

Viral Pathogens

Viruses are highly specialised intracellular parasites and they are responsible for a wide range of diseases in plants and animals (including humans). Viruses operate by utilising the host's cellular machinery to replicate new viral particles. Most are able to infect specific types of cells of only one host species. The particular **host range** is determined by the presence specific receptors on the host cell and the availability of the cellular factors needed for viral multiplication. For animal viruses, the receptor sites are on the plasma membranes of the host cells. Antiviral drugs are difficult to design because they must kill the virus without killing

the host cells. Moreover, viruses cannot be attacked when in an inert state. Antiviral drugs work by preventing entry of the virus into the host cell or by interfering with their replication. Immunisation is still regarded as the most effective way in which to control viral disease. However immunisation against viruses does not necessarily provide lifelong immunity. New viral strains develop as preexisting strains acquire mutations. These mutations allow the viruses to change their surface proteins and thus evade detection by the host's immune system. The occurrence of new strains of seasonal flu is a good example of this.

Types of Viruses Affecting Humans

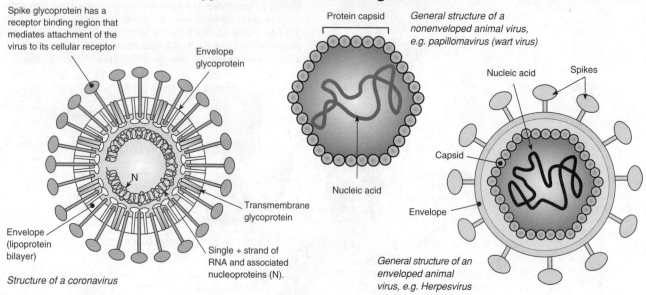

Spike glycoprotein has a receptor binding region that mediates attachment of the virus to its cellular receptor

Envelope glycoprotein

Envelope (lipoprotein bilayer)

Transmembrane glycoprotein

Single + strand of RNA and associated nucleoproteins (N).

Structure of a coronavirus

Protein capsid

General structure of a nonenveloped animal virus, e.g. papillomavirus (wart virus)

Nucleic acid

Nucleic acid

Spikes

Capsid

Envelope

General structure of an enveloped animal virus, e.g. Herpesvirus

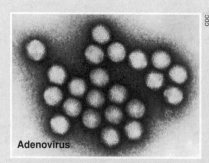

Adenovirus

Adenoviruses are medium-sized (90-100 nm), nonenveloped viruses containing double-stranded DNA. They most commonly cause respiratory illness and are unusually stable to chemical or physical agents, allowing for prolonged survival outside of the body.

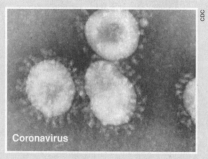

Coronavirus

Coronaviruses primarily infect the upper respiratory and gastrointestinal tracts of birds and mammals, including humans. Their name derives from the crown or corona of spikes and they have the largest genome of any of the single stranded RNA viruses.

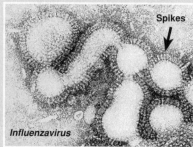

Spikes

Influenzavirus

In some viruses, the capsid is covered by an **envelope**, which protects the virus from the host's nuclease enzymes. Spikes on the envelope provide a binding site for attachment to the host. **Influenzavirus** is an enveloped virus with many glycoprotein spikes.

1. Explain the basis of host specificity in viruses: _____

_____

_____

2. (a) Describe the most effective method of controlling the spread of many viral diseases in the community:

_____

(b) Explain your answer and discuss limitations to this approach: _____

_____

_____

Related activities: The Global Threat of Disease, Antigenic Variability in Pathogens, Resistance in Pathogens **Web links:** Viral Life Cycle

Periodicals: *Viral plagues*

© Biozone International 2008-2010 **Photocopying Prohibited**

HIV and AIDS

AIDS (acquired immune deficiency syndrome) was first reported in the US in 1981. By 1983, the pathogen had been identified as a retrovirus that selectively infects **helper T cells**. It has since been established that HIV arose by the recombination of two simian viruses. It has probably been endemic in some central African regions for decades, as HIV has been found in blood samples from several African nations from as early as 1959. The disease causes a massive deficiency in the immune system due to infection with **HIV** (human immunodeficiency virus). HIV is a **retrovirus** (RNA, not DNA) and is able to splice its genes into the host cell's chromosome. As yet, there is no cure or vaccine, and the disease has taken the form of a **pandemic**, spreading to all parts of the globe and killing more than a million people each year. In southern Africa, AIDS is widespread through the heterosexual community, partly as a result of social resistance to condom use and the high incidence of risky, polygamous behaviour.

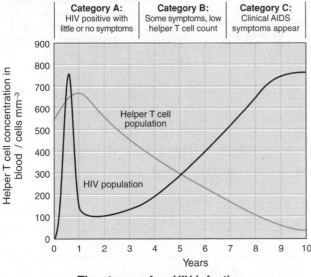

Capsid
Protein coat that protects the nucleic acids (RNA) within.

Viral envelope
A piece of the cell membrane budded off from the last human host cell.

Nucleic acid
Two identical strands of RNA contain the genetic blueprint for making more HIV viruses.

Reverse transcriptase
Two copies of this important enzyme convert the RNA into DNA once inside a host cell.

Surface proteins
These spikes allow HIV to attach to receptors on the host cells (T cells and macrophages).

The structure of HIV

HIV/AIDS

Individuals affected by the human immunodeficiency virus (HIV) may have no symptoms, while medical examination may detect swollen lymph glands. Others may experience a short-lived illness when they first become infected (resembling infectious mononucleosis). The range of symptoms resulting from HIV infection is huge, and is not the result of the HIV infection directly. The symptoms arise from an onslaught of secondary infections that gain a foothold in the body due to the suppressed immune system (due to the few helper T cells). These infections are from normally rare fungal, viral, and bacterial sources. Full blown AIDS can also feature some rare forms of cancer. Some symptoms are listed below:

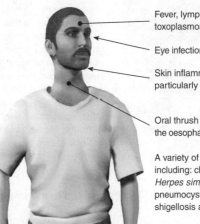

Fever, lymphoma (cancer) and toxoplasmosis of the brain, dementia.

Eye infections (*Cytomegalovirus*).

Skin inflammation (dermatitis) particularly affecting the face.

Oral thrush (*Candida albicans*) of the oesophagus, bronchi, and lungs.

A variety of opportunistic infections, including: chronic or persistent *Herpes simplex*, tuberculosis (TB), pneumocystis pneumonia, shingles, shigellosis and salmonellosis.

Diarrhoea caused by *Isospora* or *Cryptosporidium*.

Marked weight loss.
A number of autoimmune diseases, especially destruction of platelets.

Kaposi's sarcoma: a highly aggressive malignant skin tumour consisting of blue-red nodules, usually start at the feet and ankles, spreading to the rest of the body later, including respiratory and gastrointestinal tracts.

Category A:
HIV positive with little or no symptoms

Category B:
Some symptoms, low helper T cell count

Category C:
Clinical AIDS symptoms appear

Helper T cell population

HIV population

Helper T cell concentration in blood / cells mm⁻³

Years

The stages of an HIV infection

AIDS is actually only the end stage of an HIV infection. Shortly after the initial infection, HIV antibodies appear within the blood. The progress of infection has three clinical categories shown on the graph above.

1. Explain why the HIV virus has such a devastating effect on the human body's ability to fight disease:

_____

_____

_____

2. Consult the graph above showing the stages of HIV infection (remember, HIV infects and destroys helper T cells).

   (a) Describe how the virus population changes with the progression of the disease: _____

   _____

   _____

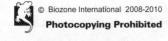

Periodicals:
AIDS, Opportunistic infections and AIDS

Related activities: Epidemiology of AIDS, Replication in HIV
Web links: HIV Interactive Animation

DA 2

Transmission, Diagnosis, Treatment, and Prevention of HIV

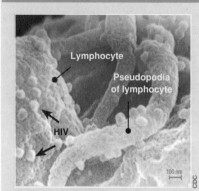

A SEM shows spherical HIV-1 virions on the surface of a human lymphocyte.

Modes of Transmission

1. HIV is transmitted in blood, vaginal secretions, semen, breast milk, and across the placenta.

2. In developed countries, blood transfusions are no longer a likely source of infection because blood is tested for HIV antibodies.

3. Historically, transmission of HIV in developed countries has been primarily through intravenous drug use and homosexual activity, but heterosexual transmission is increasing.

4. Transmission via heterosexual activity has been particularly important to the spread of HIV in Asia and southern Africa, partly because of the high prevalence of risky sexual behavior in these regions.

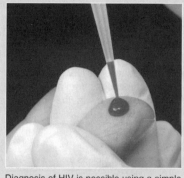

Diagnosis of HIV is possible using a simple antibody-based test on a blood sample.

HIV is easily transmitted between intravenous drug users who share needles.

Treatment and Prevention

Improving the acceptance and use of safe sex practices and condoms are crucial to reducing HIV infection rates. Condoms are protective irrespective of age, the scope of sexual networks, or the presence of other sexually transmitted infections. HIV's ability to destroy, evade, and hide inside the cells of the human immune system make it difficult to treat. Research into vaccination and chemotherapy is ongoing. The first chemotherapy drug to show promise was the nucleotide analogue AZT, which inhibits reverse transcriptase. Protease inhibitors are also used. These work by blocking the HIV protease so that HIV makes copies of itself that cannot infect other cells. An effective vaccine is still some time away, although recent vaccines based on monoclonal antibody technology appear to be promising.

A positive HIV rapid test result shows clumping (aggregation) where HIV antibodies have reacted with HIV protein-coated latex beads.

(b) Describe how the helper T cells respond to the infection: _____

_____

_____

3. Describe three common ways in which HIV can be transmitted from one person to another: _____

_____

_____

4. Explain what is meant by the term **HIV positive**: _____

_____

5. In the years immediately following the discovery of the HIV pathogen, there was a sudden appearance of AIDS cases amongst **haemophiliacs** (people with an inherited blood disorder). State why this group was being infected with HIV:

_____

_____

6. Explain why it has been so difficult to develop a **vaccine** for HIV: _____

_____

7. In a rare number of cases, people who have been HIV positive for many years still have no apparent symptoms. Explain the significance of this observation and its likely potential in the search for a cure for AIDS:

_____

_____

_____

Epidemiology of AIDS

In many urban centres of sub-Saharan Africa, Latin America, and the Caribbean, AIDS has already become the leading cause of death for both men and women aged 15 to 49 years. AIDS kills people in their most productive years and ranks as the leading cause of potential healthy life-years lost in sub-Saharan Africa. Within the next decade, crude death rates in some countries will more than double, and infant and child mortality rates will increase markedly. Perhaps the most significant impact will be seen in projected life expectancies due to the increased mortality of young adults. The AIDS pandemic has lowered the estimated world population level for the year 2050 from 9.4 billion to 8.9 billion, mostly caused by the massive toll of AIDS in Africa.

Pathogens and Human Disease

Regional HIV Statistics and Figures, end 2008

North America
People living with HIV / AIDS¶:	1.4 million
Adult prevalence rate*:	0.4%
People newly infected with HIV:	55 000
Deaths of people from AIDS:	25 000
Main modes of transmission**:	MSM, IDU, Hetero

Western & Central Europe
People living with HIV / AIDS¶:	850 000
Adult prevalence rate*:	0.3%
People newly infected with HIV:	30 000
Deaths of people from AIDS:	13 000
Main modes of transmission**:	Hetero, MSM, IDU

Eastern Europe & Central Asia
People living with HIV / AIDS¶:	1.5 million
Adult prevalence rate*:	0.7%
People newly infected with HIV:	110 000
Deaths of people from AIDS:	87 000
Main modes of transmission**:	IDU, Hetero

Caribbean
People living with HIV / AIDS¶:	240 000
Adult prevalence rate*:	1.0%
People newly infected with HIV:	20 000
Deaths of people from AIDS:	12 000
Main modes of transmission**:	Hetero, MSM, IDU

Latin America
People living with HIV / AIDS¶:	2.0 million
Adult prevalence rate*:	0.6%
People newly infected with HIV:	170 000
Deaths of people from AIDS:	77 000
Main modes of transmission**:	Hetero, MSM, IDU

North Africa and Middle East
People living with HIV / AIDS¶:	310 000
Adult prevalence rate*:	0.2%
People newly infected with HIV:	35 000
Deaths of people from AIDS:	20 000
Main modes of transmission**:	Hetero, IDU, MSM

Sub-Saharan Africa
People living with HIV / AIDS¶:	22.4 million
Adult prevalence rate*:	5.2%
People newly infected with HIV:	1.9 million
Deaths of people from AIDS:	1.4 million
Main modes of transmission**:	Hetero

Oceania
People living with HIV / AIDS¶:	59 000
Adult prevalence rate*:	0.3%
People newly infected with HIV:	3900
Deaths of people from AIDS:	2000
Main modes of transmission**:	MSM, Hetero, IDU

South & South East Asia
People living with HIV / AIDS¶:	3.8 million
Adult prevalence rate*:	0.3%
People newly infected with HIV:	280 000
Deaths of people from AIDS:	270 000
Main modes of transmission**:	Hetero, IDU

East Asia
People living with HIV / AIDS¶:	850 000
Adult prevalence rate*:	<0.1%
People newly infected with HIV:	75 000
Deaths of people from AIDS:	59 000
Main modes of transmission**:	IDU, Hetero, MSM

Estimated percentage of adults (15-49) living with HIV/AIDS

- ■ >15%
- ▨ 5 – 15%
- ☐ 0 – 5%

Source: UNAIDS, WHO

* The proportion of adults (15 to 49 years of age) living with HIV/AIDS in 2008   ¶ People includes adults & children
** Modes of transmission: **Hetero**: heterosexual sex; **IDU**: injecting drug use; **MSM**: sex between men

The Origins of HIV

AIDS researchers have confirmed that the two strains of HIV each originated from cross-species transmission (**zoonosis**) from other primates. HIV-1, responsible for the global pandemic, arose as a result of recombination between two separate strains of simian immunodeficiency virus (SIV) in infected **common chimpanzees** in west-central Africa. HIV-2 is less virulent than HIV-1 and, until recently, was restricted to West Africa. It originated from a strain of SIV found in **sooty mangabey** monkeys in that region. Killing primates for bushmeat allows the virus to transmit to human hunters when they handle infected carcasses with cuts on their hands. Such cross-species transmissions could be happening every day.

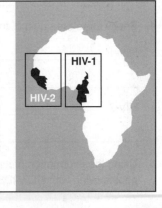

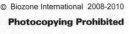

Periodicals:
Search for a cure

Related activities: HIV and AIDS, The Impact of HIV in Africa
Web links: Living with HIV/AIDS

DA 2

Factors in the spread of HIV

Epidemiologists cannot predict with certainty how rapidly a given epidemic will expand or when it will peak, although short term predictions can be made on the basis of trends in HIV spread and information on risk behaviour. Fortunately, there is strong evidence showing that countries will ultimately reduce their new infections if they carry out effective prevention programmes encouraging abstinence, or fidelity and safer sex. A crucial factor is promoting the acceptance and use of condoms, both the traditional kind and the female condom. Condoms are protective irrespective of age, the scope of sexual networks, or the presence of other sexually transmitted infections. There is evidence from around the world that many factors play a role in starting a sexually transmitted HIV epidemic or driving it to higher levels. Some of these risk factors are listed below.

In many African communities, men travel from rural settlements into the cities in search of work. These men often develop sexual networks while they are away and bring HIV with them when they return.

Social and behavioural risk factors

- Little or no condom use.
- Large proportion of the adult population with multiple partners.
- Overlapping (as opposed to serial) sexual partnerships. Individuals are highly infectious when they first acquire HIV and are more likely to infect any concurrent partners.
- Large sexual networks which are often seen in individuals who move back and forth between home and a far off work place.
- Women's economic dependence on marriage or prostitution, robbing them of control over the circumstances or safety of sex.

Biological risk factors

- High rates of sexually transmitted infections, especially those causing genital ulcers.
- Low rates of male circumcision (for poorly understood reasons, circumcised males have a reduced risk of contracting HIV).
- High viral load (HIV levels in the blood is typically highest when a person is first infected and again in the late stages of illness).

1. Comment on the social, economic, and biological factors involved in the prevalence of HIV in many of the **rural** communities of sub-Saharan Africa:

_____

_____

_____

_____

2. Describe the effects of AIDS on the countries of sub-Saharan Africa with respect to the following:

    (a) Age structure of their populations: _____

    _____

    (b) Their local economies: _____

    _____

3. Effective antiviral therapies have reduced deaths from HIV/AIDS in developed countries. Suggest why a similar reduction has not occurred in the countries of sub-Saharan Africa:

    _____

    _____

4. Briefly state the origin of the two main strains of HIV:

    HIV-1: _____

    _____

    HIV-2 _____

    _____

5. Using the information provided on the previous page and your own graph paper, plot a column graph of the number of people living with HIV/AIDS for each region. Staple the completed graph into this workbook.

Replication in HIV

Animal viruses exhibit a number of different mechanisms for **replicating**, i.e. entering a host cell and producing and releasing new virions. Enveloped viruses bud out from the host cell, whereas those without an envelope are released by rupture of the cell membrane. Three processes (attachment, penetration, and uncoating) are shared by both DNA- and RNA containing animal viruses. The example below describes replication in the retrovirus HIV, where the virus uses its own reverse transcriptase to synthesise viral DNA and produce **latent proviruses** or active, mature retroviruses.

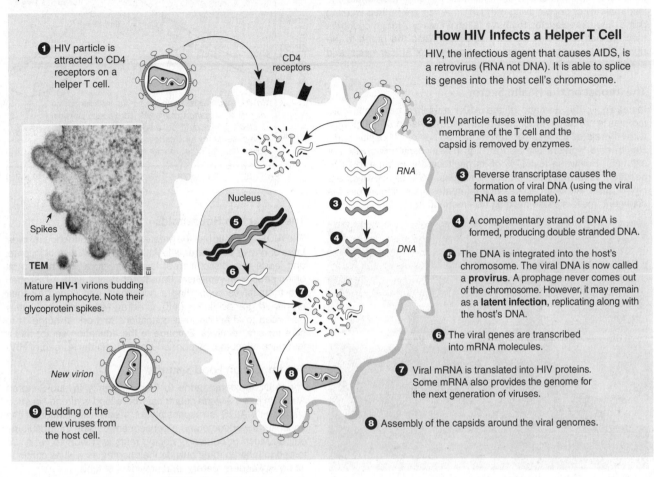

1 HIV particle is attracted to CD4 receptors on a helper T cell.

CD4 receptors

Spikes

TEM

Mature **HIV-1** virions budding from a lymphocyte. Note their glycoprotein spikes.

Nucleus

New virion

9 Budding of the new viruses from the host cell.

RNA

3

4

DNA

5

6

7

8

How HIV Infects a Helper T Cell

HIV, the infectious agent that causes AIDS, is a retrovirus (RNA not DNA). It is able to splice its genes into the host cell's chromosome.

2 HIV particle fuses with the plasma membrane of the T cell and the capsid is removed by enzymes.

3 Reverse transcriptase causes the formation of viral DNA (using the viral RNA as a template).

4 A complementary strand of DNA is formed, producing double stranded DNA.

5 The DNA is integrated into the host's chromosome. The viral DNA is now called a **provirus**. A prophage never comes out of the chromosome. However, it may remain as a **latent infection**, replicating along with the host's DNA.

6 The viral genes are transcribed into mRNA molecules.

7 Viral mRNA is translated into HIV proteins. Some mRNA also provides the genome for the next generation of viruses.

8 Assembly of the capsids around the viral genomes.

---

1. (a) Describe how an HIV particle enters a host cell: _____

_____

(b) Explain the role of the reverse transcriptase in the life cycle of a retrovirus: _____

_____

_____

(c) Explain the significance of the formation of a provirus: _____

_____

_____

2. Summarise the steps involved in invasion of a host cell by an enveloped viral particle such as *Influenzavirus*:

(a) Attachment: _____

(b) Penetration: _____

(c) Uncoating: _____

(d) Biosynthesis: _____

(e) Release: _____

3. Some of the drugs for treating HIV inhibit reverse transcriptase. Suggest why this is an effective strategy: _____

_____

© Biozone International 2008-2010

Photocopying Prohibited

Periodicals:
Are viruses alive?

Related activities: *Viral Pathogens, HIV and AIDS*
Web links: *Animation of HIV Replication, HIV Life Cycle*

A 2

The Impact of HIV/AIDS in Africa

Around 10% of the world's population lives in sub-Saharan Africa, yet this region is home to two thirds of HIV-infected people. The impact of HIV-AIDS on Africa's populations, workplaces, and economies is enormous, and is setting back Africa's economic and social progress. The effects of the disease are disproportionate; the vast majority of people living with HIV in Africa are in their working prime. Life expectancies too have fallen; in many African countries they are half what they were 15 years ago (see the graph to the right). This has been detrimental to all aspects of African social and economic structure.

The Impact on the Health Sector

Increasingly, the demand of the AIDS epidemic on health care facilities is not being adequately met. In sub-Saharan Africa, people with HIV-related diseases occupy more than half of all hospital beds. Health care workers are also at high risk of contracting HIV. For example, Botswana lost 17% of its healthcare workforce to AIDS between 1999 and 2005. In some regions, 40% of midwives are HIV positive. Although access to treatment is improving across the continent, most of those needing treatment do not receive it.

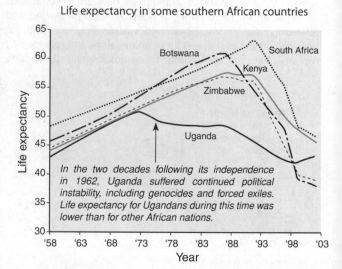

Life expectancy in some southern African countries

In the two decades following its independence in 1962, Uganda suffered continued political instability, including genocides and forced exiles. Life expectancy for Ugandans during this time was lower than for other African nations.

Photo courtesy of Karl Mueller; Flickr

This Malawi grandmother is now responsible for the care and support of her four grandchildren, whose parents both died of AIDS. The intergenerational effects of HIV/AIDS are the longest lasting and are related to how the epidemic intensifies poverty and leads to its persistence. The challenge to African countries is in achieving the sustainable development needed to respond effectively to the epidemic.

Information from various sources including UNAIDS, UNDP, and AVERT.org

The Impact on Households

The AIDS epidemic has the greatest impact on poor households. Loss of one or both parents to AIDS results in a loss of income, and leads to a dissolution of family structure. The burden on older people is immense. Parents of adults with AIDS often find themselves supporting a household and caring for their orphaned grandchildren (left). Children (especially girls) may be forced to abandon their educations to work at home or to care for sick relatives. Damage to the education sector has a feedback effect too, as poor education fuels the spread of HIV.

The Impact on Food Security

The AIDS epidemic adds to food insecurity in sub-Saharan Africa. In many African countries, where food is already in short supply, HIV/AIDS is responsible for a severe depletion of the agricultural workforce and a consequent decline in agricultural output. Much of burden of coping rests with women, who are forced to take up roles outside their homes as well as continue as housekeepers, carers, and providers of food.

The Future?

AIDS in Africa is linked to other problems, including poverty, food insecurity, and poor public infrastructure. Efforts to fight the epidemic must work within these constraints. Without control, the AIDS epidemic will continue to present the single greatest barrier to Africa's social and economic development.

1. With reference to the graph of life expectancies in southern African countries:

   (a) Describe and explain the trends in since the late 1980s: _____

   _____

   (b) Describe and give a likely explanation of the trends between 1958 and 1988: _____

   _____

   _____

2. Discuss the barriers sub-Saharan African nations will face in countering the effects of the HIV/AIDS:

   _____

   _____

   _____

   _____

   _____

The Global Threat of Disease

Emerging diseases are so named because they are diseases with no previous history in the human population. Often, as with HIV/AIDS and avian influenza (H5N1), they are **zoonoses** (animal diseases that cross to humans). Zoonoses are capable of causing highly lethal **pandemics** (world-wide epidemics) amongst an unprepared population. The increasing incidence of **multiple drug resistance** in pathogens (including those that cause tuberculosis and malaria) has lead to the **re-emergence** of diseases that were previously thought to be largely under control. In the 1940s, many common but lethal bacterial diseases (e.g. diphtheria) were conquered using antibiotics. It is now evident that antibiotics are losing efficacy and must be used wisely to reduce the risk of resistance in the target populations. The global spread of viral diseases is controlled most effectively by immunising susceptible populations against circulating strains (as occurs with seasonal flu). The challenge for the control of viral disease is in the continual development of effective vaccines against newly emerging strains.

'Swine Flu' (H1N1/09)

Pandemic H1N1/09 virus is an *Influenza A* virus subtype H1N1. It is a novel strain of flu and was responsible for the flu pandemic in 2009. Although it was first notified in Mexico, genetic analyses indicate that the strain had been circulating in pigs for some time before its transmission to humans. H1N1 seems to be the result of a reassortment of two viruses, one from North America and one from Europe. Pigs are susceptible to both avian and human flu viruses, so potentially serve as vessels in which viral reassortment can easily occur. Although H1N1 shows no sign of mutating to a more virulent form, experts are tracking its spread closely.

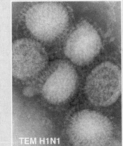

TEM H1N1 · CDC

Avian influenza A (H5N1)

In January 2004, a new strain of 'bird flu' (H5N1) spread rapidly through 8 Asian countries. Outbreaks occurred again in 2005, each time with high human mortality. It is now continuing its spread through Africa and Europe. Avian flu mutates rapidly and crosses species barriers, apparently with little difficulty. These features make it a serious public health threat. The **H5N1 virus** is pictured left. At this magnification, the stippled appearance of the protein coat encasing the virion can easily be seen.

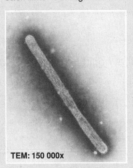

CDC · TEM: 150 000x

So-called 'swine flu' is easily transmitted by coughing and sneezing. It is normally a relatively mild disease, but can move deep into the lungs in susceptible people. In these cases, it can be fatal.

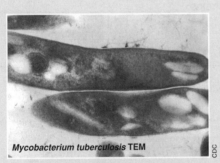

Mycobacterium tuberculosis TEM · CDC

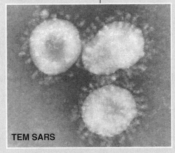

TEM SARS

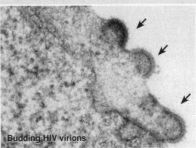

Budding HIV virions · EII

Resistant Tuberculosis (global)

The reappearance of TB as a globally significant disease is largely the result of increasing multi-drug resistance in the pathogen and reduced levels of immunisation. Globally, TB is resisting control. Even the newest TB drugs are decades old and, because TB is a disease most prevalent in poorer contries, there are few incentives to develop better diagnostic tests and new drugs.

Severe Acute Respiratory Syndrome (SARS)

The first case of this respiratory illness was reported in 2002 in China. The reservoir for the pathogen is a bat, which transmits it to other mammals. Once in the human population, SARS spread rapidly. It had a mortality of about 10%, with 50% for people aged 60+.

HIV and AIDS (global)

The AIDS pandemic is set to have a lasting effect on the world population. At least two strains are recognised: the deadly **HIV-1 virus** is more widespread than the slightly more benign **HIV-2 strain**. **HIV viruses** (arrowed) are shown above emerging from a human T cell. HIV is responsible for the massive AIDS pandemic that some claim to be species threatening.

Periodicals:
War on disease,
Preventing the next pandemic

Related activities: HIV and AIDS, Tuberculosis
Web links: Avian Flu, Flutracker

1. Using examples, explain the role of **zoonoses** in the emergence of new diseases:

_____

_____

_____

2. Explain the role of reassortment (of genetic material) in producing new viral strains: _____

_____

_____

3. (a) Explain the potential of populations of swine to act as incubators for new viral strains: _____

_____

_____

(b) Suggest precautions that could be taken by the agricultural industry to prevent viral reassortment in swine:

_____

_____

_____

4. Much criticism was directed at factory farming operations in the US and Mexico during the 2009 H1N1 pandemic. Explain why factory farming (of both pigs and poultry) has been targeted as a culprit in the emergence of new viral strains:

_____

_____

_____

5. Using an example, explain what is meant by a **re-emerging disease**: _____

_____

_____

6. Explain how drug resistance in pathogens has led to an increase in the number of re-emerging diseases:

_____

_____

_____

7. (a) Describe the responses of governments to the threat of new influenza strains: _____

_____

_____

(b) Suggest how authorities in public health and agriculture could cooperate to reduce the risk of new diseases:

_____

_____

8. HIV is another viral disease for which immunisation could offer a chance of effective disease control. Discuss the biological factors that have (thus far) prevented the development of a successful HIV vaccine:

_____

_____

_____

_____

The Control of Disease

Many factors can influence the spread of disease, including the social climate, diet, general health, and access to medical care. Human intervention and modification of behaviour can reduce the transmission rate of some diseases and inhibit their spread. Examples include the use of personal physical barriers, such as condoms, to prevent sexually transmitted infections (STIs), and the use of **quarantine** to ensure that potential carriers of disease are isolated until incubation periods have elapsed. Cleaning up the environment also lowers the incidence of disease by reducing the likelihood that pathogens or their vectors will survive. The effective control of infectious disease depends on knowing the origin of the outbreak (its natural reservoir), its mode of transmission within the population, and the methods that can be feasibly employed to contain it. Diseases are often classified according to how they behave in a given population. Any disease that spreads from one host to another, either directly or indirectly, is said to be a **communicable disease**. Those that are easily spread from one person to another, such as chicken pox or measles, are said to be **contagious**. Such diseases are a threat to **public health** and many must be notified to health authorities. **Noncommunicable diseases** are not spread from one host to another and pose less of a threat to public health. A disease that occurs only occasionally and is usually restricted in its spread is called a **sporadic disease**.

Methods for controlling the spread of disease

Transmission of disease can be prevented or reduced by adopting 'safe' behaviours. Examples include using condoms to reduce the spread of STIs, isolation of people with a specific illness (such as swine flu), or establishing quarantine procedures for people who may be infected, but are not yet ill.

The development of effective sanitation, sewage treatment, and treatment of drinking water has virtually eliminated dangerous waterborne diseases from developed countries. These practices disrupt the normal infection cycle of pathogens such as cholera and giardia.

Appropriate personal hygiene practices reduce the risk of infection and transmission. Soap may not destroy the pathogens but washing will dilute and remove them from the skin. Although popular, antibacterial soaps encourage development of strains resistant to antimicrobial agents.

The environment can be made less suitable for the growth and transmission of pathogens. For example, spraying drainage ditches and draining swamps eliminates breeding habitats for mosquitoes carrying diseases such as malaria and dengue fever.

Immunisation schedules form part of public health programmes. If most of the population is immune, 'herd immunity' limits outbreaks to sporadic cases. In such populations there are too few susceptible individuals to support the spread of an epidemic.

Disinfectants and sterilisation techniques, such as autoclaving, destroy pathogenic microbes before they have the opportunity to infect. The use of these techniques in medicine has significantly reduced post operative infections and associated deaths.

1. Distinguish between contagious and non-communicable diseases, providing an example of each:

_____

_____

2. (a) Explain the difference between **isolation** and **quarantine**: _____

_____

_____

_____

© Biozone International 2008-2010
Photocopying Prohibited

Periodicals:
New tactics against TB,
Preparing for a pandemic

Related activities: Vaccines and Vaccination, Antibiotics

RA 2

(b) Using the recent example of SARS, explain how isolation and quarantine operate to prevent the spread of disease:

_____

_____

3. Explain how the use of condoms reduces the spread of the human immunodeficiency virus (HIV) that causes AIDS:

_____

_____

4. Explain how the drainage of stagnant water in tropical regions may reduce the incidence of malaria in those countries:

_____

_____

5. Describe how each of the following methods is used to control the **growth** of disease-causing microbes:

(a) Disinfectants: _____

_____

(b) Antiseptics: _____

_____

(c) Heat: _____

_____

(d) Ionising radiation (gamma rays): _____

_____

(e) Desiccation: _____

_____

(f) Cold: _____

_____

6. The **Human Genome Project** (HGP) was launched in 1990 and completed in 2003, two years ahead of schedule. Its achieved aim was to sequence the entire human genome, but much of the research since has focussed on determining the various roles of the (expressed) gene products. It is hoped that a more complete understanding the human genome will revolutionise the treatment and prevention of disease. Briefly discuss how the HGP will facilitate:

(a) Diagnosis of disease: _____

_____

_____

(b) Treatment of disease: _____

_____

_____

7. The first measles vaccine was introduced to Britain in 1964. However, in 1993 there were 9000 cases of measles notified to the health authorities in England and Wales.

(a) Suggest why measles has not been eliminated in Britain: _____

_____

(b) Explain how vaccination interrupts the transmission of measles within a population: _____

_____

_____

Antibiotics

An **antibiotic** is a chemotherapeutic agent that inhibits or prevents microbial growth. Antibiotics are produced naturally by bacteria and fungi, but some synthetic (manufactured) **antimicrobial drugs** are also effective against microbial infections. Antimicrobial drugs interfere with the growth of microorganisms (see diagram below) by either killing microbes directly (**bactericidal**) or preventing them from growing (**bacteriostatic**). To be effective, they must often act inside the host, so their effect on the host's cells and tissues is important. The ideal antimicrobial drug has **selective toxicity**, killing the pathogen without damaging the host. Some antimicrobial drugs have a narrow **spectrum of**

activity, and affect only a limited number of microbial types. Others are **broad-spectrum drugs** and affect a large number of microbial species. When the identity of a pathogen is not known, a broad-spectrum drug may be prescribed in order to save valuable time. There is a disadvantage with this, because broad spectrum drugs target not just the pathogen, but much of the host's normal microflora also. The normal microbial community usually controls the growth of pathogens and other microbes by competing with them. By selectively removing them with drugs, certain microbes in the community that do not normally cause problems, may flourish and become **opportunistic pathogens**.

How Antimicrobial Drugs Work

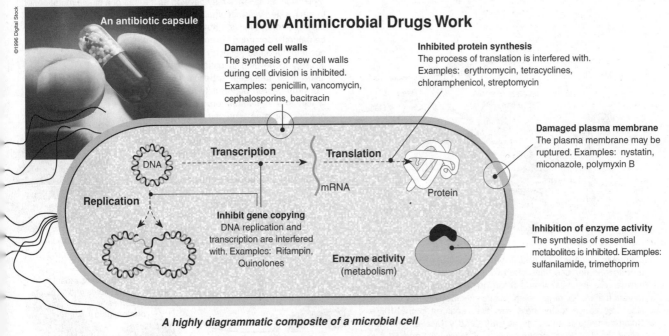

An antibiotic capsule

©1996 Digital Stock

Damaged cell walls
The synthesis of new cell walls during cell division is inhibited. Examples: penicillin, vancomycin, cephalosporins, bacitracin

Inhibited protein synthesis
The process of translation is interfered with. Examples: erythromycin, tetracyclines, chloramphenicol, streptomycin

Damaged plasma membrane
The plasma membrane may be ruptured. Examples: nystatin, miconazole, polymyxin B

DNA — Transcription — Translation — mRNA — Protein

Replication

Inhibit gene copying
DNA replication and transcription are interfered with. Examples: Rifampin, Quinolones

Enzyme activity
(metabolism)

Inhibition of enzyme activity
The synthesis of essential metabolites is inhibited. Examples: sulfanilamide, trimethoprim

A highly diagrammatic composite of a microbial cell

1. Describe one advantage and one disadvantage of using a **broad-spectrum drug** on an unidentified bacterial infection:

    (a) _____

    (b) _____

2. Describe the requirements of an "ideal" anti-microbial drug, and explain how antibiotics satisfy these requirements:

    _____

    _____

3. The diagram below shows an experiment investigating the effectiveness of different antibiotics on a pure culture of a single species of bacteria. Giving a reason, Identify which antibiotic (A-D) is most effective in controlling the bacteria:

    _____

    _____

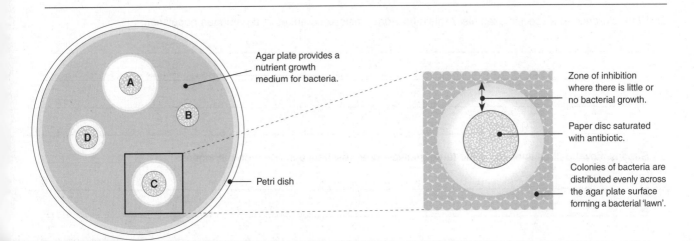

Agar plate provides a nutrient growth medium for bacteria.

Petri dish

Zone of inhibition where there is little or no bacterial growth.

Paper disc saturated with antibiotic.

Colonies of bacteria are distributed evenly across the agar plate surface forming a bacterial 'lawn'.

Periodicals:
Finding and improving
antibiotics

Related activities: The Control of Disease, Antibiotic Resistance
Web links: Antibiotics Attack

A 2

Resistance in Pathogens

Many pathogens are effectively controlled by the use of drugs and vaccines, but the emergence of drug resistant pathogens is increasingly undermining the ability to treat and control killer diseases such as HIV/AIDS, tuberculosis, and malaria. High mutation rates and short generation times in viral, bacterial, and protozoan pathogens have contributed to the rapid spread of drug resistance through populations. This is well documented for malaria (below), TB, and HIV/AIDS. Rapid evolution in pathogens is exacerbated too by the strong selection pressure created by the wide use and misuse of antimicrobial drugs, the poor quality of available drugs, and poor patient compliance. The most successful treatment for several diseases, including HIV/AIDS and TB appears to be a multi-pronged attack using a cocktail of drugs to target the pathogen at many stages.

Global Spread of Chloroquine Resistance

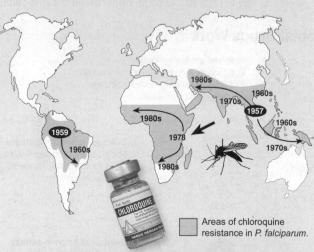

Areas of chloroquine resistance in *P. falciparum*.

Malaria in humans is caused by various species of *Plasmodium*, a protozoan parasite transmitted by *Anopheles* mosquitoes. The inexpensive antimalarial drug **chloroquine** was used successfully to treat malaria for many years, but its effectiveness has declined since resistance to the drug was first recorded in the 1960s. Chloroquine resistance has spread steadily (above) and now two of the four *Plasmodium* species, *P. falciparum* and *P. vivax* are chloroquine-resistant. *P. falciparum* alone accounts for 80% of all human malarial infections and 90% of the deaths, so this rise in resistance is of global concern. New anti-malarial drugs have been developed, but are expensive and often have undesirable side effects. Resistance to even these newer drugs is already evident, especially in *P. falciparum*, although this species is currently still susceptible to artemisinin, a derivative of the medicinal herb *Artemisia annua*.

Drug Resistance in HIV

Strains of drug-resistant HIV arise when the virus mutates during replication. Resistance may develop as a result of a single mutation, or through a step-wise accumulation of specific mutations. These mutations may alter drug binding capacity or increase viral fitness, or they may be naturally occurring polymorphisms (which occur in untreated patients). Drug resistance is likely to develop in patients who do not follow their treatment schedule closely, as the virus has an opportunity to adapt more readily to a "non-lethal" drug dose. The best practice for managing the HIV virus is to treat it with a cocktail of anti-retroviral drugs with different actions to minimise the number of viruses in the body. This minimises the replication rate, and also the chance of a drug resistant mutation being produced.

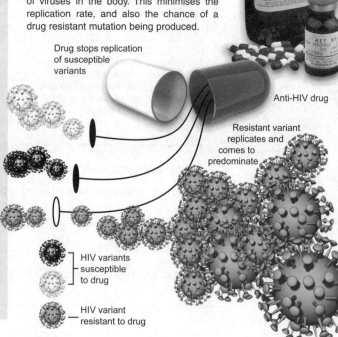

Drug stops replication of susceptible variants

Anti-HIV drug

Resistant variant replicates and comes to predominate

HIV variants susceptible to drug

HIV variant resistant to drug

1. Describe factors contributing to the rapid spread of drug resistance in pathogens: _____

_____

_____

2. With reference to a specific example, explain how drug resistance arises in a pathogen population:

_____

_____

_____

_____

3. Suggest how health authorities could target multiple drug resistance in common pathogens: _____

_____

_____

_____

KEY TERMS Crossword

Complete the crossword below, which will test your understanding of key terms in this chapter and their meanings

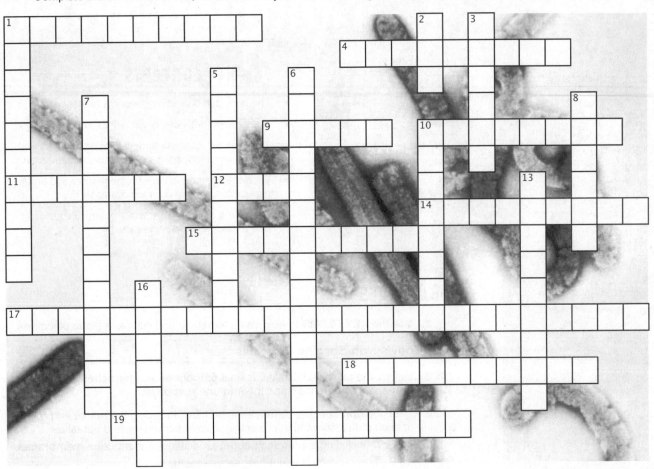

Clues Across

1. Animal-like protoctistans whose members include amoebae, intracellular parasites, and pathogenic flagellates.
4. The term describing the number of newly infected people in a population.
9. This type of white blood cell is infected by HIV (2 words: 1, 3)
10. A disease-causing organism.
11. An abnormal condition of an organism.
12. A prolonged and usually fatal syndrome characterised by a suite of diseases and resulting from immune system failure (acronym).
14. The study of the cause of a disease.
15. A type of virus that uses reverse transcriptase enzyme to make DNA from its own RNA.
17. The pathogen responsible for tuberculosis (2 words: 13, 12).
18. An antimicrobial drug used to combat bacterial infections.
19. A statistical measure of the number of years of life left at any particular age (2 words: 4, 10).

Clues Down

1. A genus of specialised protozoan parasites and the agent responsible for malaria.
2. The infectious agent for Acquired Immune Deficiency (or Immunodeficiency) Syndrome (acronym).
3. A state of general well being.
5. The term describing the number of people in a population affected by a disease.
6. A series of criteria designed to establish a causal relationship between a microbe and a disease (2 words: 5, 10).
7. A globally important disease, usually affecting the lungs, that is caused by an infectious bacterium and is spread by coughing.
8. An agent that carries a pathogen between hosts.
10. An organism that lives off a host organism and is dependent on it during at least some stage of its life cycle.
13. The death rate, often made attributable to a specific cause.
16. A disease caused by an intracellular blood parasite.

Defence and the Immune System

KEY CONCEPTS

▶ The body can defend itself against pathogens.

▶ Non-specific defences target any foreign material.

▶ The immune response targets specific antigens and has a memory for antigens previously encountered.

▶ Vaccination employs the immune response to protect populations against common diseases.

▶ Plants and microorganisms are potential sources of new medicines against disease.

KEY TERMS

active immunity
antibody (=immunoglobulin)
antigen
artificially acquired immunity
B cell (=B lymphocyte)
cell-mediated immunity
cell signalling
clonal selection
eradication
fever
humoral immunity
immune response
immunity
immunological memory
infection
inflammation
interferon
leucocyte
lymphocyte
macrophage
memory cells
naturally acquired immunity
non-specific defence
passive immunity
pathogen
phagocyte
plasma cells
primary response
secondary response
specific defence
T cell (=T lymphocyte)
thymus
vaccination (=immunisation)

OBJECTIVES

☐ 1. Use the **KEY TERMS** to help you understand and complete these objectives.

The Body's Non-Specific Defences   pages 229-233

☐ 2. Outline the primary defences against **pathogens** and **parasites**, including **non-specific defences** and the **immune response**.

☐ 3. Describe **non-specific defences** in humans, including the nature and role of each of the following in protecting against pathogens and parasites:

   (a) Skin (including **sweat** and **sebum** production) and **mucous membranes**.

   (b) Body secretions (tears, urine, saliva, gastric juice).

   (c) Natural anti-bacterial and anti-viral proteins, e.g. **interferon**.

   (d) The **inflammatory response**, **fever**, and cell death (necrosis).

   (e) **Phagocytosis** by phagocytic leucocyte (phagocytes).

The Body's Non-Specific Defences   pages 234-244

☐ 4. Describe the **immune response**, including the importance of both **specificity** and memory. Distinguish between **naturally acquired** and **artificially acquired** immunity and between **active** and **passive immunity**.

☐ 5. Describe and distinguish between the activities of the **T lymphocytes** and **B lymphocytes** in the **cell mediated** and **humoral immune responses**.

☐ 6. Describe and contrast the functional roles of **plasma cells** and **memory cells** and explain the basis for **immunological memory**.

☐ 7. Discuss the role of **cell signalling** and immunological memory in long term immunity (ability to respond quickly to previously encountered **antigens**).

☐ 8. Describe the structure of an antibody, identifying the constant and variable regions, and the antigen binding site. Explain how antibodies inactivate antigens and facilitate their destruction.

☐ 9. Describe **antibody** production, including how B cells bring about **humoral** (antibody-mediated) **immunity** to specific **antigens**.

☐ 10. Compare and contrast the **primary** and **secondary response** to infection. Explain how these responses are used to provide immunity by vaccination.

☐ 11. Explain the role of **vaccination programmes** in preventing disease. Discuss the role of vaccination in the **eradication** of infectious diseases.

☐ 12. Explain the role of health authorities in responding to the threat of new strains of pathogens as the arise (e.g. seasonal flu).

☐ 13. Describe possible sources of new medicines from microorganisms and plants and relate this to the need to preserve biodiversity.

Periodicals:
Listings for this chapter are on page 340

Weblinks:
www.biozone.co.uk/
weblink/OCR-AS-2641.html

*Teacher Resource
CD-ROM:*
Monoclonal Antibodies

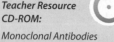

The First Line of Defence

If microorganisms never encountered resistance from the body's defence system, we would be constantly ill and would eventually die from disease. The first barrier to infection is formed by a series of obstacles which forms part of a **non-specific** defence against infection. Non-specific defences do not target any particular pathogen; they are ready to respond at all times to all pathogens, and the response is the same and maximal each time regardless of the invader. Unlike the specific immune response, no immunological memory is developed. Physical barriers (e.g. skin, mucous membranes), chemical barriers (e.g. enzymes, fatty acids, pH), and biological barriers, such as the body's natural microbiota, are the body's first line of defence against infection.

The Body's Natural Microbiota

After birth, normal and characteristic microbial populations begin to establish themselves on and in the body. A typical human body contains 1×10^{13} body cells, yet harbours 1×10^{14} bacterial cells. These microorganisms establish more or less permanent residence but, under normal conditions, do not cause disease. In fact, this normal microflora can benefit the host by preventing the overgrowth of harmful pathogens. They are not found throughout the entire body, but are located in certain regions.

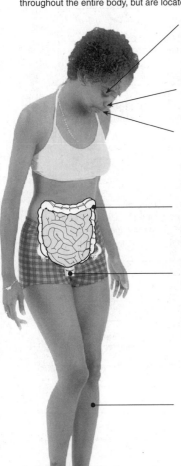

Eyes: The conjuctiva, a continuation of the skin or mucous membrane, contains a similar microbiota to the skin.

Nose and throat: Harbours a variety of microorganisms, e.g. *Staphylococcus spp.*

Mouth: Supports a large and diverse microbiota. It is an ideal microbial environment; high in moisture, warmth, and nutrient availability.

Large intestine: Contains the body's largest resident population of microbes because of its available moisture and nutrients.

Urinary and genital systems: The lower urethra in both sexes has a resident population; the vagina has a particular acid-tolerant population of microbes because of the low pH nature of its secretions.

Skin: Skin secretions prevent most of the microbes on the skin from becoming residents.

The skin is the largest organ of the body. It forms an important physical barrier against the entry of pathogens into the body. A natural population of harmless microbes live on the skin, but most other microbes find the skin inhospitable. The continual shedding of old skin cells (desquamation) physically removes bacteria from the surface of the skin. Sebaceous glands in the skin produce sebum, which has antimicrobial properties, and the slightly acidic secretions of sweat inhibit microbial growth.

Cilia line the epithelium of the nasal passage (below). Their wave-like movement sweeps foreign material out and keeps the passage free of microorganisms, preventing them from colonising the body.

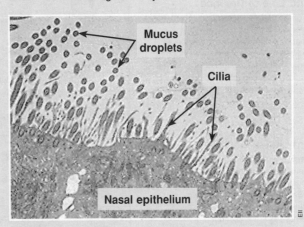

Mucus droplets

Cilia

Nasal epithelium

Antimicrobial chemicals are present in many bodily secretions. Tears, saliva, nasal secretions, and human breast milk all contain **lysozymes** and **phospholipases**. Lysozymes kill bacterial cells by catalysing the hydrolysis of cell wall linkages, whereas phospholipases hydrolyse the phospholipids in bacterial cell membranes, causing bacterial death. Low pH gastric secretions also inhibit microbial growth, and reduce the number of pathogens establishing colonies in the gastrointestinal tract.

Defence and the Immune System

1. Explain why it is healthy to have a natural population of microbes on and inside the body: _____

_____

2. Describe how the skin acts as a barrier to prevent pathogens entering the body: _____

_____

_____

3. Describe the role of each of the following in non-specific defence:

(a) Phospholipases: _____

(b) Cilia: _____

(c) Sebum: _____

The Body's Defences

The human body has a tiered system of defences to prevent or limit infection by pathogens. The first line of defence aims to keep microorganisms from entering the body. If this fails, a second line of defence targets any foreign bodies (including microbes) that manage to get inside. If microorganisms manage to evade this level of defence, the body's immune system provides a third line of (specific) defence. The ability to ward off disease through the various defence mechanisms is called **resistance**. The lack of resistance, or vulnerability to disease, is known as **susceptibility**. **Non-specific resistance** includes the first and second lines of defence. **Specific resistance** (the immune response) is specific to particular pathogens.

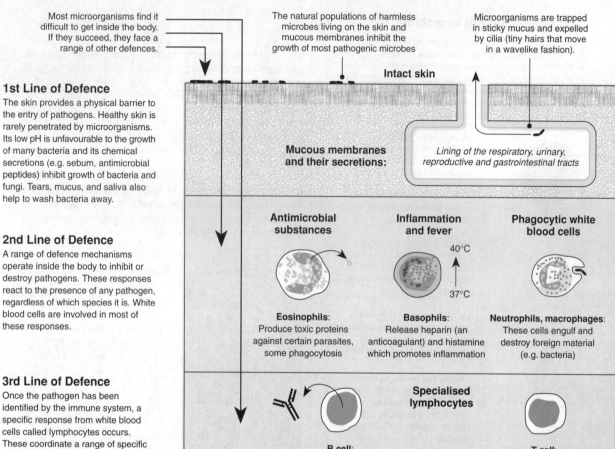

Most microorganisms find it difficult to get inside the body. If they succeed, they face a range of other defences.

The natural populations of harmless microbes living on the skin and mucous membranes inhibit the growth of most pathogenic microbes

Microorganisms are trapped in sticky mucus and expelled by cilia (tiny hairs that move in a wavelike fashion).

Intact skin

Mucous membranes and their secretions:

Lining of the respiratory, urinary, reproductive and gastrointestinal tracts

Antimicrobial substances

Eosinophils:
Produce toxic proteins against certain parasites, some phagocytosis

Inflammation and fever

40°C

37°C

Basophils:
Release heparin (an anticoagulant) and histamine which promotes inflammation

Phagocytic white blood cells

Neutrophils, macrophages:
These cells engulf and destroy foreign material (e.g. bacteria)

Specialised lymphocytes

B cell:
Antibody production

T cell:
Cell-mediated immunity

1st Line of Defence

The skin provides a physical barrier to the entry of pathogens. Healthy skin is rarely penetrated by microorganisms. Its low pH is unfavourable to the growth of many bacteria and its chemical secretions (e.g. sebum, antimicrobial peptides) inhibit growth of bacteria and fungi. Tears, mucus, and saliva also help to wash bacteria away.

2nd Line of Defence

A range of defence mechanisms operate inside the body to inhibit or destroy pathogens. These responses react to the presence of any pathogen, regardless of which species it is. White blood cells are involved in most of these responses.

3rd Line of Defence

Once the pathogen has been identified by the immune system, a specific response from white blood cells called lymphocytes occurs. These coordinate a range of specific responses to the pathogen.

1. Distinguish between specific and non-specific resistance: _____

_____

_____

_____

_____

2. Describe the functional role of each of the following defence mechanisms:

   (a) Phagocytosis by white blood cells: _____

   _____

   (b) Antimicrobial substances: _____

   _____

   (c) Antibody production: _____

   _____

3. Explain the value of a three tiered system of defence against microbial invasion: _____

_____

Related activities: *The Action of Phagocytes, Inflammation, Fever, The Immune System*   **Web links**: *Immunoanimations*

Periodicals:
Skin, scabs, and scars, Fight for your life!

© Biozone International 2008-2010
Photocopying Prohibited

The Action of Phagocytes

Human cells that ingest microbes and digest them by the process of **phagocytosis** are called **phagocytes**. All are types of white blood cells. During many kinds of infections, especially bacterial infections, the total number of white blood cells increases by two to four times the normal number. The ratio of various white blood cell types changes during the course of an infection.

How a Phagocyte Destroys Microbes

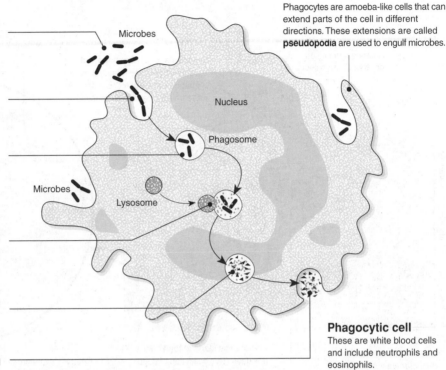

1 Detection

Phagocyte detects microbes by the chemicals they give off (chemotaxis) and sticks the microbes to its surface.

2 Ingestion

The microbe is engulfed by the phagocyte wrapping pseudopodia around it to form a vesicle.

3 Phagosome forms

A phagosome (phagocytic vesicle) is formed, which encloses the microbes in a membrane.

4 Fusion with lysosome

Phagosome fuses with a lysosome (which contains powerful enzymes that can digest the microbe).

5 Digestion

The microbes are broken down by enzymes into their chemical constituents.

6 Discharge

Indigestible material is discharged from the phagocyte cell.

Microbes

Nucleus

Phagosome

Microbes

Lysosome

Phagocytes are amoeba-like cells that can extend parts of the cell in different directions. These extensions are called **pseudopodia** are used to engulf microbes.

Phagocytic cell

These are white blood cells and include neutrophils and eosinophils.

The Interaction of Microbes and Phagocytes

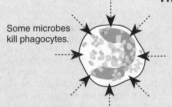

Some microbes kill phagocytes.

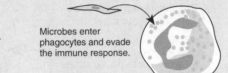

Microbes enter phagocytes and evade the immune response.

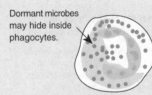

Dormant microbes may hide inside phagocytes.

Some microbes kill phagocytes

Some microbes produce toxins that can actually kill phagocytes, e.g. toxin-producing staphylococci and the dental plaque-forming bacteria *Actinobacillus*.

Microbes evade immune system

Some microbes can evade the immune system by entering phagocytes. The microbes prevent fusion of the lysosome with the phagosome and multiply inside the phagocyte, almost filling it. Examples include *Chlamydia*, *Mycobacterium tuberculosis*, *Shigella*, and malarial parasites.

Dormant microbes hide inside

Some microbes can remain dormant inside the phagocyte for months or years at a time. Examples include the microbes that cause brucellosis and tularemia.

1. Identify the white blood cells capable of phagocytosis: _____

_____

2. Describe how a blood sample from a patient may be used to determine whether they have a microbial infection (without looking for the microbes themselves):

_____

_____

3. Explain how some microbes are able to overcome phagocytic cells and use them to their advantage: _____

_____

_____

Periodicals:
Looking out for danger

Related activities: The Body's Defences, Blood
Web links: Phagocytosis and Bacterial Pathogens

RA 2

Inflammation

Damage to the body's tissues can be caused by physical agents (e.g. sharp objects, heat, radiant energy, or electricity), microbial infection, or chemical agents (e.g. gases, acids and bases). The damage triggers a defensive response called **inflammation**. It is usually characterised by four symptoms: pain, redness, heat and swelling. The inflammatory response is beneficial and has the following functions: (1) to destroy the cause of the infection and remove it and its products from the body; (2) if this fails, to limit the effects on the body by confining the infection to a small area; (3) replacing or repairing tissue damaged by the infection. The process of inflammation can be divided into three distinct stages. These are described below.

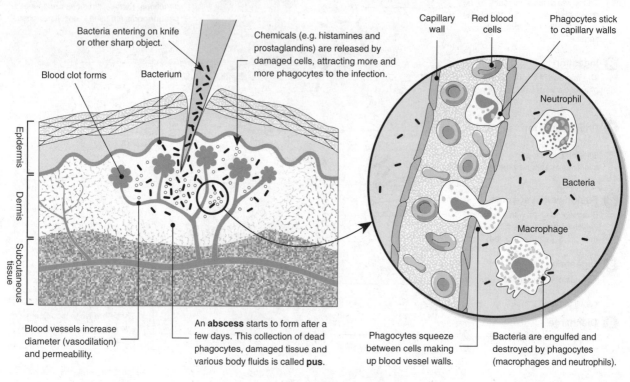

Bacteria entering on knife or other sharp object.

Chemicals (e.g. histamines and prostaglandins) are released by damaged cells, attracting more and more phagocytes to the infection.

Blood clot forms

Bacterium

Epidermis

Dermis

Subcutaneous tissue

Blood vessels increase diameter (vasodilation) and permeability.

An **abscess** starts to form after a few days. This collection of dead phagocytes, damaged tissue and various body fluids is called **pus**.

Capillary wall

Red blood cells

Phagocytes stick to capillary walls

Neutrophil

Bacteria

Macrophage

Phagocytes squeeze between cells making up blood vessel walls.

Bacteria are engulfed and destroyed by phagocytes (macrophages and neutrophils).

Stages in inflammation

Increased diameter and permeability of blood vessels	**Phagocyte migration and phagocytosis**	**Tissue repair**
Blood vessels increase their diameter and permeability in the area of damage. This increases blood flow to the area and allows defensive substances to leak into tissue spaces.	Within one hour of injury, phagocytes appear on the scene. They squeeze between cells of blood vessel walls to reach the damaged area where they destroy invading microbes.	Functioning cells or supporting connective cells create new tissue to replace dead or damaged cells. Some tissue regenerates easily (skin) while others do not at all (cardiac muscle).

1. Outline the three stages of inflammation and identify the beneficial role of each stage:

   (a) _____

   _____

   (b) _____

   _____

   (c) _____

   _____

2. Identify two features of phagocytes important in the response to microbial invasion: _____

   _____

3. State the role of histamines and prostaglandins in inflammation: _____

4. Explain why pus forms at the site of infection: _____

   _____

Fever

Fever describes a condition where the internal body temperature increases to above-normal levels. It arises because of an increase in the body's thermoregulatory set-point so that the previous "normal body temperature" is considered hypothermic. Fever is not a disease, but it is a symptom of infection and, to a point, it is beneficial, because it assists a number of the defence processes. The release of the protein **interleukin-1** helps to reset the thermostat of the body to a higher level, and increases production of **T cells** (lymphocytes). High body temperature also intensifies the effect of **interferon** (an antiviral protein) and may inhibit the growth of some bacteria and viruses. High temperatures also speed up the body's **metabolism**, so promote more rapid tissue repair. Fever also increases heart rate so that white blood cells are delivered to sites of infection more rapidly.

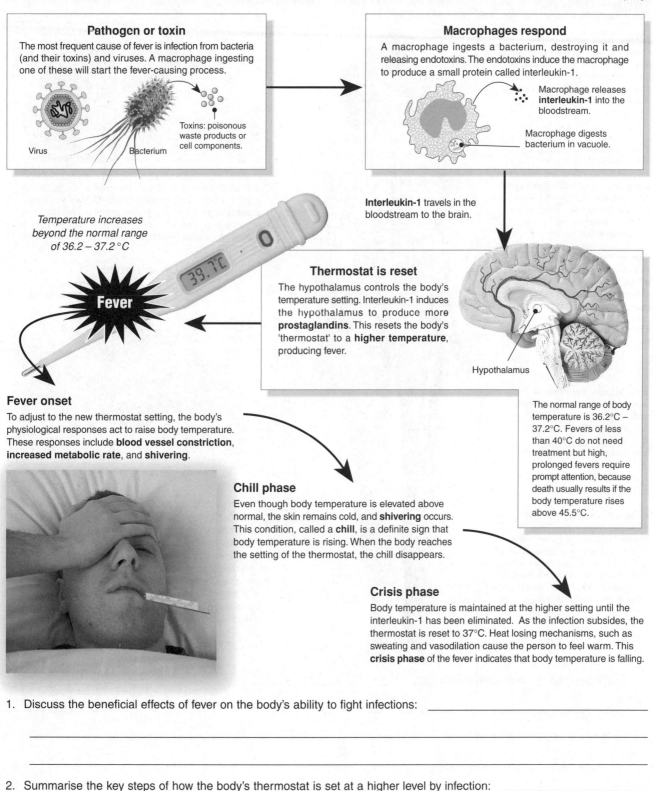

Pathogen or toxin

The most frequent cause of fever is infection from bacteria (and their toxins) and viruses. A macrophage ingesting one of these will start the fever-causing process.

Virus   Bacterium

Toxins: poisonous waste products or cell components.

Macrophages respond

A macrophage ingests a bacterium, destroying it and releasing endotoxins. The endotoxins induce the macrophage to produce a small protein called interleukin-1.

Macrophage releases **interleukin-1** into the bloodstream.

Macrophage digests bacterium in vacuole.

Interleukin-1 travels in the bloodstream to the brain.

Thermostat is reset

The hypothalamus controls the body's temperature setting. Interleukin-1 induces the hypothalamus to produce more **prostaglandins**. This resets the body's 'thermostat' to a **higher temperature**, producing fever.

Hypothalamus

Temperature increases beyond the normal range of 36.2 – 37.2 °C

Fever

Fever onset

To adjust to the new thermostat setting, the body's physiological responses act to raise body temperature. These responses include **blood vessel constriction**, **increased metabolic rate**, and **shivering**.

The normal range of body temperature is 36.2°C – 37.2°C. Fevers of less than 40°C do not need treatment but high, prolonged fevers require prompt attention, because death usually results if the body temperature rises above 45.5°C.

Chill phase

Even though body temperature is elevated above normal, the skin remains cold, and **shivering** occurs. This condition, called a **chill**, is a definite sign that body temperature is rising. When the body reaches the setting of the thermostat, the chill disappears.

Crisis phase

Body temperature is maintained at the higher setting until the interleukin-1 has been eliminated. As the infection subsides, the thermostat is reset to 37°C. Heat losing mechanisms, such as sweating and vasodilation cause the person to feel warm. This **crisis phase** of the fever indicates that body temperature is falling.

1. Discuss the beneficial effects of fever on the body's ability to fight infections: _____

_____

_____

2. Summarise the key steps of how the body's thermostat is set at a higher level by infection: _____

_____

_____

_____

Related activities: The Body's Defences

A 2

Defence and the Immune System

The Lymphatic System

Fluid leaks out from capillaries and forms the tissue fluid, which is similar in composition to plasma but lacks large proteins. This fluid bathes the tissues, supplying them with nutrients and oxygen, and removing wastes. Some of the tissue fluid returns directly into the capillaries, but some drains back into the blood circulation through a network of lymph vessels. This fluid, called **lymph**, is similar to tissue fluid, but contains more leucocytes. Apart from its circulatory role, the lymphatic system also has an important function in the immune response. Lymph nodes are the primary sites where the destruction of pathogens and other foreign substances occurs. A lymph node that is fighting an infection becomes swollen and hard as the lymph cells reproduce rapidly to increase their numbers. The thymus, spleen, and bone marrow also contribute leucocytes to the lymphatic and circulatory systems.

Tonsils: Tonsils (and adenoids) comprise a collection of large lymphatic nodules at the back of the throat. They produce lymphocytes and antibodies and are well-placed to protect against invasion of pathogens.

Thymus gland: The thymus is a two-lobed organ located close to the heart. It is prominent in infants and diminishes after puberty to a fraction of its original size. Its role in immunity is to help produce **T cells** that destroy invading microbes directly or indirectly by producing various substances.

Spleen: The oval spleen is the largest mass of lymphatic tissue in the body, measuring about 12 cm in length. It stores and releases blood in case of demand (e.g. in cases of bleeding), produces mature **B cells**, and destroys bacteria by phagocytosis.

Bone marrow: Bone marrow produces red blood cells and many kinds of leucocytes: monocytes (and macrophages), neutrophils, eosinophils, basophils, and lymphocytes (B cells and T cells).

Lymphatic vessels: When tissue fluid is picked up by lymph capillaries, it is called **lymph**. The lymph is passed along lymphatic vessels to a series of lymph nodes. These vessels contain one-way valves that move the lymph in the direction of the heart until it is reintroduced to the blood at the subclavian veins.

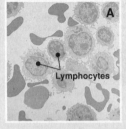
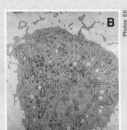

Lymphocytes

Many types of leucocytes are involved in internal defence. The photos above illustrate examples of leucocytes. **A** shows a cluster of **lymphocytes**. **B** shows a single **macrophage**: large, phagocytic cells that develop from monocytes and move from the blood to reside in many organs and tissues, including the spleen and lymph nodes.

Lymph node

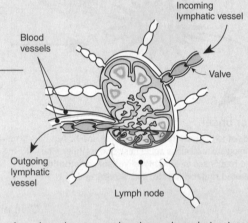

Blood vessels

Incoming lymphatic vessel

Valve

Outgoing lymphatic vessel

Lymph node

Lymph nodes are oval or bean-shaped structures, scattered throughout the body, usually in groups, along the length of lymphatic vessels. As lymph passes through the nodes, it filters foreign particles (including pathogens) by trapping them in fibres. Lymph nodes are also a "store" of **lymphocytes**, which may circulate to other parts of the body. Once trapped, macrophages destroy the foreign substances by phagocytosis. T cells may destroy them by releasing various products, and/or B cells may release antibodies that destroy them.

1. Briefly describe the composition of lymph: _____

_____

2. Discuss the various roles of lymph: _____

_____

_____

3. Describe one role of each of the following in the lymphatic system:

   (a) Lymph nodes: _____

   (b) Bone marrow: _____

Related activities: Formation of Tissue Fluid, The Immune System

The Immune System

The efficient internal defence provided by the immune system is based on its ability to respond specifically against a foreign substance and its ability to hold a memory of this response. There are two main components of the immune system: the humoral and the cell-mediated responses. They work separately and together to protect us from disease. The **humoral immune response** is associated with the serum (non-cellular part of the blood) and involves the action of **antibodies** secreted by B cell lymphocytes. Antibodies are found in extracellular fluids including lymph, plasma, and mucus secretions. The humoral response protects the body against circulating viruses, and bacteria and their toxins. The **cell-mediated immune response** is associated with the production of specialised lymphocytes called **T cells**. It is most effective against bacteria and viruses located within host cells, as well as against parasitic protozoa, fungi, and worms. This system is also an important defence against cancer, and is responsible for the rejection of transplanted tissue. Both B and T cells develop from stem cells located in the liver of foetuses and the bone marrow of adults. T cells complete their development in the thymus, whilst the B cells mature in the bone marrow.

Lymphocytes and their Functions

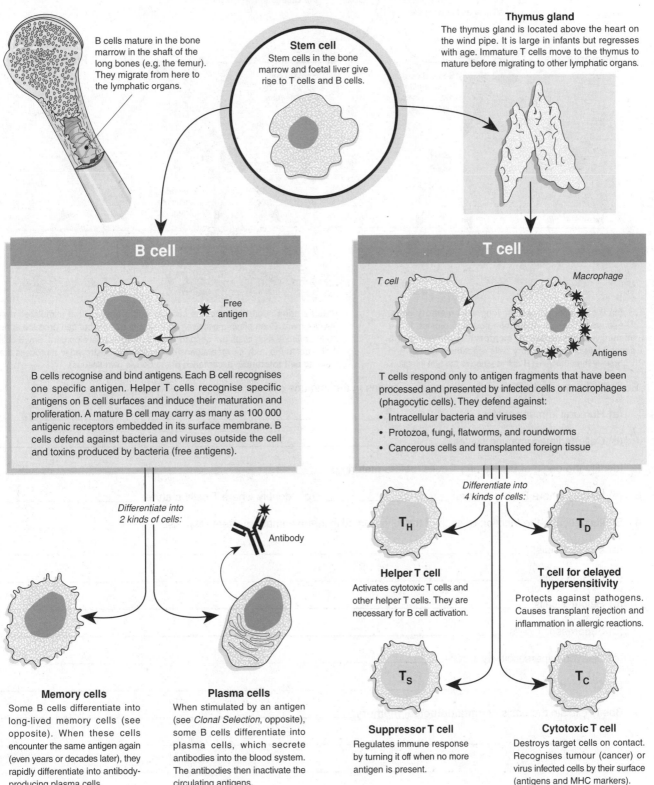

Stem cell
Stem cells in the bone marrow and foetal liver give rise to T cells and B cells.

B cells mature in the bone marrow in the shaft of the long bones (e.g. the femur). They migrate from here to the lymphatic organs.

Thymus gland
The thymus gland is located above the heart on the wind pipe. It is large in infants but regresses with age. Immature T cells move to the thymus to mature before migrating to other lymphatic organs.

B cell

Free antigen

B cells recognise and bind antigens. Each B cell recognises one specific antigen. Helper T cells recognise specific antigens on B cell surfaces and induce their maturation and proliferation. A mature B cell may carry as many as 100 000 antigenic receptors embedded in its surface membrane. B cells defend against bacteria and viruses outside the cell and toxins produced by bacteria (free antigens).

T cell

T cell   Macrophage

Antigens

T cells respond only to antigen fragments that have been processed and presented by infected cells or macrophages (phagocytic cells). They defend against:
• Intracellular bacteria and viruses
• Protozoa, fungi, flatworms, and roundworms
• Cancerous cells and transplanted foreign tissue

Differentiate into 2 kinds of cells:

Antibody

Differentiate into 4 kinds of cells:

T_H

T_D

Helper T cell
Activates cytotoxic T cells and other helper T cells. They are necessary for B cell activation.

T cell for delayed hypersensitivity
Protects against pathogens. Causes transplant rejection and inflammation in allergic reactions.

T_S

T_C

Memory cells
Some B cells differentiate into long-lived memory cells (see opposite). When these cells encounter the same antigen again (even years or decades later), they rapidly differentiate into antibody-producing plasma cells.

Plasma cells
When stimulated by an antigen (see *Clonal Selection*, opposite), some B cells differentiate into plasma cells, which secrete antibodies into the blood system. The antibodies then inactivate the circulating antigens.

Suppressor T cell
Regulates immune response by turning it off when no more antigen is present.

Cytotoxic T cell
Destroys target cells on contact. Recognises tumour (cancer) or virus infected cells by their surface (antigens and MHC markers).

Periodicals: Lymphocytes - the heart of the immune system

Related activities: The Lymphatic System, **Web links**: The Immune System Overview, The Humoral Response, Introducing Specific Immunity

A 2

Defence and the Immune System

The immune system has the ability to respond to the large and unpredictable range of potential antigens encountered in the environment. The diagram below explains how this ability is based on **clonal selection** after antigen exposure. The example illustrated is for B cell lymphocytes. In the same way, a T cell stimulated by a specific antigen will multiply and develop into different types of T cells. Clonal selection and differentiation of lymphocytes provide the basis for **immunological memory**.

Five (a-e) of the many, randomly generated B cells. Each one can recognise only one specific antigen.

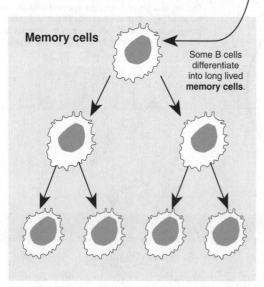

This B cell encounters and binds an antigen. It is then stimulated to proliferate.

Clonal Selection Theory

Millions of randomly generated B cells form during development. Collectively, they can recognise many antigens, including those that have never been encountered. Each B cell makes an antibody specific to the type of antigenic receptor on its surface. The receptor reacts only to that specific antigen. When a B cell encounters its antigen, it responds by proliferating and producing many clones all with the same kind of antibody. This is called **clonal selection** because the antigen selects the B cells that will proliferate.

Memory cells

Some B cells differentiate into long lived **memory cells**.

Plasma cells

Some B cells differentiate into **plasma cells**.

Antibodies inactivate antigens

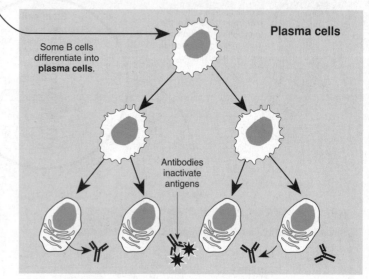

Some B cells differentiate into long lived **memory cells**. These are retained in the lymph nodes to provide future immunity (**immunological memory**). In the event of a second infection, B-memory cells react more quickly and vigorously than the initial B-cell reaction to the first infection.

Plasma cells secrete antibodies specific to the antigen that stimulated their development. Each plasma cell lives for only a few days, but can produce about 2000 antibody molecules per second. Note that during development, any B cells that react to the body's own antigens are selectively destroyed in a process that leads to **self tolerance** (acceptance of the body's own tissues).

1. State the general action of the two major divisions in the immune system:

   (a) Humoral immune system: _____

   (b) Cell-mediated immune system: _____

2. Identify the origin of B cells and T cells (before maturing): _____

3. (a) Identify where B cells mature: _____ (b) Identify where T cells mature: _____

4. Briefly describe the function of each of the following cells in the immune system response:

   (a) Memory cells: _____

   (b) Plasma cells: _____

   (c) Helper T cells: _____

   (d) Suppressor T cells: _____

   (e) Delayed hypersensitivity T cells: _____

   (f) Cytotoxic T cells: _____

5. Briefly explain the basis of **immunological memory**: _____

   _____

   _____

   _____

Antibodies

Antibodies and antigens play key roles in the response of the immune system. Antigens are foreign molecules that are able to bind to antibodies (or T cell receptors) and provoke a specific immune response. Antigens include potentially damaging microbes and their toxins (see below) as well as substances such as pollen grains, blood cell surface molecules, and the surface proteins on transplanted tissues. **Antibodies** (also called immunoglobulins) are proteins that are made in response to antigens. They are secreted into the plasma where they circulate and can recognise, bind to, and help to destroy antigens. There are five classes of **immunoglobulins**. Each plays a different

role in the immune response (including destroying protozoan parasites, enhancing phagocytosis, protecting mucous surfaces, and neutralising toxins and viruses). The human body can produce an estimated 100 million antibodies, recognising many different antigens, including those it has never encountered. Each type of antibody is highly specific to only one particular antigen. The ability of the immune system to recognise and ignore the antigenic properties of its own tissues occurs early in development and is called **self-tolerance**. Exceptions occur when the immune system malfunctions and the body attacks its own tissues, causing an **autoimmune disorder**.

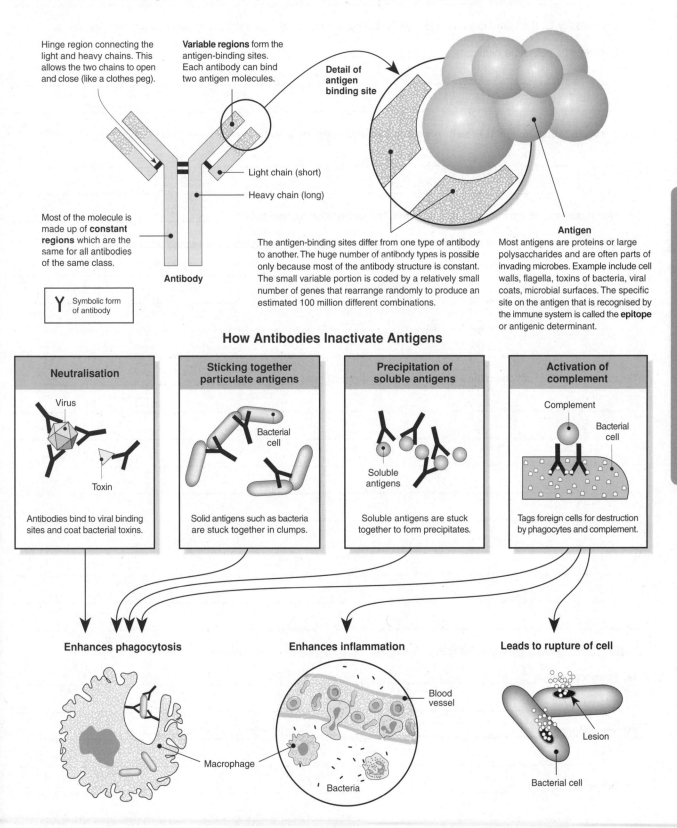

How Antibodies Inactivate Antigens

Periodicals: Antibodies

Related activities: The Immune System
Web links: How Lymphocytes Produce Antibodies

RA 2

Defence and the Immune System

238

1. Distinguish between an antibody and an antigen: _____

_____

_____

_____

_____

_____

2. It is necessary for the immune system to clearly distinguish the body's own cells and proteins from foreign ones.

 (a) Explain why this is the case: _____

_____

_____

 (b) In simple terms, explain how **self tolerance** develops (see the activity *The Immune System* if you need help):

_____

_____

 (c) Name the type of disorder that results when this recognition system fails: _____

 (d) Describe two examples of disorders that are caused in this way, identifying what happens in each case:

_____

_____

_____

_____

3. Discuss the ways in which antibodies work to inactivate antigens: _____

_____

_____

_____

_____

_____

_____

_____

_____

_____

4. Explain how antibody activity enhances or leads to:

 (a) Phagocytosis: _____

_____

 (b) Inflammation: _____

_____

 (c) Bacterial cell lysis: _____

_____

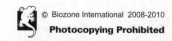

Acquired Immunity

We have natural or **innate resistance** to certain illnesses; examples include most diseases of other animal species. **Acquired immunity** refers to the protection an animal develops against certain types of microbes or foreign substances. Immunity can be acquired either passively or actively and is developed during an individual's lifetime. **Active immunity** develops when a person is exposed to microorganisms or foreign substances and the immune system responds. **Passive immunity** is acquired when antibodies are transferred from one person to another. Recipients do not make the antibodies themselves and the effect lasts only as long as the antibodies are present, usually several weeks or months. Immunity may also be **naturally acquired**, through natural exposure to microbes, or **artificially acquired** as a result of medical treatment.

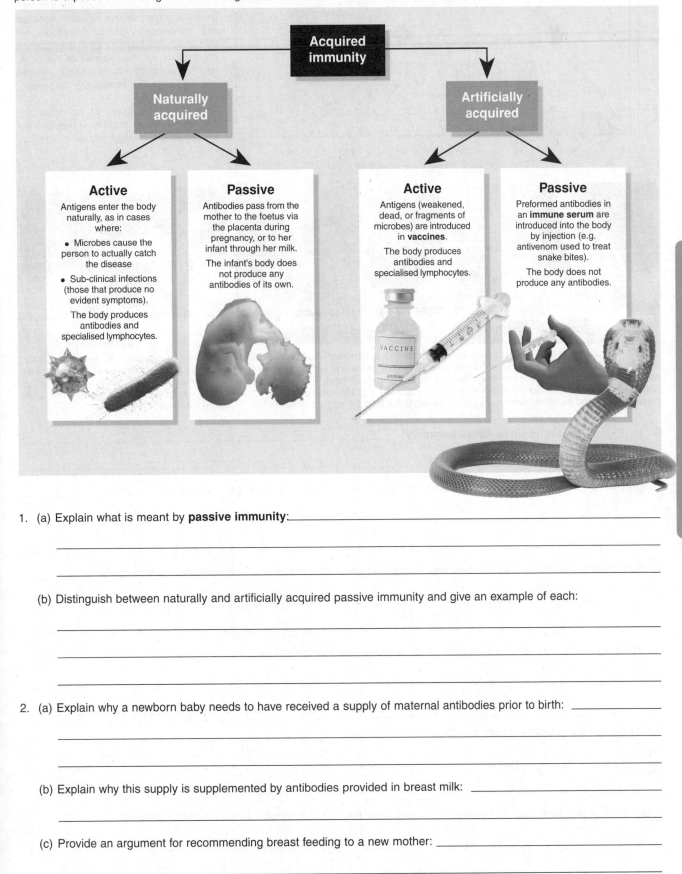

Acquired immunity

Naturally acquired

Active

Antigens enter the body naturally, as in cases where:

- Microbes cause the person to actually catch the disease
- Sub-clinical infections (those that produce no evident symptoms).

The body produces antibodies and specialised lymphocytes.

Passive

Antibodies pass from the mother to the foetus via the placenta during pregnancy, or to her infant through her milk.

The infant's body does not produce any antibodies of its own.

Artificially acquired

Active

Antigens (weakened, dead, or fragments of microbes) are introduced in **vaccines**.

The body produces antibodies and specialised lymphocytes.

Passive

Preformed antibodies in an **immune serum** are introduced into the body by injection (e.g. antivenom used to treat snake bites).

The body does not produce any antibodies.

Defence and the Immune System

1. (a) Explain what is meant by **passive immunity**: _____

_____

_____

(b) Distinguish between naturally and artificially acquired passive immunity and give an example of each:

_____

_____

_____

2. (a) Explain why a newborn baby needs to have received a supply of maternal antibodies prior to birth: _____

_____

_____

(b) Explain why this supply is supplemented by antibodies provided in breast milk: _____

_____

(c) Provide an argument for recommending breast feeding to a new mother: _____

_____

Periodicals:
Hard to swallow

Related activities: Antibodies, Vaccines and Vaccination

A 2

Primary and Secondary Responses to Antigens

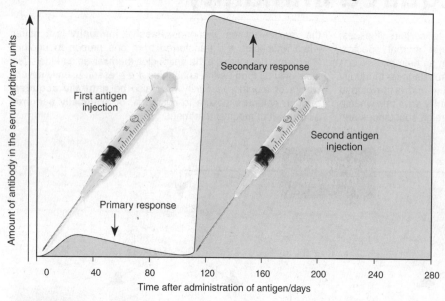

When the B cells encounter antigens and produce antibodies, the body develops **active immunity** against that antigen.

The initial response to antigenic stimulation, caused by the sudden increase in B cell clones, is called the **primary response**. Antibody levels as a result of the primary response peak a few weeks after the response begins and then decline. However, because the immune system develops an immunological memory of that antigen, it responds much more quickly and strongly when presented with the same antigen subsequently (the **secondary response**).

This forms the basis of immunisation programmes where one or more booster shots are provided following the inital vaccination.

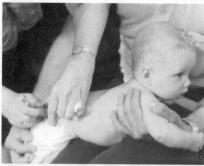

Vaccines to protect against common diseases are administered at various stages during childhood according to an immunisation schedule. Vaccination has been behind the decline of some once-common childhood diseases, such as mumps.

Many childhood diseases for which vaccination programmes exist are kept at a low level because of **herd immunity**. If most of the population is immune, those that are not immunised may be protected because the disease is uncommon.

While most vaccinations are given in childhood, adults may be vaccinated against specific diseases (e.g. tuberculosis) if they are in a high risk group or if they are travelling to a region in the world where a disease is prevalent.

3. (a) Explain what is meant by **active immunity**: _____

_____

   (b) Distinguish between naturally and artificially acquired active immunity and give an example of each: _____

_____

_____

4. (a) Describe two differences between the primary and secondary responses to presentation of an antigen: _____

_____

_____

   (b) Explain why the secondary response is so different from the primary response: _____

_____

_____

5. (a) Explain the principle of herd immunity: _____

_____

_____

   (b) Explain why health authorities are concerned when the vaccination rates for an infectious disease fall:

_____

_____

Vaccines and Vaccination

Vaccines operate on the principle that they alert the immune system to the presence of a pathogen by introducing harmless but recognisably foreign antigens against which the body can form antibodies. Thereare two basic types of vaccine: subunit vaccines and whole-agent vaccines. **Whole-agent vaccines** contain complete nonvirulent microbes, either **inactivated** (killed), or alive but **attenuated** (weakened). Attenuated viruses make very effective vaccines and often provide life-long immunity without the need for booster immunisations. Killed viruses are less effective and many vaccines of this sort have now been replaced by newer subunit vaccines. **Subunit vaccines** contain

only the parts of the pathogen that induce the immune response. They are safer than attenuated vaccines because they cannot reproduce in the recipient, and they produce fewer adverse effects because they contain little or no extra material. There are several ways to make subunit vaccines but, in all cases, the subunit vaccine loses its ability to cause disease while retaining its antigenic properties. Some of the most promising vaccines under development consist of naked DNA which is injected into the body and produces an antigenic protein. The safety of DNA vaccines is uncertain but they show promise against rapidly mutating viruses such as influenza and HIV.

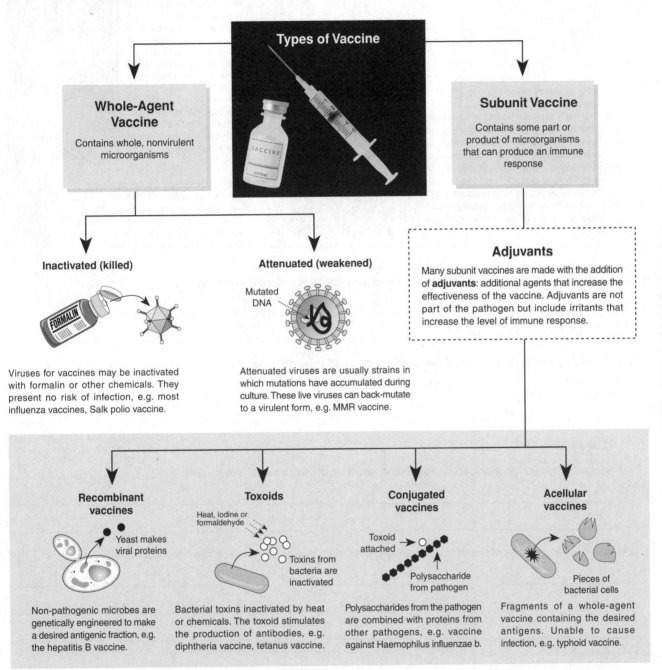

Defence and the Immune System

1. **Attenuated viruses** provide long term immunity to their recipients and generally do not require booster shots. Suggest a possible reason why attenuated viruses provide such effective long-term immunity when inactivated viruses do not:

_____

_____

_____

_____

Periodicals:
Will there ever be a
malaria vaccine?

Related activities: Acquired Immunity
Web links: Steps in Vaccine Development

RA 3

Whooping cough notifications and vaccine coverage (England and Wales) 1940-2008

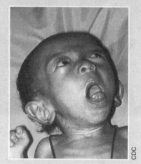

Adapted from CDSC

Notifications
Vaccination rate

Vaccination began

Whooping cough is caused by the bacterium *Bordetella pertussis*, and may last for two to three months. It is characterised by a whooping cough and painful coughing spasms, which may be followed by periods of vomiting. Infants under six months of age are most at risk of developing complications or dying because they are too young to be fully protected by the vaccine. Inclusion of the whooping cough vaccine into the UK immunisation schedule has greatly reduced the incidence rates of the disease (above).

Smallpox Eradication

Smallpox is a highly contagious disease caused by the *Variola* virus. It has two forms, the more severe of which kills about 30% of the people infected. Over several centuries, many smallpox epidemics swept across the globe, killing millions and maiming millions more. Survivors were often left blind and severeely scarred. In the late 1700s, Edward Jenner discovered that inoculation with the cowpox virus could protect humans from smallpox.

Several countries began vaccination programmes, and during the 1960s the World Health Organisation led a global programme that resulted in the complete eradication of smallpox. The key to the strategy was **surveillance** of outbreaks and then **containment**. Eradication was helped by the fact that humans were the only reservoir for infection and there were no carriers. The last natural outbreak was recorded in Somalia in 1977.

The Bangladeshi girl (above) shows the typical smallpox lesion pattern on her face and torso. The smallpox vaccine (left) contains a live vaccinia virus. Vaccina virus is similar to the virus that causes smallpox. Stocks are kept in the Netherlands.

2. (a) Explain how vaccination can help lead to the eradication of an infectious disease:

_____

_____

(b) Discuss factors in the success of the smallpox eradication programme: _____

_____

_____

_____

(c) Suggest why such an eradication programme would be difficult against tuberculosis: _____

_____

_____

3. Discuss why stocks of smallpox vaccine are maintained even though the disease has been eradicated:

_____

_____

_____

_____

4. In 1975, the UK vaccination rate for whopping cough decreased to 30% because of concerns about the vaccine's safety. Use the data in the graph above to describe what effect this had on rates of whooping cough reported:

_____

_____

_____

_____

New Medicines

One of the concerning issues in modern medicine is the need to develop new treatments for disease. Increasingly, researchers are looking to our natural biological resources for new medicines. Plants have been used for their medicinal properties for many centuries. Around half of the pharmaceuticals in use today are of plant origin. Although many plants have antimicrobial properties, the pharmaceuticals of plant origin have been obtained from relatively few species. This is generally because antibiotic production and extraction from fungi and bacteria is more straightforward. However, antimicrobial compounds from plants are being increasingly investigated as resistance to more conventional antibiotics becomes more widespread.. The advantages of using plants as starting points for drug development are compelling. There are many species whose properties are yet unknown and many could be useful medicinally. Much the same scenario applies to microorganisms, whose biodiversity is only just beginning to be discovered. Future medical treatments may depend on our conservation of this biodiversity.

Medicines from Plants

Opium poppy codeine, morphine

Aspirin
acetylsalicylic acid
200 Tablets
100 mg.

*Periwinkle plant-
anti-cancer*

Approximately 120 pure chemical substances extracted from higher plants are used in medicine throughout the world. Some, including **aspirin** (salicylic acid from willow bark) and **digitalin** (from foxglove) have been in medical use since antiquity. Others, including plant alkaloids such as **taxol** (an anticancer drug), are more recent discoveries. Most of the plant-derived medicines are now synthesised in the laboratory, but only about six are produced entirely by synthetic procedures. The rest are still extracted commercially from plants.

*The bark and needles of the Pacific yew (left) provide the anti-cancer drug **taxol**.*

White willow bark yields the active ingredient of aspirin, used to treat pain fever, and inflammation.

In recent decades, the search for new medicines has focussed on tropical plants (e.g. periwinkle and opium). However, the increased sensitivity of the new chemical screening technologies has revealed potential new drugs from plants that were not detected by previous methods. Moreover, plants can be engineered as biofactories to manufacture medicines such as vaccines and antibodies.

Digitalin, derived from foxglove, is used to treat congestive heart disease

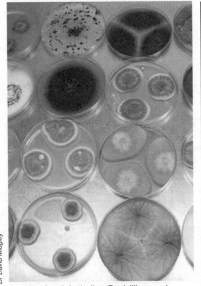

Dr David Midgley

Various fungi, including Penicillium and Aspergillus, grown in sterile (axenic) culture

The use of drugs isolated from microorganisms is a relatively recent phenomenon, which started with the discovery of the antibiotic penicillin in 1928. Now the use of microbes to produce antimicrobial drugs is a huge industry.

Some medicines come from unlikely sources. The drug **botox**, derived from the toxin of *Clostridium botulinum*, is used to treat facial neuralgia as well as for cosmetic purposes.

New approaches to microbial medicines include using **bacteriophages** (viruses that infect bacteria) to control bacterial infections in different tissues. A better understanding of microbial diversity may also provide the means to produce more effective drugs against microbial pathogens.

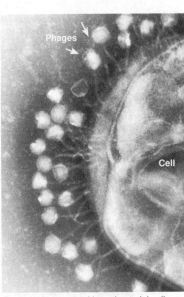

Phages

Cell

Bacteriophages attacking a bacterial cell.

Defence and the Immune System

Isolating and Testing Antimicrobial Compounds from Plants

1

A plant is chosen for testing. Plants produce a wide range of compounds, but those with aromatic rings often have medicinal properties.

2

Pulping

Solvent extraction

Extraction: Plant material is pulped, then soaked in a solvent (water, alcohol), or boiled to release and crudely fractionate the compounds.

3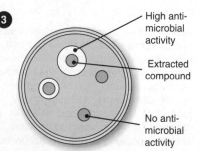

High anti-microbial activity

Extracted compound

No anti-microbial activity

Screening: Microbes are plated onto agar. Paper discs soaked with plant extract are placed on agar, or the plant extract is pipetted into wells cut into the agar. The plates are then incubated. The effectiveness of the extract can be determined by measuring the clear zone around the plant extract after incubation.

4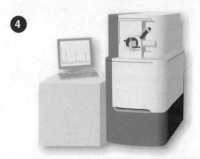

Separation and analysis: Mass spectroscopy is used to further separate and isolate compounds from the extracts that show antimicrobial activity.

5

Concentration: The compounds may be further separated, (e.g. by chromatography), before they are concentrated using freeze drying or rotary evaporation (above).

6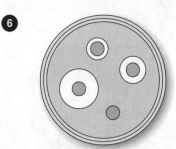

Testing: The isolated and concentrated compounds are tested for antimicrobial activity. This identifies which component of the initial extract is the antimicrobial.

The antimicrobial compound in chili peppers, is capsaicin; an antibacterial agent of relatively high potency.

The essential oil produced from the herb rosemary (above), acts as an effective general antimicrobial agent.

Phloretin is a polyphenol extracted from the leaves of apple trees. It acts as a highly potent antimicrobial.

Saponins (compounds that foam in water) extracted from ginseng are effective against a range of bacteria.

1. Describe the present and potential value of plants to modern medicine: _____

_____

_____

_____

2. Explain why a loss of tropical biodiversity could reduce the potential options for new drugs discoveries:

_____

_____

3. Explain how the effectiveness of a new antimicrobial compound can be tested: _____

_____

_____

4. Explain why research into antimicrobial plant extracts is increasing: _____

_____

_____

KEY TERMS Memory Card Game

The cards below have a keyword or term printed on one side and its definition printed on the opposite side. The aim is to win as many cards as possible from the table. To play the game.....

1) Cut out the cards and lay them definition side down on the desk. You will need one set of cards between two students.

2) Taking turns, choose a card and, BEFORE you pick it up, state your own best definition of the keyword to your opponent.

3) Check the definition on the opposite side of the card. If both you and your opponent agree that your stated definition matches, then keep the card. If your definition does not match then return the card to the desk.

4) Once your turn is over, your opponent may choose a card.

Active immunity	Antigens	Antibodies
B lymphocytes	Humoral immunity	Leucocytes
Infection	T lymphocytes	Passive immunity
Cell-mediated immunity	Clonal selection	Plasma cells
Secondary response	Fever	Vaccination
Phagocytes	Inflammation	Primary response

Defence and the Immune System

R 2

When you've finished the game keep these cutouts and use them as flash cards!

Gamma globulin proteins in the blood or other bodily fluids, which identify and neutralise foreign material, such as bacteria and viruses.

Molecules that are recognised by the immune system as foreign.

Immunity that is induced in the host itself by the antigen, and is long-lasting.

White blood cells, including lymphocytes, and macrophages and other phagocytic cells.

Immune response that is mediated by secreted antibodies.

Lymphocytes that make antibodies against specific antigens.

Immunity gained by the receipt of ready-made antibodies.

Lymphocytes responsible for cell mediated immunity.

The detrimental colonisation of a host organism by a foreign species.

B lymphocytes that have been exposed to antigen and produce and secrete large numbers of antibodies.

A model for how B and T cells are selected to target specific antigens invading the body.

Immune response involving the activation of macrophages, specific T cells, and cytokines against antigens.

The delivery by of antigenic material (the vaccine) to produce immunity to a disease.

A rise in body temperature above the normal range of as a result of an increase in the body temperature regulatory set-point

The more rapid and stronger response of the immune system to an antigen that it has encountered before.

The initial response of the immune system to exposure to an antigen.

The protective response of vascular tissues to harmful stimuli, such as irritants, pathogens, or damaged cells.

White blood cells that destroy foreign material, e.g. bacteria, by ingesting them.

Smoking: A Choice Against Health

KEY CONCEPTS

▶ Diseases of the gas exchange system can prevent air getting to the gas exchange surface.

▶ Tobacco smoke has debilitating effects on the gas exchange and cardiovascular systems that lead to chronic disease and premature death.

▶ The components of tobacco smoke cause hypertension and reduce oxygen supply to tissues.

▶ Tobacco smoking acts synergistically with other risk factors to increase the incidence of disease.

KEY TERMS

atherosclerosis

carbon monoxide

carcinogen

cardiovascular disease

chronic bronchitis

chronic obstructive
    pulmonary disease

coronary heart disease

emphysema

lung cancer

myocardial infarction

nicotine

obstructive lung disease

restrictive diseases

restrictive lung disease

risk factor

stroke

tar

tobacco

OBJECTIVES

☐ 1. Use the **KEY TERMS** to help you understand and complete these objectives.

The Nature of Respiratory Disease
pages 249-250

☐ 2. Describe the nature of non-infectious diseases affecting the human gas exchange system. Distinguish between **obstructive diseases**, such as **emphysema** and **chronic bronchitis**, and **restrictive diseases**, which occur as a result of scarring of the gas exchange surface.

☐ 3. Relate the incidence of obstructive lung diseases to the occurrence of certain behaviours, namely **tobacco smoking**. Explain how smoking can exacerbate the symptoms of diseases such as asthma, which are not related to smoking.

☐ 4. Describe the components of tobacco smoke, and their immediate and cumulative effects on the body. Include reference to **tar, nicotine, carbon monoxide**, and **carcinogens**.

Smoking and the Lungs
pages 248-252

☐ 4. Describe the effects of tobacco smoking on the human gas exchange system, with particular reference to the symptoms of:

   (a) **Chronic obstructive pulmonary disease** (COPD), including **chronic bronchitis** and **emphysema**.

   (b) **Lung cancer**

☐ 5. Evaluate the epidemiological and experimental evidence linking tobacco smoking to the incidence of lung diseases and premature death.

Smoking and the Cardiovascular System
pages 189-190, 253-254

☐ 6. Describe the physiological effects of **nicotine** and **carbon monoxide** on the cardiovascular system, including reference to the events leading to **atherosclerosis, coronary heart disease**, and **stroke**. Explain how carbon monoxide reduces the oxygen carrying capacity of the blood and how nicotine increases blood pressure.

☐ 7. Describe controllable **risk factors** for cardiovascular disease and explain ow tobacco smoke acts synergistically with other risk factors to increase the risk of a cardiovascular event (such as a **myocardial infarction**).

☐ 8. Evaluate the epidemiological and experimental evidence linking tobacco smoking to the incidence of cardiovascular diseases and premature death.

Periodicals:

Listings for this chapter are on page 340

Weblinks:

www.biozone.co.uk/
weblink/OCR-AS-2641.html

Living With Chronic Lung Disease

Activity limitation in people with and without COPD

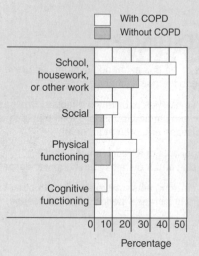

- □ With COPD
- ■ Without COPD

School, housework, or other work

Social

Physical functioning

Cognitive functioning

0 | 10 | 20 | 30 | 40 | 50

Percentage

Cognitive, physical, social and activity-related limitations are more common among people with chronic obstructive pulmonary disease.

The Impact of COPD in the UK

Chronic obstructive pulmonary disease (or COPD) includes **chronic bronchitis** and **emphysema**, which often occur together. COPD affects about one million people in the UK. Most of those affected are over the age 40 and smoking is the cause in the vast majority of cases. This relationship is clear; people who have never smoked rarely develop COPD. The symptoms of COPD and asthma are similar, but COPD causes permanent damage to the airways, and so symptoms are chronic (persistent) and treatment is limited.

COPD severely limits the capacity of sufferers to carry out even a normal daily level of activity. A survey by the American Lung Association of hundreds of people living with COPD found that nearly half became short of breath while washing, dressing, or doing light housework (left). Over 25% reported difficulty in breathing while sitting or lying still. Lack of oxygen also places those with COPD at high risk of heart failure. As the disease becomes more severe, sufferers usually require long-term oxygen therapy, in which they are more or less permanently attached to an oxygen supply.

In the UK, COPD accounts for more time off work than any other illness. The indirect costs of this chronic condition have been estimated at 24 million lost working days per annum. A 'flare-up' of COPD, during which the symptoms worsen, is one of the commonest reasons for admission to hospital and the disease places a substantial burden on health services. The number of primary care consultations for COPD is four times higher than for angina, and 30% of those admitted to hospital with COPD for the first time will be readmitted within 3 months (*NHS-UK*). At least 25,000 people die each year in the UK from the end stages of COPD, but the actual number may be higher as COPD is often present in patients who die from heart failure and stroke. Many of these people have several years of ill health before they die. Being able to breathe is something we don't often think about. What must it be like to struggle for each breath, every minute of every day for years?

A Personal Story

Deborah Ripley's message from her mother Jenny (used with permission)

"Fear, anxiety, depression, and carbon monoxide are ruining whatever life my mother has left. I posted this portrait of my Mum on the photo website Flickr because she wants to send a warning to anyone who's still smoking. I've just returned from visiting her in a nursing home where she's virtually shackled to the bed. Getting up to go to the bathroom practically kills her. She was admitted to hospital after a bout of pneumonia, which required intensive antibiotic therapy and left her hardly able to breathe. She has moderate dementia caused by a series of mini-strokes, which is aggravated by the pneumonia. She has no recollection of who has visited her or when, so consequently thinks she's alone most of the time, which is upsetting and disturbing for her.

This is all caused by damage to her brain and lungs as a result of 65 years of smoking. In those moments when she is lucid, she asks me who she can warn that this could happen to them. She said 'if people could see me lying here like this it would put them off...' None of her other known blood relatives suffered this sort of decline in their old age and, as far as I know, none of them smoked".

Thankfully, Jenny's pneumonia has since subsided and her COPD is being well managed. However, constant vigilance is important because flare-ups are common with COPD and recovery from lung infections is difficult when breathing is already compromised.

Used with permission ©deborahripley.com

1. Describe the economic impact of smoking-related diseases, such as emphysema and chronic bronchitis:

_____

_____

2. Discuss the personal costs of a smoking-related disease and comment on the value of personal testimonials such as those from Deborah's mother:

_____

_____

_____

_____

_____

Respiratory Diseases

Respiratory diseases are diseases of the gas exchange system, including diseases of the lung, bronchial tubes, trachea, and upper respiratory tract. Respiratory diseases include mild and self-limiting diseases such as the common cold, to life-threatening infections such as tuberculosis. One in seven people in the UK is affected by some form of chronic lung disease, the most common being asthma and **chronic obstructive pulmonary disease** (including emphysema and chronic bronchitis). Non-infectious respiratory diseases are categorised according to whether they prevent air reaching the alveoli (**obstructive**) or whether they affect the gas exchange tissue itself (**restrictive**). Such diseases have different causes and different symptoms (below) but all are characterised by difficulty in breathing and the end result is similar in that gas exchange rates are too low to meet metabolic requirements. Non-infectious respiratory diseases are strongly correlated with certain behaviours and are made worse by exposure to air pollutants. Obstructive diseases, such as emphysema, are associated with an inflammatory response of the lung to noxious particles or gases, most commonly **tobacco smoke**. In contrast, scarring (**fibrosis**) of the lung tissue underlies restrictive lung diseases such as **asbestosis** and **silicosis**. Such diseases are often called occupational lung diseases.

Chronic bronchitis
Excess mucus blocks airway, leading to inflammation and infection

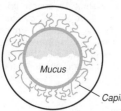

Mucus

Capillary

Asthma
Thickening of bronchiole wall and muscle hypertrophy. Bronchioles narrow.

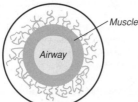

Muscle

Airway

Emphysema
Destruction of capillaries and structures supporting the small airways and lung tissue

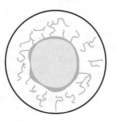

A peak flow meter is a small, hand-held device used to monitor a person's ability to breathe out air. It measures the airflow through the bronchi and thus the degree of obstruction in the airways.

Cross sections through a bronchiole with various types of obstructive lung disease

Obstructive lung disease
– passage blockage –

In obstructive lung diseases, a blockage prevents the air getting to the gas exchange surface.

The flow of air may be obstructed because of constriction of the airways (as in **asthma**), excess mucus secretion (as in **chronic bronchitis**), or because of reduced lung elasticity, which causes alveoli and small airways to collapse (as in **emphysema**). Shortness of breath is a symptom in all cases and chronic bronchitis is also associated with a persistent cough.

Chronic bronchitis and emphysema often occur together and are commonly associated with **cigarette smoking**, but can also occur with chronic exposure to air pollution.

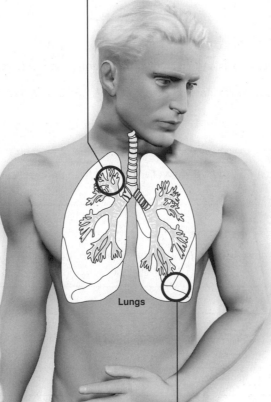

Lungs

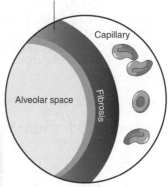

Scarring (fibrosis) makes the lung tissue stiffer and prevents adequate gas exchange

Capillary

Alveolar space

Fibrosis

Restrictive lung disease
– scarring –

Restrictive lung diseases are characterised by scarring or **fibrosis** within the gas exchange tissue of the lung (above). As a result of the scarring, the lung tissue becomes stiffer and more difficult to expand, leading to shortness of breath.

Restrictive lung diseases are usually the result of exposure to inhaled substances (especially dusts) in the environment, including **inorganic dusts** such as silica, asbestos, or coal dust, and **organic dusts**, such as those from bird droppings or mouldy hay. Like most respiratory diseases, the symptoms are exacerbated by poor air quality (such as occurs in smoggy cities).

SEM of asbestos fibres. Asbestos has different toxicity depending on the type. Some types are very friable, releasing fibres into the air, where they can be easily inhaled.

USGS

Smoking: A Choice Against Health

Related activities: Measuring Lung Function, Diseases Caused by Smoking
Web links: What is Asthma?

RA 3

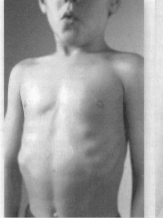  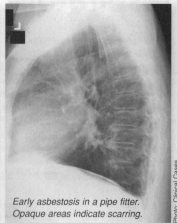 

Early asbestosis in a pipe fitter.
Opaque areas indicate scarring.

Photo: Clinical Cases

Asthma is a common disease affecting millions of people worldwide (20 million in the US alone). Asthma is the result of a hypersensitive reaction to allergens such as house dust or pollen, but attacks can be triggered by environmental factors such as cold air, exercise, or air pollutants. During an attack, sufferers show labored breathing with overexpansion of the chest cavity (above left). Asthma is treated with drugs that help to expand the airways (bronchodilators). These are usually delivered via a nebulizer or inhaler (above).

Asbestosis is a restrictive lung disease caused by breathing in asbestos fibres. The tiny fibres make their way into the alveoli where they cause damage and lead to scarring. Other occupational lung diseases include silicosis (exposure to silica dust) and coal workers' pneumoconiosis.

Chronic bronchitis is accompanied by a persistent, productive cough, where sufferers attempt to cough up the sputum or mucus which accumulates in the airways. Chronic bronchitis is indicated using **spirometry** by a reduced FEV_1/FVC ratio that is not reversed with bronchodilator therapy.

1. Distinguish between obstructive and restrictive lung diseases, and provide some examples:

_____

_____

_____

_____

2. Physicians may use spirometry to diagnosis certain types of respiratory disease. Explain the following typical results:

(a) In patients with chronic obstructive pulmonary disease, the FEV_1 / FVC ratio declines (to <70% of normal):

_____

_____

(b) Patients with asthma also have a FEV_1 / FVC ratio of <70%, but this improves following use of bronchodilators:

_____

_____

(c) In patients with restrictive lung disease, both FEV_1 and FVC are low but the FEV_1 / FVC ratio is normal to high:

_____

_____

3. Describe the mechanisms by which restrictive lung diseases reduce lung function and describe an example:

_____

_____

4. Suggest why many restrictive lung diseases are also classified as occupational lung diseases: _____

_____

_____

5. Explain why chronic bronchitis and emphysema often occur together in cigarette smokers: _____

_____

_____

_____

Smoking and the Lungs

Tobacco smoking has only recently been accepted as a major health hazard, despite its practice in developed countries for more than 400 years, and much longer elsewhere. Cigarettes became popular at the end of World War I because they were cheap, convenient, and easier to smoke than pipes and cigars. They remain popular for the further reason that they are more addictive than other forms of tobacco. The milder smoke can be more readily inhaled, allowing **nicotine** (an addictive poison) to be quickly absorbed into the bloodstream. **Lung cancer** is the most widely known harmful effect of smoking. Tobacco smoking is also directly associated with coronary artery disease and stroke, as well as obstructive lung diseases such as **chronic bronchitis** and **emphysema**. Despite indications that mortality due to smoking may be declining in the UK, one third of all deaths from cancer, including around 80% of lung cancer deaths, are linked to this cause. The damaging components of cigarette smoke include tar, carbon monoxide, nitrogen dioxide, and nitric oxide. Many of these toxins occur in greater concentrations in sidestream smoke (**passive smoking**) than in mainstream smoke (inhaled) due to the presence of a filter in the cigarette.

Effects of tobacco smoking on the gas exchange system

All forms of tobacco-smoking increase the risk of **mouth cancer**, **lip cancer**, and **cancer of the throat** (pharynx).

Lung cancer is the best known harmful effect of smoking.

Smoking is associated with obstructive lung diseases, when air flow to the gas exchange surface is prevented by excessive mucus (e.g. **chronic bronchitis**) or collapse of the airways (**emphysema**).

Short term general effects of tobacco smoking

- Reduction in capacity of the lungs.
- Increase in muscle tension and a decrease in steadiness of the hands.
- Raised blood pressure (10-30 points).
- Very sharp rise in carbon monoxide levels in the lungs contributing to breathlessness.
- Increase in pulse rate by up to 20 beats per minute.
- Surface blood vessel constriction drops skin temperature by up to 5°C.
- Dulling of appetite as well as the sense of smell and taste.

How smoking damages the lungs

Non-smoker

Normal alveoli arrangement

Cilia

Thin layer of mucus

Cells lining airways

Smoker

Coalesced alveoli

Extra mucus produced

Smoke particles

Cancerous cell

Smoke particles indirectly destroy the walls of the lung's alveoli.

Cavities lined by heavy black tar deposits.

SPECIMEN A-73-309 DATE

Gross pathology of lung tissue from a patient with emphysema. Tobacco tar deposits can be seen. Tar contains at least 17 known carcinogens.

SMOKING CAUSES LUNG CANCER
Ka mate koe i te kai hikareti
Ministry of Health Warning

Deaths from lung cancer in smokers and non-smokers (UK, 1992)

Number dying (in thousands) — Non-smokers, Smokers

Estimated percentage of deaths attributable to smoking in the UK by cause (2002 mortality data)

% of deaths attributable to smoking — Lung cancer, Upper respiratory cancer, Oesophageal cancer, Chronic obstructive lung disease, Ischaemic heart disease, Cerebrovascular disease (stroke), Aortic aneurysm, Atherosclerosis

Smoking: A Choice Against Health

Periodicals: Smoking

Related activities: Smoking and the Cardiovascular System, Respiratory Diseases **Web links:** Every Cigarette, CDC: Smoking

RDA 2

Components of Cigarette Smoke

Particulate Phase

Nicotine: a highly addictive alkaloid

Tar: composed of many chemicals

Benzene: carcinogenic hydrocarbon

Gas Phase

Carbon monoxide: a poisonous gas

Ammonia: a pungent, colourless gas

Formaldehyde: a carcinogen

Hydrogen cyanide: a highly poisonous gas

Tobacco smoke is made up of "sidestream smoke" from the burning tip and "mainstream smoke" from the filter (mouth) end. Sidestream smoke contains higher concentrations of many toxins than mainstream smoke. Tobacco smoke includes both particulate and gas phases (left), both of which contain many harmful substances.

Filter
Cellulose acetate filters trap some of the tar and smoke particles. They cool the smoke slightly, making it easier to inhale.

1. Discuss the physical changes to the lung that result from long-term smoking:

_____

_____

_____

_____

2. Describe the physiological effect of each of the following constituents of tobacco smoke when inhaled:

(a) Tar: _____

(b) Nicotine: _____

(c) Carbon monoxide: _____

3. Describe the symptoms of the following diseases associated with long-term smoking:

(a) Emphysema: _____

_____

(b) Chronic bronchitis: _____

_____

(c) Lung cancer: _____

_____

4. A long term study showed the correlation between smoking and lung cancer, providing supporting evidence for the adverse effects of smoking (right):

(a) Explain why a long term study was important:

_____

_____

(b) The study made a link between cigarette consumption and mortality from lung cancer. What else did it show?

_____

_____

_____

Cigarette consumption (men)

Lung cancer (men)

Cigarettes smoked per person per year

4000

3000

2000

1000

Lung cancer deaths per 100,000

150

100

50

Data NIH, US

1900   1920   1940   1960   1980
Year

Smoking and the Cardiovascular System

As discussed in the chapter "*Food and Health*", cardiovascular disease is an umbrella term describing a variety of diseases affecting the heart and circulatory system. A large proportion of cardiovascular diseases can be related either directly or indirectly to smoking, including coronary heart disease, peripheral vascular disease, and stroke. An estimated 13% of deaths from cardiovascular disease are attributable to smoking. Smoking increases the detrimental effects of high blood fat levels and high blood pressure to greatly increase the risk of CVD. Smoking aggravates the development of atherosclerotic plaques, at least partially through the action of nicotine and carbon monoxide, and increases heart rate and blood pressure. This it increases the body's demand for oxygen while at the same time making it harder to deliver oxygen to where it is needed.

Effects of Smoking on the Cardiovascular System

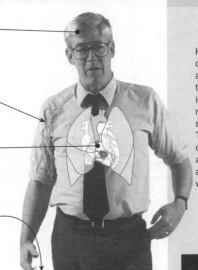

Smoking damages the arteries of the brain and may result in a **stroke**.

Nicotine is a stimulant and causes both immediate and longer term increases in blood pressure and heart rate. It also causes the body to mobilise fat stores.

In a young man who smokes 20 cigarettes a day, the risk of **coronary artery disease** is increased by about three times over that of a nonsmoker.

Smoking causes constriction of the arteries supplying blood to the extremities and leads to **peripheral vascular disease**. This can eventually lead to gangrene and the loss of digits, or even limbs.

Carbon monoxide

The blood protein haemoglobin binds and carries oxygen within the red blood cells to supply the cells and tissues of the body. However, carbon monoxide (CO) has a much higher affinity for haemoglobin than oxygen does, so when carbon monoxide is inhaled, it displaces oxygen from haemoglobin. Less oxygen is supplied to the tissues as a result.

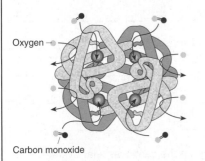

Oxygen

Carbon monoxide

Left: CO is about 200 times more likely to bind to haemoglobin than oxygen and displaces it.

Smoking and Heart Disease

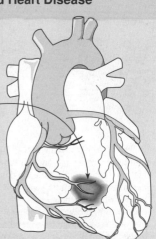

Heart attacks are usually quite sudden events. A heart attack (infarction) is normally the result of an interruption in the blood flow to the heart muscle, which becomes starved of oxygen and dies. This often follows the rupture of an unstable atherosclerotic plaque (such as that pictured below), which blocks the artery.

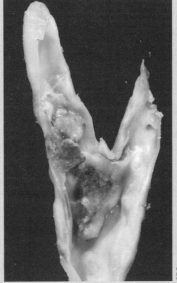

The nictoine and carbon monoxide in tobacco smoke increase the viscosity of the blood, which accelerates the rate at which unstable atherosclerotic plaques are formed in the vessels such as the coronary arteries and the carotid artery, (left). Nicotine makes the platelets stick together, and CO simulates the formation of fibrinogen, which is involved in blood clotting. Unstable clots are more likely to become dislodged and block arteries.

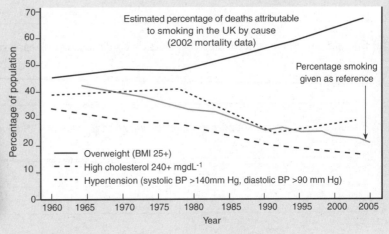

Estimated percentage of deaths attributable to smoking in the UK by cause (2002 mortality data)

Percentage smoking given as reference

- —— Overweight (BMI 25+)
- – – High cholesterol 240+ mgdL^{-1}
- ···· Hypertension (systolic BP >140mm Hg, diastolic BP >90 mm Hg)

Percentage of population

Year

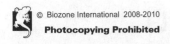

Periodicals:
Coronary heart disease

Related activities: Atherosclerosis, Cardiovascular Disease
Web links: WHO: Tobacco Warnings Database

RA 2

Smoking: A Choice Against Health

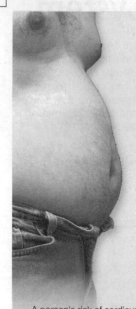

Controllable Risk Factors for Cardiovascular Disease

- Hypertension
- Cigarette smoking
- High blood cholesterol and h
- High LDL:HDL ratio
- Obesity
- Type 2 diabetes mellitus
- Sedentary lifestyle
- High achiever personality
- Environmental stress

Estimated coronary heart disease rate according to various combinations of risk factors over 10 years
(source: International Diabetes Foundation, 2001)

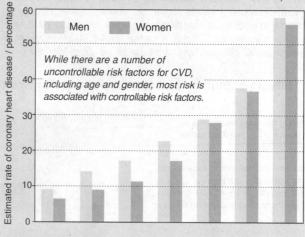

While there are a number of uncontrollable risk factors for CVD, including age and gender, most risk is associated with controllable risk factors.

A person's risk of cardiovascular disease increases markedly with an increase in the number of risk factors. This is particularly the case for smoking, because smoking acts synergistically with other risk factors, particularly hypertension (high blood pressure) and high blood lipids. This means that any given risk factor has a proportionately greater effect in a smoker than in a non-smoker.

Risk factors						
Systolic blood pressure / mm Hg 120	160	160	160	160	160	160
Cholesterol / mg dL^{-1} 220	220	259	259	259	259	259
HDL cholesterol / mg dL^{-1} 50	50	50	35	35	35	35
Diabetes –	–	–	–	+	+	+
Cigarette smoking –	–	–	–	–	+	+
Enlargement of left ventricle –	–	–	–	–	–	+

1. (a) Summarise the effects of tobacco smoking on the body's cardiovascular system: _____

_____

_____

 (b) Explain why tobacco smoking greatly increases the risk of developing cardiovascular diseases: _____

_____

_____

 (c) Evaluate the evidence supporting the link between smoking and cardiovascular disease: _____

_____

_____

_____

2. Describe the contribution of nicotine and carbon monoxide to the effects of smoking on the cardiovascular system:

_____

_____

3. (a) Suggest why some controllable risk factors for cardiovascular disease often occur together: _____

_____

_____

 (b) Evaluate the evidence supporting the observation that patients with several risk factors are at a higher risk of CVD:

_____

_____

_____

_____

Identifying Biodiversity

KEY CONCEPTS

Classification enables us to recognise and quantify the biological diversity on Earth.

▶ Organisms are assigned to taxonomic categories based on shared derived characteristics.

▶ Organisms are identified using binomial nomenclature: genus and species.

▶ Dichotomous classification keys can be used to identify unknown organisms.

KEY TERMS

Animalia
Archaebacteria (Archaea)
binomial nomenclature
class
cladistics
cladogram
classification key
common name
dichotomous
distinguishing feature
Eubacteria
Eukarya
Eukaryotae
family
five kingdom classification
Fungi
genus
kingdom
Monera
morphology
order
phylogeny
phylogenetic systematics
phylum
Plantae
Prokaryotae
Protoctista
shared derived
character (=synapomorphy)
Simpson's Index of Diversity
species
taxon (pl. taxa)
taxonomic category
three domain classification

Periodicals:
Listings for this
chapter are on page 340

Weblinks:
www.biozone.co.uk/
weblink/OCR-AS-2641.html

OBJECTIVES

☐ 1. Use the **KEY TERMS** to help you understand and complete these objectives.

Biodiversity pages 256-261, 323-324

☐ 2. Define **biodiversity** and recognise the importance of classification in recognising, appreciating, and conserving the biodiversity on Earth. Discuss current estimates of global diversity.

☐ 3. Explain how biodiversity can be recognised at several levels including **species** diversity, **habitat** diversity, and **genetic** diversity. Appreciate the significance of recognising each of these levels of biodiversity.

☐ 4. Explain the role of **sampling** in quantifying biodiversity. Appreciate the need for unbiased, quantitative information and explain how this is achieved using **random sampling**.

☐ 5. Distinguish between **species richness** and **species evenness** (distribution) and describe how these can be measured. Appreciate the importance of understanding both in the management and conservation of populations.

☐ 6. Explain how diversity indices are used to quantify species diversity in an area. Use **Simpson's Index of Diversity** (D) to calculate the diversity of a habitat. Explain the significance of different values of D.

Classification of Life pages 262-283, 287-289

☐ 7. Define the terms **classification**, **phylogeny**, and **taxonomy**, and show how they are related. Describe the principles and importance of biological classification.

☐ 8. Describe the classification of species into a taxonomic hierarchy, recognising domain, kingdom, phylum, class, order, family, genus, and species.

☐ 9. Describe the **distinguishing features** of each of the following five kingdoms: **Prokaryotae** (Monera), **Protoctista**, **Fungi**, **Plantae**, and **Animalia**.

☐ 10. Explain how **binomial nomenclature** is used to classify organisms. Appreciate the problems associated with using common names to describe organisms.

☐ 11. Explain how a dichotomous key is used to identify organisms. Use a simple taxonomic key to recognise and classify some common organisms.

New Classifications pages 284-286

☐ 12. Discuss changes to traditional classification schemes in the light of molecular evidence. Appreciate that **cladistics** emphasises relatedness and the presence of **shared derived characters**. Discuss the benefits and disadvantages of this.

☐ 13. Compare and contrast the **five kingdom** (#9 above) and **three domain** classification systems. Explain the basis and rationale for these classifications.

Global Biodiversity

The species is the basic unit by which we measure biological diversity or **biodiversity**. Biodiversity is not distributed evenly on Earth, being consistently richer in the tropics and concentrated more in some areas than in others. The simplest definition of biodiversity is as the sum of all biotic variation from the level of genes to ecosystems, but often the components of total biodiversity are distinguished. **Species diversity** describes the number of different species in an area (**species richness**), **genetic diversity** is the diversity of genes within a species, and **ecosystem diversity** refers to the diversity at the higher ecosystem level of organisation. **Habitat diversity** is also sometimes described and is essentially a subset of ecosystem diversity expressed per given unit area. Total biological diversity is often threatened because of the loss of just one of these components. Conservation International recognises 25 **biodiversity hotspots**. These are biologically diverse and ecologically distinct regions under the greatest threat of destruction. They are identified on the basis of the number of species present, the amount of **endemism**, and the extent to which the species are threatened.

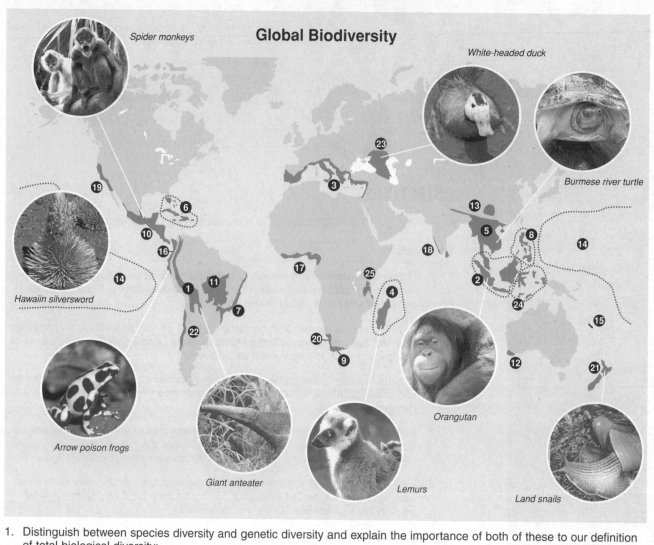

Global Biodiversity

Spider monkeys

White-headed duck

Burmese river turtle

Hawaiin silversword

Arrow poison frogs

Giant anteater

Lemurs

Orangutan

Land snails

1. Distinguish between species diversity and genetic diversity and explain the importance of both of these to our definition of total biological diversity:

_____

_____

_____

_____

2. Explain the importance of considering ecosystem (habitat) diversity when targeting regions for conservation purposes:

_____

_____

_____

3. Use your research tools (e.g. textbook, internet, or encyclopaedia) to identify each of the 25 biodiversity hotspots illustrated in the diagram above. For each region, summarise the characteristics that have resulted in it being identified as a biodiversity hotspot. Present your summary as a short report and attach it to this page of your workbook.

Related activities: Britain's Biodiversity, Loss of Biodiversity
Web links: Biodiversity Hotspots

Periodicals:
Biodiversity: Taking stock

Britain's Biodiversity

We measure biodiversity by looking at species richness. For some taxa, e.g. bacteria, the true extent of species diversity remains unidentified. Some data on species richness for the UK are shown below (note the bias towards large, conspicuous organisms). The biodiversity of the British Isles today is the result of a legacy of past climatic changes and a long history of human influence. Some of the most interesting, species-rich ecosystems, such as hedgerows,

downland turf, and woodland, are maintained as a result of human activity. Many of the species characteristic of Britain's biodiversity are also found more widely in Europe. Other species (e.g. the Scottish crossbill), or species associations (e.g. bluebell woodlands) are uniquely British. With increasing pressure on natural areas from urbanisation, roading, and other human encroachment, maintaining species diversity is paramount and should concern us all today.

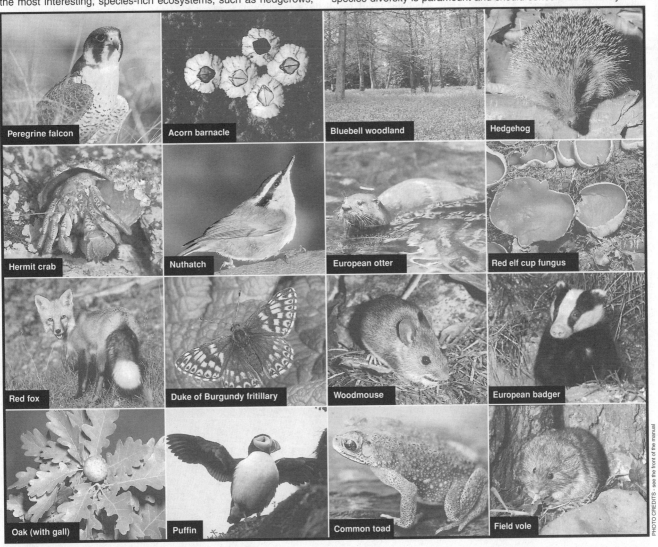

Peregrine falcon · Acorn barnacle · Bluebell woodland · Hedgehog · Hermit crab · Nuthatch · European otter · Red elf cup fungus · Red fox · Duke of Burgundy fritillary · Woodmouse · European badger · Oak (with gall) · Puffin · Common toad · Field vole

PHOTO CREDITS - see the front of the manual

Left: Fig. 1: British biodiversity, as numbers of terrestrial and freshwater species, compared with recent global estimates of described species in major taxonomic groups.

Major taxonomic group	Estimated no. of British species	Estimated no. of world species
Bacteria	unknown	> 4 000
Viruses	unknown	> 5 000
Protozoa	> 20 000	> 40 000
Algae	> 20 000	> 40 000
Fungi	> 15 000	> 70 000
Ferns and bryophytes	1 080	> 26 000
Lichens	1 500	> 17 000
Flowering plants	1 400	> 250 000
Invertebrate animals	> 28 500	> 1.28 million
Insects	22 500	> 1 million
Non-insect arthropods	> 3 000	> 190 000
All other invertebrates	> 3 000	> 90 000
Vertebrate animals	308	> 33 208
Fish (freshwater)	38	> 8 500
Amphibians	6	> 4 000
Reptiles	6	> 6 500
Birds (breeding residents)	210	9 881
Mammals	48	4 327

Source: Biodiversity: The UK Action Plan, 1994. HMSO

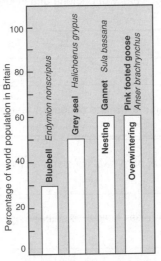

Fig. 2: Bar graph illustrating the percentage of world populations of various species permanently or temporarily resident in Britain.

(Bluebell *Endymion nonscriptus*; Grey seal *Halichoerus grypus*; Gannet *Sula bassana* – Nesting; Pink footed goose *Anser brachrynchus* – Overwintering)

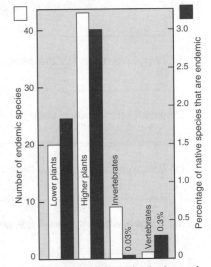

Fig 3: Bar graph illustrating the degree of endemism in Britain. Right axis indicates % endemism in relation to the number of described British native species.

(Lower plants; Higher plants; Invertebrates 0.03%; Vertebrates 0.3%)

Barn owls are predators of small mammals, birds, insects, and frogs. They are higher order consumers and, as such, have been badly affected by the bioaccumulation of pesticides in recent times. They require suitable nesting and bathing sites, and reliable sources of small prey.

Conservation of the barn owl *(Tyto alba)*

Status: The barn owl is one of the best known and widely distributed owl species in the world. It was once very common in Britain but has experienced severe declines in the last 50 years as a result of the combined impacts of habitat loss, changed farming practices, and increased sources of mortality.

Reasons for decline: Primarily, declines have been the result of changed farm management practices (e.g. increased land clearance and mechanisation) which have resulted in reduced prey abundance and fewer suitable breeding sites. Contributing factors include increases in road deaths as traffic speed and volume rises, and poorer breeding success and reduced chick survival as a result of pesticide bioaccumulation. In addition, more birds are drowned when attempting to bathe in the steep sided troughs which have increasingly replaced the more traditional shallow farm ponds.

Conservation management: A return to the population densities of 50 years ago is very unlikely, but current conservation measures have at least stabilised numbers. These involve habitat enhancement (e.g. provision of nest sites), reduction in pesticide use, and rearing of orphaned young followed by monitored release into suitable habitats.

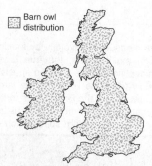

Barn owl distribution

Barn owls are widely distributed in Ireland and the UK, but numbers are not high.

Period of survey (England & Wales)	Breeding pairs (estimates)
1935	12 000
1968 – 1972	6000 – 9000
1983 – 1985	3800

1. Produce a pie graph below to show the proportions of British species in each taxonomic group (ignoring bacteria and viruses). Calculate the percentages from Fig. 1 (opposite) and tabulate the data (one has been completed for you). The chart has been marked in 5° divisions and each % point is equal to 3.6° on the pie chart. Provide a colour key in the space next to the tabulated figures. For the purposes of this exercise, use the values provided, ignoring the > sign:

Proportion of British species in different taxonomic groups

	Percentage of species in each taxon	Segment size	Key
Protozoa			
Algae			
Fungi			
Ferns and bryophytes			
Lichens			
Flowering plants			
Invertebrates	28 500 ÷ 87 788 X 100 = 32.5%	117°	
Vertebrates			

2. Comment on the proportion of biodiversity within each taxonomic group: _____

_____

3. (a) Contrast our knowledge of the biodiversity of bacteria and invertebrates with that of vertebrates:

_____

(b) Suggest a reason for the difference: _____

_____

4. Comment on the level of endemism in the UK and suggest a reason for it: _____

_____

5. (a) Calculate the percentage decline in barn owls (England and Wales) over the 50 year period 1935 – 1985:

_____

(b) Suggest why this species has been less difficult to stabilise against decline than other (more endangered) species:

_____

_____

_____

Measuring Diversity in Ecosystems

Measurements of biodiversity have essentially two components: **species richness**, which describes the number of species, and **species evenness**, which quantifies how equally the community composition is distributed. Both are important, especially when rarity is a reflection of how threatened a species is in an environment. Information about the biodiversity of ecosystems is obtained through **sampling** the ecosystem in a manner that provides a fair (unbiased) representation of the organisms present and their distribution. This is usually achieved through **random sampling**, a technique in which every possible sample of a given size has the same chance of selection. Measures of biodiversity are commonly used as the basis for making conservation decisions and different measures of biodiversity may support different solutions. Often indicator species and species diversity indices are used as a way of quantifying biodiversity. Such indicators can be particularly useful when monitoring ecosystem change and looking for causative factors in species loss.

Quantifying the Diversity of Ecosystems

Reef community:
high density, clumped distribution

Measurements of biodiversity must be appropriate to the community being investigated. Communities in which the populations are at low density and have a random or clumped distribution will require a different sampling strategy to those where the populations are uniformly distributed and at higher density. There are many sampling options (below), each with advantages and drawbacks for particular communities. How would you estimate the biodiversity of this reef community?

Random point sampling

Point sampling: systematic grid

Line and belt transects

Random quadrats

Marine ecologists use quadrat sampling to estimate biodiversity prior to works such as dredging.

Line transects are appropriate to estimate biodiversity along an environmental gradient.

Keystone Species in Ecosystems

The stability of an ecosystem refers to its apparently unchanging nature over time, something that depends partly on its ability to resist and recover from disturbance. Ecosystem stability is closely linked to biodiversity, and more biodiverse systems tend to be more stable, partly because the many species interactions that sustain them act as a buffer against change. Some species are more influential than others in the stability of an ecosystem because of their pivotal role in some ecosystem function such as nutrient recycling or productivity. Such species are called **keystone species** because of their disproportionate effect on ecosystem function.

The **European beaver**, *Castor fiber*, was originally distributed throughout most of Europe and northern Asia but populations have been decimated as a result of both hunting and habitat loss. The beaver is a keystone species; where they occur, beavers are critical to ecosystem function and a number of species depend partly or entirely on beaver ponds for survival. Their tree-felling activity is akin to a natural coppicing process and promotes vigorous regrowth, while historically they helped the spread of alder (a water-loving species) in Britain.

1. (a) Distinguish between the two measures of biodiversity: species richness and species evenness:

_____

_____

_____

(b) Explain why it is important to consider both these measures when considering species conservation:

_____

_____

_____

Related activities: Global Biodiversity, Britain's Biodiversity

RA 2

Calculation and Use of Diversity Indices

One of the best ways to determine the health of an ecosystem is to measure the variety (rather than the absolute number) of organisms living in it. Certain species, called **indicator species**, are typical of ecosystems in a particular state (e.g. polluted or pristine). An objective evaluation of an ecosystem's biodiversity can provide valuable insight into its status, particularly if the species assemblages have changed as a result of disturbance.

Diversity can be quantified using a **diversity index (DI)**. Diversity indices attempt to quantify the degree of diversity and identify indicators for environmental stress or degradation. Most indices of diversity are easy to use and they are widely used in ecological work, particularly for monitoring ecosystem change or pollution (i.e. before and after assessments). **Simpson's Index of Diversity** (below) produces values ranging between 0 and almost 1. These are more easily interpreted than some other versions of Simpson's Index because of the more limited range of values, but no single index offers the "best" measure of diversity; each is chosen on the basis of suitability to different situations.

Simpson's Index of Diversity

Simpson's Index of Diversity (D) is easily calculated using the following simple formula. Communities with a wide range of species produce a higher diversity score than communities dominated by larger numbers of only a few species.

$$D = 1 - (\Sigma (n/N)^2)$$

Where:

D = Diversity index

N = Total number of individuals (of all species) in the sample

n = Number of individuals of each species in the sample

This index ranges between 0 (low diversity) and 1 (high diversity). Indices are usually evaluated with reference to earlier measurement or a standard ecosystem measure to be properly interpreted.

Example of species diversity in a stream

The example describes the results from a survey of stream invertebrates. It is not necessary to know the species to calculate a diversity index as long as the different species can be distinguished. For the example below, Simpson's Index of Diversity using $D = 1 - (\Sigma (n/N)^2)$ is:

Species	n	n/N	$(n/N)^2$
A (backswimmer)	12	0.300	0.090
B (stonefly larva)	7	0.175	0.031
C (silver water beetle)	2	0.050	0.003
D (caddisfly larva)	6	0.150	0.023
E (water spider)	5	0.125	0.016
F (mayfly larva)	8	0.20	0.040
	$\Sigma n = 40$		$\Sigma (n/N)^2 = 0.201$

$$D = 1 - 0.201 = \mathbf{0.799}$$

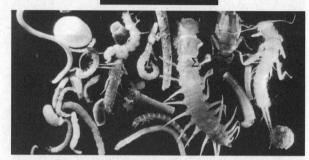

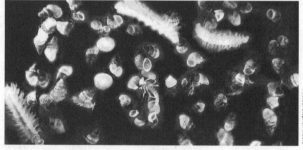

A stream community with a high macroinvertebrate diversity (above) in contrast to a low diversity stream community (below).

Photos: Stephen Moore

2. Describe two necessary considerations in attempting to make an unbiased measurement of biodiversity:

_____

_____

3. Explain why high biodiversity is generally associated with greater ecosystem stability: _____

_____

_____

4. Explain why the loss of a keystone species could be particularly disturbing for ecosystem diversity:

_____

_____

5. Describe a situation where a species diversity index may provide useful information: _____

_____

6. An area of forest floor was sampled and six invertebrate species were recorded, with counts of 7, 10, 11, 2, 4, and 3 individuals. Calculate Simpson's index of diversity for this community:

(a) DI= _____   DI = _____

(b) Comment on the diversity of this community: _____

_____

Loss of Biodiversity

More than a third of the planet's known terrestrial plant and animal species are found within the biodiversity hotspot regions which cover only 1.4% of the Earth's land area. Unfortunately, biodiversity hotspots often occur near areas of dense human habitation and rapid human population growth. Most are located in the tropics and most are forests. Background (natural) extinction rates for all organisms (including bacteria and fungi) are estimated to be 10-100 species a year. The actual extinction rate is estimated to be 100-1000 times higher, mainly due to the effects of human activity. Over 41 000 species are now on the International Union for Conservation's (IUCN) red list, and 16 000 are threatened with extinction. Loss of biodiversity reduces the stability and resilience of natural ecosystems and decreases the ability of their communities to adapt to changing environmental conditions. Humans rely heavily on the biodiversity in nature and a loss of species richness has a deleterious effect on us all.

Insects make up 80% of all known animal species. There are an estimated 6-10 million insect species on Earth, but only 900,000 have been identified. Some 44 000 species may have become extinct over the last 600 years. The Duke of Burgundy butterfly (*Hamearis lucina*), right, is an endangered British species.

Just over 5% of the 8225 reptile species are at risk. These include the two tuatara species (right) from New Zealand, which are the only living members of the order Sphenodontia, and the critically endangered blue iguana. Only about 200 blue iguanas remain, all in the Grand Caymans.

	Total number of species*	Number of IUCN listed species
Plants	310 000 - 422,000	8474
Insects	6 -10 million	622
Fish	28 000	126
Amphibians	5743	1809
Reptiles	8225	423
Birds	10 000	1133
Mammals	5400	1027

* Estimated numbers

The giant panda (above), is one of many critically endangered terrestrial mammals, with fewer than 2000 surviving in the wild. Amongst the 120 species of marine mammals, approximately 25% (including the humpback whale and Hector's dolphin) are on the ICUN's red list.

Prior to the impact of human activity on the environment, one bird species became extinct every 100 years. Today, the rate is one every year, and may increase to 10 species every year by the end of the century. Some at risk birds, such as the Hawaiian crow (right), are now found only in captivity.

Current estimates suggest as many as 47% of plant species may be endangered. Some, such as the South African cycad *Encephalartos woodii* (above), is one of the rarest plants in the world. It is extinct in the wild and all remaining specimens are clones.

Threats to Biodiversity

Rainforests in some of the most species-rich regions of the world are being destroyed at an alarming rate as world demand for tropical hardwoods increases and land is cleared for the establishment of agriculture.

Illegal trade in species (for food, body parts, or for the exotic pet trade) is pushing some species to the brink of extinction. Despite international bans on trade, illegal trade in primates, parrots, reptiles, and big cats (among others) continues.

Pollution and the pressure of human populations on natural habitats threatens biodiversity in many regions. Environmental pollutants may accumulate through food chains or cause harm directly, as with this bird trapped in oil.

1. Discuss, in general terms, the effects of loss of biodiversity on an ecosystem: _____

_____

_____

_____

_____

Periodicals:
Earth's nine lives

Related activities: Diversity, Stability, and Key Species.
Global Biodiversity

RA 2

Characteristics of Life

With each step in the hierarchy of biological order, new properties emerge that were not present at simpler levels of organisation. Life itself is associated with numerous **emergent properties**, including **metabolism** and growth. The cell is the site of life; it is the functioning unit structure from which living organisms are made. Viruses and cells are profoundly different. Viruses are non-cellular, lack the complex structures found in cells, and show only some of the properties we associate with living things. The traditional view of viruses is as a minimal particle, although the identification in 2004 of a new family of viruses, called mimiviruses, is forcing a rethink of this conservative view. Note the different scale to which the examples below are drawn. Refer to the scale bars for the comparative sizes (1000 nm = 1 µm = 0.001 mm).

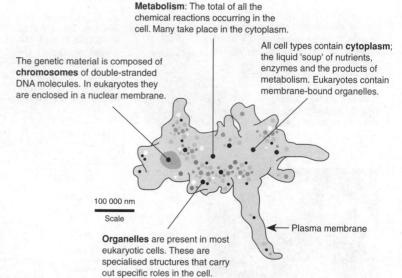

Although some viruses may contain an **enzyme**, it is incapable of working until it is inside a host cell's cytoplasm.

Single or double stranded molecule of **RNA** or **DNA**.

A **protein coat** surrounds the viral genetic material and enzyme (if present). There is no cellular membrane.

50 nm
Scale

No metabolism: The absence of cytoplasm means that a virus can not carry out any chemical reactions on its own; it is dependent upon parasitising a cell and using the cell's own machinery.

Metabolism: The total of all the chemical reactions occurring in the cell. Many take place in the cytoplasm.

The genetic material is composed of **chromosomes** of double-stranded DNA molecules. In eukaryotes they are enclosed in a nuclear membrane.

All cell types contain **cytoplasm**; the liquid 'soup' of nutrients, enzymes and the products of metabolism. Eukaryotes contain membrane-bound organelles.

100 000 nm
Scale

Organelles are present in most eukaryotic cells. These are specialised structures that carry out specific roles in the cell.

← Plasma membrane

Virus
(e.g. HIV)

Viruses cannot become active outside a living host cell. They simply exist as inert virus particles called **virions**. Only when they invade a cell and take over the cell's metabolic machinery, can the virus carry out its 'living programme'.

Cell
(e.g. Amoeba)

Cells remain alive so long as their metabolic reactions in the cytoplasm are maintained. With a few rare exceptions (that involve freezing certain types of cells) if metabolism is halted, the cell dies.

1. Identify three features that all cells have in common: _____

_____

2. Describe how cells differ from viruses in the following aspects:

(a) Size: _____

_____

(b) Metabolism: _____

_____

(c) Organelles: _____

_____

(d) Genetic material: _____

_____

(e) Life cycle: _____

_____

3. Explain why multicellular organisms are said to show emergent properties: _____

_____

_____

Related activities: Types of Living Things

Periodicals:
The living dead,
Are viruses alive?

Types of Living Things

Living things are called organisms and **cells** are the functioning unit structure from which organisms are made. Under the five kingdom system, cells can be divided into two basic kinds: the **prokaryotes**, which are simple cells without a distinct, membrane-bound nucleus, and the more complex **eukaryotes**. The eukaryotes can be further organised into broad groups according to their basic cell type: the protoctists, fungi, plants, and animals. Viruses are non-cellular and have no cellular machinery of their own. All cells must secure a source of energy if they are to survive and carry out metabolic processes. **Autotrophs** can meet their energy requirements using light or chemical energy from the physical environment. Other types of cell, called **heterotrophs**, obtain their energy from other living organisms or their dead remains.

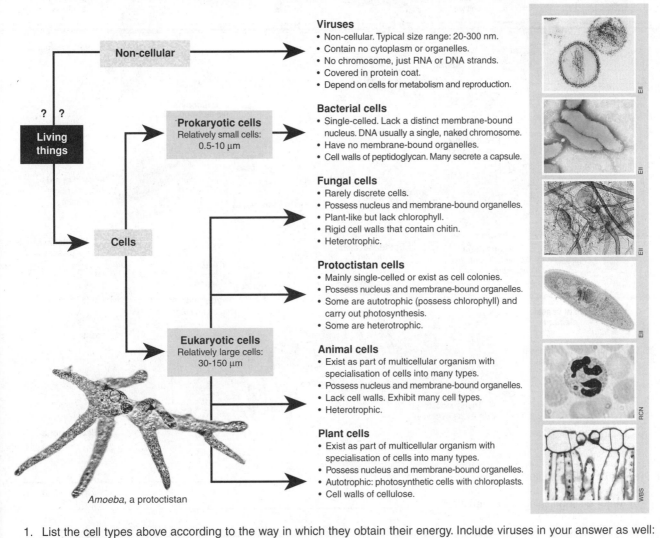

Living things → **Non-cellular**

? ?

Living things → **Cells**

Prokaryotic cells
Relatively small cells:
0.5-10 µm

Eukaryotic cells
Relatively large cells:
30-150 µm

Amoeba, a protoctistan

Viruses
- Non-cellular. Typical size range: 20-300 nm.
- Contain no cytoplasm or organelles.
- No chromosome, just RNA or DNA strands.
- Covered in protein coat.
- Depend on cells for metabolism and reproduction.

Bacterial cells
- Single-celled. Lack a distinct membrane-bound nucleus. DNA usually a single, naked chromosome.
- Have no membrane-bound organelles.
- Cell walls of peptidoglycan. Many secrete a capsule.

Fungal cells
- Rarely discrete cells.
- Possess nucleus and membrane-bound organelles.
- Plant-like but lack chlorophyll.
- Rigid cell walls that contain chitin.
- Heterotrophic.

Protoctistan cells
- Mainly single-celled or exist as cell colonies.
- Possess nucleus and membrane-bound organelles.
- Some are autotrophic (possess chlorophyll) and carry out photosynthesis.
- Some are heterotrophic.

Animal cells
- Exist as part of multicellular organism with specialisation of cells into many types.
- Possess nucleus and membrane-bound organelles.
- Lack cell walls. Exhibit many cell types.
- Heterotrophic.

Plant cells
- Exist as part of multicellular organism with specialisation of cells into many types.
- Possess nucleus and membrane-bound organelles.
- Autotrophic: photosynthetic cells with chloroplasts.
- Cell walls of cellulose.

1. List the cell types above according to the way in which they obtain their energy. Include viruses in your answer as well:

    (a) Autotrophic: _____

    (b) Heterotrophic: _____

2. Consult the diagram above and determine the two main features distinguishing **eukaryotic** cells from **prokaryotic** cells:

    (a) _____

    (b) _____

3. (a) Suggest why fungi were once classified as belonging to the plant kingdom: _____

    _____

    (b) Explain why, in terms of the distinguishing features of fungi, this classification was erroneous: _____

    _____

    _____

4. Suggest why the Protoctista have traditionally been a difficult group to classify: _____

    _____

Related activities: Plant Cells, Animal Cells, Unicellular Eukaryotes, Cell Sizes, Prokaryotic Cells **Web links**: *Types of Microbes*

Types of Cells

Cells come in a wide range of types and forms. The diagram below shows a selection of cell types from the five kingdoms. The variety that results from specialisation of undifferentiated cells is enormous. In the following exercise, identify which of the cell types belongs to each of the kingdoms and list the major distinguishing characteristics of their cells.

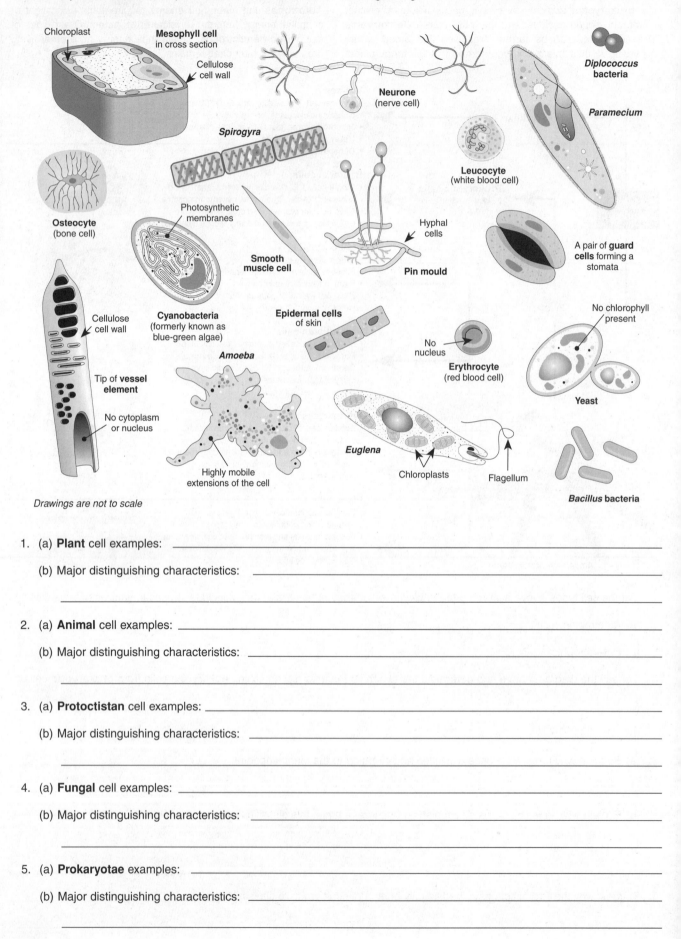

Drawings are not to scale

1. (a) **Plant** cell examples: _____

   (b) Major distinguishing characteristics: _____

   _____

2. (a) **Animal** cell examples: _____

   (b) Major distinguishing characteristics: _____

   _____

3. (a) **Protoctistan** cell examples: _____

   (b) Major distinguishing characteristics: _____

   _____

4. (a) **Fungal** cell examples: _____

   (b) Major distinguishing characteristics: _____

   _____

5. (a) **Prokaryotae** examples: _____

   (b) Major distinguishing characteristics: _____

   _____

Related activities: Plant Cells, Animal Cells, Unicellular Eukaryotes, Prokaryotic Cells

Features of Taxonomic Groups

In order to distinguish organisms, it is desirable to classify and name them (a science known as **taxonomy**). An effective classification system requires features that are distinctive to a particular group of organisms. The distinguishing features of some major taxonomic groups are provided in the following pages by means of diagrams and brief summaries. Revised classification systems, recognising three domains (rather than five kingdoms) are now recognised as better representations of the true diversity of life. However, for the purposes of describing the groups with which we are most familiar, the five kingdom system (used here) is still appropriate. Note that most animals show **bilateral symmetry** (body divisible into two halves that are mirror images). **Radial symmetry** (body divisible into equal halves through various planes) is a characteristic of cnidarians and ctenophores. Definitions of specific terms relating to features of structure or function can be found in any general biology text.

Kingdom: PROKARYOTAE (Bacteria)

- Also known as monerans or prokaryotes.
- Two major bacterial lineages are recognised: the primitive **Archaebacteria** and the more advanced **Eubacteria**.
- All have a prokaryotic cell structure: they lack the nuclei and chromosomes of eukaryotic cells, and have smaller (70S) ribosomes.
- Have a tendency to spread genetic elements across species barriers by sexual conjugation, viral transduction and other processes.
- Can reproduce rapidly by binary fission in the absence of sex.

- Have evolved a wider variety of metabolism types than eukaryotes.
- Bacteria grow and divide or aggregate into filaments or colonies of various shapes.
- They are taxonomically identified by their appearance (form) and through biochemical differences.

Species diversity: 10 000 + Bacteria are rather difficult to classify to the species level because of their relatively rampant genetic exchange, and because their reproduction is usually asexual.

Eubacteria

- Also known as 'true bacteria', they probably evolved from the more ancient Archaebacteria.
- Distinguished from Archaebacteria by differences in cell wall composition, nucleotide structure, and ribosome shape.
- Very diverse group comprises most bacteria.
- The **gram stain** provides the basis for distinguishing two broad groups of bacteria. It relies on the presence of peptidoglycan (unique to bacteria) in the cell wall. The stain is easily washed from the thin peptidoglycan layer of gram negative walls but is retained by the thick peptidoglycan layer of gram positive cells, staining them a dark violet colour.

Gram-Positive Bacteria

The walls of gram positive bacteria consist of many layers of peptidoglycan forming a thick, single-layered structure that holds the gram stain.

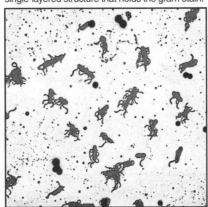

Bacillus alvei: a gram positive, flagellated bacterium. Note how the cells appear dark.

Gram-Negative Bacteria

The cell walls of gram negative bacteria contain only a small proportion of peptidoglycan, so the dark violet stain is not retained by the organisms.

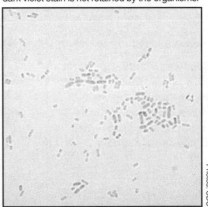

Photos: CDC

Alcaligenes odorans: a gram negative bacterium. Note how the cells appear pale.

Kingdom: FUNGI

- Heterotrophic.
- Rigid cell wall made of chitin.
- Vary from single celled to large multicellular organisms.
- Mostly saprotrophic (i.e. feeding on dead or decaying material).
- Terrestrial and immobile.

Examples:
Mushrooms/toadstools, yeasts, truffles, morels, moulds, and lichens.

Species diversity: 80 000 +

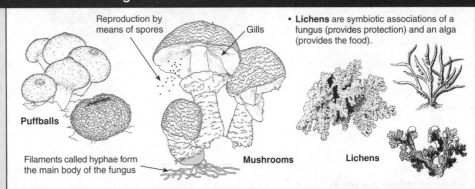

Reproduction by means of spores

Gills

- **Lichens** are symbiotic associations of a fungus (provides protection) and an alga (provides the food).

Puffballs

Filaments called hyphae form the main body of the fungus

Mushrooms

Lichens

Kingdom: PROTOCTISTA

- A diverse group of organisms that do not fit easily into other taxonomic groups.
- Unicellular or simple multicellular.
- Widespread in moist or aquatic environments.

Examples of algae: green, brown, and red algae, dinoflagellates, diatoms.

Examples of protozoa: amoebas, foraminiferans, radiolarians, ciliates.

Species diversity: 55 000 +

Algae 'plant-like' protoctists

- Autotrophic (photosynthesis)
- Characterised by the type of chlorophyll present

Cell walls of cellulose, sometimes with silica

Diatom

Protozoa 'animal-like' protoctists

- Heterotrophic nutrition and feed via ingestion
- Most are microscopic (5 μm-250 μm)

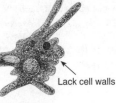

Lack cell walls

Move via projections called pseudopodia

Amoeba

Kingdom: PLANTAE

- Multicellular organisms (the majority are photosynthetic and contain chlorophyll).
- Cell walls made of cellulose; Food is stored as starch.
- Subdivided into two major divisions based on tissue structure: **Bryophytes** (non-vascular) and **Tracheophytes** (vascular) plants.

Non-Vascular Plants:

- Non vascular, lacking transport tissues (no xylem or phloem).
- They are small and restricted to moist, terrestrial environments.
- Do not possess 'true' roots, stems or leaves.

Phylum Bryophyta: Mosses, liverworts, and hornworts.

Species diversity: 18 600 +

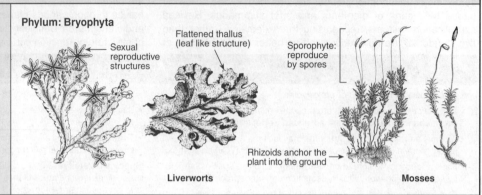

Phylum: Bryophyta

Sexual reproductive structures

Flattened thallus (leaf like structure)

Sporophyte: reproduce by spores

Rhizoids anchor the plant into the ground

Liverworts

Mosses

Vascular Plants:

- Vascular: possess transport tissues.
- Possess true roots, stems, and leaves, as well as stomata.
- Reproduce via spores, not seeds.
- Clearly defined *alternation of sporophyte and gametophyte generations*.

Seedless Plants:

Spore producing plants, includes:

Phylum Filicinophyta: Ferns
Phylum Sphenophyta: Horsetails
Phylum Lycophyta: Club mosses
Species diversity: 13 000 +

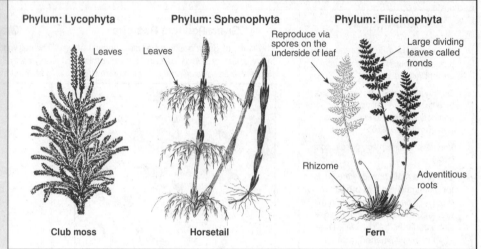

Phylum: Lycophyta

Leaves

Club moss

Phylum: Sphenophyta

Leaves

Horsetail

Phylum: Filicinophyta

Reproduce via spores on the underside of leaf

Large dividing leaves called fronds

Rhizome

Adventitious roots

Fern

Seed Plants:

Also called Spermatophyta. Produce seeds housing an embryo. Includes:

Gymnosperms

- Lack enclosed chambers in which seeds develop.
- Produce seeds in cones which are exposed to the environment.

Phylum Cycadophyta: Cycads
Phylum Ginkgophyta: Ginkgoes
Phylum Coniferophyta: Conifers
Species diversity: 730 +

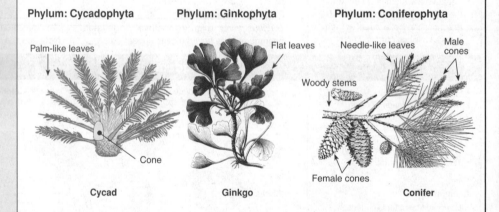

Phylum: Cycadophyta

Palm-like leaves

Cone

Cycad

Phylum: Ginkophyta

Flat leaves

Ginkgo

Phylum: Coniferophyta

Needle-like leaves

Male cones

Woody stems

Female cones

Conifer

Angiosperms

Phylum: Angiospermophyta

- Seeds in specialised reproductive structures called flowers.
- Female reproductive ovary develops into a fruit.
- Pollination usually via wind or animals.

Species diversity: 260 000 +

The phylum Angiospermophyta may be subdivided into two classes:

Class *Monocotyledoneae* (Monocots)
Class *Dicotyledoneae* (Dicots)

Angiosperms: **Monocotyledons**

Flower parts occur in multiples of 3

Leaves have parallel veins

- Only have one cotyledon (food storage organ)
- Normally herbaceous (non-woody) with no secondary growth

Lily

Examples: cereals, lilies, daffodils, palms, grasses.

Angiosperms: **Dicotyledons**

Leaves have branching veins

- Flower parts occur in multiples of 4 or 5
- Possible to have secondary growth (either herbaceous or woody)

Have two cotyledons inside the seed (acorn)

Oak

Examples: many annual plants, trees and shrubs.

Kingdom: ANIMALIA

- Over 800 000 species described in 33 existing phyla.
- Multicellular, heterotrophic organisms.
- Animal cells lack cell walls.

- Further subdivided into various major phyla on the basis of body symmetry, type of body cavity, and external and internal structures.

Phylum: Rotifera

- A diverse group of small organisms with sessile, colonial, and planktonic forms.
- Most freshwater, a few marine.
- Typically reproduce via cyclic parthenogenesis.
- Characterised by a wheel of cilia on the head used for feeding and locomotion, a large muscular pharynx (mastax) with jaw like trophi, and a foot with sticky toes.

Species diversity: 1500 +

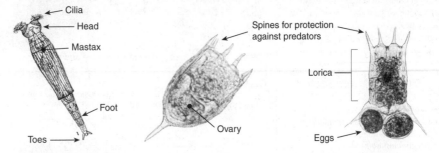

Bdelloid: non planktonic, creeping rotifer

Planktonic forms swim using their crown of cilia

Phylum: Porifera

- Lack organs.
- All are aquatic (mostly marine).
- Asexual reproduction by budding.
- Lack a nervous system.

Examples: sponges.

Species diversity: 8000 +

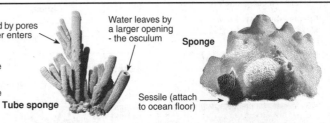

- Capable of regeneration (the replacement of lost parts)
- Possess spicules (needle-like internal structures) for support and protection

Phylum: Cnidaria

- Two basic body forms:
  Medusa: umbrella shaped and free swimming by pulsating bell.
  Polyp: cylindrical, some are sedentary, others can glide, or somersault or use tentacles as legs.
- Some species have a life cycle that alternates between a polyp stage and a medusa stage.
- All are aquatic (most are marine).

Examples: Jellyfish, sea anemones, hydras, and corals.

Species diversity: 11 000 +

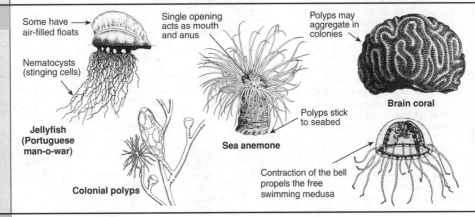

Phylum: Platyhelminthes

- Unsegmented body.
- Flattened body shape.
- Mouth, but no anus.
- Many are parasitic.

Examples: Tapeworms, planarians, flukes.

Species diversity: 20 000 +

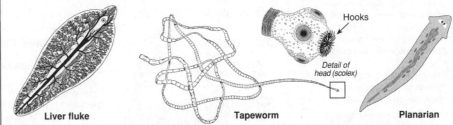

Liver fluke **Tapeworm** **Planarian**

Phylum: Nematoda

- Tiny, unsegmented roundworms.
- Many are plant/animal parasites

Examples: Hookworms, stomach worms, lung worms, filarial worms

Species diversity: 80 000 - 1 million

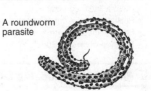

A general nematode body plan

Phylum: Annelida

- Cylindrical, segmented body with chaetae (bristles).
- Move using hydrostatic skeleton and/or parapodia (appendages).

Examples: Earthworms, leeches, polychaetes (including tubeworms).

Species diversity: 15 000 +

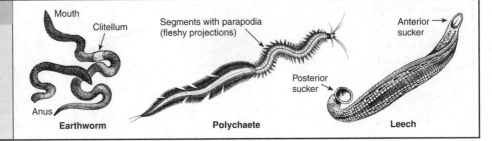

Earthworm **Polychaete** **Leech**

Kingdom: ANIMALIA (continued)

Phylum: Mollusca

- Soft bodied and unsegmented.
- Body comprises head, muscular foot, and visceral mass (organs).
- Most have radula (rasping tongue).
- Aquatic and terrestrial species.
- Aquatic species possess gills.

Examples: Snails, mussels, squid.
Species diversity: 110 000 +

Class: Bivalvia

Radula lost in bivalves

Mantle secretes shell

Muscular foot for locomotion

Two shells hinged together

Scallop

Class: Gastropoda

Mantle secretes shell

Tentacles with eyes

Head

Land snail

Class: Cephalopoda

Well developed eyes

Squid

Foot divided into tentacles

Phylum: Arthropoda

- Exoskeleton made of chitin.
- Grow in stages after moulting.
- Jointed appendages.
- Segmented bodies.
- Heart found on dorsal side of body.
- Open circulation system.
- Most have compound eyes.

Species diversity: 1 million +
Make up 75% of all living animals.

Arthropods are subdivided into the following classes:

Class: Crustacea (crustaceans)
- Mainly marine.
- Exoskeleton impregnated with mineral salts.
- Gills often present.
- Includes: Lobsters, crabs, barnacles, prawns, shrimps, isopods, amphipods
- **Species diversity:** 35 000 +

Class: Arachnida (chelicerates)
- Almost all are terrestrial.
- 2 body parts: cephalothorax and abdomen (except horseshoe crabs).
- Includes: spiders, scorpions, ticks, mites, horseshoe crabs.
- **Species diversity:** 57 000 +

Class: Insecta (insects)
- Mostly terrestrial.
- Most are capable of flight.
- 3 body parts: head, thorax, abdomen.
- Include: Locusts, dragonflies, cockroaches, butterflies, bees, ants, beetles, bugs, flies, and more
- **Species diversity:** 800 000 +

Myriapods (=many legs)
Class Diplopoda (millipedes)
- Terrestrial.
- Have a rounded body.
- Eat dead or living plants.
- **Species diversity:** 2000 +

Class Chilopoda (centipedes)
- Terrestrial.
- Have a flattened body.
- Poison claws for catching prey.
- Feed on insects, worms, and snails.
- **Species diversity:** 7000 +

Class: Crustacea

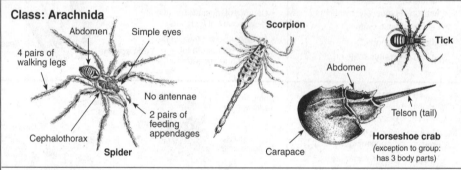

2 pairs of antennae

Cephalothorax (fusion of head and thorax)

Abdomen

Crab

3 pairs of mouthparts

Cheliped (first leg)

Shrimp

Walking legs

Swimmerets

Amphipod

Class: Arachnida

Abdomen

Simple eyes

Scorpion

Tick

4 pairs of walking legs

Abdomen

No antennae

2 pairs of feeding appendages

Telson (tail)

Cephalothorax

Spider

Carapace

Horseshoe crab
(exception to group: has 3 body parts)

Class: Insecta

1 pair of antennae

1 pair of compound eyes

Locust

Butterfly

Head

Thorax

Abdomen

2 pairs of wings

3 pairs of legs   **Honey bee**

Beetles are the largest group within the animal kingdom with more than 300 000 species.

Beetle

Class: Diplopoda

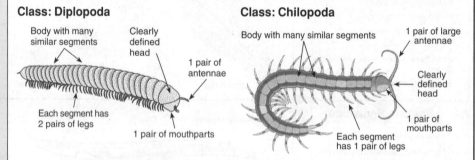

Body with many similar segments

Clearly defined head

1 pair of antennae

Each segment has 2 pairs of legs

1 pair of mouthparts

Class: Chilopoda

Body with many similar segments

1 pair of large antennae

Clearly defined head

1 pair of mouthparts

Each segment has 1 pair of legs

Phylum: Echinodermata

- Rigid body wall, internal skeleton made of calcareous plates.
- Many possess spines.
- Ventral mouth, dorsal anus.
- External fertilisation.
- Unsegmented, marine organisms.
- Tube feet for locomotion.
- Water vascular system.

Examples: Starfish, brittlestars, feather stars, sea urchins, sea lilies.
Species diversity: 6000 +

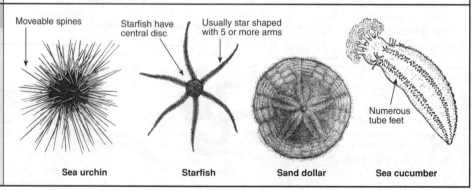

Moveable spines

Starfish have central disc

Usually star shaped with 5 or more arms

Numerous tube feet

Sea urchin

Starfish

Sand dollar

Sea cucumber

Kingdom: ANIMALIA (continued)

Phylum: Chordata

- Dorsal notochord (flexible, supporting rod) present at some stage in the life history.
- Post-anal tail present at some stage in their development.
- Dorsal, tubular nerve cord.
- Pharyngeal slits present.
- Circulation system closed in most.
- Heart positioned on ventral side.

Species diversity: 48 000 +

- A very diverse group with several sub-phyla:
  - Urochordata (sea squirts, salps)
  - Cephalochordata (lancelet)
  - Craniata (vertebrates)

Sub-Phylum Craniata (vertebrates)
- Internal skeleton of cartilage or bone.
- Well developed nervous system.
- Vertebral column replaces notochord.
- Two pairs of appendages (fins or limbs) attached to girdles.

Further subdivided into:

Class: Chondrichthyes (cartilaginous fish)
- Skeleton of cartilage (not bone).
- No swim bladder.
- All aquatic (mostly marine).
- Include: Sharks, rays, and skates.

Species diversity: 850 +

Class: Osteichthyes (bony fish)
- Swim bladder present.
- All aquatic (marine and fresh water).

Species diversity: 21 000 +

Class: Amphibia (amphibians)
- Lungs in adult, juveniles may have gills (retained in some adults).
- Gas exchange also through skin.
- Aquatic and terrestrial (limited to damp environments).
- Include: Frogs, toads, salamanders, and newts.

Species diversity: 3900 +

Class Reptilia (reptiles)
- Ectotherms with no larval stages.
- Teeth are all the same type.
- Eggs with soft leathery shell.
- Mostly terrestrial.
- Include: Snakes, lizards, crocodiles, turtles, and tortoises.

Species diversity: 7000 +

Class: Aves (birds)
- Terrestrial endotherms.
- Eggs with hard, calcareous shell.
- Strong, light skeleton.
- High metabolic rate.
- Gas exchange assisted by air sacs.

Species diversity: 8600 +

Class: Mammalia (mammals)
- Endotherms with hair or fur.
- Mammary glands produce milk.
- Glandular skin with hair or fur.
- External ear present.
- Teeth are of different types.
- Diaphragm between thorax/abdomen.

Species diversity: 4500 +
Subdivided into three subclasses:
Monotremes, marsupials, placentals.

Class: Chondrichthyes (cartilaginous fish)

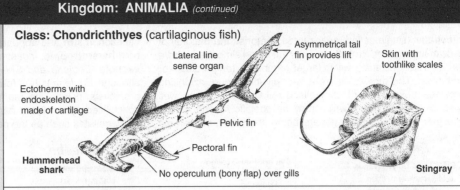

- Lateral line sense organ
- Asymmetrical tail fin provides lift
- Skin with toothlike scales
- Ectotherms with endoskeleton made of cartilage
- Pelvic fin
- Pectoral fin
- No operculum (bony flap) over gills

Hammerhead shark

Stingray

Class: Osteichthyes (bony fish)

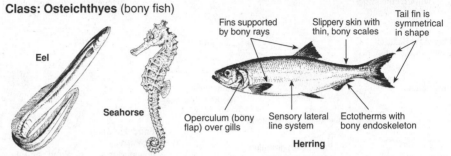

- Fins supported by bony rays
- Slippery skin with thin, bony scales
- Tail fin is symmetrical in shape

Eel

Seahorse

- Operculum (bony flap) over gills
- Sensory lateral line system
- Ectotherms with bony endoskeleton

Herring

Class: Amphibia

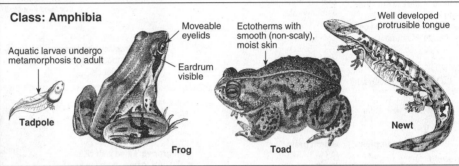

- Aquatic larvae undergo metamorphosis to adult
- Moveable eyelids
- Ectotherms with smooth (non-scaly), moist skin
- Well developed protrusible tongue
- Eardrum visible

Tadpole

Frog

Toad

Newt

Class: Reptilia

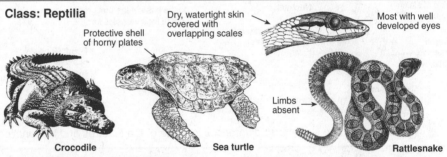

- Protective shell of horny plates
- Dry, watertight skin covered with overlapping scales
- Most with well developed eyes
- Limbs absent

Crocodile

Sea turtle

Rattlesnake

Class: Aves

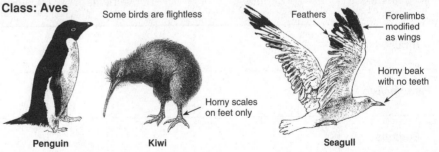

- Some birds are flightless
- Feathers
- Forelimbs modified as wings
- Horny scales on feet only
- Horny beak with no teeth

Penguin

Kiwi

Seagull

Class: Mammalia

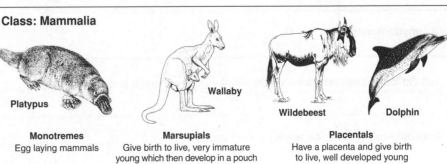

Platypus

Wallaby

Wildebeest

Dolphin

Monotremes
Egg laying mammals

Marsupials
Give birth to live, very immature young which then develop in a pouch

Placentals
Have a placenta and give birth to live, well developed young

Unicellular Eukaryotes

Unicellular (single-celled) **eukaryotes** comprise the majority of the diverse kingdom, Protoctista. They are found almost anywhere there is water, including within larger organisms (as parasites or symbionts). The protoctists are a very diverse group, exhibiting some features typical of generalised eukaryotic cells, as well as specialised features, which may be specific to one genus. Note that even within the genera below there is considerable variation in size and appearance. *Amoeba* and *Paramecium* are both **heterotrophic**, ingesting food, which accumulates inside a **vacuole**. *Euglena* and *Chlamydomonas* are autotrophic algae, although *Euglena* is heterotrophic when deprived of light. Other protoctists include the marine foraminiferans and radiolarians, specialised intracellular parasites such as *Plasmodium*, and zooflagellates such as the parasites *Trypanosoma* and *Giardia*.

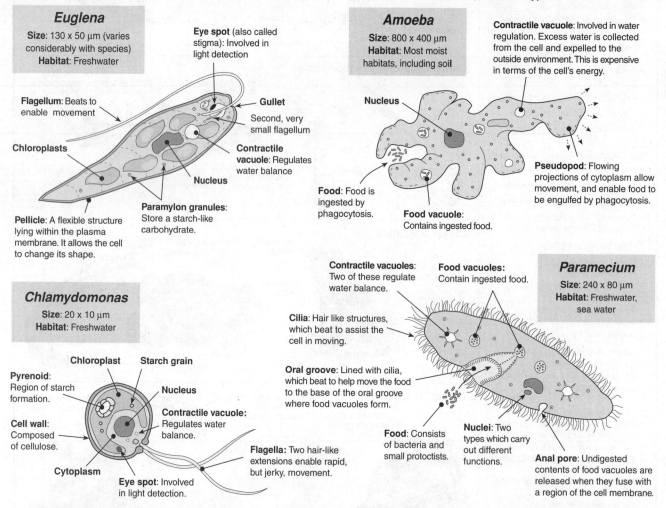

Euglena

Size: 130 x 50 μm (varies considerably with species)
Habitat: Freshwater

Eye spot (also called stigma): Involved in light detection

Flagellum: Beats to enable movement

Gullet Second, very small flagellum

Chloroplasts

Contractile vacuole: Regulates water balance

Nucleus

Pellicle: A flexible structure lying within the plasma membrane. It allows the cell to change its shape.

Paramylon granules: Store a starch-like carbohydrate.

Amoeba

Size: 800 x 400 μm
Habitat: Most moist habitats, including soil

Contractile vacuole: Involved in water regulation. Excess water is collected from the cell and expelled to the outside environment. This is expensive in terms of the cell's energy.

Nucleus

Pseudopod: Flowing projections of cytoplasm allow movement, and enable food to be engulfed by phagocytosis.

Food: Food is ingested by phagocytosis.

Food vacuole: Contains ingested food.

Chlamydomonas

Size: 20 x 10 μm
Habitat: Freshwater

Chloroplast **Starch grain**

Pyrenoid: Region of starch formation.

Nucleus

Cell wall: Composed of cellulose.

Contractile vacuole: Regulates water balance.

Cytoplasm

Eye spot: Involved in light detection.

Flagella: Two hair-like extensions enable rapid, but jerky, movement.

Contractile vacuoles: Two of these regulate water balance.

Food vacuoles: Contain ingested food.

Paramecium

Size: 240 x 80 μm
Habitat: Freshwater, sea water

Cilia: Hair like structures, which beat to assist the cell in moving.

Oral groove: Lined with cilia, which beat to help move the food to the base of the oral groove where food vacuoles form.

Food: Consists of bacteria and small protoctists.

Nuclei: Two types which carry out different functions.

Anal pore: Undigested contents of food vacuoles are released when they fuse with a region of the cell membrane.

1. Fill in the table below to summarise differences in some of the features and life functions of the protoctists shown above:

Organism	Nutrition	Movement	Osmoregulation	Eye spot present / absent	Cell wall present / absent
Amoeba					
Paramecium					
Euglena					
Chlamydomonas					

2. List the four organisms shown above in order of size (largest first): _____

_____

3. Suggest why an autotroph would have an eye spot: _____

_____

Related activities: Cell Structures and Organelles, Types of Cells

Classification System

The classification of organisms is designed to reflect how they are related to each other. The fundamental unit of classification of living things is the **species**. Its members are so alike genetically that they can interbreed. This genetic similarity also means that they are almost identical in their physical and other characteristics. Species are classified further into larger, more comprehensive categories (higher taxa). It must be emphasised that all such higher classifications are human inventions to suit a particular purpose.

1. The table below shows part of the classification for humans using the seven major levels of classification. For this question, use the example of the classification of the European hedgehog, on the next page, as a guide.

   (a) Complete the list of the classification levels on the left hand side of the table below:

	Classification level	Human classification
1.	_____	_____
2.	_____	_____
3.	_____	_____
4.	_____	_____
5.	Family	Hominidae
6.	_____	_____
7.	_____	_____

   (b) The name of the Family that humans belong to has already been entered into the space provided. Complete the classification for humans (*Homo sapiens*) on the table above.

2. Describe the two-part scientific naming system (called the **binomial system**) that is used to name organisms:

   _____

   _____

3. Give two reasons why the classification of organisms is important:

   (a) _____

   _____

   (b) _____

   _____

4. Traditionally, the classification of organisms has been based largely on similarities in physical appearance. More recently, new methods involving biochemical comparisons have been used to provide new insights into how species are related. Describe an example of a biochemical method for comparing how species are related:

   _____

   _____

   _____

5. As an example of physical features being used to classify organisms, mammals have been divided into three major sub-classes: monotremes, marsupials, and placentals. Describe the main physical feature distinguishing each of these taxa:

   (a) Monotreme: _____

   _____

   (b) Marsupial: _____

   _____

   (c) Placental: _____

   _____

Periodicals: A passion for order

Related activities: New Classification Schemes, Classification Keys, Features of Taxonomic Groups

Classification of the European Hedgehog

Below is the classification for the **European hedgehog**. Only one of each group is subdivided in this chart showing the levels that can be used in classifying an organism. Not all possible subdivisions have been shown here. For example, it is possible to indicate such categories as **super-class** and **sub-family**. The only natural category is the **species**, often separated into geographical **races**, or **sub-species**, which generally differ in appearance.

Kingdom: **Animalia**
Animals: one of five kingdoms

Phylum: **Chordata**
Animals with a notochord (supporting rod of cells along the upper surface).
tunicates, salps, lancelets, and vertebrates

23 other phyla

Sub-phylum: **Vertebrata**
Animals with backbones.
fish, amphibians, reptiles, birds, mammals

Class: **Mammalia**
Animals that suckle their young on milk from mammary glands.
placentals, marsupials, monotremes

Sub-class: **Eutheria**
Mammals whose young develop for some time in the female's reproductive tract gaining nourishment from a placenta.
placentals

Order: **Insectivora**
Insect eating mammals.
An order of over 300 species of primitive, small mammals that feed mainly on insects and other small invertebrates.

17 other orders

Sub-order: **Erinaceomorpha**
The hedgehog-type insectivores. One of the three suborders of insectivores. The other suborders include the tenrec-like insectivores (*tenrecs and golden moles*) and the shrew-like insectivores (*shrews, moles, desmans, and solenodons*).

Family: **Erinaceidae**
The only family within this suborder. Comprises two subfamilies: the true or spiny hedgehogs and the moonrats (gymnures). Representatives in the family include the desert hedgehog, long-eared hedgehog, and the greater and lesser moonrats.

Genus: *Erinaceus*
One of eight genera in this family. The genus *Erinaceus* includes four Eurasian species and another three in Africa.

7 other genera

Species: *europaeus*
The European hedgehog. Among the largest of the spiny hedgehogs. Characterised by a dense covering of spines on the back, the presence of a big toe (hallux) and 36 teeth.

6 other species

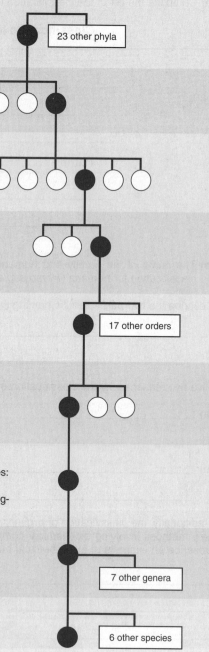

The order *Insectivora* was first introduced to group together shrews, moles, and hedgehogs. It was later extended to include tenrecs, golden moles, desmans, tree shrews, and elephant shrews, and the taxonomy of the group became very confused. Recent reclassification of the elephant shrews and tree shrews into their own separate orders has made the Insectivora a more cohesive group taxonomically.

European hedgehog
Erinaceus europaeus

Features of the Five Kingdoms

The classification of organisms into taxonomic groups is based on how biologists believe they are related in an evolutionary sense. Organisms in a taxonomic group share features which set them apart from other groups. By identifying these features, it is possible to develop an understanding of the evolutionary history of the group. The focus of this activity is to summarise the **distinguishing features** of each of the five kingdoms in the five kingdom classification system.

1. Distinguishing features of Kingdom **Prokaryotae**:

   _____

   _____

   _____

   _____

   _____

2. Distinguishing features of Kingdom **Protoctista**:

   _____

   _____

   _____

   _____

   _____

3. Distinguishing features of Kingdom **Fungi**:

   _____

   _____

   _____

   _____

   _____

4. Distinguishing features of Kingdom **Plantae**:

   _____

   _____

   _____

   _____

   _____

5. Distinguishing features of Kingdom **Animalia**:

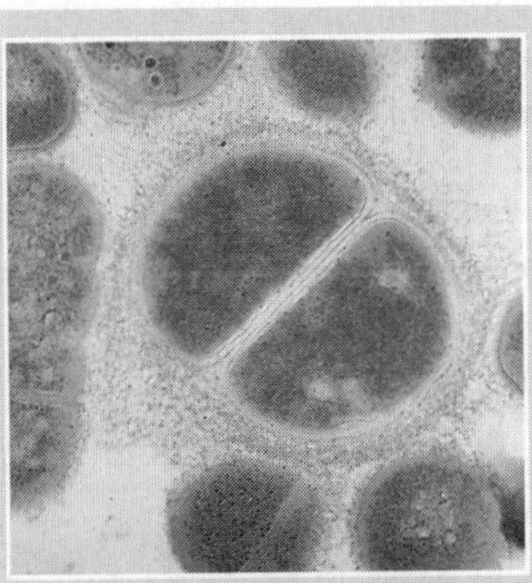

   _____

   _____

   _____

   _____

   _____

Staphylococcus dividing

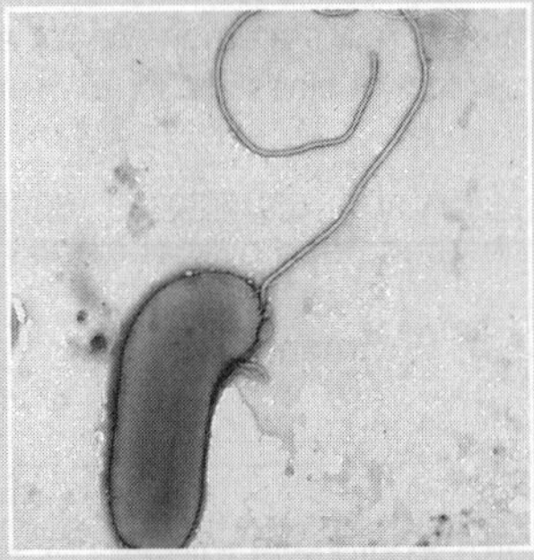

Helicobacter pylori

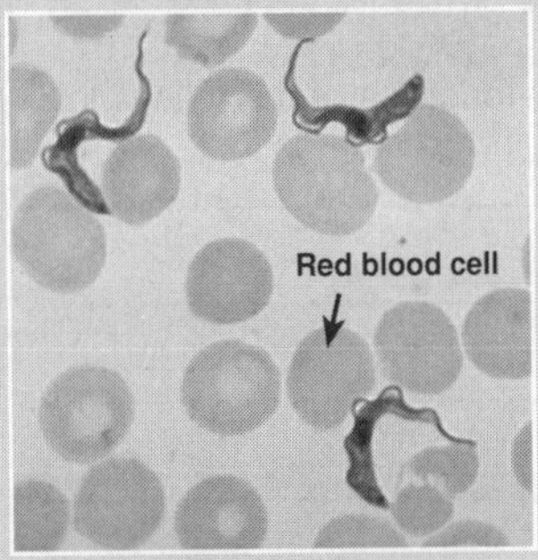

Red blood cell

Trypanosoma parasite

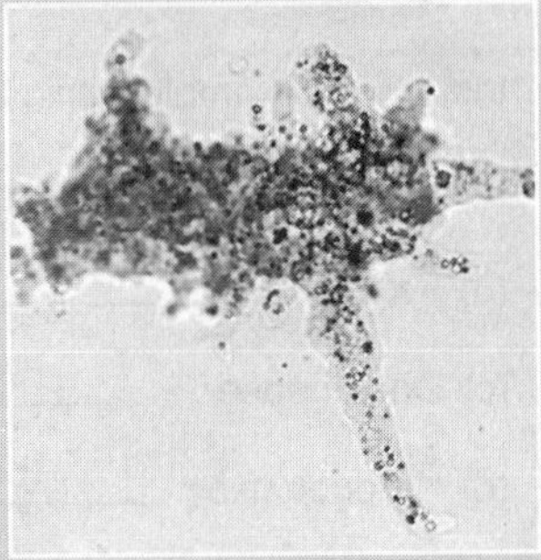

Amoeba

Mushrooms

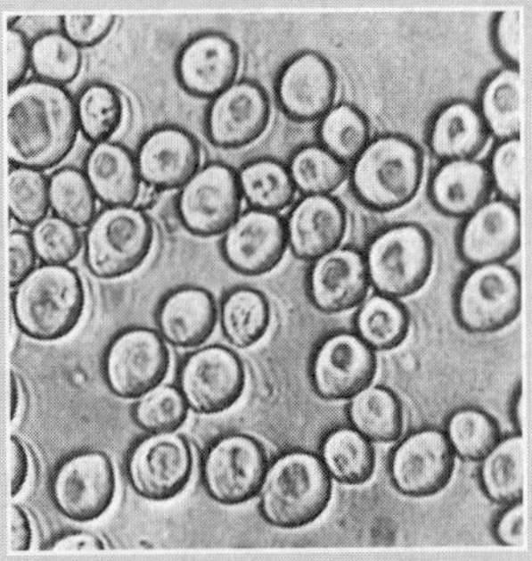

Yeast cells in solution

Maple seeds

Pea plants

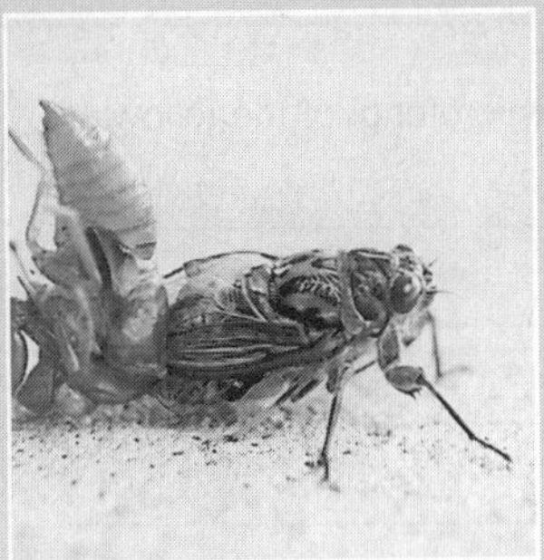

Cicada moulting

Gibbon

Features of Microbial Groups

A microorganism (or microbe) is literally a microscopic organism. The term is usually reserved for the organisms studied in microbiology: bacteria, fungi, microscopic protoctistans, and viruses. The first three of these represent three of the five kingdoms for which you described distinguishing features in an earlier activity (viruses are non-cellular and therefore not included in the five-kingdom classification). Most microbial taxa, but particularly the fungi, also have macroscopic representatives. The distinction between a macrofungus and a microfungus is an artificial but convenient one. Unlike microfungi, which are made conspicuous by the diseases or decay they cause, macrofungi are most likely to be observed with the naked eye. The microfungi include yeasts and pathogenic species. Macrofungi, e.g. mushrooms, toadstools, and lichens, are illustrated in *Features of Macrofungi and Plants*.

1. Describe aspects of each of the following for the bacteria and cyanobacteria (Kingdom Prokaryotae):

   (a) Environmental range: _____

   _____

   (b) Ecological role: _____

   _____

2. Identify an example within the bacteria of the following:

   (a) Photosynthetic: _____

   (b) Pathogen: _____

   (c) Decomposer: _____

   (d) Nitrogen fixer: _____

3. Describe aspects of each of the following for the microscopic protoctistans (Kingdom Protoctista):

   (a) Environmental range: _____

   _____

   (b) Ecological role: _____

   _____

4. Identify an example in the protoctists of the following:

   (a) Photosynthetic: _____

   (b) Pathogen: _____

   (c) Biological indicator: _____

5. Describe aspects of each of the following for the microfungi (Kingdom Fungi):

   (a) Environmental range: _____

   _____

   (b) Ecological role: _____

   _____

6. Identify examples within the microfungi of the following:

   (a) Animal pathogen: _____

   _____

   (b) Plant pathogens: _____

   _____

   _____

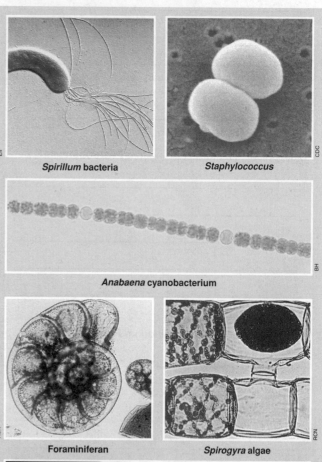

Spirillum bacteria   *Staphylococcus*

Anabaena cyanobacterium

Foraminiferan   *Spirogyra* algae

Diatoms: *Pleurosigma*

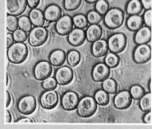

Curvularia sp. conidiophore   Yeast cells in solution

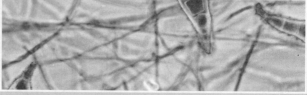

Microsporum distortum (a pathogenic fungus)

Related activities: The New Tree of Life, Features of Taxonomic Groups, Features of Macrofungi and Plants

Features of Animal Taxa

The animal kingdom is classified into about 35 major **phyla**. Representatives of the more familiar taxa are illustrated below: **cnidarians** (includes jellyfish, sea anemones, and corals), **annelids** (segmented worms), **arthropods** (insects, crustaceans, spiders, scorpions, centipedes and millipedes), **molluscs** (snails, bivalve shellfish, squid and octopus), **echinoderms** (starfish and sea urchins), **vertebrates** from the phylum **chordates** (fish, amphibians, reptiles, birds, and mammals). The **arthropods** and the **vertebrates** have been represented in more detail, giving the **classes** for each of these **phyla**. This activity asks you to describe the **distinguishing features** of each of the taxa represented below.

Sea anemones

Jellyfish

Tubeworms

Earthworm

Long-horned beetle

Butterfly

Crab

Woodlouse

Scorpion

Spider

Centipede

Millipede

1. **Cnidarian** features: _____

_____

_____

_____

_____

2. **Annelid** features: _____

_____

_____

_____

_____

3. **Insect** features: _____

_____

_____

_____

_____

4. **Crustacean** features: _____

_____

_____

_____

_____

5. **Arachnid** features: _____

_____

_____

_____

_____

6. **Myriapod** (class Chilopoda and Diplopoda) features: _____

_____

_____

_____

_____

Nautilus

Abalone

Sea urchin

Starfish

Grouper

Shark

Frog

Salamander

Iguana

Rattlesnake

Penguin

Pelican

Horse

Bear

7. **Mollusc** features: _____
_____
_____
_____
_____

8. **Echinoderm** features: _____
_____
_____
_____
_____
_____

9. **Fish** features: _____
_____
_____
_____
_____
_____

10. **Amphibian** features: _____
_____
_____
_____
_____
_____

11. **Reptile** features: _____
_____
_____
_____
_____
_____

12. **Bird** features: _____
_____
_____
_____
_____
_____

13. **Mammal** features: _____
_____
_____
_____
_____
_____

Features of Macrofungi and Plants

Although plants and fungi are some of the most familiar organisms in our environment, their classification has not always been straightforward. We know now that the plant kingdom is monophyletic, meaning that it is derived from a common ancestor. The variety we see in plant taxa today is a result of their enormous diversification from the first plants. Although the fungi were once grouped together with the plants, they are unique organisms that differ from other eukaryotes in their mode of nutrition, structural organisation, growth, and reproduction. The focus of this activity is to summarise the features of the fungal kingdom, the major divisions of the plant kingdom, and the two classes of flowering plants (angiosperms).

Lichen

Bracket fungus

Liverwort

Moss

Fern frond

Ground fern

Pine tree cone

Cycad

Coconut palms

Wheat plants

Deciduous tree

Flowering plant

1. **Macrofungi** features:

_____

_____

_____

_____

2. **Moss and liverwort** features:

_____

_____

_____

_____

3. **Fern** features:

_____

_____

_____

_____

4. **Gymnosperm** features:

_____

_____

_____

_____

5. **Monocot angiosperm** features:

_____

_____

_____

_____

6. **Dicot angiosperm** features:

_____

_____

_____

_____

Periodicals:
World flowers bloom
after recount

Related activities: The New Tree of Life,
Features of Taxonomic Groups

R 1

The Classification of Life

For this activity, cut away the two pages of diagrams that follow from your book. The five kingdoms that all living things are grouped into, are listed on this page and the following page.

1. Cut out all of the images of different living organisms (cut around each shape closely, taking care to include their names).

2. Sort them into their classification groups by placing them into the spaces provided on this and the following page.

3. To fix the images in place, first use a temporary method, so that you can easily reposition them if you need to. Make a permanent fixture when you are completely satisfied with your placements on the page.

Kingdom Prokaryotae (Monera)

Kingdom Protoctista

Kingdom Fungi

Kingdom Plantae

Phylum Bryophyta

Phylum Filicinophyta

Phylum Angiospermophyta

Class Monocotyledoneae   Class Dicotyledoneae

Phylum Cycadophyta

Phylum Coniferophyta

Related activities: Features of Taxonomic Groups

Cut out the organisms on this page and paste them into the spaces provided at the start of this activity. Organisms are not to scale.

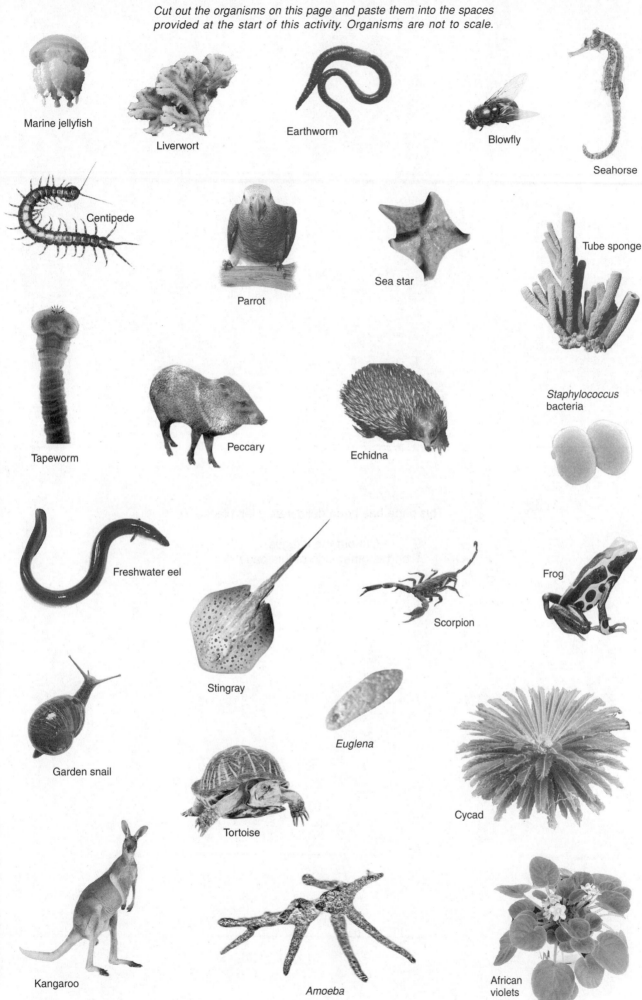

Marine jellyfish

Liverwort

Earthworm

Blowfly

Seahorse

Centipede

Parrot

Sea star

Tube sponge

Tapeworm

Peccary

Echidna

Staphylococcus bacteria

Freshwater eel

Stingray

Scorpion

Frog

Garden snail

Euglena

Cycad

Tortoise

Kangaroo

Amoeba

African violets

This page has been deliberately left blank

Cut out the images
on the other side of this page

Cut out the organisms on this page and paste them into the spaces provided at the start of this activity. Organisms are not to scale.

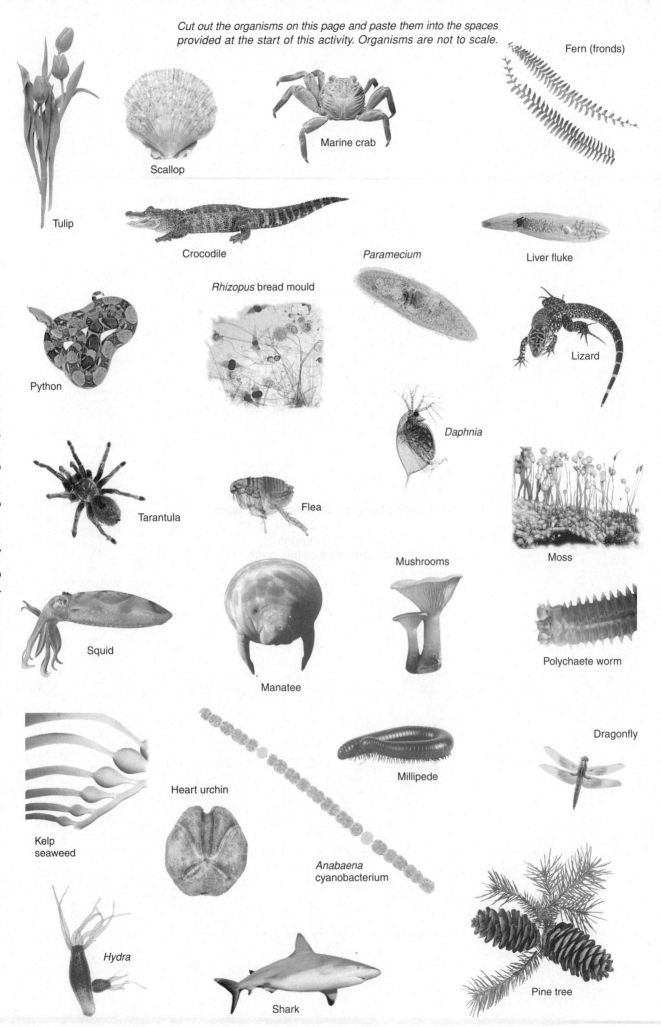

Fern (fronds)

Scallop

Marine crab

Tulip

Crocodile

Paramecium

Liver fluke

Rhizopus bread mould

Python

Lizard

Daphnia

Tarantula

Flea

Mushrooms

Moss

Squid

Manatee

Polychaete worm

Kelp seaweed

Heart urchin

Anabaena cyanobacterium

Millipede

Dragonfly

Hydra

Shark

Pine tree

This page has been deliberately left blank

Cut out the images
on the other side of this page

Kingdom Animalia

Phylum Porifera	Phylum Cnidaria	Phylum Platyhelminthes	Phylum Annelida

Phylum Mollusca

Class Gastropoda   Class Bivalvia   Class Cephalopoda

Phylum Echinodermata

Phylum Arthropoda
Superclass Crustacea   Classes Chilopoda/Diplopoda   Class Arachnida   Class Insecta

Class Chondrichthyes   **Phylum Chordata**   Class Osteichthyes   Class Amphibia

Class Reptilia
Order Squamata   Order Crocodilia   Order Chelonia   Class Aves

Class Mammalia
Subclass Prototheria   Subclass Metatheria   Subclass Eutheria

The New Tree of Life

Taxonomy is the science of classification and, like all science, it is a dynamic field, subject to change as new information comes to light. With the advent of DNA sequencing technology, scientists began to analyse the genomes of many bacteria. In 1996, the results of a scientific collaboration examining DNA evidence confirmed the proposal that life comprises three major evolutionary lineages (domains) and not two as was the convention. The recognised lineages were the **Eubacteria**, the **Eukarya** and the **Archaea** (formerly the Archaebacteria). The new classification reflects the fact that there are very large differences between the Archaea and the Eubacteria. All three domains probably had a distant common ancestor.

A Five (or Six) Kingdom World (right)

The diagram (right) represents the **five kingdom system** of classification commonly represented in many biology texts. It recognises two basic cell types: prokaryote and eukaryote. The domain Prokaryota includes all bacteria and cyanobacteria. Domain Eukaryota includes protoctists, fungi, plants, and animals. More recently, based on 16S ribosomal RNA sequence comparisons, Carl Woese divided the prokaryotes into two kingdoms, the Eubacteria and Archaebacteria. Such **six-kingdom systems** are also commonly recognised in texts.

A New View of the World (below)

In 1996, scientists deciphered the full DNA sequence of an unusual bacterium called *Methanococcus jannaschii*. An **extremophile**, this methane-producing archaebacterium lives at 85°C; a temperature lethal for most bacteria as well as eukaryotes. The DNA sequence confirmed that life consists of three major evolutionary lineages, not the two that have been routinely described. Only 44% of this archaebacterium's genes resemble those in bacteria or eukaryotes, or both.

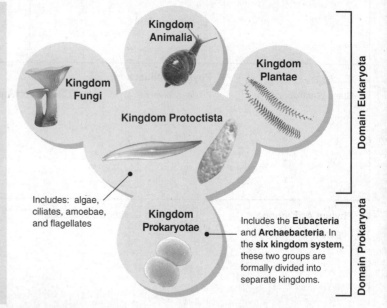

Includes: algae, ciliates, amoebae, and flagellates

Includes the **Eubacteria** and **Archaebacteria**. In the **six kingdom system**, these two groups are formally divided into separate kingdoms.

Domain Eubacteria

Lack a distinct nucleus and cell organelles. Generally prefer less extreme environments than Archaea. Includes well-known pathogens, many harmless and beneficial species, and the cyanobacteria (photosynthetic bacteria containing the pigments chlorophyll *a* and phycocyanin).

Domain Archaea

Closely resemble eubacteria in many ways but cell wall composition and aspects of metabolism are very different. Live in extreme environments similar to those on primeval Earth. They may utilise sulfur, methane, or halogens (chlorine, fluorine), and many tolerate extremes of temperature, salinity, or pH.

Domain Eukarya

Complex cell structure with organelles and nucleus. This group contains four of the kingdoms classified under the more traditional system. Note that Kingdom Protoctista is separated into distinct groups: e.g. amoebae, ciliates, flagellates.

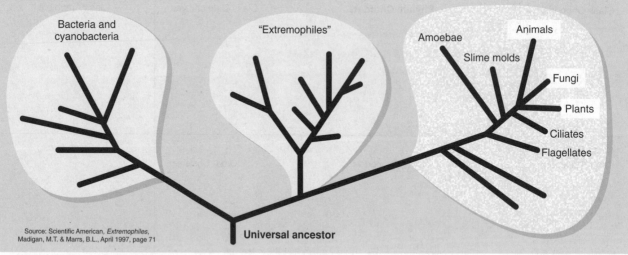

Source: Scientific American, *Extremophiles*, Madigan, M.T. & Marrs, B.L., April 1997, page 71

1. Explain why some scientists have recommended that the conventional classification of life be revised so that the Archaea, Eubacteria and Eukarya are three separate domains:

_____

_____

2. Describe one feature of the three domain system that is very different from the five kingdom classification:

_____

_____

3. Describe one way in which the three domain system and the six kingdom classification are alike: _____

_____

© Biozone International 2008-2010

A 2

Related activities: *Features of Taxonomic Groups*
Web links: *Types of Microbes, Introduction to the Archaea*

Periodicals:
What is a species?

Phylogenetic Systematics

The aim of classification is to organise species in a way that most accurately reflects their evolutionary history (**phylogeny**). Each successive group in the taxonomic hierarchy should represent finer and finer branching from a common ancestor. Traditional classification systems emphasise morphological similarities in order to group species into genera and other higher level taxa. In contrast, **cladistic analysis** relies on **shared derived characteristics** (**synapomorphies**), and emphasises features that are the result of shared ancestry (homologies), rather than

convergent evolution. Technology has assisted taxonomy by providing biochemical evidence for the relatedness of species. Traditional and cladistic schemes do not necessarily conflict, but there have been reclassifications of some taxa (notably the primates, but also the reptiles, dinosaurs, and birds). Popular classifications will probably continue to reflect similarities and differences in appearance, rather than a strict evolutionary history. In this respect, they are a compromise between phylogeny and the need for a convenient filing system for species diversity.

Constructing a Simple Cladogram

A table listing the features for comparison allows us to identify where we should make branches in the **cladogram**. An outgroup (one which is known to have no or little relationship to the other organisms) is used as a basis for comparison.

Comparative features	Jawless fish (outgroup)	Bony fish	Amphibians	Lizards	Birds	Mammals
Vertebral column	✔	✔	✔	✔	✔	✔
Jaws	✘	✔	✔	✔	✔	✔
Four supporting limbs	✘	✘	✔	✔	✔	✔
Amniotic egg	✘	✘	✘	✔	✔	✔
Diapsid skull	✘	✘	✘	✔	✔	✘
Feathers	✘	✘	✘	✘	✔	✘
Hair	✘	✘	✘	✘	✘	✔

Taxa (column header above table)

The table above lists features shared by selected taxa. The outgroup (jawless fish) shares just one feature (vertebral column), so it gives a reference for comparison and the first branch of the cladogram (tree).

As the number of taxa in the table increases, the number of possible trees that could be drawn increases exponentially. To determine the most likely relationships, the rule of **parsimony** is used. This assumes that the tree with the least number of evolutionary events is most likely to show the correct evolutionary relationship.

Three possible cladograms are shown on the right. The top cladogram requires six events while the other two require seven events. Applying the rule of parsimony, the top cladogram must be taken as correct.

Parsimony can lead to some confusion. Some evolutionary events have occurred multiple times. An example is the evolution of the four chambered heart, which occurred separately in both birds and mammals. The use of fossil evidence and DNA analysis can help to solve problems like this.

Possible Cladograms

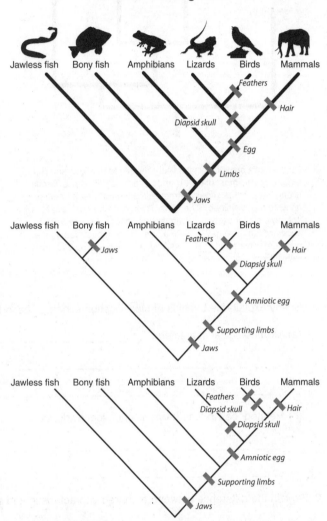

Using DNA Data

DNA analysis has allowed scientists to confirm many phylogenies and refute or redraw others. In a similar way to morphological differences, DNA sequences can be tabulated and analysed. The ancestry of whales has been in debate since Darwin. The radically different morphologies of whales and other mammals makes it difficult work out the correct phylogenetic tree. However recently discovered fossil ankle bones, as well as DNA studies, show whales are more closely related to hippopotami than to any other mammal. Coupled with molecular clocks, DNA data can also give the time between each split in the lineage.

The DNA sequences on the right show part of a the nucleotide subset 141-200 and some of the matching nucleotides used to draw the cladogram. Although whales were once thought most closely related to pigs, based on the DNA analysis the most parsimonious tree disputes this.

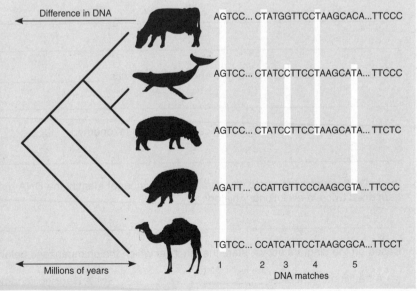

AGTCC... CTATGGTTCCTAAGCACA...TTCCC

AGTCC... CTATCCTTCCTAAGCATA... TTCCC

AGTCC... CTATCCTTCCTAAGCATA... TTCTC

AGATT... CCATTGTTCCCAAGCGTA...TTCCC

TGTCC... CCATCATTCCTAAGCGCA...TTCCT

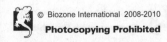
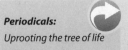

Periodicals: Uprooting the tree of life

Related activities: Classification System, The New Tree of Life

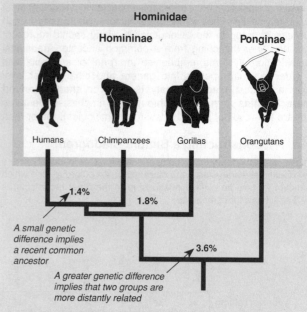

A Classical Taxonomic View

Hominidae | **Pongidae**
The 'great apes'

Humans | Chimpanzees | Gorillas | Orangutans

On the basis of overall anatomical similarity (e.g. bones and limb length, teeth, musculature), apes are grouped into a family (Pongidae) that is separate from humans and their immediate ancestors (Hominidae). The family Pongidae (the great apes) is not monophyletic (of one phylogeny), because it stems from an ancestor that also gave rise to a species in another family (i.e. humans). This traditional classification scheme is now at odds with schemes derived after considering genetic evidence.

A Cladistic View

Hominidae

Homininae | **Ponginae**

Humans | Chimpanzees | Gorillas | Orangutans

1.4%
A small genetic difference implies a recent common ancestor

1.8%

3.6%
A greater genetic difference implies that two groups are more distantly related

Based on the evidence of genetic differences (% values above), chimpanzees and gorillas are more closely related to humans than to orangutans, and chimpanzees are more closely related to humans than they are to gorillas. Under this scheme there is no true family of great apes. The family Hominidae includes two subfamilies: Ponginae and Homininae (humans, chimpanzees, and gorillas). This classification is monophyletic: the Hominidae includes all the species that arise from a common ancestor.

1. Briefly explain the benefits of classification schemes based on:

   (a) Morphological characters: _____

   _____

   _____

   (b) Relatedness in time (from biochemical evidence): _____

   _____

   _____

2. Explain the difference between a shared characteristic and a shared derived characteristic: _____

   _____

   _____

   _____

3. Explain how the rule of parsimony is applied to cladistics: _____

   _____

4. Describe the contribution of biochemical evidence to taxonomy: _____

   _____

5. In the DNA data for the whale cladogram (previous page) identify the DNA match that shows a mutation event must have happened twice in evolutionary history:

   _____

6. Based on the diagram above, state the family to which the chimpanzees belong under:

   (a) A traditional scheme: _____ (b) A cladistic scheme: _____

Classification Keys

Classification systems provide biologists with a way in which to identify species. They also indicate how closely related, in an evolutionary sense, each species is to others. An organism's classification should include a clear, unambiguous **description**, an accurate **diagram**, and its unique name, denoted by the **genus** and **species**. Classification keys are used to identify an organism and assign it to the correct species (assuming that the organism has already been formally classified and is included in the key) Typically, keys are **dichotomous** and involve a series of linked steps. At each step, a choice is made between two features; each alternative leads to another question until an identification is made. If the organism cannot be identified, it may be a new species or the key may need revision. Two examples of **dichotomous keys** are provided here. The first (below) describes features for identifying the larvae of various genera within the order Trichoptera (caddisflies). From this key you should be able to assign a generic name to each of the caddisfly larvae pictured. The key on the next page identifies aquatic insect orders.

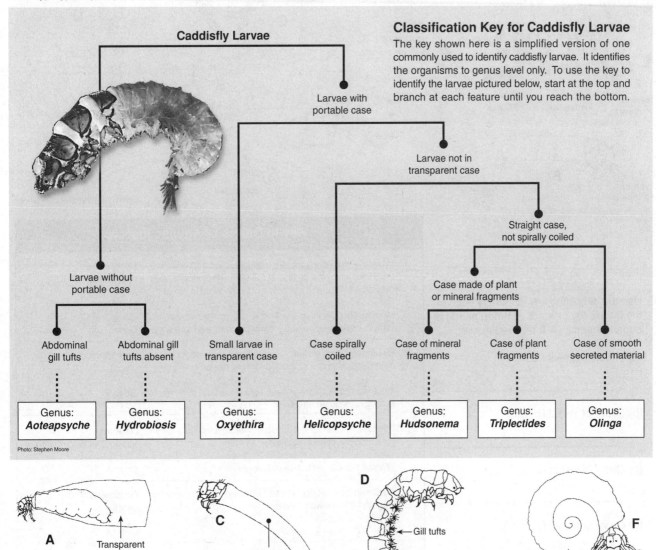

Classification Key for Caddisfly Larvae

The key shown here is a simplified version of one commonly used to identify caddisfly larvae. It identifies the organisms to genus level only. To use the key to identify the larvae pictured below, start at the top and branch at each feature until you reach the bottom.

Photo: Stephen Moore

1. Describe the main feature used to distinguish the genera in the key above: _____

_____

2. Use the key above to assign each of the caddisfly larvae (**A-G**) to its correct genus:

A: _____ D: _____ G: _____

B: _____ E: _____

C: _____ F: _____

Related activities: Keying Out Plant Species
Web links: What is the Key to Classification?

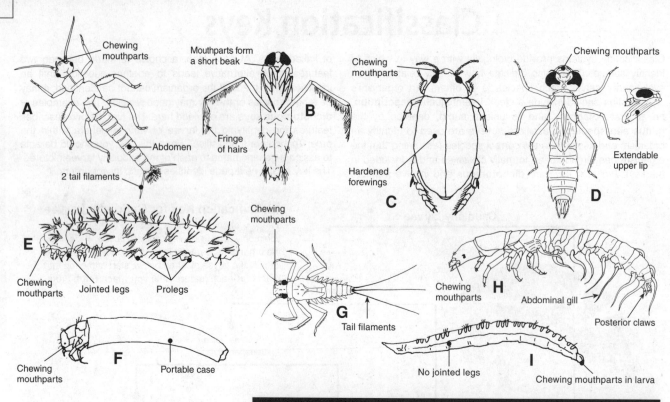

A — Chewing mouthparts — Abdomen — 2 tail filaments

B — Mouthparts form a short beak — Fringe of hairs

C — Chewing mouthparts — Hardened forewings

D — Chewing mouthparts — Extendable upper lip

E — Chewing mouthparts — Jointed legs — Prolegs

F — Chewing mouthparts — Portable case

G — Chewing mouthparts — Tail filaments

H — Chewing mouthparts — Abdominal gill — Posterior claws

I — No jointed legs — Chewing mouthparts in larva

Key to Orders of Aquatic Insects

1	Insects with chewing mouthparts; forewings are hardened and meet along the midline of the body when at rest (they may cover the entire abdomen or be reduced in length).	**Coleoptera** (beetles)
	Mouthparts piercing or sucking and form a pointed cone	*Go to 2*
	With chewing mouthparts, but without hardened forewings	*Go to 3*
2	Mouthparts form a short, pointed beak; legs fringed for swimming or long and spaced for suspension on water.	**Hemiptera** (bugs)
	Mouthparts do not form a beak; legs (if present) not fringed or long, or spaced apart.	*Go to 3*
3	Prominent upper lip (labium) extendable, forming a food capturing structure longer than the head.	**Odonata** (dragonflies & damselflies)
	Without a prominent, extendable labium	*Go to 4*
4	Abdomen terminating in three tail filaments which may be long and thin, or with fringes of hairs.	**Ephemeroptera** (mayflies)
	Without three tail filaments	*Go to 5*
5	Abdomen terminating in two tail filaments	**Plecoptera** (stoneflies)
	Without long tail filaments	*Go to 6*
6	With three pairs of jointed legs on thorax	*Go to 7*
	Without jointed, thoracic legs (although non-segmented prolegs or false legs may be present).	**Diptera** (true flies)
7	Abdomen with pairs of non-segmented prolegs bearing rows of fine hooks.	**Lepidoptera** (moths and butterflies)
	Without pairs of abdominal prolegs	*Go to 8*
8	With eight pairs of finger-like abdominal gills; abdomen with two pairs of posterior claws.	**Megaloptera** (dobsonflies)
	Either, without paired, abdominal gills, or, if such gills are present, without posterior claws.	*Go to 9*
9	Abdomen with a pair of posterior prolegs bearing claws with subsidiary hooks; sometimes a portable case.	**Trichoptera** (caddisflies)

3. Use the simplified key to identify each of the orders (by order or common name) of aquatic insects (**A-I**) pictured above:

(a) Order of insect A:

_____

(b) Order of insect B:

_____

(c) Order of insect C:

_____

(d) Order of insect D:

_____

(e) Order of insect E:

_____

(f) Order of insect F:

_____

(g) Order of insect G:

_____

(h) Order of insect H:

_____

(i) Order of insect I:

_____

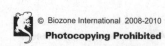

Keying Out Plant Species

Dichotomous keys are a useful tool in biology and can enable identification to the species level provided the characteristics chosen are appropriate for separating species. Keys are extensively used by botanists as they are quick and easy to use in the field, although they sometime rely on the presence of particular plant parts such as fruits or flowers. Some also require some specialist knowledge of plant biology. The following simple activity requires you to identify five species of the genus *Acer* from illustrations of the leaves. It provides valuable practice in using characteristic features to identify plants to species level.

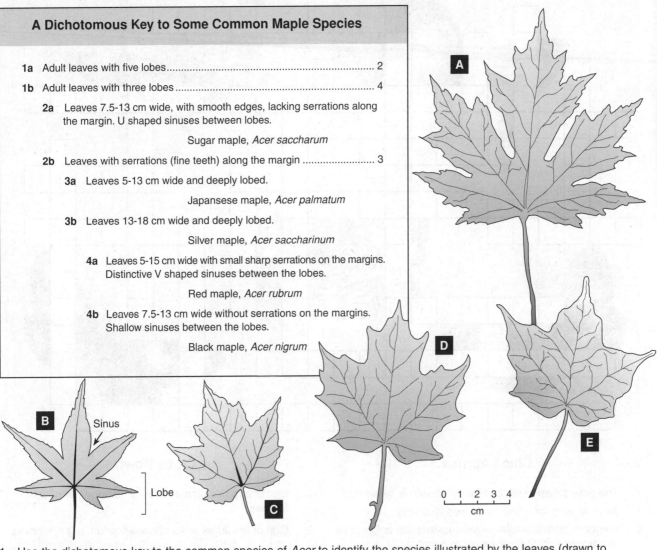

A Dichotomous Key to Some Common Maple Species

1a Adult leaves with five lobes.. 2

1b Adult leaves with three lobes 4

 2a Leaves 7.5-13 cm wide, with smooth edges, lacking serrations along the margin. U shaped sinuses between lobes.

 Sugar maple, *Acer saccharum*

 2b Leaves with serrations (fine teeth) along the margin 3

  3a Leaves 5-13 cm wide and deeply lobed.

  Japansese maple, *Acer palmatum*

  3b Leaves 13-18 cm wide and deeply lobed.

  Silver maple, *Acer saccharinum*

   4a Leaves 5-15 cm wide with small sharp serrations on the margins. Distinctive V shaped sinuses between the lobes.

   Red maple, *Acer rubrum*

   4b Leaves 7.5-13 cm wide without serrations on the margins. Shallow sinuses between the lobes.

   Black maple, *Acer nigrum*

1. Use the dichotomous key to the common species of *Acer* to identify the species illustrated by the leaves (drawn to scale). Begin at the top of the key and make a choice as to which of the illustrations best fits the description:

 (a) Species A: _____

 (b) Species B: _____

 (c) Species C: _____

 (d) Species D: _____

 (e) Species E: _____

2. Identify a feature that could be used to identify maple species when leaves are absent: _____

3. Suggest why it is usually necessary to consider a number of different features in order to classify plants to species level:

 _____
 _____

4. When identifying a plant, suggest what you should be sure of before using a key to classify it to species level:

 _____
 _____

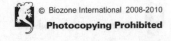
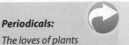
Related activities: *Classification Keys*
Web links: *Tree ID*

A 2

KEY TERMS Crossword

Complete the crossword below, which will test your understanding of key terms in this chapter and their meanings

Clues Across

3. The physical appearance of an organism is called this.

7. Another term for a shared derived character.

9. Kingdom of multicellular organisms that use chlorophyll to harness light to produce carbohydrates.

11. A unit of classification. A group of related species.

12. Key that gives two options at each step of the identification process.

15. The largest, least specific unit of classification within a kingdom.

16. A member of the sub phylum of the kingdom Animalia. Possess backbones, normally made of bone, that protect a dorsal nerve cord.

17. A unit of classification; a group of related genera.

18. A phylogenetic tree based on shared derived characters.

20. Ancient line of bacteria distinct from the eubacteria. It includes a number of extremophiles.

21. A calculated measure by which to assess species richness (2 words: 9, 5).

22. A characteristic that is shared by all members of a (taxonomic) group and is absent in members of other such taxa (2 words: 14, 7)

Clues Down

1. Unit of classification used to group together related families.

2. One of the larger units of classification, e.g. mammals.

4. A group of organisms that possess a chitinous cell wall, feed using hyphae, and reproduce by spores.

5. Diverse group of eukaryotes, either unicellular or multicellular but lacking specialised tissues.

7. Historically, the highest rank in biological taxonomy.

8. Kingdom including organisms with cells that lack organelles and have a single circular chromosome.

10. A member of the phylum whose name literally means "jointed foot".

13. The smallest, most precise unit of classification; a closely related group of organisms able to interbreed.

14. A large and diverse group of prokaryotes, separate in evolutionary terms from the Archaea.

19. Group of multicellular organisms that feed off other living organisms. Cells do not possess cell walls.

Evolution

KEY CONCEPTS

▶ Variation in species can be continuous or discontinuous. Both genes and environment contribute to variation.

▶ Evolution, through adaptation, equips species for survival. Natural selection sorts the variability in gene pools and establishes adaptive phenotypes.

▶ The evidence for evolution is overwhelming. Modern examples show that evolution is occurring today.

KEY TERMS

adaptation
analogy
behavioural adaptation
continuous variation
directional selection
discontinuous variation
disruptive selection
drug resistance
evolution
fitness
fossil
genotype
homologous structure
homology
natural selection
pesticide resistance
phenotype
physiological adaptation
population
speciation
species
stabilising selection
structural adaptation
transitional fossil
variation
vestigial organ

OBJECTIVES

☐ 1. Use the **KEY TERMS** to help you understand and complete these objectives.

Variation in Species   pages 292-296

☐ 2. Define **variation**, and discuss how variation occurs both within and between species. Explain both the genetic and environmental causes of variation.

☐ 3. Explain what is meant by **fitness** and explain how evolution, through **adaptation**, equips species for survival.

☐ 4. Describe examples of **behavioural**, **physiological**, and **structural** (anatomical or morphological) **adaptation**.

☐ 5. Distinguish between **continuous variation** and **discontinuous variation** and explain the basis of each. Describe examples of discontinuous and continuous variation in the characteristics of plants, animals, and microorganisms.

Evolution in Populations   pages 297-298

☐ 6. Describe Darwin's four observations on which he based his "*Theory of evolution by natural selection*" and describe the consequences of these.

☐ 7. Define the terms **speciation** and **evolution**, explaining how evolution is a feature of populations and not of individuals. Explain how variation, adaptation, and selection are major components of evolution.

The Evidence for Evolution   pages 299-320

☐ 8. Discuss the evidence in support of evolutionary theory, including:

(a) **Fossil evidence**, including the significance of **transitional fossils** and the development of accurate dating methods for fossil material.

(b) **Molecular evidence** from DNA, amino acids, and protein structures for the common ancestry of living organisms. Examples include comparisons of DNA, amino acid sequences, or immunological proteins.

(c) Evidence provided by comparative anatomy, e.g. **homologous structures**.

☐ 9. Explain how **natural selection** is responsible for most evolutionary change by selectively reducing or changing genetic variation through differential survival and reproduction. Interpret data to explain how natural selection produces change within a population.

☐ 10. Recognise three types of natural selection: **stabilising**, **directional**, and **disruptive selection**. If required, describe the outcome of each type of selection in a population exhibiting a normal curve of phenotypic variation.

☐ 11. Describe examples of evolution by natural selection. Examples should include:

(a) Evolution of drug resistance in microorganisms.

(b) Evolution of pesticide resistance in insects.

Periodicals:

Listings for this chapter are on page 341

Weblinks:

www.biozone.co.uk/
weblink/OCR-AS-2641.html

Teacher Resource CD-ROM:

Small Flies & Giant Buttercups

Variation in Species

Variation is a characteristic of all living organisms; we see it not only between species but between individuals of the same species. Under the biological species definition, individuals belong to the same species if they can interbreed to produce fertile offspring. The highly variable phenotypes of some species has arisen as a result of centuries of selective breeding under domestication. Many plants (e.g. brassicas) and animals (e.g. dogs) are examples of this. All breeds of dog are members of the same species, **Canis familiaris**. This species is descended from a single wild species, the grey wolf **Canis lupus**, over 15,000 years ago. There are now over 400 different dog breeds and all can interbreed, despite the considerable phenotypic variation between them. Five ancient breeds are recognised, from which all other breeds are thought to have descended by selective breeding. Some more closely resemble the original wolf-wild type phenotype than others.

Grey wolf *Canis lupus pallipes*

The grey wolf is distributed throughout Europe, North America, and Asia. Amongst members of this species, there is a lot of variation in coat coloration. This accounts for the large variation in coat colours of dogs today.

The Ancestor of Domestic Dogs

Until recently, it was unclear whether the ancestor to the modern domestic dogs was the desert wolf of the Middle East, the woolly wolf of central Asia, or the grey wolf of Northern Hemisphere. Recent genetic studies (mitochondrial DNA comparisons) now provide strong evidence that the ancestor of domestic dogs throughout the world is the grey wolf. It seems likely that this evolutionary change took place in a single region, most probably China.

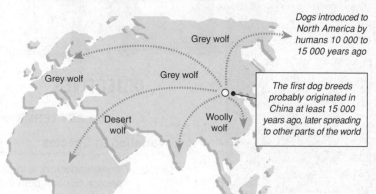

Dogs introduced to North America by humans 10 000 to 15 000 years ago

Grey wolf

Grey wolf

Grey wolf

Desert wolf

Woolly wolf

The first dog breeds probably originated in China at least 15 000 years ago, later spreading to other parts of the world

Mastiff-type
Canis familiaris inostranzevi
Originally from Tibet, the first records of this breed of dog go back to the Stoneage.

Greyhound
Canis familiaris leineri
Drawings of this breed on pottery dated from 8000 years ago in the Middle East make it one of the oldest.

Pointer-type
Canis familiaris intermedius
Probably derived from the greyhound breed for the purpose of hunting small game.

Sheepdog
Canis familiaris metris optimae
Originating in Europe, this breed has been used to guard flocks from predators for thousands of years.

Wolf-like
Canis familiaris palustris
Found in snow covered habitats in northern Europe, Asia (Siberia), and North America (Alaska).

1. Explain why all dog breeds belong to the same species: _____

   _____

   _____

2. (a) Explain how the natural variation in the grey wolf provided the basis for the phenotypic variability we see in dog breeds today. Provide an example to support your explanation:

   _____

   _____

   _____

   _____

   (b) Explain how selective breeding was used to produce this phenotypic variation: _____

   _____

   _____

   _____

Related activities: *Natural Selection*
Web links: *Dog and More Dogs*

Sources of Variation

The genetic variability within species is due mostly to a **shuffling** of the existing genetic material into new combinations as genetic information is passed from generation to generation. In addition to this, **mutation** creates new alleles in individuals. While most mutations are harmful, some are 'silent' (without visible effect on the phenotype), and some may even be beneficial. Depending on the nature of the inheritance pattern, variation in a population can be continuous or discontinuous. Traits determined by a single gene (e.g. ABO blood groups) show **discontinuous variation**, with a very limited number of variants present in the population. In contrast, traits determined by a large number of genes (e.g. skin colour) show **continuous variation**, and the number of phenotypic variations is exceedingly large. Environmental influences (differences in diet for example) also contribute to the observable variation in a population, helping or hindering the expression of an individual's full genetic potential.

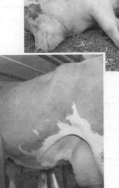

Albinism (above) is the result of the inheritance of recessive alleles for melanin production. Those with the albino phenotype lack melanin pigment in the eyes, skin, and hair.

Comb shape in poultry is a **qualitative trait** and birds have one of four phenotypes depending on which combination of four alleles they inherit. The dash (missing allele) indicates that the allele may be recessive or dominant.

Quantitative traits are characterised by **continuous variation**, with individuals falling somewhere on a normal distribution curve of the phenotypic range. Typical examples include skin colour and height in humans (left), grain yield in corn (above), growth in pigs (above, left), and milk production in cattle (far left). Quantitiative traits are determined by genes at many loci (polygenic) but most are also influenced by environmental factors.

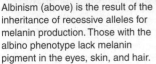

Single comb	Walnut comb	Pea comb	Rose comb
rrpp	**R_P_**	**rrP_**	**R_pp**

Flower colour in snapdragons (right) is also a **qualitative trait** determined by two alleles. (red and white) The alleles show incomplete dominance and the heterozygote (C^{R}C^W) exhibits an intermediate phenotype between the two homozygotes.

C^{R}C^R

C^{W}C^W

Sources of Variation in Organisms

Sources of genetic variation
- Dominant alleles
- Recessive alleles
- Mutations
- Crossing over
- Independent assortment
- Gene interactions

Provides inheritable variation

Combine in their effects

Phenotype

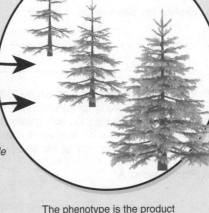

Sources of environmentally induced variation
- Climate and altitude
- Water availability
- Acidity (pH)
- Soil type
- Light
- Predation
- Competition

Provides non-inheritable variation

The phenotype is the product of the genotype and the environment

Periodicals:
What is variation?

Related activities: Descriptive Statistics, Interpreting Sample Variability

RA 2

The Effects of Environment on Phenotype

Severe stunting (krummholz)

Growth to genetic potential

Cline

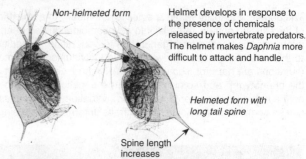

Non-helmeted form

Helmet develops in response to the presence of chemicals released by invertebrate predators. The helmet makes *Daphnia* more difficult to attack and handle.

Helmeted form with long tail spine

Spine length increases

Altitude and achievement of genetic potential in plants
Increasing altitude can stunt the phenotype of plants with the same genotype. In some conifers, e.g. **Engelmann spruce**, plants at low altitude grow to their full genetic potential, but become progressively more stunted as elevation increases, forming gnarled growth forms (krummholz) at the highest elevations. Continuous gradation in a phenotypic character within a species, associated with a change in an environmental variable, is called a **cline**.

Phenotypic response to predation in zooplankton
Some organisms respond to the presence of other, potentially harmful, organisms by changing their morphology or body shape. Invertebrates such as *Daphnia* will grow a large helmet when a predatory midge larva (*Chaoborus*) is present. Such responses are usually mediated through the action of chemicals produced by the predator (or competitor), and are common in plants as well as animals.

1. Describe the differences between **continuous** and **discontinuous** variation, giving examples to illustrate your answer:

_____

_____

_____

_____

_____

2. Identify each of the following phenotypic traits as continuous (quantitative) or discontinuous (qualitative):

(a) Wool production in sheep: _____   (d) Albinism in mammals: _____

(b) Hand span in humans: _____   (e) Body weight in mice: _____

(c) Blood groups in humans: _____   (f) Flower colour in snapdragons: _____

3. In the examples above, identify those in which an environmental influence on phenotype could be expected:

_____

4. From a sample of no less than 30 adults, collect data (by request or measurement) for one continuous variable (e.g. height, weight, shoe size, or hand span). On a separate sheet, record your results, produce a tally chart, and then plot a frequency histogram of the data, Staple the sheet into your workbook:

(a) Describe the pattern of the distribution: _____

(b) Explain the basis of this distribution: _____

_____

_____

5. On a windswept portion of a coast, two different species of plant (species A and species B) were found growing together. Both had a low growing (prostrate) phenotype. One of each plant type was transferred to a greenhouse where "ideal" conditions were provided to allow maximum growth. In this controlled environment, species B continued to grow in its original prostrate form, but species A changed its growing pattern and became erect in form. Identify the **cause** of the prostrate phenotype in each of the coastal grown plant species and explain your answer:

(a) Plant species A: _____

_____

Plant species B: _____

_____

(b) Identify which of these species (A or B) would be most likely to exhibit clinal variation: _____

Adaptations and Fitness

An **adaptation**, is any heritable trait that suits an organism to its natural function in the environment (its niche). These traits may be structural, physiological, or behavioural. The idea is important for evolutionary theory because adaptive features promote fitness. **Fitness** is a measure of an organism's ability to maximise the numbers of offspring surviving to reproductive age. Adaptations are distinct from properties which, although they may be striking, cannot be described as adaptive unless they are shown to be functional in the organism's natural habitat. Genetic adaptation must not be confused with **physiological adjustment** (acclimatisation), which refers to an organism's ability to adapt during its lifetime to changing environmental conditions (e.g. a person's acclimatisation to altitude). Examples of adaptive features arising through evolution are illustrated below.

Ear Length in Rabbits and Hares

The external ears of many mammals are used as important organs to assist in thermoregulation (controlling loss and gain of body heat). The ears of rabbits and hares native to hot, dry climates, such as the jack rabbit of south-western USA and northern Mexico, are relatively very large. The Arctic hare lives in the tundra zone of Alaska, northern Canada and Greenland, and has ears that are relatively short. This reduction in the size of the extremities (ears, limbs, and noses) is typical of cold adapted species.

Arctic hare: *Lepus arcticus*

Black-tail jackrabbit: *Lepus californicus*

Body Size in Relation to Climate

Regulation of body temperature requires a large amount of energy and mammals exhibit a variety of structural and physiological adaptations to increase the effectiveness of this process. Heat production in any endotherm depends on body volume (heat generating metabolism), whereas the rate of heat loss depends on surface area. Increasing body size minimises heat loss to the environment by reducing the surface area to volume ratio. Animals in colder regions therefore tend to be larger overall than those living in hot climates. This relationship is know as **Bergman's rule** and it is well documented in many mammalian species. Cold adapted species also tend to have more compact bodies and shorter extremities than related species in hot climates.

Fennec fox

Arctic fox

The **fennec fox** of the Sahara illustrates the adaptations typical of mammals living in hot climates: a small body size and lightweight fur, and long ears, legs, and nose. These features facilitate heat dissipation and reduce heat gain.

The **Arctic fox** shows the physical characteristics typical of cold-adapted mammals: a stocky, compact body shape with small ears, short legs and nose, and dense fur. These features reduce heat loss to the environment.

Number of Horns in Rhinoceroses

Not all differences between species can be convincingly interpreted as adaptations to particular environments. Rhinoceroses charge rival males and predators, and the horn(s), when combined with the head-down posture, add effectiveness to this behaviour. Horns are obviously adaptive, but it is not clear that the possession of one (Indian rhino) or two (black rhino) horns is necessarily related directly to the environment in which those animals live.

Great Indian rhino

African black rhino

Evolution

1. Distinguish between adaptive features (genetic) and acclimatisation: _____

_____

_____

2. Explain the nature of the relationship between the length of extremities (such as limbs and ears) and climate:

_____

_____

3. Explain the adaptive value of a larger body size at high latitude: _____

_____

_____

Periodicals:
Optimality

Related activities: Darwin's Theory, Darwin's Finches

A 2

Snow Bunting
(Plectrophenax nivalis)

The snow bunting is a small ground feeding bird that lives and breeds in the Arctic and sub-Arctic islands. Although migratory, snow buntings do not move to traditional winter homes but prefer winter habitats that resemble their Arctic breeding grounds, such as bleak shores or open fields of northern Britain and the eastern United States. Snow buntings have the unique ability to molt very rapidly after breeding. During the warmer months, the buntings are a brown color, changing to white in winter (right). They must complete this color change quickly, so that they have a new set of feathers before the onset of winter and before migration. In order to achieve this, snow buntings lose as many as four or five of their main flight wing feathers at once, as opposed to most birds, which lose only one or two.

Very few small birds breed in the Arctic, because most small birds lose more heat than larger ones. In addition, birds that breed in the brief Arctic summer must migrate before the onset of winter, often traveling over large expanses of water. Large, long winged birds are better able to do this. However, the snow bunting is superbly adapted to survive in the extreme cold of the Arctic region.

White feathers are hollow and filled with air, which acts as an insulator. In the dark colored feathers the internal spaces are filled with pigmented cells.

Less heat is lost from white plumage compared to dark plumage.

Snow buntings, on average, lay one or two more eggs than equivalent species further south. They are able to rear more young because the continuous daylight and the abundance of insects at high latitudes enables them to feed their chicks around the clock.

During snow storms or periods of high wind, snow buntings will burrow into snowdrifts for shelter.

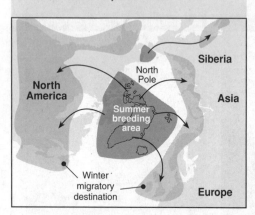

Habitat and ecology: Widespread throughout Arctic and sub-Arctic Islands. Active throughout the day and night, resting for only 2-3 hours in any 24 hour period. Snow buntings may migrate up to 6000 km but are always found at high latitudes. **Reproduction and behavior**: The nest, which is concealed amongst stones, is made from dead grass, moss, and lichen. The male bird feeds his mate during the incubation period and helps to feed the young.

4. Describe a structural, physiological, and behavioral adaptation of the **snow bunting**, explaining how each adaptation assists survival:

(a) Structural adaptation: _____

_____

_____

(b) Physiological adaptation: _____

_____

_____

(c) Behavioral adaptation: _____

_____

_____

5. Examples of adaptations are listed below. Identify them as predominantly structural, physiological, and/or behavioral:

(a) Relationship of body size and shape to latitude (tropical or Arctic): _____

(b) The production of concentrated urine in desert dwelling mammals: _____

(c) The summer and winter migratory patterns in birds and mammals: _____

(d) The C4 photosynthetic pathway and CAM metabolism of plants: _____

(e) The thick leaves and sunken stomata of desert plants: _____

(f) Hibernation or torpor in small mammals over winter: _____

(g) Basking in lizards and snakes: _____

The Modern Theory of Evolution

Although **Charles Darwin** is credited with the development of the theory of evolution by natural selection, there were many people that contributed ideas upon which he built his own. Since Darwin first proposed his theory, aspects that were problematic (such as the mechanism of inheritance) have now been explained. The development of the modern theory of evolution has a history going back at least two centuries. The diagram below illustrates the way in which some of the major contributors helped to form the currently accepted model, or **new synthesis**. Understanding of evolutionary processes continued to grow through the 1980s and 1990s as comparative molecular sequence data were amassed and understanding of the molecular basis of developmental mechanisms improved. Most recently, in the exciting new area of evolutionary developmental biology (**evo-devo**), biologists have been exploring how developmental gene expression patterns explain how groups of organisms evolved.

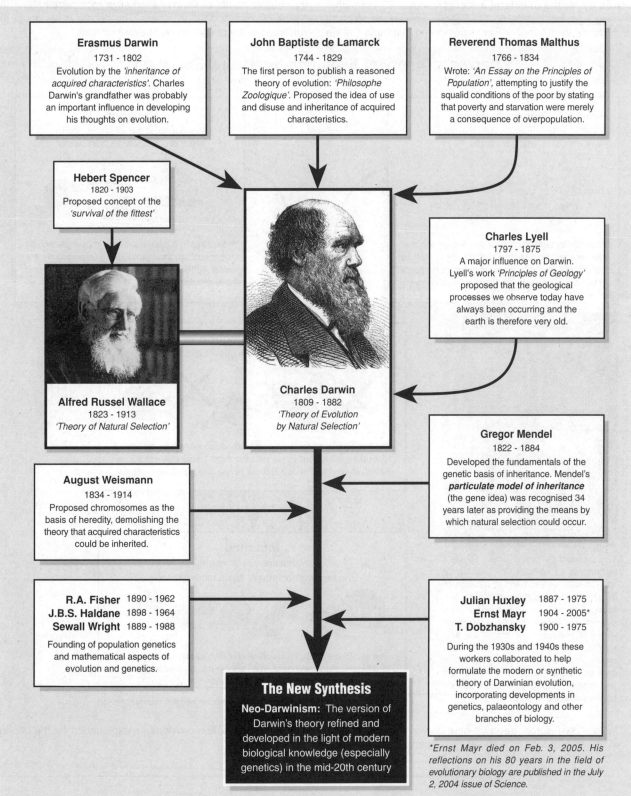

Erasmus Darwin
1731 - 1802
Evolution by the *'inheritance of acquired characteristics'*. Charles Darwin's grandfather was probably an important influence in developing his thoughts on evolution.

John Baptiste de Lamarck
1744 - 1829
The first person to publish a reasoned theory of evolution: *'Philosophe Zoologique'*. Proposed the idea of use and disuse and inheritance of acquired characteristics.

Reverend Thomas Malthus
1766 - 1834
Wrote: *'An Essay on the Principles of Population'*, attempting to justify the squalid conditions of the poor by stating that poverty and starvation were merely a consequence of overpopulation.

Hebert Spencer
1820 - 1903
Proposed concept of the *'survival of the fittest'*

Charles Lyell
1797 - 1875
A major influence on Darwin. Lyell's work *'Principles of Geology'* proposed that the geological processes we observe today have always been occurring and the earth is therefore very old.

Alfred Russel Wallace
1823 - 1913
'Theory of Natural Selection'

Charles Darwin
1809 - 1882
'Theory of Evolution by Natural Selection'

Gregor Mendel
1822 - 1884
Developed the fundamentals of the genetic basis of inheritance. Mendel's **particulate model of inheritance** (the gene idea) was recognised 34 years later as providing the means by which natural selection could occur.

August Weismann
1834 - 1914
Proposed chromosomes as the basis of heredity, demolishing the theory that acquired characteristics could be inherited.

R.A. Fisher 1890 - 1962
J.B.S. Haldane 1898 - 1964
Sewall Wright 1889 - 1988

Founding of population genetics and mathematical aspects of evolution and genetics.

Julian Huxley 1887 - 1975
Ernst Mayr 1904 - 2005*
T. Dobzhansky 1900 - 1975

During the 1930s and 1940s these workers collaborated to help formulate the modern or synthetic theory of Darwinian evolution, incorporating developments in genetics, palaeontology and other branches of biology.

The New Synthesis

Neo-Darwinism: The version of Darwin's theory refined and developed in the light of modern biological knowledge (especially genetics) in the mid-20th century

Ernst Mayr died on Feb. 3, 2005. His reflections on his 80 years in the field of evolutionary biology are published in the July 2, 2004 issue of Science.

Evolution

1. From the diagram above, choose one of the contributors to the development of evolutionary theory (excluding Charles Darwin himself), and write a few paragraphs discussing their role in contributing to Darwin's ideas. You may need to consult an encyclopaedia or other reference to assist you.

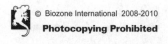

Periodicals:
Evolution: Five big questions

Related activities: Variation, Darwin's Theory

A 2

Darwin's Theory

In 1859, Darwin and Wallace jointly proposed that new species could develop by a process of natural selection. Natural selection is the term given to the mechanism by which better adapted organisms survive to produce a greater number of viable offspring. This has the effect of increasing their proportion in the population so that they become more common. It is Darwin who is best remembered for the theory of evolution by natural selection through his famous book: '**On the origin of species by means of natural selection**', written 23 years after returning from his voyage on the Beagle, from which much of the evidence for his theory was accumulated. Although Darwin could not explain the origin of variation nor the mechanism of its transmission (this was provided later by Mendel's work), his basic theory of evolution by natural selection (outlined below) is widely accepted today. The study of population genetics has greatly improved our understanding of evolutionary processes, which are now seen largely as a (frequently gradual) change in allele frequencies within a population. Students should be aware that scientific debate on the subject of evolution centres around the relative merits of various alternative hypotheses about the nature of evolutionary processes. The debate is not about the existence of the phenomenon of evolution itself.

Darwin's Theory of Evolution by Natural Selection

Overproduction
Populations produce too many young: many must die

Populations tend to produce more offspring than are needed to replace the parents. Natural populations normally maintain constant numbers. There must therefore be a certain number dying.

Variation
Individuals show variation: some are more favourable than others

Individuals in a population vary in their phenotype and therefore, their genotype. Some variants are better suited in the prevailing environment and have greater survival and reproductive success.

Natural Selection
Natural selection favours the best suited at the time

The struggle for survival amongst individuals competing for limited resources will favour those with the most favourable variations. Relatively more of those without favourable variations will die.

Inherited
Variations are Inherited.
The best suited variants leave more offspring.

The variations (both favourable and unfavourable) are passed on to offspring. Each new generation will contain proportionally more descendents of individuals with favourable characters.

Andrew Dunn www.andrewdunnphoto.com

The banded or grove snail, *Cepaea nemoralis*, is famous for the highly variable colours and banding patterns of its shell. These **polymorphisms** are thought to have a role in differential survival in different regions, associated with both the risk of predation and maintenance of body temperature. Dark brown grove snails are more abundant in dark woodlands, whilst snails with light yellow shells and thin banding are more commonly found in grasslands.

1. In your own words, describe how Darwin's theory of evolution by natural selection provides an explanation for the change in the appearance of a species over time:

_____

_____

_____

_____

Related activities: The Modern Theory of Evolution, Darwin's Finches
Web links: Variation : Snails

Periodicals:
Was Darwin wrong?

Fossil Formation

Fossils are the remains of long-dead organisms that have escaped decay and have, after many years, become part of the Earth's crust. A fossil may be the preserved remains of the organism itself, the impression of it in the sediment (called trace fossils). For fossilisation to occur, rapid burial of the organism is required (usually in water-borne sediment). This is followed by chemical alteration, where minerals are added or removed. Fossilisation requires the normal processes of decay to be permanently arrested. This can occur if the remains are isolated from the air or water and decomposing microbes are prevented from breaking them down. Fossils provide a record of the appearance and extinction of organisms, from species to whole taxonomic groups. Once this record is calibrated against a time scale (by using a broad range of dating techniques), it is possible to build up a picture of the evolutionary changes that have taken place.

Modes of Preservation

Silicification: Silica from weathered volcanic ash is gradually incorporated into partly decayed wood (also called petrification).

Phosphatisation: Bones and teeth are preserved in phosphate deposits.

Pyritisation: Iron pyrite replaces hard remains of the dead organism.

Tar pit: Animals fall into and are trapped in mixture of tar and sand.

Trapped in amber: Gum from conifers traps insects and then hardens.

Limestone: Calcium carbonate from the remains of marine plankton is deposited as a sediment that traps the remains of other sea creatures.

Brachiopod (lamp shell), Jurassic (New Zealand)

Mould: This impression of a lamp shell is all that is left after the original shell material was dissolved after fossilisation.

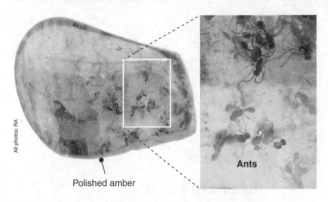

All photos: RA

Polished amber

Ants

Insects in amber: The fossilised resin or gum produced by some ancient conifers trapped these insects (including the ants visible in the enlargement) about 25 million years ago (Madagascar).

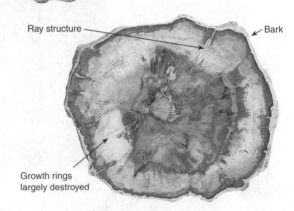

Ray structure — Bark

Growth rings largely destroyed

Petrified wood: A cross-section of a limb from a coniferous tree (Madagascar).

Rock phosphate matrix

Shell

Stone interior

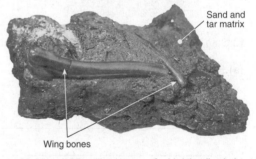

Sand and tar matrix

Wing bones

Shark tooth: The tooth of a shark *Lamna obliqua* from phosphate beds, Eocene (Khouribga, Morocco).

Ammonite: This ammonite still has a layer of the original shell covering the stone interior, Jurassic (Madagascar).

Bird bones: Fossilised bones of a bird that lived about 5 million years ago and became stuck in the tar pits at la Brea, Los Angeles, USA.

Shell and chambers replaced by iron pyrite

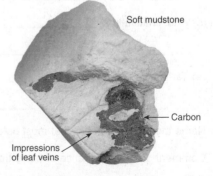

Soft mudstone

Carbon

Impressions of leaf veins

Cast: This ammonite has been preserved by a process called pyritisation, late Cretaceous (Charmouth, England).

Fossil fern: This compression fossil of a fern frond shows traces of carbon and wax from the original plant, Carboniferous (USA).

Sub-fossil: Leaf impression in soft mudstone (can be broken easily with fingers) with some of the remains of the leaf still intact (a few thousand years old, New Zealand).

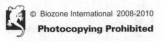
Periodicals: Meet your ancestor, The quick and the dead

Related activities: The Fossil Record
Web links: Getting into the Fossil Record

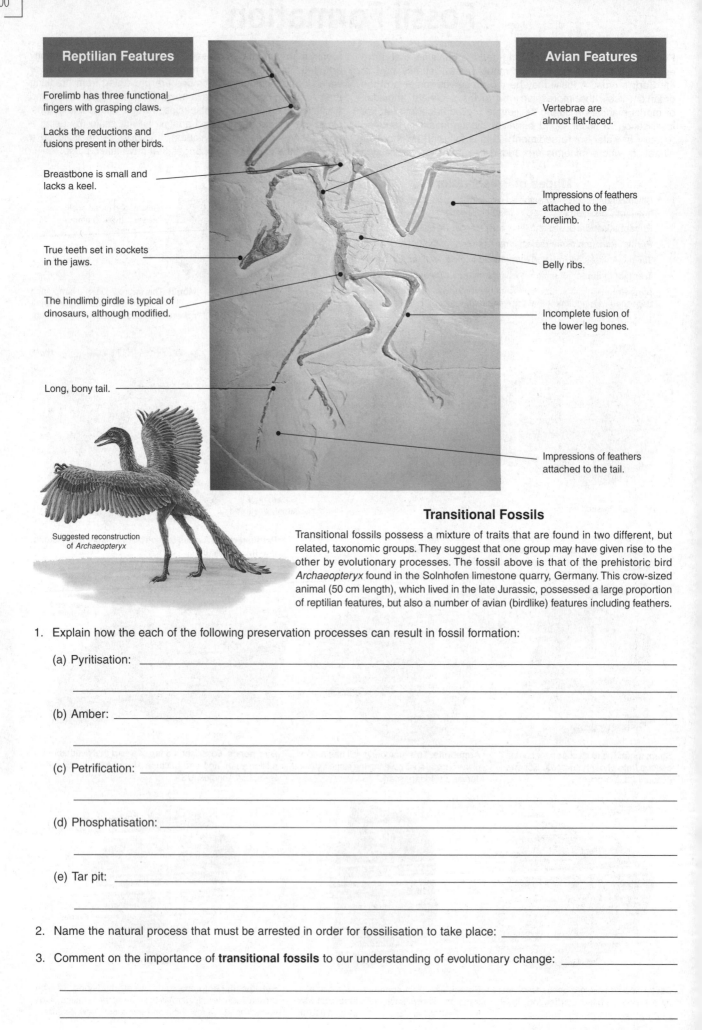

Reptilian Features

Forelimb has three functional fingers with grasping claws.

Lacks the reductions and fusions present in other birds.

Breastbone is small and lacks a keel.

True teeth set in sockets in the jaws.

The hindlimb girdle is typical of dinosaurs, although modified.

Long, bony tail.

Suggested reconstruction of *Archaeopteryx*

Avian Features

Vertebrae are almost flat-faced.

Impressions of feathers attached to the forelimb.

Belly ribs.

Incomplete fusion of the lower leg bones.

Impressions of feathers attached to the tail.

Transitional Fossils

Transitional fossils possess a mixture of traits that are found in two different, but related, taxonomic groups. They suggest that one group may have given rise to the other by evolutionary processes. The fossil above is that of the prehistoric bird *Archaeopteryx* found in the Solnhofen limestone quarry, Germany. This crow-sized animal (50 cm length), which lived in the late Jurassic, possessed a large proportion of reptilian features, but also a number of avian (birdlike) features including feathers.

1. Explain how the each of the following preservation processes can result in fossil formation:

   (a) Pyritisation: _____

   _____

   (b) Amber: _____

   _____

   (c) Petrification: _____

   _____

   (d) Phosphatisation: _____

   _____

   (e) Tar pit: _____

   _____

2. Name the natural process that must be arrested in order for fossilisation to take place: _____

3. Comment on the importance of **transitional fossils** to our understanding of evolutionary change: _____

   _____

   _____

The Fossil Record

The diagram below represents a cutting into the earth revealing the layers of rock. Some of these layers may have been laid down by water (sedimentary rocks) or by volcanic activity (volcanic rocks). Fossils are the actual remains or impressions of plants or animals that become trapped in the sediments after their death. Layers of sedimentary rock are arranged in the order that they were deposited, with the most recent layers near the surface (unless they have been disturbed).

Profile with Sedimentary Rocks Containing Fossils

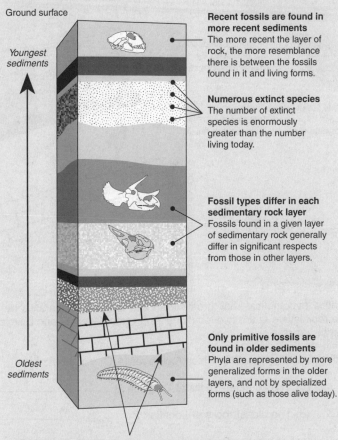

Ground surface

Youngest sediments

Recent fossils are found in more recent sediments
The more recent the layer of rock, the more resemblance there is between the fossils found in it and living forms.

Numerous extinct species
The number of extinct species is enormously greater than the number living today.

Fossil types differ in each sedimentary rock layer
Fossils found in a given layer of sedimentary rock generally differ in significant respects from those in other layers.

Only primitive fossils are found in older sediments
Phyla are represented by more generalized forms in the older layers, and not by specialized forms (such as those alive today).

Oldest sediments

New fossil types mark changes in environment
In the rocks marking the end of one geological period, it is common to find many new fossils that become dominant in the next. Each geological period had an environment very different from those before and after. Their boundaries coincided with drastic environmental changes and the appearance of new niches. These produced new selection pressures resulting in new adaptive features in the surviving species, as they responded to the changes.

The rate of evolution can vary

According to the fossil record, rates of evolutionary change seem to vary. There are bursts of species formation and long periods of relative stability within species (stasis). The occasional rapid evolution of new forms apparent in the fossil record, is probably a response to a changing environment. During periods of stable environmental conditions, evolutionary change may slow down.

The Fossil Record of Proboscidea

African and Indian elephants have descended from a diverse group of animals known as **proboscideans** (named for their long trunks). The first pig-sized, trunkless members of this group lived in Africa 40 million years ago. From Africa, their descendants invaded all continents except Antarctica and Australia. As the group evolved, they became larger; an effective evolutionary response to deter predators. Examples of extinct members of this group are illustrated below:

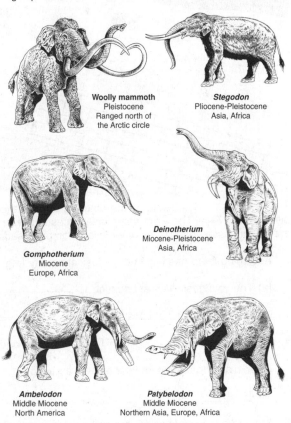

Woolly mammoth
Pleistocene
Ranged north of the Arctic circle

Stegodon
Pliocene-Pleistocene
Asia, Africa

Deinotherium
Miocene-Pleistocene
Asia, Africa

Gomphotherium
Miocene
Europe, Africa

Ambelodon
Middle Miocene
North America

Patybelodon
Middle Miocene
Northern Asia, Europe, Africa

- **Modern day species can be traced:** The evolution of many present-day species can be very well reconstructed. For instance, the evolutionary history of the modern elephants is exceedingly well documented for the last 40 million years. The modern horse also has a well understood fossil record spanning the last 50 million years.

- **Fossil species are similar to but differ from today's species:** Most fossil animals and plants belong to the same major taxonomic groups as organisms living today. However, they do differ from the living species in many features.

1. Name an animal or plant taxon (e.g. family, genus, or species) that has:

   (a) A good fossil record of evolutionary development: _____

   (b) Appeared to have changed very little over the last 100 million years or so: _____

2. Discuss the importance of **fossils** as a record of evolutionary change over time: _____
   _____
   _____
   _____

Periodicals:
The accidental discovery of a feathered giant dinosaur

Related activities: *Fossil Formation*
Web links: *29+ Evidences for Macroevolution*

RA 2

Evolution

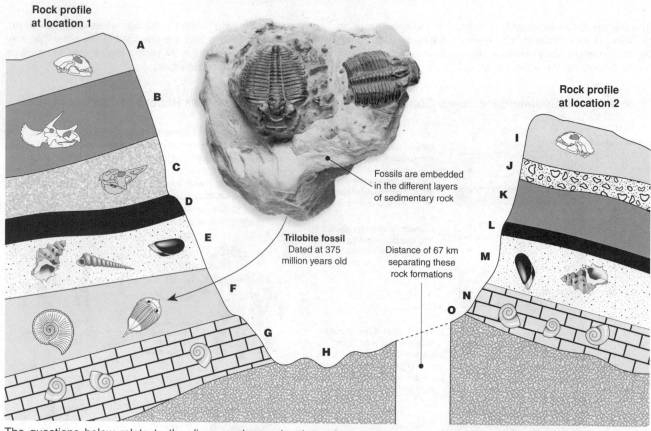

Rock profile at location 1

A
B
C
D
E
F
G
H

Rock profile at location 2

I
J
K
L
M
N
O

Fossils are embedded in the different layers of sedimentary rock

Trilobite fossil
Dated at 375 million years old

Distance of 67 km separating these rock formations

The questions below relate to the diagram above, showing a hypothetical rock profile from two locations separated by a distance of 67 km. There are some differences between the rock layers at the two locations. Apart from layers D and L which are volcanic ash deposits, all other layers comprise sedimentary rock.

3. Assuming there has been no geological activity (e.g. tilting or folding), state in which rock layer (A-O) you would find:

    (a) The youngest rocks at Location 1: _____ (c) The youngest rocks at Location 2: _____

    (b) The oldest rocks at Location 1: _____ (d) The oldest rocks at Location 2: _____

4. (a) State which layer at location 1 is of the same age as layer M at location 2: _____

    (b) Explain the reason for your answer above: _____

    _____

5. The rocks in layer H and O are sedimentary rocks. Explain why there are no visible fossils in layers:

    _____

    _____

6. (a) State which layers present at location 1 are missing at location 2: _____

    (b) State which layers present at location 2 are missing at location 1: _____

7. Describe three methods of dating rocks: _____

    _____

    _____

8. Using radiometric dating, the trilobite fossil was determined to be approximately 375 million years old. The volcanic rock layer (D) was dated at 270 million years old, while rock layer B was dated at 80 million years old. Give the approximate **age range** (i.e. greater than, less than or between given dates) of the rock layers listed below:

    (a) Layer A: _____ (d) Layer G: _____

    (b) Layer C: _____ (e) Layer L: _____

    (c) Layer E: _____ (f) Layer O: _____

Dating a Fossil Site

The diagram below shows a rock shelter typical of those found in the Dordogne Valley of Southwest France. Such shelters have yielded a rich source of Neanderthal and modern human remains. It illustrates the way human activity is revealed at archaeological excavations. Occupation sites included shallow caves or rocky overhangs of limestone. The floors of these caves accumulated the debris of natural rockfalls, together with the detritus of human occupation at various layers, called **occupation horizons**. A wide array of techniques can be used for dating, some of which show a high degree of reliability (see the table below). The use of several appropriate techniques to date material improves the reliability of the date determined.

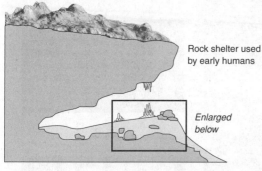

Rock shelter used by early humans

Enlarged below

Dating method	Dating range (years ago)	Datable materials
Radiocarbon (^{14}C)	1000 - 50 000+	Bone, shell, charcoal
Potassium-argon (K/Ar)	10 000 - 100 million	Volcanic rocks and minerals
Uranium series decay	less than 1 million	Marine carbonate, coral, shell
Thermoluminescence	less than 200 000	Ceramics (burnt clay)
Fission track	1000 - 100 million	Volcanic rock, glass, pottery
Electron spin resonance	2000 - 500 000	Bone, teeth, loess, burnt flint

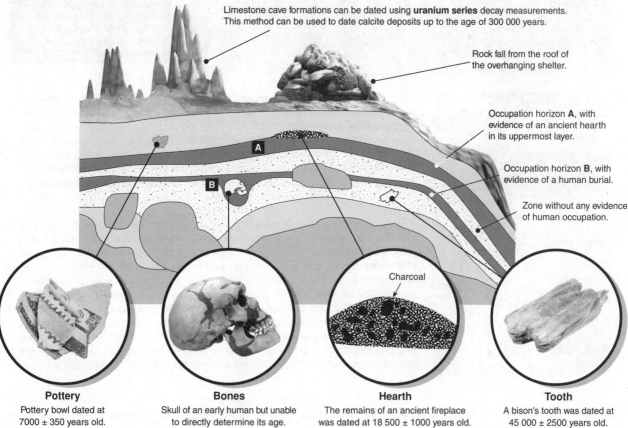

Limestone cave formations can be dated using **uranium series** decay measurements. This method can be used to date calcite deposits up to the age of 300 000 years.

Rock fall from the roof of the overhanging shelter.

Occupation horizon **A**, with evidence of an ancient hearth in its uppermost layer.

Occupation horizon **B**, with evidence of a human burial.

Zone without any evidence of human occupation.

Charcoal

Pottery
Pottery bowl dated at 7000 ± 350 years old.

Bones
Skull of an early human but unable to directly determine its age.

Hearth
The remains of an ancient fireplace was dated at 18 500 ± 1000 years old.

Tooth
A bison's tooth was dated at 45 000 ± 2500 years old.

Evolution

1. Discuss the significance of **occupation horizons**: _____

_____

_____

_____

2. Determine the approximate date range for the items below (Hint: take into account layers/artifacts with known dates):

   (a) The skull at point B: _____

   (b) Occupation horizon A: _____

3. Name the dating methods that could have been used to date each of the following, at the site above:

   (a) Pottery bowl: _____   (c) Hearth: _____

   (b) Skull: _____   (d) Tooth: _____

Periodicals:
How old is...?

Related activities: Fossil Formation
Web links: Neanderthals: Dig and Deduce

RDA 2

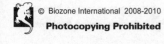

Interpreting Fossil Sites

Human skull

Charcoal fragments (possible evidence of fire use and excellent for radiocarbon dating).

Bones from a large mammal with evidence of butchering (cut and scrape marks from stone tools). These provide information on the past ecology and environment of the hominins in question.

Excavation through rock strata (layers). The individual layers can be dated using both chronometric (absolute) and relative dating methods.

Stone tools

Photo: RA

istock

Searching for ancient human remains, including the evidence of culture, is the work of **palaeoanthropologists**. Organic materials, such as bones and teeth, are examined and analysed by physical anthropologists, while cultural materials, such as tools, weapons, shelters, and artworks, are examined by archaeologists. Both these disciplines, **palaeoanthropology** and **archaeology**, are closely associated with other scientific disciplines, including **geochemistry** (for **chronometric dates**), **geology** (for reconstructions of past physical landscapes), and **palaeontology** (for knowledge of the past species assemblages).

The reconstruction of a **dig site**, pictured above, illustrates some of the features that may be present at a site of hominin activity. Naturally, the type of information recovered from a site will depend on several factors, including the original nature of the site and its contents, the past and recent site environment, and earlier disturbance by people or animals. During its period of occupation, a site represents an interplay between additive and subtractive processes; building vs destruction, growth vs decay. Organic matter decays, and other features of the site, such as tools, can be disarranged, weathered, or broken down. The archaeologists goal is to maximise the recovery of information, and recent trends have been to excavate and process artifacts immediately, and sometimes to leave part of the site intact so that future work, perhaps involving better methodologies, is still possible.

4. Explain why palaeoanthropologists date and interpret all of the remains at a particular site of interest (e.g. animal bones, pollen, and vegetation, as well as hominin remains):

_____

_____

_____

_____

5. Discuss the importance of involving several scientific disciplines when interpreting a site of hominin activity:

_____

_____

_____

_____

_____

The Evolution of Horses

The evolution of the horse from the ancestral *Hyracotherium* to modern *Equus* is well documented in the fossil record. For this reason it is often used to illustrate the process of evolution. The rich fossil record, which includes numerous **transitional fossils**, has enabled scientists to develop a robust model of horse phylogeny. Although the evolution of the line was once considered to be a gradual straight line process, it has been radically revised to a complex tree-like lineage with many divergences (below). It showed no inherent direction, and a diverse array of species coexisted for some time over the 55 million year evolutionary period. The environmental transition from forest to grasslands drove many of the changes observed in the equid fossil record. These include reduction in toe number, increased size of cheek teeth, lengthening of the face, and increasing body size.

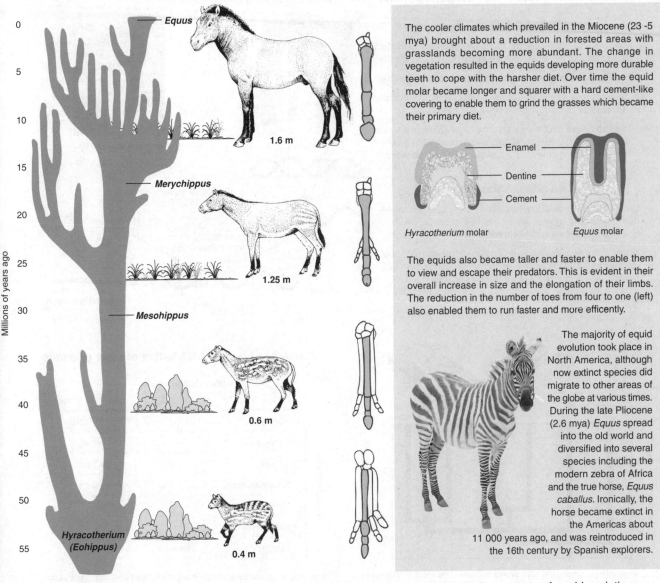

The cooler climates which prevailed in the Miocene (23 -5 mya) brought about a reduction in forested areas with grasslands becoming more abundant. The change in vegetation resulted in the equids developing more durable teeth to cope with the harsher diet. Over time the equid molar became longer and squarer with a hard cement-like covering to enable them to grind the grasses which became their primary diet.

Enamel
Dentine
Cement

Hyracotherium molar   *Equus* molar

The equids also became taller and faster to enable them to view and escape their predators. This is evident in their overall increase in size and the elongation of their limbs. The reduction in the number of toes from four to one (left) also enabled them to run faster and more efficently.

The majority of equid evolution took place in North America, although now extinct species did migrate to other areas of the globe at various times. During the late Pliocene (2.6 mya) *Equus* spread into the old world and diversified into several species including the modern zebra of Africa and the true horse, *Equus caballus*. Ironically, the horse became extinct in the Americas about 11 000 years ago, and was reintroduced in the 16th century by Spanish explorers.

Millions of years ago

Equus — 1.6 m
Merychippus — 1.25 m
Mesohippus — 0.6 m
Hyracotherium (Eohippus) — 0.4 m

Evolution

1. Explain how the environmental change from forest to grassland influenced the following aspects of equid evolution:

   (a) Change in tooth structure: _____

   _____

   (b) Limb length: _____

   _____

   (c) Reduction in number of toes: _____

   _____

2. Explain why the equid fossil record provides a good example of the evolutionary process: _____

   _____

   _____

   _____

DNA Homologies

Establishing a phylogeny on the basis of homology in a protein, such as cytochrome c, is valuable, but it is also analogous to trying to see a complete picture through a small window. The technique of **DNA-DNA hybridisation** provides a way to compare the total genomes of different species by measuring the degree of genetic similarity between pools of DNA sequences. It is usually used to determine the genetic distance between two species; the more closely two species are related, the fewer differences there will be between their genomes. This is because there has been less time for the point mutations that will bring about these differences to occur. This technique gives a measure of 'relatedness', and can be calibrated as a **molecular clock** against known fossil dates. It has been applied to primate DNA samples to help determine the approximate date of human divergence from the apes, which has been estimated to be between 10 and 5 million years ago.

DNA Hybridisation

1. DNA from the two species to be compared is extracted, purified and cut into short fragments (e.g. 600-800 base pairs).

2. The DNA of one species is mixed with the DNA of another.

3. The mixture is incubated to allow DNA strands to dissociate and reanneal, forming hybrid double-stranded DNA.

4. The hybridised sequences that are highly similar will bind more firmly. A measure of the heat energy required to separate the hybrid strands provides a measure of DNA relatedness.

DNA Homologies Today

DNA-DNA hybridisation has been criticised because duplicated sequences within a single genome make it unreliable for comparisons between closely related species.

Today, DNA sequencing and computed comparisons are more widely used to compare genomes, although DNA-DNA hybridisation is still used to help identify bacteria.

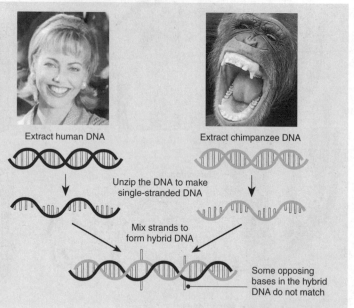

Extract human DNA

Extract chimpanzee DNA

Unzip the DNA to make single-stranded DNA

Mix strands to form hybrid DNA

Some opposing bases in the hybrid DNA do not match

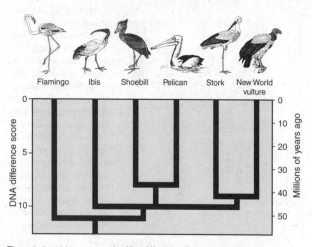

Flamingo   Ibis   Shoebill   Pelican   Stork   New World vulture

The relationships among the **New World vultures** and **storks** have been determined using DNA hybridisation. It has been possible to estimate how long ago various members of the group shared a common ancestor.

Similarity of human DNA to that of other primates

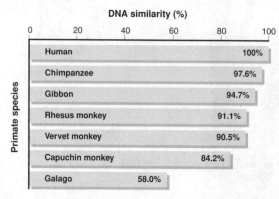

DNA similarity (%)

Primate species	DNA similarity
Human	100%
Chimpanzee	97.6%
Gibbon	94.7%
Rhesus monkey	91.1%
Vervet monkey	90.5%
Capuchin monkey	84.2%
Galago	58.0%

The genetic relationships among the **primates** has been investigated using DNA hybridisation. Human DNA was compared with that of the other primates. It largely confirmed what was suspected from anatomical evidence.

1. Explain how **DNA hybridisation** can give a measure of genetic relatedness between species:

_____

_____

_____

2. Study the graph showing the results of a DNA hybridisation between human DNA and that of other primates.

   (a) Identify which is the most closely related primate to humans: _____

   (b) Identify which is the most distantly related primate to humans: _____

3. State the DNA difference score for: (a) Shoebills and pelicans: _____   (b) Storks and flamingos: _____

4. On the basis of DNA hybridisation, state how long ago the ibises and New World vultures shared a common ancestor:

_____

Protein Homologies

Traditionally, phylogenies were based largely on anatomical or behavioural traits and biologists attempted to determine the relationships between organisms based on overall degree of similarity or by tracing the appearance of key characteristics. With the advent of molecular techniques, homologies can now be studied at the molecular level as well and these can be compared to the phylogenies established using other methods. Protein sequencing provides an excellent tool for establishing **homologies** (similarities resulting from shared ancestry). Each protein has a specific number of amino acids arranged in a specific order. Any differences in the sequence reflect changes in the DNA sequence. Commonly studied proteins include blood proteins, such as **haemoglobin** (below), and the respiratory protein **cytochrome c** (overleaf). Many of these proteins are highly conserved, meaning they change very little over time, presumably because mutations would be detrimental to basic function. Conservation of protein sequences is indicated by the identical amino acid residues at corresponding parts of proteins.

Amino Acid Differences in Haemoglobin

Human beta chain	0
Chimpanzee	0
Gorilla	1
Gibbon	2
Rhesus monkey	8
Squirrel monkey	9
Dog	15
Horse, cow	25
Mouse	27
Grey kangaroo	38
Chicken	45
Frog	67

When the sequence of the **beta haemoglobin chain** (right), which is 146 amino acids long, is compared between humans, five other primates, and six other vertebrates, the results support the phylogenies established using other methods. The numbers in the table (left) represent the number of amino acid differences between the beta chain of humans and those of other species. In general, the number of amino acid differences between the haemoglobins of different vertebrates is inversely proportional to genetic relatedness.

Shading indicates (from top) primates, non-primate placental mammals, marsupials, and non-mammals.

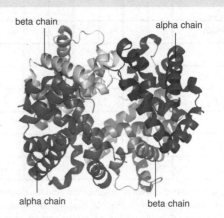

In most vertebrates, the oxygen-transporting blood protein haemoglobin is composed of four polypeptide chains, two alpha chains and two beta chains. Haemoglobin is derived from myoglobin, and ancestral species had just myoglobin for oxygen transport. When the amino acid sequences of myoglobin, the haemoglobin alpha chain, and the haemoglobin beta chain are compared, there are several amino acids that remain conserved between all three. These amino acid sequences must be essential for function because they have remained unchanged throughout evolution.

Using Immunology to Determine Phylogeny

The immune system of one species will recognise the blood proteins of another species as foreign and form antibodies against them. This property can be used to determine the extent of homology between species. Blood proteins, such as albumins, are used to prepare **antiserum** in rabbits. The antiserum contains antibodies against the test blood proteins (e.g. human) and will react to those proteins in any blood sample they are mixed with. The extent of the reaction indicates how different the proteins are; the greater the reaction, the greater the homology. This principle is illustrated (right) for antiserum produced to human blood and its reaction with the blood of other primates and a rat.

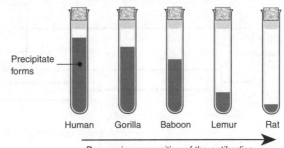

Decreasing recognition of the antibodies against human blood proteins

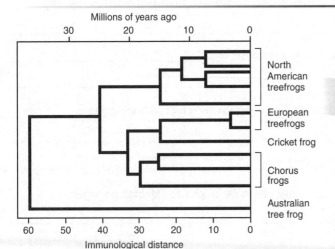

The relationships among tree frogs have been established by immunological studies based on blood proteins such as immunoglobulins and albumins. The **immunological distance** is a measure of the number of amino acid substitutions between two groups. This, in turn, has been calibrated to provide a time scale showing when the various related groups diverged.

Evolution

Related activities: DNA Homologies
Web links: Evidence for Evolution

DA 2

Cytochrome c and the Molecular Clock Theory

Evolutionary change at the molecular level occurs primarily through fixation of neutral mutations by genetic drift. The rate at which one neutral mutation replaces another depends on the mutation rate, which is fairly constant for any particular gene.

If the rate at which a protein evolves is roughly constant over time, the amount of molecular change that a protein shows can be used as a molecular clock to date evolutionary events, such as the divergence of species.

The molecular clock for each species, and each protein, may run at different rates, so scientists calibrate the molecular clock data with other evidence (morphological, molecular) to confirm phylogenetic relationships.

For example, 20 amino acid substitutions in a protein since two organisms diverged from a known common ancestor 400 mya indicates an average substitution rate of 5 substitutions per 100 my.

		1					6				10				14			17	18		20		
Human		Gly	Asp	Val	Glu	Lys	Gly	Lys	Lys	Ile	Phe	Ile	Met	Lys	Cys	Ser	Gln	Cys	His	Thr	Val	Glu	Lys
Pig												Val	Gln			Ala								
Chicken				Ile								Val	Gln											
Dogfish										Val		Val	Gln			Ala								Asn
Drosophila	<<									Leu		Val	Gln	Arg		Ala								Ala
Wheat	<<		Asn	Pro	Asp	Ala		Ala				Lys	Thr			Ala							Asp	Ala
Yeast	<<		Ser	Ala	Lys			Ala	Thr	Leu		Lys	Thr	Arg		Glu	Leu							

This table shows the N-terminal 22 amino acid residues of human cytochrome c, with corresponding sequences from other organisms aligned beneath. Sequences are aligned to give the most position matches. A shaded square indicates no change. In every case, the cytochrome's heme group is attached to the Cys-14 and Cys-17. In *Drosophila*, wheat, and yeast, arrows indicate that several amino acids precede the sequence shown.

The sequence homology of cytochrome c (right), a respiratory protein, has been used to construct a phylogenetic tree for species across a wide range of taxa. Overall, the phylogeny aligns well to other evolutionary data, although the tree indicates that primates branched off before the marsupials diverged from other placental mammals, which is incorrect based on a variety of other evidence. As indicated by the table above, cytochrome c is highly conserved, which means that its sequence changes very little despite speciation.

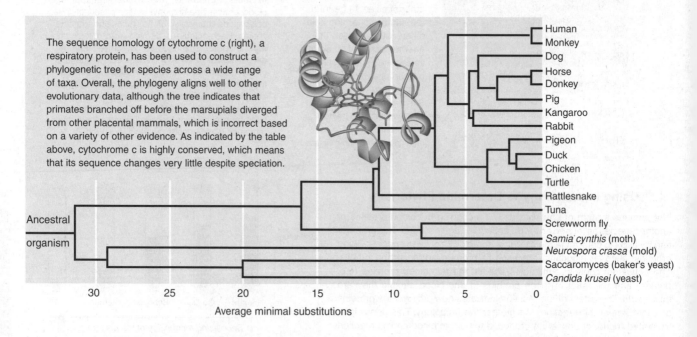

1. Explain why chimpanzees and gorillas are considered most closely related to humans, while monkeys are less so:

   _____

2. (a) Explain why a respiratory protein like cytochrome C would be highly conserved: _____

   _____

   (b) Suggest why highly conserved proteins are good candidates for use in establishing protein homologies

   _____

   _____

3. Discuss some of the limitations of using protein homology, specifically molecular clocks, to establish phylogeny:

   _____

Homologous Structures

The evolutionary relationships between groups of organisms is determined mainly by structural similarities called **homologous structures** (homologies), which suggest that they all descended from a common ancestor with that feature. The bones of the forelimb of air-breathing vertebrates are composed of similar bones arranged in a comparable pattern. This is indicative of a common ancestry. The early land vertebrates were amphibians and possessed a limb structure called the **pentadactyl limb**: a limb with 5 fingers or toes (below left). All vertebrates that descended from these early amphibians, including reptiles, birds and mammals, have limbs that have evolved from this same basic pentadactyl pattern. They also illustrate the phenomenon known as **adaptive radiation**, since the basic limb plan has been adapted to meet the requirements of different niches.

Generalised Pentadactyl Limb

The forelimbs and hind limbs have the same arrangement of bones but they have different names. In many cases bones in different parts of the limb have been highly modified to give it a specialised locomotory function.

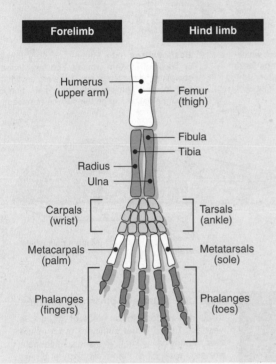

Specialisations of Pentadactyl Limbs

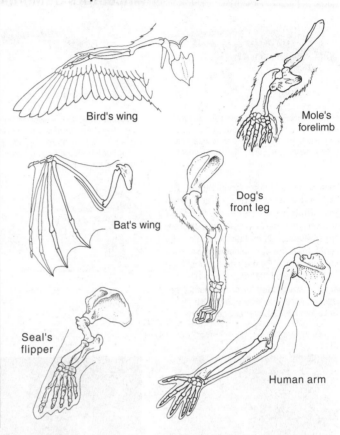

1. Briefly describe the purpose of the major anatomical change that has taken place in each of the limb examples above:

   (a) Bird wing: <u>Highly modified for flight. Forelimb is shaped for aerodynamic lift and feather attachment.</u>

   (b) Human arm: _____

   (c) Seal flipper: _____

   (d) Dog foot: _____

   (e) Mole forelimb: _____

   (f) Bat wing: _____

2. Describe how **homology** in the pentadactyl limb is evidence for adaptive radiation: _____

   _____

   _____

   _____

3. Homology in the behaviour of animals (for example, sharing similar courtship or nesting rituals) is sometimes used to indicate the degree of relatedness between groups. Suggest how behaviour could be used in this way:

   _____

   _____

   _____

Evolution

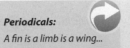

Periodicals:

A fin is a limb is a wing...

Related activities: DNA Homologies
Web links: All in the Family

A 2

Vestigial Organs

Some classes of characters are more valuable than others as reliable indicators of common ancestry. Often, the less any part of an animal is used for specialised purposes, the more important it becomes for classification. This is because common ancestry is easier to detect if a particular feature is unaffected by specific adaptations arising later during the evolution of the species. Vestigial organs are an example of this because, if they have no clear function and they are no longer subject to natural selection, they will remain unchanged through a lineage. It is sometimes argued that some vestigial organs are not truly vestigial, i.e. they may perform some small function. While this may be true in some cases, the features can still be considered vestigial if their new role is a minor one, unrelated to their original function.

Ancestors of Modern Whales

1.8 m long

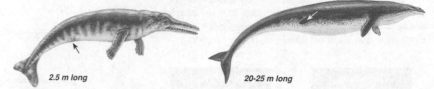

2.5 m long

20-25 m long

Pakicetus (early Eocene) a carnivorous, four limbed, early Eocene whale ancestor, probably rather like a large otter. It was still partly terrestrial and not fully adapted for aquatic life.

Protocetus (mid Eocene). Much more whale-like than *Pakicetus*. The hind limbs were greatly reduced and although they still protruded from the body (arrowed), they were useless for swimming.

Basilosaurus (late Eocene). A very large ancestor of modern whales. The hind limbs contained all the leg bones, but were vestigial and located entirely within the main body, leaving a tissue flap on the surface (arrowed).

Vestigial organs are common in nature. The vestigial hind limbs of modern whales (right) provide anatomical evidence for their evolution from a carnivorous, four footed, terrestrial ancestor. The oldest known whale, *Pakicetus*, from the early Eocene (~54 mya) still had four limbs. By the late Eocene (~40 mya), whales were fully marine and had lost almost all traces of their former terrestrial life. For fossil evidence, see *Whale Origins* at: www.neoucom.edu/Depts/Anat/whaleorigins.htm

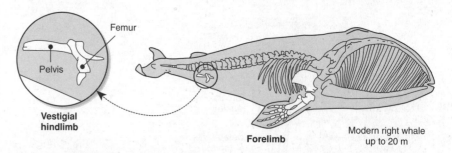

Femur

Pelvis

Vestigial hindlimb

Forelimb

Modern right whale up to 20 m

RM-DoC

Vestigial organs in birds and reptiles

In all snakes (far left), one lobe of the lung is vestigial (there is not sufficient room in the narrow body cavity for it). In some snakes there are also vestiges of the pelvic girdle and hind limbs of their walking ancestors. Like all ratites, kiwis (left) are flightless. However, more than in other ratites, the wings of kiwis are reduced to tiny vestiges. Kiwis evolved in the absence of predators to a totally ground dwelling existence.

1. In terms of natural selection explain how structures, that were once useful to an organism, could become vestigial:

_____

_____

_____

_____

2. Suggest why a vestigial structure, once it has been reduced to a certain size, may not disappear altogether:

_____

_____

3. Whale evolution shows the presence of **transitional forms** (fossils that are intermediate between modern forms and very early ancestors). Suggest how vestigial structures indicate the common ancestry of these forms:

_____

_____

_____

Related activities: Natural Selection

Periodicals:
A waste of space

Oceanic Island Colonisers

The distribution of organisms around the world lends powerful support to the idea that modern forms evolved from ancestral populations. **Biogeography** is the study of the geographical distribution of species, both present-day and extinct. It stresses the role of dispersal of species from a point of origin across pre-existing barriers. Studies from the island populations (below) indicate that flora and fauna of different islands are more closely related to adjacent continental species than to each other.

Galapagos and Cape Verde islands

Galapagos Is — South America — 800 km — Pacific Ocean

Cape Verde Is — Western Africa — 450 km — Atlantic Ocean

Biologists did not fully appreciate the uniqueness and diversity of tropical island biota until explorers began to bring back samples of flora and fauna from their expeditions in the 19th Century. The Galapagos Islands, the oldest of which arose 3-4 million years ago, had species similar to but distinct from those on the South American mainland. Similarly, in the Cape Verde Islands, species had close relatives on the West Africa mainland. This suggested to biologists that ancestral forms found their way from the mainland to the islands where they then underwent evolutionary changes.

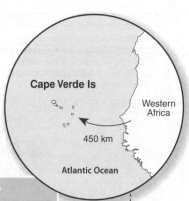

Tristan da Cunha

South America — 4500 km — South Atlantic Ocean — 3000 km — Africa — Tristan da Cunha

The island of Tristan da Cunha in the South Atlantic Ocean is a great distance from any other land mass. Even though it is closer to Africa, there are more species closely related to South American species found there (see table on right). This is probably due to the predominant westerly trade winds from the direction of South America. The flowering plants of universal origin are found in both Africa and South America and could have been introduced from either land mass.

South American origin	
7	Flowering plants
5	Ferns
30	Liverworts

African origin	
2	Flowering plants
2	Ferns
5	Liverworts

Universal origin	
19	Flowering plants

The flightless cormorant is one of a number of bird species that lost the power of flight after becoming resident on an island. Giant tortoises, such as the 11 subspecies remaining on the Galapagos today (center) were, until relatively recently, characteristic of many islands in the Indian Ocean including the Seychelles archipelago, Reunion,

Mauritius, Farquhar, and Diego Rodriguez. These were almost completely exterminated by early Western sailors, although a small population remains on the island of Aldabra. Another feature of oceanic islands is the adaptive radiation of colonising species into different specialist forms. The three species of Galapagos iguana almost certainly arose,

through speciation, from a hardy traveler from the South American mainland. The marine iguana (above) feeds on shoreline seaweeds and is an adept swimmer. The two species of land iguana (not pictured) feed on cacti, which are numerous. The second of these (the pink iguana) was identified as a separate species only in 2009.

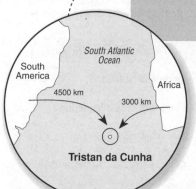

Evolution

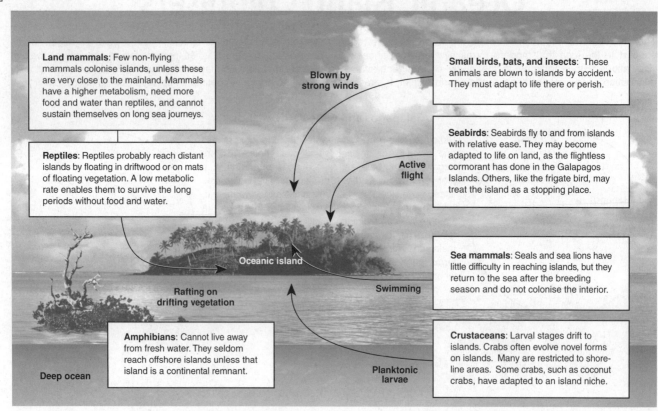

Land mammals: Few non-flying mammals colonise islands, unless these are very close to the mainland. Mammals have a higher metabolism, need more food and water than reptiles, and cannot sustain themselves on long sea journeys.

Reptiles: Reptiles probably reach distant islands by floating in driftwood or on mats of floating vegetation. A low metabolic rate enables them to survive the long periods without food and water.

Blown by strong winds

Small birds, bats, and insects: These animals are blown to islands by accident. They must adapt to life there or perish.

Active flight

Seabirds: Seabirds fly to and from islands with relative ease. They may become adapted to life on land, as the flightless cormorant has done in the Galapagos Islands. Others, like the frigate bird, may treat the island as a stopping place.

Oceanic island

Rafting on drifting vegetation

Swimming

Sea mammals: Seals and sea lions have little difficulty in reaching islands, but they return to the sea after the breeding season and do not colonise the interior.

Deep ocean

Amphibians: Cannot live away from fresh water. They seldom reach offshore islands unless that island is a continental remnant.

Planktonic larvae

Crustaceans: Larval stages drift to islands. Crabs often evolve novel forms on islands. Many are restricted to shore-line areas. Some crabs, such as coconut crabs, have adapted to an island niche.

The diversity and uniqueness of biota of islands is determined by migration to and from the island and extinctions and diversifications following colonisation. These events are themselves affected by a number of other factors (see table, below right). The animals that successfully colonise oceanic islands have to be marine in habit, or able to survive long periods at sea or in the air. This precludes large numbers from ever reaching distant islands.

Plants also have limited capacity to reach distant islands. Only some have fruits and seeds that are salt tolerant. Many plants are transferred to the islands by wind or migrating birds. The biota of the **Galapagos islands** provide a good example of the results of such a colonisation process. For example, all the subspecies of giant tortoise evolved in Galápagos from a common ancestor that arrived from the mainland, floating with the ocean currents.

1. The Galapagos and the Cape Verde Islands are both tropical islands close to the equator, yet their biotas are quite different. Explain why this is the case:

_____

_____

_____

2. Explain why the majority of the plant species found on Tristan da Cunha originated from South America, despite its greater distance from the island:

_____

_____

3. The table (right) identifies some of the factors influencing the composition of island biota. Explain how each of the three following might affect the diversity and uniqueness of the biota found on an oceanic island:

(a) Large island area: _____

_____

(b) Long period of isolation from other land masses: _____

_____

(c) Relatively close to a continental land mass: _____

_____

4. Describe one feature typical of an oceanic island coloniser and explain its significance: _____

_____

Factors affecting final biota
Degree of isolation
Length of time of isolation
Size of island
Climate (tropical/Arctic, arid/humid)
Location relative to ocean currents
Initial plant and animal composition
The species composition of earliest arrivals (if always isolated)
Serendipity (chance arrivals)

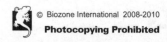

Natural Selection

Natural selection operates on the phenotypes of individuals, produced by their particular combinations of alleles. In natural populations, the allele combinations of some individuals are perpetuated at the expense of other genotypes. This differential survival of some genotypes over others is called **natural selection**. The effect of natural selection can vary; it can act to maintain the genotype of a species or to change it.

Stabilising selection maintains the established favourable characteristics and is associated with stable environments. In contrast, **directional selection** favours phenotypes at one extreme of the phenotypic range and is associated with gradually changing environments. **Disruptive selection** is a much rarer form of selection favouring two phenotypic extremes, and is a feature of fluctuating environments.

Stabilising Selection

Extreme variations are culled from the population (there is selection against them). Those with the established (middle range) adaptive phenotype are retained in greater numbers. This reduces the variation for the phenotypic character. In the example right, light and dark snails are eliminated, leaving medium coloured snails. Stabilising selection can be seen in the selection pressures on human birth weights.

Directional Selection

Directional selection is associated with gradually changing conditions, where the adaptive phenotype is shifted in one direction and one aspect of a trait becomes emphasised (e.g. colouration). In the example right, light coloured snails are eliminated and the population becomes darker. Directional selection was observed in peppered moths in England during the Industrial Revolution. They responded to the air pollution of industrialisation by increasing the frequency of darker, melanic forms.

Disruptive or Diversifying Selection

Disruptive selection favours two extremes of a characteristic at the expense of intermediate forms. It is associated with a fluctuating environment (for example periodic droughts). In the example right, there is selection against medium coloured snails, which are eliminated. Disruptive selection is not the same as balanced selection, although the outcomes often appear similar. Balanced selection occurs when different morphs persist because of heterozygous advantage or frequency dependent selection, and it is not associated with phenotypic extremes.

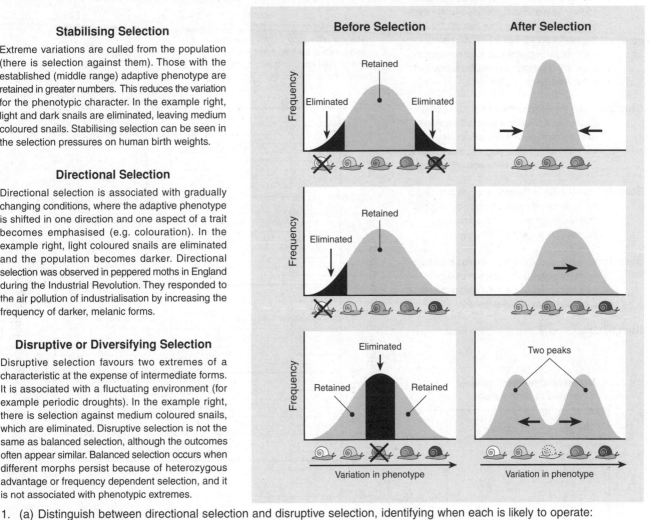

1. (a) Distinguish between directional selection and disruptive selection, identifying when each is likely to operate:

_____

_____

_____

(b) Identify which of the three types of selection described above will lead to evolution, and explain why:

_____

_____

2. Explain how a change in environment may result in selection becoming directional rather than stabilising:

_____

_____

3. Explain how, in a population of snails, through natural selection, shell colour could change from light to dark over time:

_____

_____

Related activities: Darwin's Theory, Selection for Human Birth Weight
Web links: Natural Selection in Populations, Variation : Snails

A 2

Selection for Human Birth Weight

Selection pressures operate on populations in such a way as to reduce mortality. For humans, giving birth is a special, but often traumatic, event. In a study of human birth weights it is possible to observe the effect of selection pressures operating to constrain human birth weight within certain limits. This is a good example of **stabilising selection**. This activity explores the selection pressures acting on the birth weight of human babies. Carry out the steps below:

Step 1: Collect the birth weights from 100 birth notices from your local newspaper (or 50 if you are having difficulty getting enough; this should involve looking back through the last 2-3 weeks of birth notices). If you cannot obtain birth weights in your local newspaper, a set of 100 sample birth weights is provided in the Model Answers booklet.

Step 2: Group the weights into each of the 12 weight classes (of 0.5 kg increments). Determine what percentage (of the total sample) fall into each weight class (e.g. 17 babies weigh 2.5-3.0 kg out of the 100 sampled = 17%)

Step 3: Graph these in the form of a histogram for the 12 weight classes (use the graphing grid provided right). Be sure to use the scale provided on the left vertical (y) axis.

Step 4: Create a second graph by plotting percentage mortality of newborn babies in relation to their birth weight. Use the scale on the right y axis and data provided (below).

Step 5: Draw a line of 'best fit' through these points.

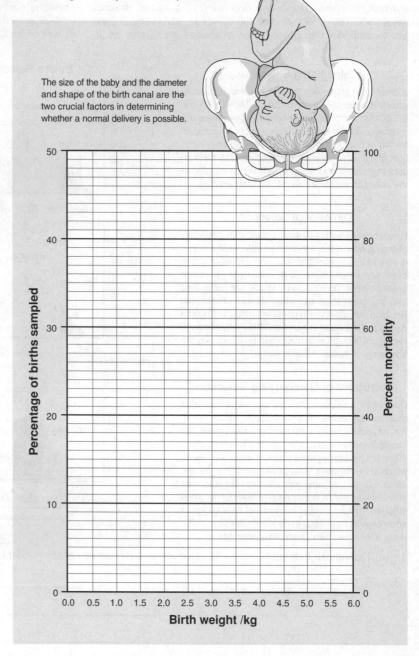

The size of the baby and the diameter and shape of the birth canal are the two crucial factors in determining whether a normal delivery is possible.

Mortality of newborn babies related to birth weight

Weight /kg	Mortality /%
1.0	80
1.5	30
2.0	12
2.5	4
3.0	3
3.5	2
4.0	3
4.5	7
5.0	15

Source: Biology: The Unity & Diversity of Life (4th ed), by Starr and Taggart

1. Describe the shape of the histogram for birth weights: _____

2. State the optimum birth weight in terms of the lowest newborn mortality: _____

3. Describe the relationship between newborn mortality and birth weight: _____

_____

4. Describe the selection pressures that are operating to control the range of birth weight: _____

_____

_____

5. Describe how medical intervention methods during pregnancy and childbirth may have altered these selection pressures:

_____

_____

Adaptation & Evolution: Darwin's Finches

The Galápagos Islands, off the West coast of Ecuador, comprise 16 main islands and six smaller islands. They are home to a unique range of organisms, including 13 species of finches, each of which is thought to have evolved from a single species of grassquit. After colonising the islands, the grassquits underwent adaptive radiation in response to the availability of unexploited feeding niches on the islands. This adaptive radiation is most evident in the present beak shape of each species. The beaks are adapted for different purposes such as crushing seeds, pecking wood, or probing flowers for nectar. Current consensus groups the finches into ground finches, tree finches, warbler finches, and the Cocos Island finches. Between them, the 13 species of this endemic group fill the roles of seven different families of South American mainland birds. DNA analyses have confirmed Darwin's insight and have shown that all 13 species evolved from a flock of about 30 birds arriving a million years ago.

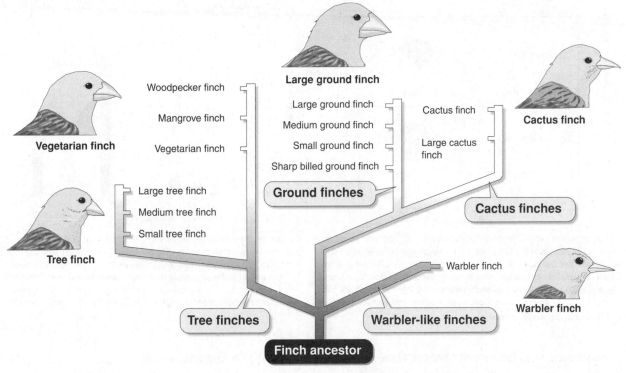

Woodpecker finch
Mangrove finch
Vegetarian finch

Vegetarian finch

Large ground finch

Large ground finch
Medium ground finch
Small ground finch
Sharp billed ground finch

Ground finches

Cactus finch
Large cactus finch

Cactus finches

Cactus finch

Large tree finch
Medium tree finch
Small tree finch

Tree finch

Tree finches

Warbler finch

Warbler-like finches

Warbler finch

Finch ancestor

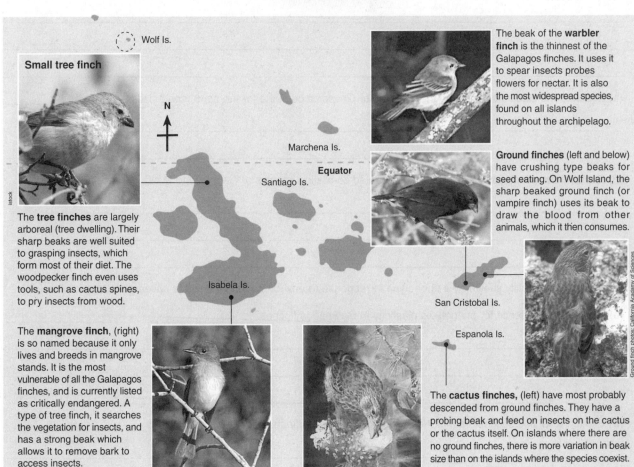

Wolf Is.

Small tree finch

N

Marchena Is.

Equator

Santiago Is.

Isabela Is.

San Cristobal Is.

Espanola Is.

The beak of the **warbler finch** is the thinnest of the Galapagos finches. It uses it to spear insects probes flowers for nectar. It is also the most widespread species, found on all islands throughout the archipelago.

Ground finches (left and below) have crushing type beaks for seed eating. On Wolf Island, the sharp beaked ground finch (or vampire finch) uses its beak to draw the blood from other animals, which it then consumes.

The **tree finches** are largely arboreal (tree dwelling). Their sharp beaks are well suited to grasping insects, which form most of their diet. The woodpecker finch even uses tools, such as cactus spines, to pry insects from wood.

The **mangrove finch**, (right) is so named because it only lives and breeds in mangrove stands. It is the most vulnerable of all the Galapagos finches, and is currently listed as critically endangered. A type of tree finch, it searches the vegetation for insects, and has a strong beak which allows it to remove bark to access insects.

The **cactus finches,** (left) have most probably descended from ground finches. They have a probing beak and feed on insects on the cactus or the cactus itself. On islands where there are no ground finches, there is more variation in beak size than on the islands where the species coexist.

Evolution

Related activities: Darwin's Theory, Adaptations and Fitness
Web links: Darwin's Finches

A 2

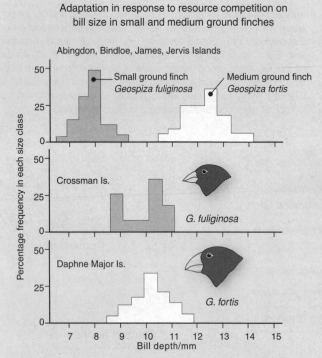

Adaptation in response to resource competition on bill size in small and medium ground finches

Abingdon, Bindloe, James, Jervis Islands

Small ground finch *Geospiza fuliginosa*

Medium ground finch *Geospiza fortis*

Crossman Is.

G. fuliginosa

Daphne Major Is.

G. fortis

Percentage frequency in each size class

Bill depth/mm

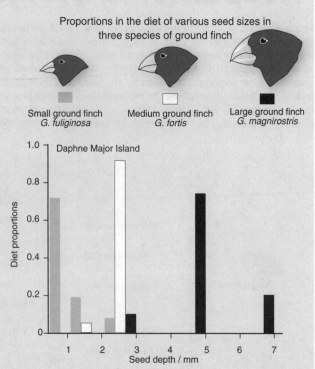

Proportions in the diet of various seed sizes in three species of ground finch

Small ground finch *G. fuliginosa*

Medium ground finch *G. fortis*

Large ground finch *G. magnirostris*

Daphne Major Island

Diet proportions

Seed depth / mm

Two species of ground finch (*G. fuliginosa* and *G. fortis*) are found on a number of islands in the Galapagos. On islands where the species occur together, the bill sizes of the two species are quite different and they feed on different sized seeds, thus avoiding direct competition. On islands where each of these species occurs alone, and there is no competition, the bill sizes of both species move to an intermediate range.

Data based on an adaptation by Strickberger (2000)

Ground finches feed on seeds, but the upper limit of seed size they can handle is constrained by the bill size. Even though small seeds are accessible to all, the birds concentrate on the largest seeds available to them because these provide the most energy for the least handling effort. For example, the large ground finch can easily open smaller seeds, but concentrates on large seeds for their high energy rewards.

1. Describe the main factors that have contributed to the adaptive radiation of Darwin's finches: _____

_____

_____

_____

2. (a) Describe the evidence indicating that species of *Geospiza* compete for the same seed sizes: _____

_____

_____

(b) Explain how adaptations in bill size have enabled coexisting species of *Geospiza* to avoid resource competition:

_____

_____

_____

3. The range of variability shown by a phenotype in response to environmental variation is called **phenotypic plasticity**.

(a) Discuss the evidence for phenotypic plasticity in Galapagos finches: _____

_____

_____

_____

(b) Explain what this suggests about the biology of the original finch ancestor: _____

_____

_____

What is Speciation?

The diagram below represents a possible sequence of genetic events involved in **speciation**, i.e. the origin of two new species from an ancestral population. As time progresses (from top to bottom of the diagram) the amount of genetic divergence increases and each group becomes increasingly isolated from the other. The mechanisms that operate to keep the two gene pools isolated from one another may begin with physical isolation via **geographical barriers**. This is usually followed by **prezygotic** mechanisms, such as behavioural or morphological differences, which further isolate each gene pool. A longer period of isolation may lead to **postzygotic** mechanisms (such as hybrid sterility) that further isolate the two gene pools. As the two gene pools become increasingly isolated they are progressively labelled: population, race, subspecies, and species.

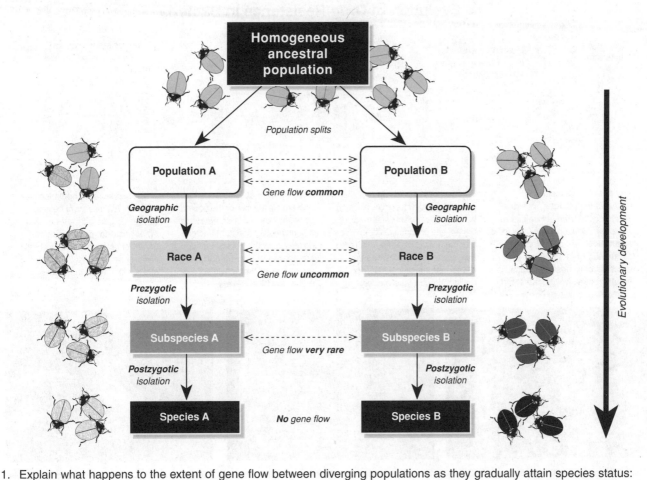

1. Explain what happens to the extent of gene flow between diverging populations as they gradually attain species status:

   _____

   _____

2. Early human populations about 500 000 ya were scattered across Africa, Europe, and Asia. This was a time of many regional variants, collectively called archaic *Homo sapiens*. The fossil skulls from different regions showed mixtures of characteristics, some modern and some 'primitive'. These regional populations are generally given subspecies status. Suggest reasons why gene flow between these populations may have been rare, but still occasionally occurred:

   _____

   _____

   _____

   _____

3. In the southern hemisphere, the native grey duck and the introduced mallard duck (from the Northern hemisphere) are undergoing 'species breakdown'. These two closely related species can interbreed to form hybrids.

   (a) Describe the factor preventing the two species interbreeding before the introduction of the mallards:

   _____

   (b) Describe the factor that may be deterring some of the ducks from interbreeding with the other species:

   _____

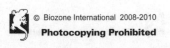
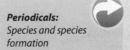

Periodicals:
Species and species
formation

Related activities: Variation in Species
Web links: Mechanisms of Speciation

A 3

Evolution

Antibiotic Resistance

Antibiotics are drugs that inhibit bacterial growth and are used to treat bacterial infections. Resistance to drugs results from an adaptive response that allows microbes to tolerate levels of antibiotic that would normally inhibit their growth. This resistance may arise spontaneously as the result of mutation, or by transfer of genetic material between microbes. Over the years, more and more bacteria have developed resistance to once-effective antibiotics. Methicillin resistant strains of the common bacterium *Staphylococcus aureus* (MRSA) have acquired genes that confer antibiotic resistance to all penicillins, including **methicillin** and other narrow-spectrum pencillin-type drugs. Such strains, called "superbugs", were discovered in the UK in 1961 and are now widespread, and the infections they cause are exceedingly difficult to treat.

The Evolution of Drug Resistance in Bacteria

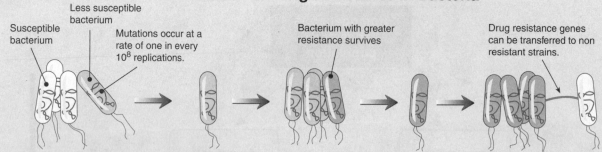

Susceptible bacterium — Less susceptible bacterium — Mutations occur at a rate of one in every 10^8 replications. — Bacterium with greater resistance survives — Drug resistance genes can be transferred to non resistant strains.

Any population, including bacterial populations, includes variants with unusual traits, in this case reduced sensitivity to an antibiotic. These variants arise as a result of mutations in the bacterial chromosome. Such mutations are well documented.

When a person takes an antibiotic, only the most susceptible bacteria will die. The more resistant cells remain and continue dividing. Note that the antibiotic does not create the resistance; it provides the environment in which selection for resistance can take place.

If the amount of antibiotic delivered is too low, or the course of antibiotics is not completed, a population of resistant bacteria develops. Within this population too, there will be variation in susceptibility. Some will survive higher antibiotic levels.

A highly resistant population has evolved. The resistant cells can exchange genetic material with other bacteria, passing on the genes for resistance. The antibiotic initially used against this bacterial strain will now be ineffective.

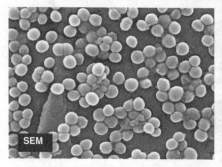

SEM

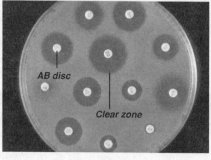

AB disc — Clear zone

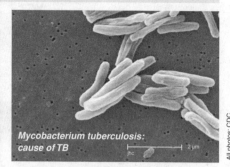

Mycobacterium tuberculosis: cause of TB — 2 µm

All photos: CDC

Staphylococcus aureus is a common bacterium responsible various minor skin infections in humans. MRSA (above) is variant strain that has evolved resistance to penicillin and related antibiotics. MRSA is troublesome in hospital-associated infections where patients with open wounds, invasive devices (e.g. catheters), and weakened immune systems are at greater risk for infection than the general public.

The photo above shows an antibiogram plate culture of *Enterobacter sakazakii*, a rare cause of invasive infections in infants. An antibiogram measures the biological resistance of disease-causing organisms to antibiotic agents. The bacterial lawn (growth) on the agar plate is treated with antibiotic discs, and the sensitivity to various antibiotics is measured by the extent of the clearance zone in the bacterial lawn.

TB is a disease that has experienced spectacular ups and downs. Drugs were developed to treat it, but then people became complacent when they thought the disease was beaten. TB has since resurged because patients stop their medication too soon and infect others. Today, one in seven new TB cases is resistant to the two drugs most commonly used as treatments, and 5% of these patients die.

1. (a) Explain how antibiotic resistance arises in a bacterial population: _____

   _____

   (b) Describe two ways in which antibiotic resistance can become widespread: _____

   _____

   _____

2. With reference to a specific example, discuss the implications to humans of widespread antibiotic resistance:

   _____

   _____

   _____

   _____

   _____

Related activities: *Resistance in Pathogens, Insecticide Resistance*
Web links: *Why Evolution Matters Now, The Rise of Antibiotic Resistance*

Periodicals:
MRSA: A hospital superbug

Insecticide Resistance

Insecticides are pesticides used to control insects considered harmful to humans, their livelihood, or environment. Insecticides have been used for hundreds of years, but their use has proliferated since the advent of synthetic insecticides (e.g. DDT) in the 1940s. When **insecticide resistance** develops the control agent will no longer control the target species. Insecticide resistance can arise through a combination of behavioural, anatomical, biochemical, and physiological mechanisms, but the underlying process is a form of **natural selection**, in which the most resistant organisms survive to pass on their genes to their offspring. To combat increasing resistance, higher doses of more potent pesticides are sometimes used. This drives the selection process, so that increasingly higher dose rates are required to combat rising resistance. This cycle is made worse by the development of multiple resistance in some pest species. High application rates may also kill non-target species, and persistent chemicals may remain in the environment and accumulate in food chains. These concerns have led to some insecticides being banned (DDT has been banned in most developed countries since the 1970s). Insecticides are used in medical, agricultural, and environmental applications, so the development of resistance has serious environmental and economic consequences.

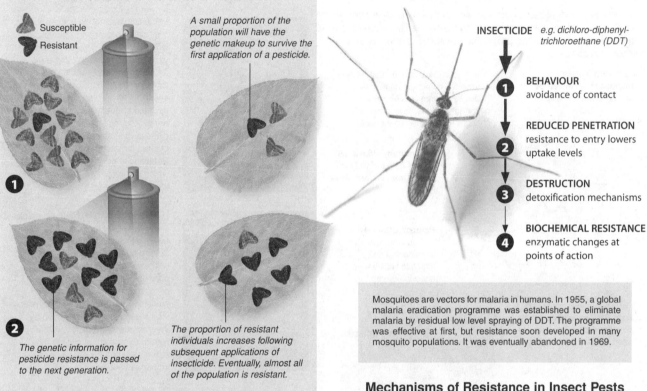

Susceptible

Resistant

A small proportion of the population will have the genetic makeup to survive the first application of a pesticide.

1

2

The genetic information for pesticide resistance is passed to the next generation.

The proportion of resistant individuals increases following subsequent applications of insecticide. Eventually, almost all of the population is resistant.

INSECTICIDE *e.g. dichloro-diphenyl-trichloroethane (DDT)*

1 **BEHAVIOUR**
avoidance of contact

2 **REDUCED PENETRATION**
resistance to entry lowers uptake levels

3 **DESTRUCTION**
detoxification mechanisms

4 **BIOCHEMICAL RESISTANCE**
enzymatic changes at points of action

Mosquitoes are vectors for malaria in humans. In 1955, a global malaria eradication programme was established to eliminate malaria by residual low level spraying of DDT. The programme was effective at first, but resistance soon developed in many mosquito populations. It was eventually abandoned in 1969.

The Development of Resistance

The application of an insecticide can act as a potent selection pressure for resistance in pest insects. The insecticide acts as a selective agent, and only individuals with greater natural resistance survive the application to pass on their genes to the next generation. These genes (or combination of genes) may spread through all subsequent populations.

Mechanisms of Resistance in Insect Pests

Insecticide resistance in insects can arise through a combination of mechanisms. (1) Increased sensitivity to an insecticide will cause the pest to avoid a treated area. (2) Certain genes (e.g. the *PEN* gene) confer stronger physical barriers, decreasing the rate at which the chemical penetrates the cuticle. (3) Detoxification by enzymes within the insect's body can render the pesticide harmless, and (4) structural changes to the target enzymes make the pesticide ineffective. No single mechanism provides total immunity, but together they transform the effect from potentially lethal to insignificant.

1. Give two reasons why widespread insecticide resistance can develop very rapidly in insect populations:

    (a) _____

    (b) _____

2. Explain how repeated insecticide applications acts as a selective agent for evolutionary change in insect populations:

    _____

    _____

3. With reference to synthetic insecticides, discuss the implications of insecticide resistance to human populations:

    _____

    _____

    _____

Antigenic Variability in Pathogens

Influenza (flu) is a disease of the upper respiratory tract caused by the viral genus *Influenzavirus*. Globally, up to 500 000 people die from influenza every year. It is estimated that up to 20% of Britons are affected by the flu every year, and a small number of deaths occur as a result. Three types of *Influenzavirus* affect humans. They are simply named *Influenzavirus* A, B, and C. The most common and most virulent of these strains is *Influenzavirus* A, which is discussed in more detail below. Influenza viruses are constantly undergoing genetic changes. **Antigenic drifts** are small changes in the virus which happen continually over time.

Such changes mean that the influenza vaccine must be adjusted each year to include the most recently circulating influenza viruses. **Antigenic shift** occurs when two or more different viral strains (or different viruses) combine to form a new subtype. The changes are large and sudden and most people lack immunity to the new subtype. New influenza viruses arising from antigenic shift have caused influenza pandemics that have killed millions people over the last century. *Influenzavirus* A is considered the most dangerous to human health because it is capable of antigenic shift.

Structure of *Influenzavirus*

Viral strains are identified by the variation in their H and N surface antigens. Viruses are able to combine and readily rearrange their RNA segments, which alters the protein composition of their H and N glycoprotein spikes.

The *influenzavirus* is surrounded by an **envelope** containing protein and lipids.

The genetic material is actually closely surrounded by protein capsomeres (these have been omitted here and below right in order to illustrate the changes in the RNA more clearly).

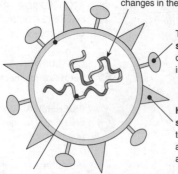

The **neuraminidase (N) spikes** help the virus to detach from the cell after infection.

Hemagglutinin (H) spikes allow the virus to recognise and attach to cells before attacking them.

The viral genome is contained on **eight RNA segments**, which enables the exchange of genes between different viral strains.

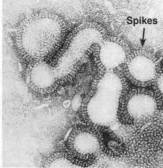

Spikes

Photo right: *Electron micrograph of Influenzavirus showing the glycoprotein spikes projecting from the viral envelope*

Antigenic Shift in *Influenzavirus*

CDC

Influenza vaccination is the primary method for preventing influenza and is 75% effective. The ability of the virus to recombine its RNA enables it to change each year, so that different strains dominate in any one season. The 'flu' vaccination is updated annually to incorporate the antigenic properties of currently circulating strains. Three strains are chosen for each year's vaccination. Selection is based on estimates of which strains will be predominant in the following year.

CDC

H1N1, H1N2, and H3N2 (below) are the known *Influenza A* viral subtypes currently circulating among humans. Although the body will have acquired antibodies from previous flu strains, the new combination of N and H spikes is sufficiently different to enable new viral strains to avoid detection by the immune system. The World Health Organisation coordinates strain selection for each year's influenza vaccine.

| H1N1 | H1N2 | H3N2 |

1. The *Influenzavirus* is able to mutate readily and alter the composition of H and N spikes on its surface.

    (a) Explain why this is the case: _____

    _____

    _____

    (b) Explain how this affects the ability of the immune system to recognise the virus and launch an attack:

    _____

    _____

    _____

2. Discuss why a virus capable of antigenic shift is more dangerous to humans than a virus undergoing antigenic drift:

    _____

    _____

    _____

Related activities: The Global Threat of Disease, Resistance in pathogens

Periodicals: Tracking the next killer flu

KEY TERMS: Mix and Match

INSTRUCTIONS: *Test your vocab by matching each term to its correct definition, as identified by its preceding letter code.*

ADAPTATION

BEHAVIOURAL ADAPTATION

CONTINUOUS VARIATION

DIRECTIONAL SELECTION

DISCONTINUOUS VARIATION

DISRUPTIVE SELECTION

DRUG RESISTANCE

EVOLUTION

FITNESS

FOSSIL

GENOTYPE

HOMOLOGOUS STRUCTURE

HOMOLOGY

NATURAL SELECTION

PESTICIDE RESISTANCE

PHENOTYPE

PHYSIOLOGICAL ADAPTATION

POPULATION

SPECIES

STABILISING SELECTION

STRUCTURAL ADAPTATION

TRANSITIONAL FOSSIL

VARIATION

VESTIGIAL ORGAN

A Selection in which a single phenotype and therefore allele frequency is favoured and the phenotypic norm shifts in one direction.

B The observable properties of an organism that result from both genes and environment.

C The genetic make-up of an organism.

D Reduction in effectiveness of, or response to, a drug.

E A feature of an organism's physiology that contributes to evolutionary fitness.

F Changes in population genetics in which intermediate (most common) values for a characteristic are favoured over more extreme forms.

G Variation within a population of a characteristic that falls into two or more discrete categories.

H A group of organisms capable of breeding together to produce viable offspring.

I The fossilised remains of intermediate forms of life that show an evolutionary transition.

J The preserved remains or traces of animals, plants, and other organisms from the remote past.

K Changes in population genetics in which extreme values for a characteristic are favoured over intermediate forms.

L Variation within a population of a characteristic that is the result of the contribution of many genes and the environment.

M A body part that has the same basic structure and embryonic origin as that of another organism as a result of shared ancestry.

N Degenerate or rudimentary structures that are usually largely or entirely functionless, and a legacy of an organism's evolutionary past.

O The evolutionary process by which a species becomes better suited to its habitat *(verb)*. A feature that results from this process *(noun)*.

P Reduction in effectiveness of, or response to, a pesticide.

Q Variation in alleles of genes, occuring both within and among species.

R In evolutionary terms, the capability of an individual of certain genotype to reproduce. It is usually equal to the proportion of the individual's genes in all the genes of the next generation.

S The process by which certain heritable traits that improve fitness in the prevailing environment become more common in a population over successive generations.

T Similarity as a result of shared ancestry.

U A feature of an organism's behaviour that contributes to evolutionary fitness.

V A group of organisms of one species that interbreed and live in the same place at the same time.

W A cumulative change in the characteristics and gene pool of a population over many successive generations.

X A feature of an organism's structure (anatomy) that contributes to evolutionary fitness.

Maintaining
Biodiversity

KEY CONCEPTS

▶ High biodiversity contributes to ecosystem stability and resilience.

▶ Climate change will have implications for biodiversity, agriculture, and incidence of disease.

▶ Agriculture can benefit from the implementation of practices that maintain or enhance biodiversity.

▶ *In-situ* and *ex-situ* methods both have a role in the conservation of threatened species.

KEY TERMS

biodiversity
botanic gardens
captive breeding
CITES
ecosystem restoration
endangered
environmental impact assessment
ex-situ conservation
gene bank
global warming
greenhouse effect
in-situ conservation
keystone species
resilience
Rio Convention on Biodiversity
seedbank
stability
The Hedgerows Regulations
threatened
vulnerable
zoos

Periodicals:
Listings for this chapter are on page 341

Weblinks:
www.biozone.co.uk/
weblink/OCR-AS-2641.html

OBJECTIVES

☐ 1. Use the **KEY TERMS** to help you understand and complete these objectives.

Why Conserve Biodiversity?   pages 256-258, 323-324

☐ 2. Recognise that areas of high **biodiversity** are under increasing pressure as human populations expand.

☐ 3. EXTENSION: Recognise the relationship between **diversity** and **ecosystem stability** and resilience. Describe the role of **keystone species** in ecosystem function and the possible consequences of removing these species.

☐ 4. Using an appropriate example (e.g. conservation of hedgerows in Britain), discuss the ethical, ecological, economic, and aesthetic reasons for the conservation of biodiversity.

Agriculture, Climate, & Biodiversity   pages 201-202, 259-260, 325-336

☐ 5. BACKGROUND/EXTENSION: Describe and explain the causes and consequences of the enhanced **greenhouse effect (global warming)**. Include an analysis of the changes in concentration of atmospheric CO_2 as documented by historical records.

☐ 6. Discuss the consequences of global **climate change** on global and local biodiversity. Include in your discussion reference to changing patterns of agriculture and changes in the patterns and incidence of disease.

☐ 7. Recall the differences between intensive and traditional agricultural practices. Using an appropriate example, e.g. conservation of hedgerows or sustainable farming, describe and explain how agriculture can benefit by maintaining or enhancing biodiversity.

☐ 8. Describe *in-situ* conservation (e.g. **ecosystem restoration**) and *ex-situ* conservation (e.g. **captive breeding** in **zoos**). Discuss the benefits and problems of each approach in the conservation of endangered species.

☐ 9. Using examples, discuss the role of **botanic gardens** and seedbanks (a type of **gene bank**) in the *ex-situ* conservation of plant species that are rare or extinct in the wild.

☐ 10. Discuss the role of international cooperation in the management of endangered species and the conservation of biodiversity. Include reference to **CITES** and the **Rio Convention on Biodiversity**.

☐ 11. Recognise the impact of local developments on biodiversity. Discuss the importance of **environmental impact assessments** (including estimates of biodiversity) when planning decisions are made by local authorities. Use an appropriate example, e.g. the Hedgerows Regulations 1997.

Diversity, Stability, and Key Species

Ecological theory suggests that all species in an ecosystem contribute in some way to ecosystem function. Therefore, species loss past a certain point is likely to have a detrimental effect on the functioning of the ecosystem and on its ability to resist change (its stability). Although many species still await discovery, we do know that the rate of species extinction is increasing. Scientists estimate that human destruction of natural habitat is implicated in the extinction of up to 100 000 species every year. This substantial loss of biodiversity has serious implications for the long term stability of many ecosystems.

The Concept of Ecosystem Stability

The stability of an ecosystem refers to its apparently unchanging nature over time. Ecosystem stability has various components, including **inertia** (the ability to resist disturbance) and **resilience** (ability to recover from external disturbances). Ecosystem stability is closely linked to the biodiversity of the system, although it is difficult to predict which factors will stress an ecosystem beyond its range of tolerance.

It was once thought that the most stable ecosystems were those with the greatest number of species, since these systems had the greatest number of biotic interactions operating to buffer them against change. This assumption is supported by experimental evidence but there is uncertainty over what level of biodiversity provides an insurance against catastrophe.

Monoculture / Natural grassland

Rainforest

Deforestation

Single species crops (monocultures), such as the soy bean crop (above, left), represent low diversity systems that can be vulnerable to disease, pests, and disturbance. In contrast, natural grasslands (above, right) may appear homogeneous, but contain many species which vary in their predominance seasonally. Although they may be easily disturbed (e.g. by burning) they are very resilient and usually recover quickly.

Tropical rainforests (above, left) represent the highest diversity systems on Earth. Whilst these ecosystems are generally resistant to disturbance, once degraded, (above, right) they have little ability to recover. The biodiversity of ecosystems at low latitudes is generally higher than that at high latitudes, where climates are harsher, niches are broader, and systems may be dependent on a small number of key species.

Community Response to Environmental Change

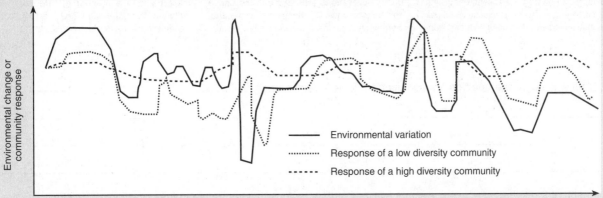

— Environmental variation
···· Response of a low diversity community
- - - Response of a high diversity community

Modified from Biol. Sci. Rev., March 1999 (p. 22)

Time or space

In models of ecosystem function, higher species diversity increases the stability of ecosystem functions such as productivity and nutrient cycling. In the graph above, note how the low diversity system varies more consistently with the environmental variation, whereas the high diversity system is buffered against major fluctuations. In any one ecosystem, some species may be more influential than others in the stability of the system. Such **keystone (key) species** have a disproportionate effect on ecosystem function due to their pivotal role in some ecosystem function such as nutrient recycling or production of plant biomass.

Elephants can change the entire vegetation structure of areas into which they migrate. Their pattern of grazing on taller plant species promotes a predominance of lower growing grasses with small leaves.

Termites are amongst the few larger soil organisms able to break down plant cellulose. They shift large quantities of soil and plant matter and have a profound effect on the rates of nutrient processing in tropical environments.

The starfish *Pisaster* is found along the coasts of North America where it feeds on mussels. If it is removed, the mussels dominate, crowding out most algae and leading to a decrease in the number of herbivore species.

Maintaining Biodiversity

Keystone Species in the Northen Hemisphere

The importance of keystone species is usually brought to attention when they disappear from an ecosystem and there is a dramatic shift in community structure and function as a result. The phenomenon has been observed in a wide range of ecosystems and for a wide range of organisms. The concept of the keystone species has become very popular in conservation biology and has been used as an argument for the reintroduction of some species (such as grey wolves and European beavers) into parts of their former range.

Grey wolf

Beaver, *Castor canadensis*

Sea otter, *Enhydra lutris*

Scots pine

Grey or timber wolves (Canis lupus) are a keystone predator and were once widespread through North America, Europe, and Eurasia. Historically, they have been eliminated because of their perceived threat to humans and livestock and now occupy only a small fraction of their former range. As a top predator, the wolf is a keystone species in the ecosystems to which it belongs. When they are absent, populations of their prey (e.g. red deer) increase to the point that they adversely affect other flora and fauna.

Two smaller mammals are also important keystone species in North America. **Beavers** (top) play a crucial role in biodiversity and many species, including 43% of North America's endangered species, depend partly or entirely on beaver ponds. Californian **sea otters** are also critical to ecosystem function. When their numbers were decimated by the fur trade, sea urchin populations exploded and the dense kelp forests, on which many species depend, were destroyed, leaving a desert-like terrain.

Scots pine (Pinus sylvestris) is the most widely distributed conifer in the world. In the Scots pine forests in Scotland, this species occupies a unique position, both because of the absence of other native conifers and because it directly or indirectly supports so many other species. Among those dependent on Scots pine for survival are blaeberries, wood ants, pine martens, and a number of bird species, including the capercallie and the UK's only endemic bird, the Scottish crossbill.

1. Explain why **keystone species** are so important to ecosystem function: _____

   _____

   _____

   _____

2. For each of the following species, discuss features of their biology that contribute to their position as keystone species:

   (a) Grey wolf: _____

   _____

   (b) Californian sea otter: _____

   _____

   (c) Scots pine: _____

   _____

3. Using an illustrative example if possible, suggest why the concept of keystone species has been used to argue for the reintroduction of species into regions where they have been removed (e.g. by hunting):

   _____

   _____

   _____

   _____

Global Warming

The Earth's atmosphere comprises a mixture of gases including nitrogen, oxygen, and water vapour. Also present are small quantities of carbon dioxide (CO_2), methane, and a number of other "trace" gases. In the past, our climate has shifted between periods of stable warm conditions to cycles of glacials and interglacials. The current period of warming climate is partly explained by the recovery after the most recent ice age that finished 10 000 years ago. Eight of the ten warmest years on record (records kept since the mid-1800s) were in the 1980s and 1990s. Global surface temperatures in 1998 set a new record by a wide margin, exceeding those of the previous record year, 1995. Many researchers believe the current warming trend has been compounded by human activity, in particular, the release of certain gases into the atmosphere. The term '**greenhouse effect**' describes a process of global climate warming caused by the release of 'greenhouse gases', which act as a thermal blanket in the atmosphere, letting in sunlight, but trapping the heat that would normally radiate back into space. About three-quarters of the natural greenhouse effect is due to water vapour. The next most significant agent is CO_2. Since the industrial revolution and expansion of agriculture about 200 years ago, additional CO_2 has been pumped into the atmosphere. The effect of global warming on agriculture, other human activities, and the biosphere in general, is likely to be considerable.

Maintaining Biodiversity

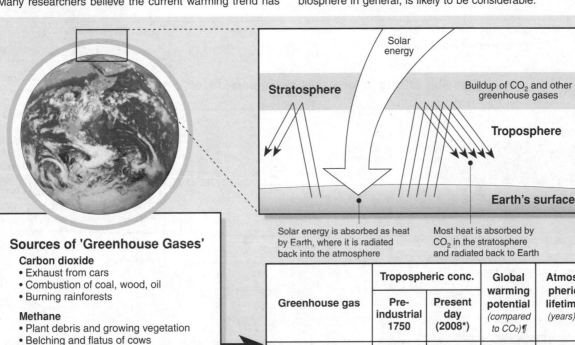

Sources of 'Greenhouse Gases'

Carbon dioxide
- Exhaust from cars
- Combustion of coal, wood, oil
- Burning rainforests

Methane
- Plant debris and growing vegetation
- Belching and flatus of cows

Chloro-fluoro-carbons (CFCs)
- Leaking coolant from refrigerators
- Leaking coolant from air conditioners

Nitrous oxide
- Car exhaust

Tropospheric ozone*
- Triggered by car exhaust (smog)

*Tropospheric ozone is found in the lower atmosphere (not to be confused with ozone in the stratosphere)

Solar energy is absorbed as heat by Earth, where it is radiated back into the atmosphere

Most heat is absorbed by CO_2 in the stratosphere and radiated back to Earth

Greenhouse gas	Tropospheric conc.		Global warming potential (compared to CO_2)¶	Atmospheric lifetime (years)§
	Pre-industrial 1750	Present day (2008*)		
Carbon dioxide	280 ppm	383.9 ppm	1	120
Methane	700 ppb	1796 ppb	25	12
Nitrous oxide	270 ppb	320.5 ppb	310	120
CFCs	0 ppb	0.39 ppb	4000+	50-100
HFCs‡	0 ppb	0.045 ppb	1430	14
Tropospheric ozone	25 ppb	34 ppb	17	hours

ppm = parts per million; **ppb** = parts per billion; ‡Hydrofluorcarbons were introduced in the last decade to replace CFCs as refrigerants; * Data from July 2007-June 2008. ¶ Figures contrast the radiative effect of different greenhouse gases relative to CO_2 over 100 years, e.g. over 100 years, methane is 25 times more potent as a greenhouse gas than CO_2 § How long the gas persists in the atmosphere. Source: CO_2 Information Analysis Centre, Oak Ridge National Laboratory, USA.

The graph on the right shows how the mean temperature for each year from 1860 until 2008 (grey bars) compared with the average temperature between 1961 and 1990. The thick black line represents the mathematically fitted curve and shows the general trend indicated by the annual data. Most anomalies since 1977 have been above normal; warmer than the long term mean, indicating that global temperatures are tracking upwards. In 1998 the global temperature exceeded that of the previous record year, 1995, by about 0.2°C.

Source: Hadley Center for Prediction and Research

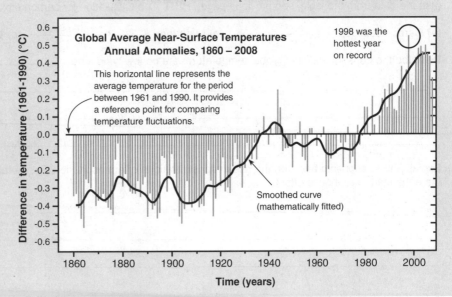

© Biozone International 2008-2010
Photocopying Prohibited

Periodicals: Global warming

Related activities: Biodiversity and Global Warming
Web links: The Greenhouse Effect, NRDC: Global Warming

DA 2

Effects of increases in temperature on crop yields

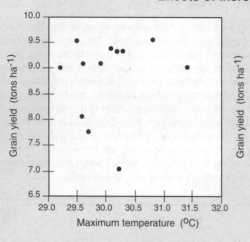

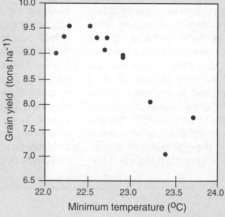

Studies on the grain production of rice have shown that maximum daytime temperatures have little effect on crop yield. However minimum night time temperatures lower crop yield by as much as 5% for every 0.5°C increase in temperature.

Source: Peng S. *et.al.* PNAS 2004

Possible effects of increases in temperature on crop damage

Source: Currano *et.al.* PNAS 2007

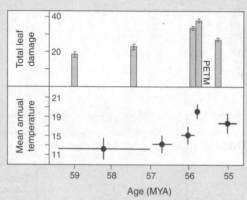

The fossil record shows that global temperatures rose sharply around 56 million years ago. Studies of fossil leaves with insect browse damage indicate that leaf damage peaked at the same time as the Paleocene Eocene Thermal Maximum (PETM). This gives some historical evidence that as temperatures increase, plant damage caused by insects also rises. This could have implications for agricultural crops.

Under a warming climate, it is likely that there will be a northwards shift in natural habitats and agricultural zones of 50-80 km per decade. Climate models predict significant impacts on soils, water resources, and agriculture in the south of England, which will be at increased risk of drought. In the north, growing seasons may increase.

Predicted rises in sea levels will threaten low-lying coastal areas of the world, such as the Netherlands and Bangladesh, with inundation. In Britain, East Anglia, the Thames Estuary and low-lying cities will be at risk from the rising sea and more frequent tidal surges. Estimates vary, but rises may be between 30 and 50 cm by 2100.

1. Calculate the increase (as a %) in the 'greenhouse gases' between the pre-industrial era and the 2008 measurements (use the data from the table, see previous page). **HINT**: The calculation for carbon dioxide is: $(383.9 - 280) \div 280 \times 100 =$

   (a) Carbon dioxide: _____ (b) Methane: _____ (c) Nitrous oxide: _____

2. Describe the consequences of global temperature rise on low lying land: _____

3. Explain the relationship between the rise in concentrations of atmospheric CO_2, methane, and oxides of nitrogen, and the enhanced greenhouse effect:

Biodiversity and Global Warming

Since the last significant period of climate change at the end of the ice age 10,000 years ago, plants and animals have adapted to survive in their current habitats. Accelerated global warming is again changing the habitats that plants and animals live in and this could have significant effects on the biodiversity of specific regions as well as on the planet overall. As temperatures rise, organisms will be forced to move to new areas where temperatures are similar to their current level. Those that cannot move face extinction, as temperatures move outside their limits of tolerance. Changes in precipitation as a result of climate change also affect where organisms can live. Long term changes in climate could see the contraction of many organisms' habitats while at the same time the expansion of others. Habitat migration, the movement of a habitat from its current region into another, will also become more frequent. Already there are a number of cases showing the effects of climate change on a range of organisms.

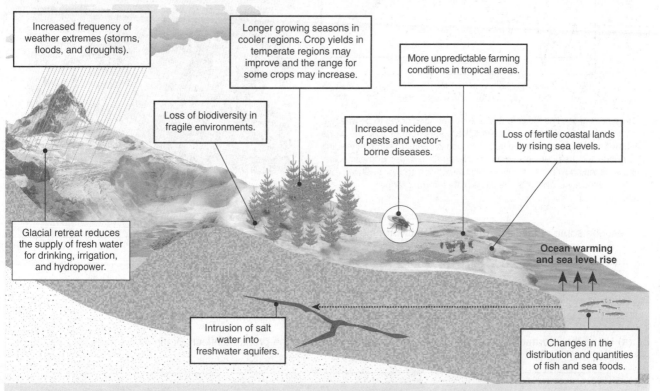

Increased frequency of weather extremes (storms, floods, and droughts).

Longer growing seasons in cooler regions. Crop yields in temperate regions may improve and the range for some crops may increase.

More unpredictable farming conditions in tropical areas.

Loss of biodiversity in fragile environments.

Increased incidence of pests and vector-borne diseases.

Loss of fertile coastal lands by rising sea levels.

Glacial retreat reduces the supply of fresh water for drinking, irrigation, and hydropower.

Ocean warming and sea level rise

Intrusion of salt water into freshwater aquifers.

Changes in the distribution and quantities of fish and sea foods.

Rothiemurchus Caledonian forest

Scotland's ancient Caledonian pinewood forests once covered thousands of kilometres of the highlands. Now they are restricted to 180 km^2. at only 4 sites. A warmer, wetter climate is predicted for Scotland, and the diversity of lowland flora is expected to increase as southern species move northwards. Species typical of the Caledonian forest, such as one-flowered wintergreen, will be threatened if the forest converts to broadleaved woodland, so conservation of pines will be a priority.

Butterfly populations in the UK are in steep decline as the climate warms and their capacity to move north is limited by habitat fragmentation. For some species, e.g. the marbled white, translocation to cooler regions has enabled viable populations to thrive.

Photo: Walter Siegmund

Studies of sea life along the Californian coast have shown that between 1931 and 1996, shoreline ocean temperatures increased by 0.79°C and populations of invertebrates including sea stars, limpets and snails moved northward in their distributions.

An Australian study in 2004 found the centre of distribution for the AdhS gene in *Drosophila*, which helps survival in hot and dry conditions, had shifted 400 kilometres south in the last twenty years.

A 2009 study of 200 million year old plant fossils from Greenland has provided evidence of a sudden collapse in biodiversity that is correlated with, and appears to be caused by, a very slight rise in CO$_2$ levels.

Periodicals:
Climate change
and biodiversity

Related activities: Global Warming
Web links: Climate Change, Global Warming Ecocentre

A 2

Effects of increases in temperature on animal populations

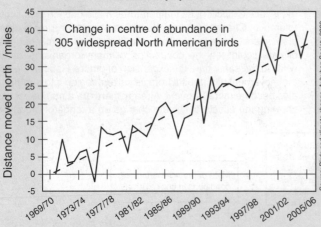

Change in centre of abundance in 305 widespread North American birds

Distance moved north / miles

45
40
35
30
25
20
15
10
5
0
-5

1969/70 1973/74 1977/78 1981/82 1985/86 1989/90 1993/94 1997/98 2001/02 2005/06

Source: Birds and climate change. Aububon Society 2009

A number of studies indicate that animals are beginning to be affected by increases in global temperatures. Data sets from around the world show that birds are migrating up to two weeks earlier to summer feeding grounds and are often not migrating as far south in winter. Some species, such as the snow bunting, which relies directly on snow for habitat, may become locally extinct in some parts of their former range.

Animals living at altitude are also affected by warming climates and are being forced to shift their normal range. As temperatures increase, the snow line increases in altitude pushing alpine animals to higher altitudes. In some areas of North America this has resulting the local extinction of the North American pika (*Ochotona princeps*).

Photo: sevenstar

1. Describe some of the likely effects of global warming on physical aspects of the environment: _____

   _____

   _____

   _____

2. (a) Using the information on this and the previous activity, discuss the probable effects of global warming on plant crops:

   _____

   _____

   (b) Suggest how farmers might be able to adjust to these changes: _____

   _____

   _____

   _____

3. Discuss the evidence that insect populations are affected by global temperature: _____

   _____

   _____

4. (a) Describe how increases in global temperatures have affected some migratory birds: _____

   _____

   _____

   (b) Explain how these changes in migratory patterns might affect food availability for these populations: _____

   _____

   _____

5. Explain how global warming could lead to the local extinction of some alpine species: _____

   _____

   _____

   _____

Agriculture and Diversity

The English countryside has been shaped by many hundreds of years of agriculture. The landscape has changed as farming practices evolved through critical stages. Farming has always had an impact on Britain's rich biodiversity (generally in a negative manner). Modern farming practices, such as increasing mechanisation and the move away from mixed farming operations, have greatly accelerated this decline. In recent years active steps to conserve the countryside, such as **hedgerow legislation**, policies to increase woodland cover, and schemes to promote environmentally sensitive farming practices are slowly meeting their objectives. Since 1990, expenditure on agri-environmental measures has increased, the area of land in organic farming has increased, and the overall volume of inorganic fertilisers and pesticides has decreased. The challenge facing farmers, and those concerned about the countryside, is to achieve a balance between the goals of production and conservation of diversity.

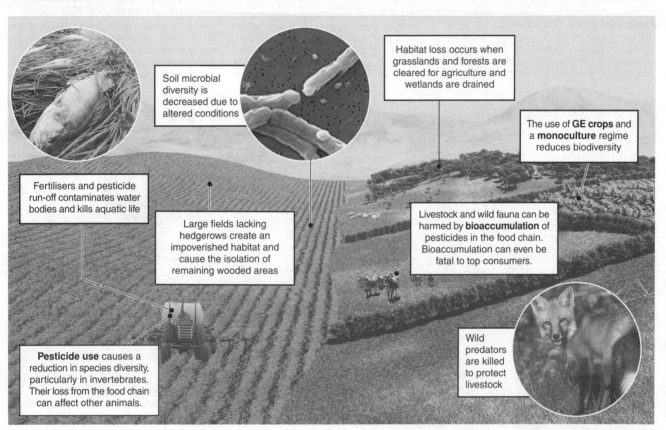

Soil microbial diversity is decreased due to altered conditions

Habitat loss occurs when grasslands and forests are cleared for agriculture and wetlands are drained

The use of **GE crops** and a **monoculture** regime reduces biodiversity

Fertilisers and pesticide run-off contaminates water bodies and kills aquatic life

Large fields lacking hedgerows create an impoverished habitat and cause the isolation of remaining wooded areas

Livestock and wild fauna can be harmed by **bioaccumulation** of pesticides in the food chain. Bioaccumulation can even be fatal to top consumers.

Pesticide use causes a reduction in species diversity, particularly in invertebrates. Their loss from the food chain can affect other animals.

Wild predators are killed to protect livestock

Maintaining Biodiversity

Natural grasslands are diverse and productive ecosystems. Ancient meadows may have contained 80-100 plant species, in contrast to currently cultivated grasslands, which may contain as few as three species. Unfortunately, many of the management practices that promote grassland species diversity conflict with modern farming methods. For example, the extensive use of fertilisers and selective herbicides on pastures favours aggressive species, such as nettles and docks, which outcompete ecologically important species such as orchids and cowslips. Appropriate management can help to conserve grassland ecosystems while maintaining their viability for agriculture.

1. One solution to the conflicting needs of conserving biodiversity and productivity is to intensively farm designated areas, leaving other areas for conservation. From the farmer's perspective, outline two advantages of this approach:

(a) _____

_____

(b) _____

_____

(c) Describe a disadvantage of this management approach: _____

_____

_____

© Biozone International 2008-2010
Photocopying Prohibited

Periodicals:
Down on the farm

Related activities: Hedgerows: An Ancient Tradition
Web links: Conservation Grazing, Meadows

RA 2

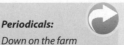

An increase in urban sprawl and the pressure on farmers to increase productivity are having a dramatic impact on the once common flowering plants of Britain's grasslands. Diversity be maintained only through careful management and conservation of existing ecosystems.

Conservation of grasslands is not only important for maintaining plant diversity Many birds, reptiles, invertebrates, and mammals rely on these ecosystems for food and shelter. A reduction in the diversity of grassland plant species translates to a reduction in the diversity of other species.

This woodland in Yorkshire, England, is home to numerous species of organisms. Clearing land for agriculture reduces both biodiversity and the ability of the community to adapt to changing environmental conditions. Natural ecosystem stability is decreased as a result.

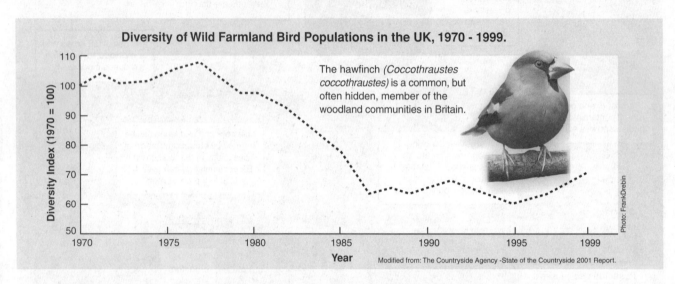

Diversity of Wild Farmland Bird Populations in the UK, 1970 - 1999.

The hawfinch (Coccothraustes coccothraustes) is a common, but often hidden, member of the woodland communities in Britain.

Modified from: The Countryside Agency -State of the Countryside 2001 Report.

2. Populations of wild farmland bird species, because of their wide distribution and position near the top of the food chain, provide good indicators of the state of other wildlife species and of environmental health in general. Over the last 25 years, there has been a marked net decline in the diversity of farmland bird populations (above). However, since 1986, diversity has ceased to decline further and, in recent years, has actually showed an increase.

Suggest two possible reasons for this decline in the diversity of farmland birds (also see the activity on hedgerows):

(a) _____

(b) _____

3. (a) Describe three initiatives local and national government have implemented in an attempt to reverse this decline:

(b) Discuss the role of environmental impact assessments and biodiversity estimates when planning such initiatives:

4. Provide an argument for retaining areas of uncultivated meadow alongside more intensively managed pasture:

Hedgerows: An Ancient Tradition

A significant feature of landscape change in England since WWII has been the amalgamation of fields and the removal of traditional hedgerows. Many traditional, mixed farms, which required hedgerows to contain livestock, have been converted to arable farms, and fields have become larger to accommodate modern machinery. In Britain, this conversion has resulted in the loss of thousands of kilometres of hedgerows each year. Deliberate removal is not the only threat to hedgerows either. Neglect and improper management have been responsible for almost half of lost hedgerows every year. Hedgerows require maintenance and management in order to remain viable, yet hedge-laying and trimming skills are rapidly becoming lost. In 1997, legislation was introduced to control the destruction of hedgerows in rural settings. In England and Wales, landowners must now apply to the local authority for permission to remove a hedgerow of greater than 20 metres in length, and this can be refused if the hedge is shown to be significant in terms of its age, environmental, or historical importance.

Hedgerows are important because...

- Hedges may support up to 80% of England's birds, 50% of its mammals, and 30% of its butterflies.
- The ditches and banks associated with hedgerows provide habitat for amphibians and reptiles.
- Hedges provide habitat, nesting material and food for birds and mammals.
- Some small mammals, e.g. dormice, once used hay ricks as overwintering habitat. With the loss of hay ricks, hedgerows are virtually their only alternative.
- They act as corridors, along which animals (e.g. pheasants) can safely move.
- They provide overwintering habitat for predatory insects which move into crops to control pest insects in spring.
- Hedges provide shelter for stock and crops and reduce wind speed, which prevents erosion.
- Hedges act as barriers for windborne pests.

Photo courtesy, Kimberley Mallady

Bjorn Schulz

Hazel dormouse

Hedgerows commonly comprise hawthorn, blackthorn, field maple, hazel, and bramble. A hedgerow is essentially a linear wood and many of the associated plants are woodland species. At least 30 bird species nest in hedges. Hedgerows of different heights are preferred by different bird species, so management to provide a range of hedge heights and tree densities provides the best option for increasing diversity. For example, bullfinches prefer well-treed hedgerows over 4 m tall, whereas whitethroats, linnets, and yellowhammers favour shorter hedgerows (2-3 m) with fewer trees. The hedge base is important for ground-nesting species like the grey partridge. Hedgerows are important habitat for dormice and are used as dispersal corridors linking copses that are too small to support a viable populations on their own. Crucially they also support breeding populations independent of other habitats.

1. From an environmental perspective, describe three benefits of hedgerows to biodiversity:

   (a) _____

   (b) _____

   (c) _____

2. Explain why hedgerows might be regarded as undesirable from the perspective of a modern farmer: _____

   _____

   _____

3. Outline a brief argument to convince a farmer to retain and manage hedgerows, rather than remove them:

   _____

   _____

   _____

   _____

   _____

Periodicals:
Insecticides and the
conservation of hedgerows

Related activities: Britain's Biodiversity, Agriculture and Diversity
Web links: Web links: Defra: Hedgerows, NatureNet: Hedgerows

In-Situ Conservation

One of the concerns facing conservationists today is the rapidly accelerating rate at which species are being lost. In 1992, the **Convention on Biological Diversity** was adopted in Rio de Janeiro. It is an international treaty and its aims are to conserve biodiversity, use biodiversity in a sustainable way, and ensure that the benefits of genetic resources are shared equitably. Various strategies are available to protect at-risk species and help the recovery of those that are threatened. *In-situ* methods focus on ecological restoration and legislation to protect ecosystems of special value (below left). Ecological restoration is a long term process and usually involves collaborative work between institutions with scientific expertise and the local communities involved.

A Case Study in *In-Situ* Conservation
East Midland Ancient Woodland Project

The East Midland Ancient Woodland Project is a government project to restore ancient woodland sites in the Northants Forest District in central England. England was once extensively wooded, but much native woodland was cleared or replanted last century, mostly with Norway spruce. The project, which was launched in 2000, aims to restore ancient woodland sites (those dating back to 1600 or before) to native woodland. The project will focus on key species.

The objectives

- ▶ To identify key features of ancient woodlands.
- ▶ To draw up design plans (with input from the public) for existing woodlands.
- ▶ To establish groups to monitor and review progress.
- ▶ To promote the value of woodlands for recreation, ecology, and heritage.

How the project will proceed

- ▶ Restore plantation sof ancient woodland sites to semi-natural woodland.
- ▶ Remove conifers and exotic broadleaf species.
- ▶ Expand the size and range of key species.
- ▶ Restore and manage the woodland to reflect its culture and history, while recognising current uses.
- ▶ To generate income to contribute to the costs of long term management.

Source: UK Clearing House Mechanisms for Biodiversity

Bluebell woodland in Buckinghamshire

Keith Hulbert and Paul Zarucki

Woodland-pond restoration (UK)

Orangutan (endangered species)

Confiscated ivory, Kenya

Habitat Protection and Restoration

Most countries have a system of reserve lands focused on ecosystem conservation. These areas aim to protect and restore habitats of special importance and they may be intensively managed through pest and weed control, revegetation, reintroduction of threatened species, and site specific management practices (such as coppicing).

Convention on International Trade in Endangered Species (CITES)

CITES is an international agreement between governments which aims to ensure that trade in species of wild animals and plants does not threaten their survival. CITES comprises more than 150 member nations and includes virtually all important wildlife producing and consuming countries. Trade in over 40,000 species is controlled or prohibited depending on their level of threat. In 1989, CITES imposed a global ban on the international trade in ivory and ivory products (above) in a move that has helped enormously in reducing the slaughter of elephants.

1. Explain why *in-situ* conservation commonly involves both ecosystem restoration and legislation to protect species:

_____

_____

_____

_____

Related activities: Ex-Situ Conservation
Web links: Protected Areas and In-Situ Conservation

Periodicals:
Conflicted conservation

Ex-Situ Conservation

Ex-situ conservation methods operate away from the natural environment and are particularly useful where species are critically endangered. Zoos, aquaria, and botanical gardens are the most conventional vehicles for *ex-situ* conservation. They house and protect specimens for breeding and, where necessary and possible, they reintroduce them into the wild to restore natural populations. Many UK-based organisations, such as Twycross Zoo, the Durrell Wildlife Conservation Trust, and Kew Gardens, are involved in collaborative projects, both within the UK and internationally, in both *in-situ* and *ex-situ* conservation efforts. The maintenance of seedbanks by botanic gardens and breeding registers by zoos has been particularly important in ensuring that efforts to conserve species are not impaired by problems of inbreeding. Some animal species respond more favourably to captive breeding and relocation efforts than others. Those implementing captive breeding programmes are faced with balancing the needs of the animals in captivity, while not inadvertently selecting for features that make survival in the wild less likely.

Captive Breeding and Relocation

Individuals are captured and bred under protected conditions. If breeding programs are successful and there is suitable habitat available, captive individuals may be relocated to the wild where they can establish natural populations. Zoos now have an active role in captive breeding. There are problems with captive breeding; individuals are inadvertently selected for fitness in a captive environment and their survival in the wild may be compromised. this is especially so for marine species. However, for some taxa, such as reptiles, birds, and small mammals, captive rearing is very successful.

The Important Role of Zoos and Aquaria

As well as keeping their role in captive breeding programmes and as custodians of rare species, zoos have a major role in public education. They raise awareness of the threats facing species in their natural environments and engender public empathy for conservation work. Modern zoos tend to concentrate on particular species and are part of global programmes that work together to help retain genetic diversity in captive bred animals.

Above: England is home to a rare sub-species of sand lizard (Lacerta agilis). it is restricted to southern heathlands and the coastal sand dunes of north west England. The UK Herpetological Conservation Trust is the lead partner in the action plan for this species and Chester Zoo hosts a captive breeding colony.

Right: A puppet 'mother' shelters a takahe chick. Takahe, a rare rail species native to New Zealand, were brought back from the brink of extinction through a successful captive breeding programme.

In New Zealand, introduced predatory mammals, including weasels and stoats, have decimated native bird life. Relocation of birds on to predator-free islands or into areas that have been cleared of predators has been instrumental in the recovery of some species such as the North Island kokako. Sadly, others have been lost forever.

Above: The okapi is a species of rare forest antelope related to giraffes. Okapi are only found naturally in the Ituri Forest, in the northeastern rainforests of the Democratic Republic of Congo (DRC), Africa, an area at the front line of an ongoing civil war. A okapi calf was born to Bristol Zoo Gardens in 2009, one of only about 100 okapi in captivity.

1. Describe the key features of *ex-situ* conservation methods: _____

_____

_____

_____

2. Explain why some animal species are more well suited to *ex-situ* conservation efforts than others:

_____

_____

_____

Maintaining Biodiversity

333

Related activities: Loss of Biodiversity, Biodiversity and Global Warming

RA 2

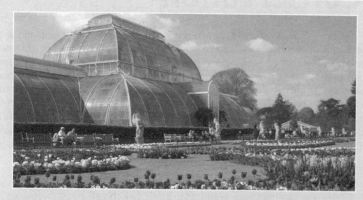

The Role of Botanic Gardens

Botanic gardens have years of collective expertise and resources and play a critical role in plant conservation. They maintain seed banks, nurture rare species, maintain a living collection of plants, and help to conserve indigenous plant knowledge. They also have an important role in both research and education. The Royal Botanic Gardens at Kew (above) contain an estimated 25,000 species, 2700 of which are classified by the ICUN as rare, threatened, or endangered. Kew Gardens are involved in both national and international projects associated with the conservation of botanical diversity and are the primary advisors to CITES on threatened plant species. Kew's Millennium Seed Bank partnership is the largest *ex situ* plant conservation project in the world; working with a network in over 50 countries they have banked 10% of the world's wild plant species.

Seedbanks and Gene Banks

Seedbanks and gene banks around the world have a role in preserving the genetic diversity of species. A seedbank (above) stores seeds as a source for future planting in case seed reserves elsewhere are lost. The seeds may be from rare species whose genetic diversity is at risk, or they may be the seeds of crop plants, in some cases of ancient varieties no longer used in commercial production.

3. Describe three key roles of zoos and aquaria and explain the importance of each:

(a) _____

_____

(b) _____

_____

(c) _____

_____

4. Explain the importance of gene and seed banks, both to conservation and to agriculture: _____

_____

_____

_____

5. Compare and contrast *in-situ* and *ex-situ* methods of conservation, including reference to the advantages and disadvantages of each approach:

_____

_____

_____

_____

_____

_____

_____

_____

_____

_____

National Conservation

The UK has a highly modified natural environment that has resulted from a legacy of human exploitation reaching far back into prehistory. Few areas have escaped modification. The conservation problems faced by the UK are typical of many other developed nations: loss of biodiversity, natural habitat loss, pollution, waste disposal, and inadequate recycling. The main government agencies for conservation in the UK are **English Nature** (in England), **Scottish** **Nature Heritage** (Scotland), and **The Countryside Council for Wales** (Wales). These agencies are supported in their roles by a number of voluntary organisations that provide an additional source of expertise, labour, and finance for assisting conservation work. Conservation involves not just preservation of habitats in their existing state but also the restoration of damaged areas that previously had high conservation value.

Protected Habitats in the United Kingdom

National Parks: There are currently 7 National Parks in England and 3 in Wales. While Scotland and Northern Ireland do not have National Parks, they do have essentially equivalent areas in the form of Regional Parks (Scotland) and Areas of Outstanding Beauty (Northern Ireland). Legislation permits some farming, forestry and quarrying within these parks.

Sites of Special Scientific Interest (SSSIs): These are notified by the government agency because of their plants, animals, or geological or physiographical features. In England, about 40% are owned or managed by public bodies or by the Crown (e.g. Ministry of Defence).

Environmentally Sensitive Areas (ESAs): These are areas in the UK whose environmental significance is a result of particular farming practices. If these methods change, then the ecological value of the area will decline. To preserve these areas, restrictions are imposed on the practices allowed.

National Nature Reserves (NNRs): These are sites which have been assigned as reserves under government legislation. They are either owned or controlled by government agencies or held by approved non-governmental organisations.

Marine Nature Reserves (MNRs): In England, these are declared by the Secretary of State for the Environment. At present there two: one in England and one in Wales.

National Parks in the United Kingdom

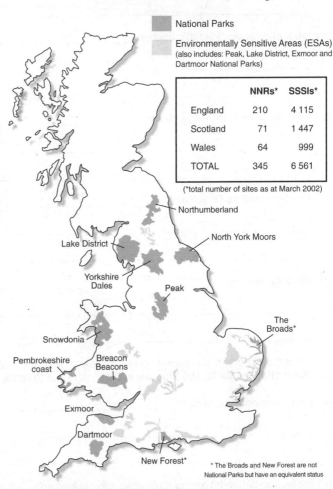

- ▉ National Parks
- ▉ Environmentally Sensitive Areas (ESAs) (also includes: Peak, Lake District, Exmoor and Dartmoor National Parks)

	NNRs*	SSSIs*
England	210	4 115
Scotland	71	1 447
Wales	64	999
TOTAL	345	6 561

(*total number of sites as at March 2002)

Northumberland
Lake District
North York Moors
Yorkshire Dales
Peak
The Broads*
Snowdonia
Pembrokeshire coast
Breacon Beacons
Exmoor
Dartmoor
New Forest*

* The Broads and New Forest are not National Parks but have an equivalent status

Condition of SSSIs (March 2002)

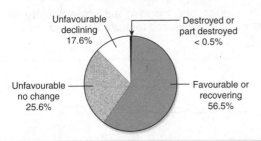

- Unfavourable declining 17.6%
- Destroyed or part destroyed < 0.5%
- Unfavourable no change 25.6%
- Favourable or recovering 56.5%

The European Union Habitats Directive

In 1992, the Council of the European Communities adopted a directive for the conservation of natural habitats and wild flora and fauna, known as the **Habitats Directive**. The global objective of the Habitats Directive is "to contribute towards ensuring biodiversity through the conservation of natural habitats and of wild fauna and flora in the European territory of the Member States to which the Treaty applies". Within the Habitats Directive, is the ecological network of special areas of conservation called **Natura 2000**. Natura 2000 areas aim to conserve natural habitats and species of plants and animals that are rare, endangered, or vulnerable in the European Community. The Natura 2000 network will include two types of areas:

Special Areas of Conservation (SAC): areas with rare, endangered, or vulnerable natural habitats, and plant or animal species (other than birds).

Special Protection Areas (SPAs): areas with significant numbers of wild birds and their habitats.

Areas of very great importance on land and sea may become both SAC and SPA sites.

Environmental cleanup (pond, Glasgow)

High value habitat: woodland and lakes

Paper recycling

Recovery of birds affected by oil spills

Any conservation or restoration programme must be multifaceted: preserving or restoring valuable habitat, repairing damage and aiding species recovery, and educating people to consider environmentally friendly options (e.g. recycling and reuse) in their general lives.

Maintaining Biodiversity

1. Explain the purpose of the following areas in the conservation of habitats and species diversity in the UK:

(a) National Parks: _____

_____

_____

(b) National Nature Reserves: _____

_____

_____

(c) Sites of Specific Scientific Interest: _____

_____

_____

(d) Environmentally Sensitive Areas: _____

_____

_____

(e) The EU Habitats Directive: _____

_____

_____

(f) Natura 2000: _____

_____

_____

2. Describe the contributions made to conservation by non governmental organisations, including the **Royal Society for the Protection of Birds**, **the Woodland Trust**, and **the Wildlife Trusts**. Include reference to each of the following:

(a) Conservation of biodiversity: _____

_____

_____

(b) Protection of habitat and unique geographical features: _____

_____

_____

(c) Environmental restoration: _____

_____

_____

(d) Education and promotion of conservation aims: _____

_____

_____

3. Find out about the EU **set-aside scheme**, and answer the following:

(a) Describe what the set-aside scheme offers to farmers: _____

_____

(b) Describe the benefits of the scheme to farmers: _____

_____

KEY TERMS: Word Find

Use the clues below to find the relevant key terms in the WORD FIND grid

```
M T F J J B N X H S C B H B B I P N S T N C T A W
E U J P I K U B S K Q G D L P H B H P S V V Z R M
M I G R X Y L C N B O W D V V J D U W H U T P G C
P J A C T A L X F S K E Y S T O N E S P E C I E S
E C O S Y S T E M R E S T O R A T I O N W S W T K
S J G E N E B A N K D N N R V X C P G R O H S H S
T R T Z I N S I T U C O N S E R V A T I O N V M E
A P H B I O D I V E R S I T Y K P V Q U Z O O S E
B Q R R D R E S I L I E N C E G P V O A C N Z M D
I P E B H U L A X S D W K P Q L Z Q E K Y P X A B
L O A M X V X N H O B O T A N I C G A R D E N S A
I K T X U T C E V L B Y D F V I I M O H A F O G N
T G E E X S I T U C O N S E R V A T I O N V A I K
Y U N Q E A T G T U H O M S E N D A N G E R E D C
N H E A H A E Y M J G L O B A L W A R M I N G S C
P W D Y W D S K G R E E N H O U S E E F F E C T S
R E E P U H S E C A P T I V E B R E E D I N G R V
```

Maintaining Biodiversity

Breeding and educational facilities in which animals are confined and displayed to the public.

These species fall into three categories: vulnerable, endangered, and critically endangered.

The ability of an ecosystem to resist change.

The increase in the average temperature of Earth's near-surface air and oceans since the mid-20th century and its projected continuation (2 words: 6, 7).

The ability of an ecosystem to recover from disturbance.

A species that has a disproportionate effect on its environment relative to its biomass (2 words: 8, 7)

A repository of plant or animal genetic material (2 words: 4, 4).

The process of protecting an endangered species in its natural habitat, either by protecting or restoring the habitat itself, or by defending the species from loss (2 words: 6, 12).

The warming effect of the radiative energy absorbed by the atmosphere (2 words: 10, 6).

Protecting an endangered species of plant or animal outside of its natural habitat (2 words: 6, 12).

A population of organisms which is at risk of becoming extinct is called this.

An assessment of the possible impact (positive or negative) of a proposed project on the environment (abbrev).

The return of a damaged ecological system to a stable, healthy, and sustainable state.

An international agreement to regulate the trade in endangered species (acronym).

The process of breeding animals in human controlled environments such as a zoo (2 words: 7, 8).

Well tended parks containing a range of plants, many of which may be rare or endangered (2 words: 7, 7)

The variation of life forms within a given ecosystem, biome, or on the entire Earth.

A repository of preserved seeds.

Appendix 1

SKILLS IN BIOLOGY

▶ **The Truth Is Out There**

New Scientist, 26 February 2000 (Inside Science). *The philosophy of scientific method: starting with an idea, formulating a hypothesis, and following the process to theory.*

▶ **Experiments**

Biol. Sci. Rev., 14(3) February 2002, pp. 11-13. *The basics of experimental design and execution: determining variables, measuring them, and establishing a control.*

▶ **Descriptive Statistics**

Biol. Sci. Rev., 13(5) May 2001, pp. 36-37. *An account of descriptive statistics using text, tables and graphs.*

▶ **Percentages**

Biol. Sci. Rev., 17(2) Nov. 2004, pp. 28-29. *The calculation of percentage and the appropriate uses of this important transformation.*

▶ **It's a Plot!**

Biol. Sci. Rev., 22 (2) Nov. 2009, pp. 16-19. *Using graphs to evaluate and explain data.*

▶ **Dealing with Data**

Biol. Sci. Rev., 12 (4) March 2000, pp. 6-8. *A short account of the best ways in which to deal with the interpretation of graphically presented data in examinations.*

▶ **Drawing Graphs**

Biol. Sci. Rev., 19(3) Feb. 2007, pp. 10-13. *A guide to creating graphs. The use of different graphs for different tasks is explained and there are a number of pertinent examples described to illustrate points.*

▶ **Size Does Matter**

Biol. Sci. Rev., 17 (3) February 2005, pp. 10-13. *Measuring the size of organisms and calculating magnification and scale.*

▶ **Describing the Normal Distribution**

Biol. Sci. Rev., 13(2) Nov. 2000, pp. 40-41. *The normal distribution: data spread, mean, median, variance, and standard deviation.*

▶ **Estimating the Mean and Standard Deviation**

Biol. Sci. Rev., 13(3) January 2001, pp. 40-41. *Simple statistical analysis. Includes formulae for calculating sample mean and standard deviation.*

▶ **The Variability of Samples**

Biol. Sci. Rev., 13(4) March 2001, pp. 34-35. *The variability of sample data*

and the use of sample statistics as estimators for population parameters.

CELL STRUCTURE

▶ **Size Does Matter**

Biol. Sci. Rev., 17 (3) Feb. 2005, pp. 10-13. *Measuring the size of organisms and calculating magnification and scale.*

▶ **Light Microscopy**

Biol. Sci. Rev., 13(1) Sept. 2000, pp. 36-38. *An excellent account of the basis and various techniques of light microscopy.*

▶ **Transmission Electron Microscopy**

Biol. Sci. Rev., 19(4) April 2007, pp. 6-9. *An account of the techniques and applications of TEM. Includes an excellent diagram comparing features of TEM and light microscopy.*

▶ **Scanning Electron Microscopy**

Biol. Sci. Rev., 13(3) Jan. 2001, pp. 6-9. *An account of the techniques and applications of SEM. Includes details of specimen preparation and recent advancements in the technology.*

▶ **Bacteria**

National Geographic, 184(2) August 1993, pp. 36-61. *The structure and diversity of bacteria; our most abundant and useful organisms.*

▶ **Cellular Factories**

New Scientist, 23 November 1996 (Inside Science). *The role of different organelles in plant and animal cells*

▶ **Chloroplasts: Biosynthetic Powerhouses**

Biol. Sci. Rev., 21(4) April 2009, pp. 25-27. *The features of chloroplasts and their origin from proplastids.*

▶ **The Beat Goes On: Cilia and Flagella**

Biol. Sci. Rev., 18(4) April 2006, pp. 2-6. *The structure and function of cilia and flagella and their important roles in biological systems.*

▶ **The Power Behind an Electron Microscopist**

Biol. Sci. Rev., 18(1) Sept. 2005, pp. 16-20. *The use of TEMs to obtain greater resolution of finer details than is possible from optical microscopes.*

CELL MEMBRANES AND TRANSPORT

▶ **Cellular Factories**

New Scientist, 23 November 1996 (Inside Science). *The role of different organelles in plant and animal cells*

▶ **Border Control**

New Scientist, 15 July 2000 (Inside Science). *The role of the plasma membrane in cell function: membrane structure, transport processes, and the role of receptors on the cell membrane.*

▶ **The Fluid-Mosaic Model for Membranes**

Biol. Sci. Rev., 22(2), Nov. 2009, pp. 20-21. *Diagrammatic revision of membrane structure and function.*

▶ **Getting in and Out**

Biol. Sci. Rev., 20(3), Feb. 2008, pp. 14-16. *Diffusion: some adaptations and some common misunderstandings.*

▶ **How Biological Membranes Achieve Selective Transport**

Biol. Sci. Rev., 21(4), April 2009, pp. 32-36. *The structure of the plasma membrane and the proteins that enable the selective transport of molecules.*

▶ **What is Endocytosis?**

Biol. Sci. Rev., 22(3), Feb. 2010, pp. 38-41. *The mechanisms of endocytosis and the role of membrane receptors in concentrating important molecules before ingestion.*

CELL DIVISION AND ORGANISATION

▶ **The Cell Cycle and Mitosis**

Biol. Sci. Rev., 14(4) April 2002, pp. 37-41. *Cell growth and division, stages in the cell cycle, and the complex control over different stages of mitosis.*

▶ **Rebels Without a Cause**

New Scientist, 13 July 2002, (Inside Science). *The causes of cancer: the uncontrolled division of cells that results in tumour formation. Breast cancer is a case example.*

▶ **What is a Stem Cell?**

Biol. Sci. Rev., 16(2) Nov. 2003, pp. 22-23. *The nature of stem cells and their therapeutic applications.*

▶ **Cell Differentiation**

Biol. Sci. Rev., 20(4), April 2008, pp. 10-13. *How tissues arise through the control of cellular differentiation during development. The example provided is the differentiation of blood cells.*

EXCHANGE SURFACES AND BREATHING

▶ **Getting in and Out**

Biol. Sci. Rev., 20(3), Feb. 2008, pp. 14-16. *Diffusion: some adaptations and some common misunderstandings*

Appendix 1

PERIODICAL REFERENCES

Gas Exchange in the Lungs

Bio. Sci. Rev. 16(1) Sept. 2003, pp. 36-38. *The structure and function of the alveoli of the lungs, with an account of respiratory problems and diseases.*

TRANSPORT IN ANIMALS

▶ Venous Disease

Biol. Sci. Rev., 19(3), Feb. 2007, pp. 15-17. *Valves in the deep veins of the legs assist venous return but when these are damaged, superficial veins are put under more pressure and circulation is compromised.*

▶ The Heart

Bio. Sci. Rev. 18(2) Nov. 2005, pp. 34-37. *The structure and physiology of the heart.*

▶ Keeping Pace - Cardiac Muscle & Heartbeat

Biol. Sci. Rev., 19(3), Feb. 2007, pp. 21-24. *The structure and properties of cardiac muscle.*

▶ Blood Pressure

Biol. Sci. Rev., 12(5) May 2000, pp. 9-12. *Blood pressure: its control, measurement, and significance to diagnosis.*

▶ Cunning Plumbing

New Scientist, 6 Feb. 1999, pp. 32-37. *The arteries can actively respond to changes in blood flow, spreading the effects of mechanical stresses to avoid extremes.*

A Fair Exchange

Biol. Sci. Rev., 13(1), Sept. 2000, pp. 2-5. *The role of tissue fluid in the body and how it is produced and reabsorbed.*

▶ Red Blood Cells

Bio. Sci. Rev. 11(2) Nov. 1998, pp. 2-4. *The structure and function of erythrocytes, including details of oxygen transport.*

▶ Humans with Altitude

New Scientist, 2 Nov. 2002, pp. 36-39. *The short term adjustments and long term adaptations to life at altitude.*

TRANSPORT IN PLANTS

▶ How Trees Lift Water

Biol. Sci. Rev., 18(1), Sept. 2005, pp. 33-37. *Cohesion-tension theory and others on how trees lift water.*

▶ Cacti

Biol. Sci. Rev., 20(1), Sept. 2007, pp. 26-30. *The growth forms and structural and physiological adaptations of cacti.*

▶ High Tension

Biol. Sci. Rev., 13(1), Sept. 2000, pp. 14-18. *Cell specialisation and transport in plants: an excellent account of the mechanisms by which plants transport water and solutes.*

BIOLOGICAL MOLECULES

▶ Water, Life, and Hydrogen Bonding

Biol. Sci. Rev., 21(2) Nov. 2008, pp. 18-20. *The molecules of life and the important role of hydrogen bonding.*

▶ Glucose & Glucose-Containing Carbohydrates

Biol. Sci. Rev., 19(1) Sept. 2006, pp. 12-15. *The structure of glucose and its polymers.*

▶ Designer Starches

Biol. Sci. Rev., 19(3) Feb. 2007, pp. 18-20. *The composition of starch, and an excellent account of its properties and functions.*

▶ What is Tertiary Structure?

Biol. Sci. Rev., 21(1) Sept. 2008, pp. 10-13. *How amino acid chains fold into the functional shape of a protein.*

▶ Enzymes: Nature's Catalytic Machines

Biol. Sci. Rev., 22(2) Nov. 2009, pp. 22-25. *Enzymes as catalysts: a very up-to-date description of enzyme specificity and binding, how enzymes work, and how they overcome the energy barriers for a reaction. Some well known enzymes are described.*

▶ Enzymes: Fast and Flexible

Biol. Sci. Rev., 19(1) Sept. 2006, pp. 2-5. *The structure of enzymes and how they work so efficiently at relatively low temperatures.*

NUCLEIC ACIDS

▶ DNA: 50 Years of the Double Helix

New Scientist, 15 March 2003, pp. 35-51. *A special issue on DNA: structure and function, repair, the new-found role of histones, and the functional significance of chromosome position in the nucleus.*

▶ DNA Polymerase

Biol. Sci. Rev., 22(4) April 2010, pp. 38-41. *How DNA polymerase operates during DNA replication in the cell and how the enzyme is used in PCR.*

▶ What is a Gene?

Biol. Sci. Rev. 21(4) April 2009, pp. 10-12. *The molecular basis of genes, gene transcription, and production of a functional mRNA by removal of introns.*

FOOD AND HEALTH

▶ Feast and Famine

Scientific American, Sept. 2007, (special issue). *This issue covers recent developments in health and nutrition science, including the best diet for good health, the role of exercise in weight management, what controls obesity, and the role of globalisation in the impoverished nutrition those in the world who live in poverty.*

▶ The Good, the Fad and the Unhealthy

New Scientist, 27 Sept. 2006, pp. 42-49. *The facts and myths of good nutrition.*

▶ The Alphabet Soup of Vitamins

Biol. Sci. Rev., 20 (3) Feb. 2008, pp. 31-34. *The role of vitamins in a balanced diet, and the consequences of vitamin deficiencies to health.*

▶ Why are we so Fat?

National Geographic, 206(2), August 2004, pp. 46-61. *The occurrence and health problems associated with obesity.*

▶ Coronary Heart Disease

Biol. Sci. Rev., 18(1) Sept. 2005, pp. 21-24. *An account of cardiovascular disease, including risk factors and treatments.*

▶ Atherosclerosis: The New View

Sci. American, May 2002, pp. 28-37. *The pathological development and rupture of plaques in atherosclerosis.*

▶ Heart Disease and Cholesterol

Biol. Sci. Rev., 13(2) Nov. 2000, pp. 2-5. *The links between dietary fat, cholesterol level, and heart disease.*

▶ What Price More Food?

New Scientist, 14 June 2008, pp. 28-33. *A discussion of the looming world food crisis, precipitated by the shortage of grain and the lack of the research and development needed to increase agricultural yields.*

▶ The Adaptations of Cereals

Biol. Sci. Rev., 13(3) Jan. 2001, pp. 30-33. *The world's major cereal crops: production and adaptations.*

▶ The Microbiology of Cheese

Biol. Sci. Rev., 15(4) April 2003, pp. 37-41. *The microbiology behind cheese-making: different cheeses and how they are made, pasteurisation, and microbial culture and fermentation.*

▶ Poultry Farming

Biol. Sci. Rev., 11(4) March 1999, pp. 14-17. *Producing poultry in intensive farming situations requires close attention to nutrition and welfare.*

Appendices and Index

Appendix 1

PERIODICAL REFERENCES

PATHOGENS AND HUMAN DISEASE

► **Koch's Postulates**

Biol. Sci. Rev., 15(3) February 2003, pp. 24-25. *Koch's postulates and the diagnosis of infectious disease.*

► **The White Plague**

New Scientist (Inside Science), 9 Nov. 2002. *The causes and nature of TB, its global incidence, and a discussion of the implications of drug resistance to TB treatment.*

► **Tuberculosis**

Biol. Sci. Rev., 14(1) Sept. 2001, pp. 30-33. *Despite vaccination, TB has become more common recently. Why has it returned?*

► **Malaria**

Biol. Sci. Rev., 15(1) Sept. 2002, pp. 29-33. *An account of the world's most important parasitic infection of humans. The parasite's life cycle, disease symptoms, control and prevention, and future treatments are all discussed.*

► **Beating the Bloodsuckers**

Biol. Sci. Rev., 16(3) Feb. 2004, pp. 31-35. *The global distribution of malaria, the current state of malaria research, and an account of the biology of the Plasmodium parasite and the body's immune response to it.*

► **Viral Plagues**

Biol. Sci. Rev., 17(3) Feb. 2005, pp. 37-41. *Viruses: nature and transmission.*

► **AIDS**

Biol. Sci. Rev., 20(1) Sept. 2007, pp. 30-12. *The HIV virus can evade the immune system and acquire drug resistance. This has prevented effective cures from being developed.*

► **Opportunistic Infections & AIDS**

Biol. Sci. Rev., 14 (4) April 2002, pp. 21-24. *An account of the suite of infections characterising AIDS (good).*

► **Search for a Cure**

National Geographic, 201(2) February 2002, pp. 32-43. *An account of the status of the AIDS epidemic and the measures to stop it.*

► **Are Viruses Alive?**

Scientific American, Dec. 2004, pp. 77-81. *This account covers the nature of viruses, including viral replication and an evaluation of the status of viruses in the world.*

► **War on Disease**

National Geographic, 201(2) February 2002, pp. 4-31. *An excellent account on the global importance of a range of infectious diseases. A great overview.*

► **Preventing the Next Pandemic**

Scientific American, April 2009, pp. 60-65. *More than half of human infectious diseases, past and present, originated in animals. What can be done to stop the spread of zoonoses?*

► **New Tactics Against TB**

Scientific American, March 2009, pp. 56-63. *Strategies for halting the growing pandemic of drug-resistant tuberculosis.*

► **Preparing for a Pandemic**

Scientific American, Nov. 2005, pp. 22-31. *A predicted global epidemic caused by some newly evolved strain of influenza may be temporarily contained with antiviral drugs.*

► **Finding and Improving Antibiotics**

Biol. Sci. Rev. 12(1) Sept. 1999, pp. 36-38. *Antibiotics, their production & testing, and the search for new drugs.*

DEFENCE AND THE IMMUNE SYSTEM

► **What is the Human Microbiome?**

Biol. Sci. Rev., 22(2) Nov. 2009, pp. 38-41. *An informative acount of the nature and role of the microbes that inhabit our bodies.*

► **Skin, Scabs and Scars**

Biol. Sci. Rev., 17(3) Feb. 2005, pp. 2-6. *The roles of skin, including its role in wound healing and the processes involved in its repair when damaged.*

► **Fight for Your Life!**

Biol. Sci. Rev., 18(1) September 2005, pp. 2-6. *The mechanisms by which we recognise pathogens and defend ourselves against them (overview).*

► **Looking Out for Danger: How White Blood Cells Protect Us**

Biol. Sci. Rev., 19 (4) April 2007, pp. 34-37. *The various types of leucocytes and they work together to protect the body against infection.*

► **Inflammation**

Biol. Sci. Rev., 17(1) Sept. 2004, pp. 18-20. *The role of this nonspecific defence response to tissue injury and infection. The processes involved in inflammation are discussed.*

► **Immunology**

Biol. Sci. Rev., 22(4) April 2010, pp. 20-21. *A pictorial but information-packed review of the basic of internal defense functions.*

► **Antibodies**

Biol. Sci. Rev., 11(3) January 1999, pp. 34-35. *The operation of the immune system and the production of antibodies*

(including procedures for producing monoclonal antibodies).

► **Lymphocytes - The Heart of the Immune System**

Biol. Sci. Rev., 12 (1) Sept. 1999 pp. 32-35. *An excellent account of the role of lymphocytes in the immune response (includes the types and actions of different lymphocytes).*

► **Hard to Swallow**

New Scientist, 26 Jan. 2008, pp. 37-39. *Many people fear that vaccines are unsafe and cause health problems. Particular reference to the polio and measles vaccines.*

► **Will there ever be a Malaria Vaccine?**

Biol. Sci. Rev., 19(1), Sept. 2006, pp. 24-28. *An outline of the categories of malarial vaccine development.*

SMOKING: A CHOICE AGAINST HEALTH

► **Smoking**

Biol. Sci. Rev., 10(1), Sept. 1999, pp. 14-16. *Tobacco as a drug and the nature of emphysema and other smoking-related diseases such as cardiovascular disease and cancer.*

► **Coronary Heart Disease**

Biol. Sci. Rev., 18(1) Sept. 2005, pp. 21-24. *An account of cardiovascular disease, including the influence of risk factors and treatment options.*

IDENTIFYING BIODIVERSITY

► **Biodiversity: Taking Stock**

National Geographic, 195(2) Feb. 1999 (entire issue). *Special issue exploring the Earth's biodiversity and what we can do to preserve it.*

► **Earth's Nine Lives**

New Scientist, 27 Feb. 2010, pp. 31-35. *How much can we push the Earth's support systems. This account examine the human interaction with nine critical global functions.*

► **The Living Dead**

New Scientist, 13 October 2001, (Inside Science). *The non-cellular nature of viruses and how they operate.*

► **Are Viruses Alive?**

Scientific American, December 2004, pp. 77-81. *Although viruses challenge our concept of what "living" means, they are vital members of the web of life. This account covers the nature of viruses, including viral replication and a critical evaluation of the status of viruses in the natural world.*

Appendix 1

PERIODICAL REFERENCES

▶ **A Passion for Order**

National Geographic, 211(6) June 2007, pp. 73-87. *The history of Carl Linnaeus and the classification of plant species.*

▶ **The Family Line - The Human-Cat Connection**

National Geographic, 191(6) June 1997, pp. 77-85. *An examination of the genetic diversity and lineages within the felidae. A good context within which to study classification.*

▶ **World Flowers Bloom after Recount**

New Scientist, 29 June 2002, p. 11. *A systematic study of flowering plants indicates more species than expected, especially in regions of high biodiversity such as South American and Asia.*

▶ **What is a Species?**

Scientific American June 2008, pp. 48-55. *The science of classification; modern and traditional approaches, the value of each, and the importance of taxonomy to identifying and recognising diversity. Excellent.*

▶ **Uprooting the Tree of Life**

Scientific American Feb. 2000, pp. 72-77. *Using molecular techniques to redefine phylogeny and divulge the path of evolution.*

▶ **The Loves of the Plants**

Scientific American, Feb. 1996, pp. 98-103. *The classification of plants and the development of keys to plant identification.*

EVOLUTION

▶ **What is Variation?**

Biol. Sci. Rev., 13(1) Sept. 2000, pp. 30-31. *The nature of continuous and discontinuous variation. The distribution pattern of traits that show continuous variation is discussed.*

▶ **Optimality**

Biol. Sci. Rev., 17(4), April 2005, pp. 2-5. *Environmental stability and optimality of structure and function can explain evolutionary stasis in animals. Examples are described.*

▶ **Evolution: Five Big Questions**

New Scientist, 14 June 2003, pp. 32-39, 48-51. *A synopsis of the five most common points of discussion regarding evolution and the mechanisms by which it occurs.*

▶ **Was Darwin Wrong?**

National Geographic, 206(5) Nov. 2004, pp. 2-35. *An account of the scientific evidence for evolution. A good way to remind students that the scientific debate around evolutionary theory is associated with the mechanisms by which evolution occurs, not the fact of evolution itself.*

▶ **Meet your Ancestor**

New Scientist, 9 Sept. 2006, pp. 35-39. *The significance of a recent fossil find: the missing link between fish and tetrapods.*

▶ **The Quick and the Dead**

New Scientist, 5 June 1999, pp. 44-48. *The formation of fossils: fossil types and preservation in different environments.*

▶ **The Accidental Discovery of a Feathered Giant Dinosaur**

Biol. Sci. Rev., 20(4), April 2008, pp. 18-20. *How scientists piece together and interpret sometimes confusing fossil evidence.*

▶ **How Old is...**

Nat. Geographic, 200(3) Sept. 2001, pp. 79-101. *A comprehensive discussion of dating methods and their application.*

▶ **A Fin is a Limb is a Wing-How Evolution Fashioned its Masterworks**

National Geographic, 210(5) Nov. 2006, pp. 110-135. *An excellent account of the role of developmental genes in the evolution of complex organs and structures in animals. Beautifully illustrated, compelling evidence for the mechanisms of evolutionary change.*

▶ **A Waste of Space**

New Scientist, 25 April 1998, pp. 38-39. *Vestigial organs: how they arise in an evolutionary sense and what role they may play.*

▶ **Species and Species Formation**

Biol. Sci. Rev., 20(3), Feb. 2008, pp. 36-39. *A feature covering the definition of species and how new species come into being through speciation.*

▶ **MRSA: A Hospital Superbug**

Biol. Sci. Rev., 19(4) April 2007, pp. 30-33. *An excellent account of how the evolution of MRSA has been driven by the misuse of antibiotics.*

▶ **Tracking the Next Killer Flu**

National Geographic, 208(4) Oct. 2005, pp. 4-31. *Discussion on flu viruses and how they spread.*

MAINTAINING DIVERSITY

▶ **Biodiversity and Ecosystems**

Biol. Sci. Rev., 11(4) March 1999, pp. 18-23. *Ecosystem diversity and its relationship to ecosystem stability.*

▶ **Global Warming**

Time, special issue, 2007. *A. engaging special issue on global warming: the causes, perils, solutions, and actions.*

▶ **Climate Change and Biodiversity**

Biol. Sci. Rev., 16(1) Sept. 2003, pp. 10-14. *While the focus of this account is on climate change, it provides useful coverage of ecosystem structure and processes and how these are studied.*

▶ **Ecology & Nature Conservation**

Biol. Sci. Rev., 18(1) Sept. 2005, pp. 11-15. *The prnciples of conservation and restoration ecology.*

▶ **Down on the Farm: The Decline in Farmland Birds**

Biol. Sci. Rev., 16(4) April 2004, pp. 17-20. *Factors in the decline of bird populations in the UK.*

▶ **Insecticides and the Conservation of Hedgerows**

Biol. Sci. Rev., 16(4) April 2004, pp. 28-31. *Well managed hedgerows can reduce the need for insecticides in adjacent crops by encouraging natural pest control agents.*

▶ **Conflicted Conservation**

Scientific American, Sept. 2009, pp. 10-11. *Measures to protect biodiversity could force indigenous peoples off their land into poverty.*

Appendix 2

INDEX OF LATIN AND GREEK ROOTS, PREFIXES, AND SUFFIXES

Many biological terms have a Latin or Greek origin. Understanding the meaning of these components in a word will help you to understand and remember its meaning and predict the probable meaning of new words. Recognizing some common roots, suffixes, and prefixes will make learning and understanding biological vocabulary easier.

The following terms are identified, together with an example illustrating their use in biology.

a(n)- without .... anoxic
ab- away from .... abductor
ad- towards............................ adductor
affer- carrying to........................afferent
amphi- both............................amphibian
amyl- starch .... amylase
anemo- wind ....anemometer
ante- before .... antenatal
anthro- human....................anthropology
anti- against, opposite .... antibiotic
apo- separate, from ....apoenzyme
aqua- water................................aquatic
arach- spider ....arachnoid
arbor- tree ....arboreal
arch(ae/i)- ancient.................... Archaea
arthro- joint............................arthropod
artic- jointed........................articulation
artio- even-numbered .... artiodactyl
auto- self ....autologous
avi- bird....................................... avian
axi- axis....................................axillary

blast- germ............................blastopore
brachy- short....................brachycardia
brady- slow .... bradycardia
branch- gill...............................branchial
bronch- windpipe....................bronchial
bucca- mouth cavity..................buccal

caec- blind...............................caecum
card- heart ....cardiac
cauda- tail...................................caudal
centi- hundred ....centimorgan
ceph(al)- head ....cephalothorax
cera(s)(t)- horn ....ceratopsian
cerebro- brain ....cerebrospinal

cerv- neck .... cervix
chrom- color.......................chromoplast
chym- juice .... chyme
cili- eyelash..................................... cilia
cloaca- sewer .... cloacal
coel- hollow..........................coelomate
contra- opposite..............contraception
cotyl- cup .... hypocotyl
crani- skull...............................cranium
crypt- hidden ....crptic
cyan- blue ....cyanobacteria
cyt- cell .... cytoplasm

dactyl- finger.....................polydactylic
deci-(a) ten .... decibel, decapod
dendr- tree ....dendrogram
dent- tooth...............................edentate
derm- skin ....pachyderm
di- two ....dihybrid
dors- back.................................dorsal
dur- hard .... dura mater

echino- spiny ....echinoderm
ecto- outside.......................... ectoderm
effer- carrying away ....efferent
endo- inside.....................endoparasite
equi- horse, equal...............equilibrium
erect- upright .... Homo erectus
erythr- red ....erthyrocyte
eu- well, very........................eukaryote
eury- wide ....eurythermal
ex- out of..................................explant
exo- outside ....exoskeleton
extra- beyond................extraperitoneal

foramen- opening ..... foramen magnum

gast(e)r- stomach, pouch ..... gastropod
gymn- naked.....................gymnosperm

hal- salty ....halophyte
haplo- single, simple..................hapolid
holo- complete, whole............. holozoic
hydr- water..........................hydrophyte
hyper- above..........................hypertoic
hypo- beneath.......................hypotonic

infra- under ....infrared
inter- between ....interspecific
intra- within.....................intraspecific
iso- equal .... isotonic

kilo- thousand .... kilogram

labi- lips .... labial palps
lacuna- space .... lacunae
lamella- leaf, layer...........lamellar bone
leuc- white.........................leu(k)cocyte
lip- fat...................................lipoprotein
lith- stone .... Palaeolithic
lumen- cavity...............................lumen
lute- yellow..................corpus luteum
lymph- clear water .... lymphatic

magni- large.................... magnification
mamma- breast.......................mammal
mat(e)ri- mother ....maternal
mega- large ....megakaryocyte
melan- black .... melanocyte
meso- middle ....Mesolithic
meta- after....................metamorphosis
micro- small ....microorganism
milli- thousand.....................millimetre
mirabile- wonderful ....rete mirabile
mono- one ....monohybrid
morph- form ....morphology
motor- mover.....................motor nerve
multi- many ....multicellular
myo- muscle ....myofibril

necro- dead .... necrosis
neo- new ....Neolithic
nephr- kidney............................nephro
neur- nerve...............................neural
notho- southern...................Nothofagus
noto- back, south...................notochord

oecious- house of .... monoecius
oed- swollen.............................oedema
olfact- smelling.......................olfactory
opistho- behind.................opisthosoma
os(s/t)- bone...........................osteocyte
ovo- egg..........................ovoviviparous

pachy- thick .... pachyderm
palae- old............................Palaeocene
pect(or)- chest ....pectoral fin
ped- foot ....quadraped
pent- five.........................pentose sugar
per(i)- through, beyond..........peristalsis
peri- around ....periosteum
phaeo- dark ....phaeomelanin
phag- eat ....phagosome
phyll- leaf ....sclerophyll

Appendix 3

MULTIPLES AND SI UNITS

physio- nature .... physiology
phyto- plant.....................phytohormone
pisc- fish ....piscivorous
plagio- oblique................. plagioclimax
pneu(mo/st)- air, lung...........pneumonia
pod- foot ....sauropod
poly- many .... polydactyly
pre- before ....premolar
pro- in front of ....Prokaryote
prot- first............................... protandry
pseud- false .... pseudopodia
pter- wing, fern....................Pterophyta
pulmo- lung........................pulmonary

radi- root ....radicle
ren- kidney....................................renal
retic- network ....reticulated
rhin- nose, snout ....rhinoceros
rostr- beak, prow........................rostrum

sacchar- sugar.............. polysaccharide

schizo- split.................schizocoelomate
scler- hard ....sclerophyll
seba- tallow, wax .... sebaceous
semi- half ....semi-conservative
sept- seven, wall.......................septum
soma- body.........opisthosoma, somatic
sperm- seed...................spermatophyte
sphinct- closing ....sphincter
stereo- solid ....stereocilia
stom- mouth ....stoma
strat- layer........................ stratification
sub- below ....subtidal
sucr- sugar ....sucrase
sulc- furrow ....sulci
super- beyond.........................superior
supra- above............supracoracoideus
sym- with................................symbiosis
syn- with................................synapsis

tact- touch .... tactile
tachy- fast ....tachycardia

taenia- ribbon...........Taenia (tapeworm)
trans- across .... transmembrane
tri- three............................... triploblastic
trich- hair ....trichome

ultra- above.......................... ultraviolet
un- one .... unicellular
uro- tail ....urodele

vas- vessel ....vascular
ven- vein .... venous
ventr- belly ....ventral
vern- spring............................. vernal
visc- organs of body cavity ....viscera
vitr- glass ....in vitro

xanth- yellow ....xanthophyll
xen- stranger ....xenotransplant
xer- dry .... xerophyte
xyl- wood......................................xylem
zo- animal ....zoological

TEST YOURSELF!

Use the index of Latin and Greek terms to deduce the meaning of the following terms:

1. *Sclerophyll*: _____

2. *Osteocyte*: _____

3. *Polydactyly*: _____

4. *Gymnosperm*: _____

5. *Tachycardia*: _____

MULTIPLES

MULTIPLE	PREFIX	SYMBOL	EXAMPLE
10^9	giga	G	gigawatt (GW)
10^6	mega	M	megawatt (MW)
10^3	kilo	k	kilogram (kg)
10^2	hecto	h	hectare (ha)
10^{-1}	deci	d	decimetre (dm)
10^{-2}	centi	c	centimetre (cm)
10^{-3}	milli	m	milliimetre (mm)
10^{-6}	micro	μ	microsecond (μs)
10^{-9}	nano	n	nanometre (nm)
10^{-12}	pico	p	picosecond (ps)

INTERNATIONAL SYSTEM OF UNITS (SI)

Examples of SI derived units

DERIVED QUANTITY	NAME	SYMBOL
area	square metre	m^2
volume	cubic metre	m^3
speed, velocity	metre per second	ms^{-1}
acceleration	metre per second squared	ms^{-2}
mass density	kilogram per cubic metre	kgm^{-3}
specific volume	cubic meter per kilogram	m^3kg^{-1}
amount-of-substance concentration	mole per cubic meter	$molm^{-3}$
luminance	candela per square meter	cdm^{-2}

Appendix 4

COMMAND WORDS

Questions come in a variety of forms. Whether you are studying for an exam or writing an essay, it is important to understand exactly what the question is asking. A question has two parts to it: one part of the question will provide you with information, the second part of the question will provide you with instructions as to how to answer the question. Following these instructions is most important. Often students in examinations know the material but fail to follow instructions and do not answer the question appropriately. Examiners often use certain key words to introduce questions. Look out for them and be clear as to what they mean. Below is a description of terms commonly used when asking questions in biology.

Commonly used Terms in Biology

The following terms are frequently used when asking questions in examinations and assessments. Students should have a clear understanding of each of the following terms and use this understanding to answer questions appropriately.

Account for: Provide a satisfactory explanation or reason for an observation.

Analyse: Interpret data to reach stated conclusions.

Annotate: Add **brief** notes to a diagram, drawing or graph.

Apply: Use an idea, equation, principle, theory, or law in a new situation.

Appreciate: To understand the meaning or relevance of a particular situation.

Calculate: Find an answer using mathematical methods. Show the working unless instructed not to.

Compare: Give an account of similarities and differences between two or more items, referring to both (or all) of them throughout. Comparisons can be given using a table. Comparisons generally ask for similarities more than differences (see contrast).

Construct: Represent or develop in graphical form.

Contrast: Show differences. Set in opposition.

Deduce: Reach a conclusion from information given.

Define: Give the precise meaning of a word or phrase as concisely as possible.

Derive: Manipulate a mathematical equation to give a new equation or result.

Describe: Give a detailed account, including all the relevant information.

Design: Produce a plan, object, simulation or model.

Determine: Find the only possible answer.

Discuss: Give an account including, where possible, a range of arguments, assessments of the relative importance of various factors, or comparison of alternative hypotheses.

Distinguish: Give the difference(s) between two or more different items.

Draw: Represent by means of pencil lines. Add labels unless told not to do so.

Estimate: Find an approximate value for an unknown quantity, based on the information provided and application of scientific knowledge.

Evaluate: Assess the implications and limitations.

Explain: Give a clear account including causes, reasons, or mechanisms.

Identify: Find an answer from a number of possibilities.

Illustrate: Give concrete examples. Explain clearly by using comparisons or examples.

Interpret: Comment upon, give examples, describe relationships. Describe, then evaluate.

List: Give a sequence of names or other brief answers with no elaboration. Each one should be clearly distinguishable from the others.

Measure: Find a value for a quantity.

Outline: Give a brief account or summary. Include essential information only.

Predict: Give an expected result.

Solve: Obtain an answer using algebraic and/or numerical methods.

State: Give a specific name, value, or other answer. No supporting argument or calculation is necessary.

Suggest: Propose a hypothesis or other possible explanation.

Summarise: Give a brief, condensed account. Include conclusions and avoid unnecessary details.

In Conclusion

Students should familiarise themselves with this list of terms and, where necessary throughout the course, they should refer back to them when answering questions. The list of terms mentioned above is not exhaustive and students should compare this list with past examination papers / essays etc. and add any new terms (and their meaning) to the list above. The aim is to become familiar with interpreting the question and answering it appropriately.

Index

Acquired immunity 239-240
Active immunity 239
Active site, of enzymes 155
Active transport 67, 69
  - in phloem transport 136
Adaptations 295-296
  - Darwin's finches 315-316
  - of xerophytes 133-134
Africa, impact of HIV 220
Agriculture, and diversity 329-330
Agriculture, types of 182
AIDS 217-218
Allostery, in enzymes 159
Altitude, effects of 119
Alveolar-capillary membrane 94
Alveolus, structure of 93-94
Amino acid coding table 177
Amino acids 143-144
Amoeba, structure of 270
Amphibians, gas exchange in 88
Anabolic reactions 155
Animal cells 48
Animal circulatory systems 103
Animal classification key 287
Animal tissues 83
Animal transport systems 100
Animals, features of 267-269
Antibiotics 225
Antibodies 236
Antigenic shift 320
Apoplast pathway in roots 130
Arterial system 104
Arteries 110
Artificial selection, in crop plants 194
Artificial selection, livestock 198
Assumptions 13
Atherosclerosis 190
Atrioventricular node 107
Attenuated vaccine 241

B lymphocyte 235
Bacterial cells 45
Bacterial disease 207
Balanced diet 183
Bar graphs 20
Barn owl, conservation 258
Base pairing, of DNA 169
Beak size, Darwin's finches 316
Benedict's test 161
Biochemical nature of the cell 140
Biochemical tests 161
Biodiversity hotspots 256
Biodiversity 256
  - Britain 257-258
  - grasslands 329-330
  - impact of global warming 326-328
  - loss of 261
Biogeography 311-312
Biological drawings 29-30
Biological molecules, portraying 142
Birds, gas exchange in 88
Birth weight, selection for 314
Biuret test for proteins 161
Blood cell production 77
Blood cells 115
Blood gases 117
Blood pressure 106
Blood vessels 110-113
Blood, composition of 115
BMI 187
Body mass index 187
Body size, mammals 295
Bohr effect 118
Botanic gardens, role of 334
Brassica, domestication 196
Breathing in humans 91
Bundle of His 107

Calibration curve, glucose 161
Cancer 75-76
Cancer cells 75
Capillaries 112-113
Capsid 215
Captive breeding 334
Carbohydrate isomerism 149
Carbohydrates 140
Carbohydrates, reactions of 149
Carbohydrates, types of 148, 150-152
Cardiac cycle 108
Cardiac output 108
Cardiovascular disease 191-192
  - and cholesterol 191-192
  - and smoking 253-254
  - risk factors 189, 191, 253
Carcinogen 75
Catabolic reactions 155
Catalase, experiments with 14, 35
Cattle, artificial selection of 198
Cell cycle 73
Cell cytoskeleton 49
Cell division 72
Cell organelles 50-53
Cell signalling 63
Cells 264
  - animal 48
  - bacterial 45
  - plant 47
  - sizes 40
  - specialisation 78, 80
Cellular membranes, structure of 59
Cellular respiration 87
Cellulose 151
Chlamydomonas, structure of 270
Chloroquine resistance 226
Cholera 211
Cholesterol in membranes 59
Cholesterol, and CVD 191-192
Chromosome structure 165-166
Cigarette smoke, components of 252
Ciliated epithelium 83
Circulatory system, human 104
CITES 332
Cladistics 285-286
Cladogram 285
Classification exercise 278-283
Classification key 287-289
Classification of human 286
Classification systems 271-272
Clonal selection theory 236
Closed circulatory system 102
Codon 176
Coenzymes 158
Colorimetery 161
Communicable disease 223
Compartmentation, in organelles 61
Competitive inhibition 159
Complex plant tissue 84
Concentration gradient 64
Condensation reactions 144, 149, 168
Conservation methods 332-335
Conservation, of the barn owl 258
Continuous variation 293
Control of heartbeat 108
Control, experimental 11
Controlled variables 13
COPD 248-249
Crop plants, improvement of 194
Cytochrome c, molecular clock 308
Cytokinesis 73
Cytoskeleton 49

Darwin, Charles 297-298
Darwin's finches 303-304
Data 16
  - analysis 20-28, 31-34
  - distribution of 31-33

  - manipulation 17
  - measuring spread in32-34
  - presentation 18-19
Deficiency disease 183, 185
Denaturation of proteins 145
Dependent variable13
Descriptive statistics 31-34
Dicots, structure of 125-127
Diet, balanced 183
Diet, role in disease 185-188, 191-192
Differentation of cells 77, 79
Diffusion 64, 100
Directional selection 313
Disaccharides 148
Discontinuous variation 293
Disease transmission 207-210, 216
Diseases 206-207
  - bacterial 207, 209-211, 221
  - communicable 223
  - control of 223
  - emerging 221
  - of lungs 248-250
  - protozoan 212-213
  - smoking related 248
  - viral 214-221
Disruptive selection 313
Distribution of data 31-33
Diversity indices 260
Diversity, and ecosystem stability 323
DNA analysis 178
DNA homologies 306
DNA hybridisation 306
DNA model 164
DNA molecules 164
DNA molecules, sizes of 164
DNA replication 173-175
DNA, in phylogenetics 285
Dog, variation in 292
Domestication of animals 197
Drawings, biological 29-30
Drug development 244
Drug resistance 226
  - in bacteria 318
  - in viruses 320
Drugs, antimicrobial 225

Ear length, rabbits 295
ECG 108-109
Ecosystem restoration 333
Ecosystem stability 323
Electron micrographs 54-54
Electron microscopes 43
Emerging disease 221
Emulsion test for lipids 161
Endocytosis 68
Environment, and phenotype 293-294
Enzyme action 156
Enzyme cofactors 158
Enzyme inhibitors 159-160
Enzyme reaction rates 157
Enzyme structure 155
Enzymes, in DNA replication 174
Epidemiology of AIDS 217-218
Epithelial tissues 83
Euglena, structure of 270
Eukaryotes, unicellular 270
Evaluating quantitative work 35-36
Evolution of horses 305
Evolution, development of theory 297
Evolution
  - evidence 299-312, 315-16, 318-20
  - modern theories of 297
  - of drug resistance 318
Ex-situ conservation 333-334
Exercise
  - effect on blood flow 120
  - effect on gas exchange system 92
Exocytosis 68

Facilitated diffusion 64
Farming, environmental impact 329
Fats 153
Fatty acids, types of 153
Fever 233
Fibrous protein 146
Fick's law 64, 87
Finches, Darwin 315-316
Fish, gas exchange in 88
Fitness 295
Five kingdom classification 284
Fluid mosaic model 59
Food preservation techniques 203
Food production 199-202
Food pyramid 183
Food technology 199-200
Formula, of molecules 142
Fossil record 301-302
Fossil sites, dating 303-304
Fossils, formation of 299
Freeze-fracture studies 57
Freezing, for preservation 203
Freshwater, gas exchange in 88

Gametogenesis 72
Gas exchange 87-88
  - in humans 93-94, 97
  - plants 127
Gas exchange surfaces 87, 94
Gas transport, in humans 117
Gel electrophoresis analysis 178
Gene banks 334
Gene expression 176
Genes to proteins 176
Genetic code 177
Genetic gain 198
Genotype 293
Gills, in fish 88
Global warming 325-326
Global warming 325
  - effects on biodiversity 326-328
Globular protein 146
Graphs, constructing 19
Graphs, interpretation 27
Graphs, types of 20-28
Grasslands and biodiversity 329-330
Green revolution 193
Greenhouse effect 325
Ground tissue 84
Guard cell 127

Habitats Directive 335
Haemoglobin 117
Halophytes, adaptations of 133-134
Health 206
Heart activity, control of 107
Heart structure 105-106
Hedgerow 331
High density lipoprotein 187
Histograms 21
HIV 215-216
  - drug resistance 226
  - in Africa 220
  - replication in 219
  - transmission 216
  - treatment 216
Homologies, biochemical 306-308
Homologies, structural 309
Homologous chromosome 73
Homologous structures 309
Horses, evolution of 305
Human cells, differentiation 77-78
Human chromosomes 165
Human classification 286
Human disease 206
Human heart, review 109
Human nutrition 182

Index

Hydrolysis reactions 144, 149, 168
Hypotheses 9-10

Immune system 235-236
Immunisation 241-242
Immunity, acquired 239-240
Immunoglobulins 237
Immunology, and evolution 307
In-situ conservation 332
Inactivated vaccine 241
Independent variable 13
Induced fit model 156
Infection and disease 207
Inflammation 232
Inorganic ions, role of 140
Insecticide resistance 319
Insects, gas exchange in 88
Intensive farming 201, 329
Internal transport 100
Iodine test for starch 161
Ion pumps 67
Irradiation for preservation 203
Island colonisation 311-312
Isomers, of amino acids 144

Keystone species 323-324
Kingdoms, features of 265-269, 273
Kite graphs 23
Koch's postulates 207
Kwashiorkor 185

Leaf structure 127
LDL: HDL ratio 187
L-form of amino acids 144
Life characteristics of 262
Line graphs 24-26
Lipids 140, 153
    - test for 161
Living things, types of 263
Lock and key model 156
Low density lipoprotein 187
Lung disease 248-250
Lung function, overview 97
Lung volume 95
Lungs, human 91, 93, 95
Lungs, mammalian 88, 93
Lungs, effect of smoking 251
Lymph 234
Lymphatic system 234

Macrofungi, features of 277
Macromolecules 142
Malaria 213
Malnutrition 182
    - and obesity 187
Mammalian transport 100
Marasmus 185
Mean (average) of data 31
Median of data 31
Medicines, new 243-244
Meiosis 72
Membrane receptors 63
Membranes, role of 61
Microbial groups, features of 274
Microorganisms
    - in food production 199-200
    - medicines from 243-244
Microscope drawings 29-30
Microscopes
    - optical   41
    - electron 43
    - resolution of 44
Mineral deficiency 186
Mineral uptake in plants 130
Mitosis 73-74
Mode of data 31
Modification of proteins 147
Molecular clock 308

Molecular formula 142
Monosaccharides 148
mRNA 177
mRNA amino acid table 177
Multidrug resistant TB 210, 221
Multiple drug resistance 21
Multipotent stem cell 77
Mutation, role in variation 293
Myocardial infarction 189
Myoglobin 117

National conservation 335
Natura 2000 335
Natural selection 298, 313
Non-competitive inhibition 159
Notation, scientific 11
Nucleic acids 140, 167-168
Nucleotide structure 167
Nucleotide condensation 168
Nucleotide hydrolysis 168
Null hypothesis 9
Nutrient deficiency 186
Nutritional guidelines, UK 184

Obesity 187-188
Okazaki fragments 173
Oncogenes 75
Open circulatory system 101
Optical microscopes 41
Organelles, cellular 50-53
Organic molecules 142
Organisation of tissues 82-84
Origins of cancer 75
Osmosis 64-65
Oxygen transport 117-118
Oxygen-haemoglobin dissociation 118

Pacemaker, of heart 107
Packaging DNA 165
Paramecium, structure of 270
Passive immunity 239
Passive transport 64
Pathogens, types of 207
Pentadactyl limb 309
Peptide bonds 143
Phagocytes, action of 231
Phagocytosis 68
Phenotype 293
Phenotypic variation, shifts on 313
Phloem 129
Phospholipids, in membranes 59
Phospholipids, structure of 154
Phylogenies, constructing 285
Pickling, for preservation 203
Pie graphs 22
Pinocytosis 68
Plant adaptations 133-134
Plant cells 47
    - specialisation of 80
Plant classification key 289
Plant tissues 84
Plants, features of 277
Plants, medicines from 243-244
Plants, transport in 124
Plasma membrane, role of 61-62
Plasma membrane, structure of 59
Plasmolysis, in plant cells 66
Polysaccharide 150
Portal system 104
Potency, of stem cells 77
Potometer 132
Prediction 9
Pressure flow, sucrose transport 136
Pressure potential 65
Prokaryotic cells 45
Protein 140
Protein deficiency 185
Protein homologies 307-308

Protein modification 147
Protein secretion 53
Protein synthesis 176
Proteins 145-146
    - test for 161
Proton pumps, phloem transport 136
Protozoan disease 212
Purkyne fibres, of heart 107

Qualitative data 16
Qualitative data, collecting 12
Quantitative data 16
Quantitative data, collecting 13
Quarantine 223
Range, of data 31
Raw data, manipulating 17
Reaction rates, enzymes 157
Receptors 63
Recording results 15
Reducing sugars, test for 12, 161
Reference nutrient intakes 184
Replication of DNA 173-175
Replication, in HIV 219
Resilience, ecosystem 323
Resistance
    - antibiotic 318
    - insecticide 319
    - specific 239
    - to drugs 226
Resolution vs magnification 41
Resolution of microscopes 44
Respiratory disease 249-250
Respiratory pigments 118
Root cell development 81
Root hairs 125
Root structure 125, 130

Salting, for preservation 203
Sampling, diversity 259
Scatter plots 28
Scientific method 9
Scientific notation 11
Seedbanks 334
Selection, human birth weight 314
Selective breeding, crops 194-195
Selective breeding, livestock 197-198
Signalling, cell 63
Simple plant tissue 84
Simpson's index of diversity 260
Sinoatrial node 107
Sinusoids 112
Six kingdom classification 284
Smoking and the heart 253-254
Smoking and COPD 248
Smoking, and the lungs 251
Solute potential 65
Specialisation, of cells 78, 80
Speciation 317
Species development 317
Species diversity 256
    - measuring 259-260
Species, variation in 292
Spirometry 95-96
Squamous epithelium 83
Stabilising selection 313
Stains, in microscopy 42
Standard deviation 32
Starch 151
    - test for 161
Statistics, descriptive 31-34
Stem cells 77
Stem structure 125
Steroids, structure of 154
Stomata 127
Structural formula 142
Subunit vaccine 241
Sucrose, transport in phloem 136
Sugar syrup, for preservation 203

Surface area: volume ratio 89-90
Sustainable food production 202
Symplast pathway in roots 130

T lymphocyte 235
Tables, constructing 18
Tables, of results 15
Tables, presentation 18
Taxonomic groups, features of 265-269
Terms, scientific 11
The body's defences 229-230
Thrombosis 190
Tissue fluid 114
Tissues 82-84
    - animal 83
    - plant 84
Totipotent stem cell 77
Tracheal tubes, for gas exchange
Transcription 176
Transcription unit 176
Transitional fossils 300, 305
Translation 176
Translocation 135
Transpiration 131
Transport in plants 124
Transport, active 67, 69
Transport, passive 64
Tree of life 284
Triplet 176
Tuberculosis 210
Turgor, in plant cells 66

UK, conservation issues 335
Unicellular eukaryotes 270
Uptake of water in roots 130

Vaccination 241-242
Vaccine, types of 241
Vacuolar pathway in roots 130
Variables, identifying 13
Variation, sources of 293-294
Vascular tissue 84
Veins 111
Venous system 111
Vestigial organ 310
Viral disease 214
Viral pathogens 214
Viral replication 219
Viruses, evolution in 320
Vitamin deficiency 185

Water potential, of cells 65
Water potential, in plant transport 131
Water uptake in plants 130
Water, properties of 141
Wheat evolution 195
Whole agent vaccine 241

Xerophytes, adaptations of 133-134
Xylem 128

Yeast cell cycle 74

Zoonoses 221
Zoos, role in conservation 334
Zygote 72